MACRO-PROJECTS: ENVIRONMENT AND TECHNOLOGY

MACRO-PROJECTS: ENVIRONMENT AND TECHNOLOGY

ALEXANDER BOLONKIN
AND
RICHARD BROOK CATHCART

Nova Science Publishers, Inc.
New York

For permission to use material from this book please contact us:
Telephone 631-231-7269; Fax 631-231-8175
Web Site: http://www.novapublishers.com

LIBRARY OF CONGRESS CATALOGING-IN-PUBLICATION DATA

Bolonkin, Alexander.
Macro-projects : environment and technology / Alexander Bolonkin and Richard Brook Cathcart (editor).
p. cm.
ISBN 978-1-60456-998-8 (hardcover)
1. Environmental engineering--Technological innovations. 2. Sustainable engineering--Technological innovations. 3. Large scale systems. I. Cathcart, Richard Brook. II. Title.

TA170.B65 2009
620'.41--dc22

2008037245

Published by Nova Science Publishers, Inc. ✦ New York

CONTENTS

About the Authors

Bolonkin, Alexander Alexandrovich (1933-)

Alexander A. Bolonkin was burn in former USSR. He holds doctoral degree in Aviation Engineering from Moscow Aviation Institute and post-doctoral degree in Aerospace Engineering from Leningrad Polytechnic University. His positions included Senior Engineer in the Antonov Aircraft Design Company and Chairman of the Reliability Department in the Clushko Rocket Design Company. He also lectured in the Moscow Aviation Universities. He arrived in the USA 1988, and has lectured at the New Jersey Institute of Technology and worked as a Senior Researcher in NASA and US Air Force Research Laboratories.

Bolonkin is author more then 170 scientific articles and books and 17 inventions.

Main Bolonkin's books (in English): The Development of Soviet Rocket Engine, Delphic Ass., Inc., Washington , 1991, 122 pgs.; Non-Rocket Space Launch and Flight, Elsevier, 2006, 488 pgs.; New Concepts, Ideas, Innovation in Aerospace, Technology and Human Life, NOVA, 2008, 400 pgs.; Human Immortality and Electronic Civilization, 3-rd Edition, Lulu, 2007, 66 pgs.

Cathcart, Richard Brook (1943-)

A geographer specializing in macroengineering. Richard Cathcart is the author of many papers and articles dealing with novel aspects of superscale engineering and astroengineering projects. He was awarded an MA in Geography and founded GEOGRAPHOS, a Burbank, California, consultancy, in 1969. His current research interests include the history of anthropogeomorphology and terraforming; in 1995 he completed a 600-page textbook on these subjects which emphasizes the geopolitical aspects of famous as well as little-known macroproject proposals.

ABSTRACT

In recent years of the 21st Century the authors of this book and other scientists as well, have instigated and described many new macro-projects, USA and other countries patented concepts, speculative Macro-engineering ideas, and other general innovations in technology and environment change. These all hold the enticing promise for a true revolution in the lives of humans everywhere in the Solar System.

Here, the authors include and review new methods for travel in outer space, promising means to increase the safety of aviation, comfortable permanent settlements on the Moon and Mars, as well as for Earth's hazardous polar regions, control of local and global weather conditions, new methods of irrigation "without water", conversion of cold and hot deserts and Earth's non-twin polar regions to 'evergreen' gardens, protection against forecasted hurricane storm surge waves and inundations, unpredictable tsunami, and other weather-related floods, cheap protection of cities against nuclear warheads and aviation bombs, magnetic aircraft, channels for free traveling in outer space, thermonuclear reactors, wind power stations, along with many others.

Here we succinctly summarize some of these revolutionary macro-projects, concepts, ideas, innovations, and methods for scientists, engineers, technical students, and the world public. We do seek future attention from the general public, other macro-engineers, inventors, as well as scientists of all persuasions for these presented innovations. And, naturally, we fervently hope the popular news media, various governments and the large international aerospace and other engineering-focused corporations will, as well, increase their respective observation, R&D activity in the technologies for living and the surrounding human environment.

PREFACE

New macro-projects, concepts, ideas, methods, and innovations are explored here, but hardly developed. There remain many problems that must be researched, modeled, and tested before these summarized research ideas can be practically designed, built, and utilized—that is, fully developed and utilized.

Most ideas in our book are described in the following way: 1) Description of current state in a given field of endeavor. A brief explanation of the idea researched, including its advantages and short comings; 2) Then methods, estimation and computations of the main system parameters are listed, and 3) A brief description of possible applications—candidate macro-projects, including estimations of the main physical parameters of such economic developmental undertakings.

The first and third parts are in a popular form accessible to the wider reading public, the second part of this book will require some mathematical and scientific knowledge, such as may be found amongst technical school graduate students. Our book gives the main physical data and technical equations in attachments which will help researchers, engineers, dedicated students and enthusiastic readers make estimations for their own macro-projects. Also, inventors will find an extensive field of inventions and innovations revealed in our book.

The authors have published many new ideas and articles and proposed macro-projects in recent years (see: General References). Our book is useful as an archive of material from the authors' own articles published during the last few years.

Acknowledgement

1. Some data in this work is garnered from Wikipedia under the Creative Commons License. 2. The authors wish to acknowledge Joseph Friedlander for help in editing of this book.

PART I. TECHNOLOGY

Chapter 1

AB METHOD OF IRRIGATION WITHOUT WATER (CLOSED-LOOP WATER CYCLE)*

ABSTRACT

Authors methodically researched a revolutionary Macro-engineering idea for a closed-loop freshwater irrigation and in this chapter it is unveiled in some useful detail. We offer to cover a given site by a thin, enclosure film (with controlled heat conductivity and clarity) located at an altitude of 50 – 300 m. The film is supported, at its working altitude, by small additional induced air over-pressuring, and anchored to the ground by thin cables. We show that this closed dome allows full control of the weather within at a given planetary surface region (the day is always fine, it will rain only at night, no strong winds). The average Earth (white cloudy) reflectance equals 0.3 - 0.5. Consequently, Earth does lose about 0.3 - 0.5 of the maximum potential incoming solar energy. The dome (having control of the clarity of film and heat conductivity) converts the cold regions to controlled subtropics, hot deserts and desolate wildernesses to prosperous regions blessed temperate climate. This is, today, a realistic and cheap method of evaporation-economical irrigation and virtual weather control on Earth!

Keywords: Global weather control, gigantic film dome, converting a cold region to subtropics, converting desolate wilderness to a prosperous region

INTRODUCTION

1. Precipitation

1. General Information about Precipitation

The mass of water in Earth's hydrosphere during the current era is taken to be a physical constant of our home planet. The average annual layer of Earth's precipitation is about 1000 mm or 511,000 km^3. Twenty-one percent of this (108,000 km^3) falls on land and 79%

---

* Presented in http://arxiv.org in 2007.

(403,000 km^3) on the ocean. Most of it falls between geographical latitudes 20° North and 20° South. Both polar zones collect only 4% of Earth's precipitation. ("Precipitation" is scientifically defined as including both rainfall and snowfall.) The evaporation from the ocean equals 1250 mm (450,000 km^3). About 1120 mm returns to the ocean as precipitation on it and 130 mm by river inflows. Evaporation from land equals 410 mm (61,000 km^3) while the precipitation on land is 720 mm. The land loses 310 mm as river flows to the ocean (47,000 km^3). These are average data. In some regions the precipitation is very different, distinctly non-average.

2. A Desert

A desert is a landscape form or region that receives very little precipitation. Deserts are defined as areas that receive, an average, an annual precipitation of less than 250 mm. In the commonly used Köppen Climate Classification, deserts are classed as (BW).

Deserts take up at least one-third of the Earth's land surface. They usually have a large diurnal and seasonal temperature range, with high daytime temperatures (during summertime up to 45 °C), and low night-time temperatures (in winter down to 0 °C) due to extremely low air humidity. Water acts to trap infrared radiation from both the Sun and emitted by the heated ground, and dry desert air is incapable of blocking sunshine during the day or trapping heat after nightfall. Thus, during daylight, all of the Sun's energy reaches the desert surface. As soon as the Sun sets, the desert cools quickly by radiating its heat to outer space above. Urban areas in deserts lack large (>14 °C) daily temperature ranges, partially due to the "urban heat island effect".

Many of the Earth's deserts are located on the lee-side of blocking mountains, the "rain shadows", the mountains having wrung water as precipitation from the passing wind masses on the windward side. Deserts are often composed of almost vegetation-less rock and sandy ground surfaces.

Bottomlands—low valley places which, sometimes, contain watercourses—may be salt-covered flats. Winds and flashfloods are major factors in shaping all desert landscapes. Polar region deserts have similar landscapes but the main form of precipitation is snow rather than rain. The continent of Antarctica is the world's largest cold desert. It is composed of about 98% kilometers-thick icesheet and ~2% barren rock. The largest hot desert is the Sahara of northern Africa. Deserts, like other Earthly locales, sometimes do contain valuable mineral deposits that were probably formed in the arid environment, or were exposed by water flows and wind erosion.

Rain *does* fall occasionally in deserts, and desert storms are often violent. A record 44 millimeters of rain once fell within three hours on the Sahara. Large Saharan storms may deliver up to one millimeter per minute. Normally dry stream channels there, called wadis, can quickly fill after heavy rains, and devastating flashfloods make these channels dangerous places.

Though little rain falls in deserts, deserts receive runoff from ephemeral, or short-lived, streams fed considerable quantities of sediment for a day or two. Although most deserts are in basins with closed or interior drainage, a few deserts are crossed by 'exotic' rivers (such as Africa's Nile River) that derive their water from outside the desert traversed by the watercourse. Such rivers infiltrate soils and evaporate large amounts of water on their journeys through the deserts, but their volumes are such that they maintain their continuity. Deserts may also have underground springs, rivers, or reservoirs that lay close to the surface, or deep underground. Plants that have not completely adapted to sporadic rainfalls in a desert environment may tap into underground water sources that can be reached by the extensive and deep their tapping roots.

Lakes form where rainfall or snow meltwater in interior drainage basins is sufficient or excesssive. Desert lakes are generally shallow, temporary, and salty. Because these lakes are shallow and have a low bottom gradient, wind stress may cause the lake waters to move over many square kilometers. When small lakes dry up, they leave a salt crust. The flat area of clay, silt, or sand encrusted with salt that forms is known as a playa. Most are relics of large lakes that existed during Earth's most recent Ice Age occurring some 12,000 or more years before the present time. Because playas are arid landforms from a wetter past, they contain useful clues to climatic change.

When the occasional precipitation does occur, it erodes the desert rocks quickly. Winds are the other affective factor that erodes deserts—they are often slow, yet constant.

A desert is a hostile, potentially deadly environment for unprepared humans. The high temperature causes rapid loss of water due to sweating, which can result in dehydration and death eventual within just a few days. In addition, unprotected humans are also at risk from heatstroke. Despite this, some cultures have made deserts their home for thousands of years. Modern technology, including advanced land irrigation systems, desalination and aira conditioning technology have made deserts much more settleable. In the USA and Israel, profitable commercial desert farming takes place.

Australia's Great Sandy Desert receives nearly all of its rain from seasonal monsoonal thunderstorms, or the occasional tropical cyclone. Thunderstorm days average 20-30 annually (Burbidge 1983). Although the desert has fairly high precipitation rates due to the high rates of evaporation, this region remains an unimproved arid environment featured by vast areas of sand landforms.

Other regions of the world's drylands, which get these rare precipitation events, are northwestern Mexico, southwestern South America, and southwestern Asia. In North America, the Southwest's desert has received some tropical rainfall at infrequent times. Tropical activity is rare in all deserts, but what rain does fall is important to the existing ecosystem.

3. Arid

In general terms, the climate regime of a specified region is assumed to be arid when it is characterized by a severe lack of available freshwater, to the extent of hindering or even preventing the growth and normally-paced lifetime development of plants and animals. As a result, places on land subject to arid climates tend to lack vegetation and are called xeric or desertic environments.

The expression 'available water' refers to water in the soil in excess to the wilting point. The air over a hot desert may actually contain substantial amounts of water vapor but that water may not be generally accessible to plants, except for very specialized organisms (lichen, for example). 'Lack of water' refers to use by plants. The freshwater that is actually present in the environment may be sufficient for some species or usage, and grossly insufficient for others usages. Aridity, the characteristic nature of dryland climates, may thus depend on the land's natural and artificial use. For life, what is more important than the amount of rain is the fraction of precipitation that is not quickly lost by evaporation or runoff. Attempts to quantitatively describe the degree of aridity of a place have led to the development of indicative aridity indexes. There is no universal agreement on the precise boundaries between desert classes such as "hyper-arid", "arid", "semi-arid".

Figure 1. Mahktesh Gadol, an erosional basin in the Negev Desert of southern Israel.

While different classification schemes and global or regional maps differ in their details, there is a general agreement about the fact that large regions of our Earth are recognized as indisputably "arid". These include the hot deserts, located broadly in sub-tropical regions, where the accumulation of freshwater is largely prevented by either low precipitation, or high evaporation, or both, as well as cold deserts near the planet's poles, where freshwater may be permanently represented by snow and ice. Other arid regions include regions located in the rain shadows of major high mountain ranges or along coastal regions affected by significant ocean water upwellings (such as South America's Atacama Desert).

The distribution of aridity observed at any one place over time is largely the result of the general circulation of our Earth's atmosphere. The latter does change significantly over time through global climate regime change. In addition, changes caused by altered land use can result in greater plant demands on soil water and, thus, induce a higher degree of prevalent aridity.

4. Drought

Drought is an extended period of months or years when a region notes a deficiency in its commonly expected freshwater supply. Generally, this occurs when a region receives consistently below-average precipitation. It can have a substantial impact on the natural ecosystem and the agriculture of the settled region affected. Although droughts can persist for several years, even a short, intense drought can cause significant damage and do economic harm the local economy.

Drought is a normal, recurring feature of the climate regimes prevailing in most parts of the world. Having adequate drought mitigation strategies can greatly reduce the impact of droughts. Recurring or long-term drought can bring about desertification. Recurring droughts in Africa have created grave ecological problems, prompting massive food shortages. Some have suggested that long-term drought in Sudan's Darfur region, also affecting adjacent Chad, is the cause of human conflict there. A combination of drought, desertification and over-population are among the causes of the infamous Darful social conflict because the nomads searching for freshwater supplies for their grazing livestock have had to drive their herds further south than usual and into farmland mainly occupied by settled people.

The Himalayan glaciers that are the sources of Asia's biggest rivers – the Ganges, Indus, Brahmaputra, Yangtze, Mekong, Salween and Yellow rivers – could, according to some reported climatological scenarios, disappear by 2035 as global air temperature rise. Of our Earth's 6.4 billion persons, approximately 2.4 billion live in the drainage basins of the Himalayan rivers. Paradoxically, during coming years of the 21st Century, India, China, Pakistan, Bangladesh, Nepal and Myanmar could experience river floods followed by droughts. Drought in India affecting the Ganges River is of particular concern, as it provides drinking and agricultural irrigation freshwater for more than 500 million people in that country.

In 2005, parts of the Amazon River Basin experienced the worst drought in a century. Scientists at the Brazilian National Institute of Amazonian Research allege that the current drought response, coupled with the alteration effects of deforestation on the region's climate, are pushing the Amazon River Basin's rainforest towards a "tipping point", where it would irreversibly start to die. Researchers there suggest that the tropical rainforest is on the cusp of becoming a savvana or grassland region, with catastrophic consequences for the world's climate. The combination of abrupt climate change and deforestation increases the drying effect of dead trees and dead trees can become fuel for widespread Amazon River Basin forests fires, which then release carbon dioxide gas into the planet's atomosphere, possibly amplifying enchanced global warming.

5. Tundra

Tundra is a land region where tree growth is markedly hindered by low air temperatures and short growing seasons. There are two types of tundra: Arctic tundra (which also occurs in Antarctica on its few places without permanent icesheet), and alpine tundra. In tundra, the vegetation is composed of dwarf shrubs, sedges and grasses, mosses, and lichens. Scattered trees grow on some tundra landscapes. The ecological boundary between the tundra and the traditional forest is the "tree line" or, on mountain-sides, the "timberline".

Arctic tundra occurs in the far Northern Hemisphere, north of the taiga belt. The word "tundra" usually refers only to the areas where the subsoil is permafrost, or permanently frozen soil. Permafrost tundra includes vast areas of northern Russia and northern Canada. The polar tundra is home to several peoples who are mostly nomadic reindeer herders.

The Arctic tundra is a vast area of starkly barren landscape, which is frozen most of the year. The soil there is frozen from 25-90 cm down, and it is impossible for normal-size trees to grow. Instead, bare and sometimes rocky land can only support low growing plants such as moss, heath, and lichen. There are two main seasons, winter and summer, in the polar Tundra regions. During the wintertime it is very cold and dark, with the average temperature around -28 °C, sometimes dipping as low as -50 °C. However, extreme cold temperatures on the tundra do not get as low as those occurring in taiga regions further south. (Russia and Canada's lowest temperatures were recorded in locations south of the treeline). During summer, air temperatures rise, and the topmost layer of the permafrost melts, leaving the ground very soggy. The tundra is covered with marshes, lakes, bogs and streams during the warmest months. Generally daytime temperatures during the summer rise to about 12°C but can often drop to 3°C, or below freezing. Arctic tundras are sometimes the object of habitat conservationprograms. In Canada and Russia, for instance, many of these regions are protected by national biodiversity laws.

The tundra is a very windy area, with air often blowing upwards at 48–97 km/h. However, in terms of precipitation, it is desert-like, with only about 15–25 cm falling yearly; summertime is, typically, the season of maximum precipitation. During summer, the permafrost thaws only enough to let plants grow and reproduce, but because the ground below this thawed soil zone remains frozen, the freshwater cannot sink any lower, and so the freshwater forms lakes and marshes that exist during the summer. Although precipitation is low in quantity, evaporation is also relatively minimal.

The biodiversity of the tundras is low: 1,700 species of vascular plants and only 48 land mammals can be found, although thousands of insects and birds migrate there each year to the marshes. There are also a few fish. There are few species, however, with really large populations. Notable animals in the Arctic tundra include caribou (reindeer), musk ox, arctic hare, arctic fox, snowy owl, lemmings, and polar bears that live mainly on the flow ice of the Arctic Ocean.

Due to the harsh climate of the Arctic's tundra, regions of this kind have seen minimal human activity, even though they are sometimes rich in natural subterranean resources such as oil, natural gas, diamonds and uranium. Global warming evidently poses a severe threat to the tundras, specifically to the permafrost. Permafrost is essentially a frozen bog – during the brief summertime, only its surface melts. The melting of the permafrost in a given region, on human a time scale of decades or centuries, could radically change which species can survive in the Arctic's tundra.

Figure 2. Permafrost.

Another concern is that about one third of the world's soil-bound carbon is in taiga and tundra regions. If and when the finally permafrost melts, it is likely to release carbon dioxide gas, which is a greenhouse gas. The expected effect has already been observed in northern Alaska. The world's tundra was once a carbon sink, but during the 21st Century, in all liklihood, it will become a significant source of carbon when it is released by possibly permanent tundra melting.

6. Permafrost

65% of all Russian Federation territory is classed as permafrost. Permafrost or permafrost soil is soil at or below the freezing point of water (0 °C) for two or more years. Ice is not always present, as may be in the case of non-porous bedrock, but it frequently occurs and it may be in amounts exceeding the potential hydraulic saturation of the ground material. Most permafrost is located in high latitudes (e.g., Earth's North and South Poles), but alpine permafrost exists at high altitudes in mountains.

The extent of permafrost can vary as the Earth's climate changes over time. Today, approximately 20% of the Earth's landmass is covered by permafrost (including discontinuous permafrost) or glacial ice. Overlying permafrost is a thin *active layer* that seasonally thaws during the summer. Plant life can be supported only within the active layer since growth can occur only in soil that is fully thawed for some part of the year. Thickness of the active layer varies by year and location, but is typically 0.6 – 4 m thick. In regions of continuous permafrost and harsh winters, the depth of the permafrost can be as much as 1493 m in the northern Lena River and Yana River watersheds of Siberia.

2. Irrigation

The reader learns, then, that more than 50% of the Earth's land surface is in dry or cold regions which are, by definition of these growing conditions unsuitable for agriculture. Many dry regions and deserts are not used because no freshwater is available for plant irrigation agriculture. In the rare cases where it is practical, humanity spends a lot of money and expends a lot of costly energy to obtain freshwater, digging irrigation canals, constructing water pumping stations and freshwater distribution systems—and in the long term, still more effort, time and expense is applied to fighting the inevitable silting of canals which literally can require rebuilding such works generation by generation.

1. Irrigation

Irrigation is the artificial application of freshwater to the potentially fertile soil, usually for assisting the growth of crops of food and fiber. In crop production, it is mainly used to replace absent rainfall during periods of drought, but also to protect plants against sudden frosts. Additionally, irrigation suppresses weed growth in rice-fields. In contrast, agriculture that relies only on direct rainfall is sometimes referred to as dryland farming or as rain-fed farming. Irrigation techniques are generally studied in conjunction with drainage techniques, which are the natural or artificial removal of surface and sub-surface water from a selected land region.

Various types of irrigation techniques differ in how the freshwater obtained from a source is ultimately distributed within the cropped field. In general, the farmer's goal is to supply the entire field uniformly with freshwater, so that each plant has the amount of freshwater it actually needs to flourish.

By the mid-20th Century, the advent of diesel engines and electric motors led, for the first time, to irrigation systems that could pump groundwater from out of major deep-lying aquifers faster than it was naturally recharged. This can lead to permanent a loss of aquifer capacity caused by compression of the rock formation holding the freshwater, decreased freshwater quality, ground subsidence at the land surface, and other problems. The future of food and fiber production in such world-famous farming regions as the North China Plain, the India-Pakistan Punjab, and the USA's Great Plains is threatened by over-pumping of groundwater resources.

At the global scale, by the year 2000 AD, 2,788,000 sq km of agricultural land was equipped with irrigation infrastructure. About 68 % of the landscape equipped for freshwater irrigation is located in Asia, 17 % in North America, 9 % in Europe, 5 % in Africa and 1 % in Oceania. The largest contiguous region of high irrigation density are found in North India and Pakistan along the rivers Ganges and Indus, in the Hai He, Huang He and Yangtze watersheds of China, along the Nile River in Egypt and the Sudan, in the Mississippi-Missouri river basins and in parts of California. Smaller irrigation areas are spread across almost all populated parts of our world.

Irrigation gives high-stability harvests which are ~ 3 – 5 times more than conventional agriculture would provide humanity. Sources of irrigation freshwater can be groundwater extracted from springs or by using bored wells, surface water withdrawn from rivers, lakes or reservoirs or non-conventional sources like treated wastewater, desalinated water or drainage water. A special form of irrigation, using surface water, is spate irrigation, also called floodwater harvesting. In case of a flood (spate) water is diverted to normally dry river beds (wadis) using a network of dams, gates and channels and spread over large land regions. The moisture stored in the soil will be used thereafter to grow crops. Spate irrigation regioms are, in particular, located in semi-arid or arid valleys of mountain regions. While floodwater harvesting belongs to the accepted irrigation methods, rainwater harvesting is usually not considered as a form of irrigation. Rainwater harvesting is the collection of runoff water from roofs or unused land and the concentration of this collected water on cultivated land.

Problems in Conventional Irrigation

- Competition for surface water rights.
- Depletion of underground aquifers.
- Ground subsidence.
- Build-up of toxic salts on soil surface in place of high evaporation. (This requires either leaching to remove these salts and a method of drainage to carry the salts away or use of mulch to minimize evaporation.)
- Over-irrigation because of non-uniformity of water distribution or poor management of waste water. Chemicals may lead to water pollution.

2. Hydroponics

Hydroponics is a method for growing selected useful plants using mineral nutrient solutions instead of natural soil. Terrestrial plants may be grown with their roots in the mineral nutrient solution only, or in an inert stabilizing soil-like medium, such as perlite, fine gravel or rockwool.

19th Century plant physiologistsd discovered that green growing plants absorb essential mineral nutrients as inorganic ions in water. In natural conditions, the holding soil acts as a mineral nutrient reservoir but the soil itself is not essential to plant growth. When the mineral nutrients in the soil dissolve in water, plant roots are able to absorb them. When the required mineral nutrients are introduced into a plant's water supply artificially, soil is no longer required for the plant to thrive. Almost any terrestrial plant will grow with hydroponics, but some species do better than other species. It is also very easy as farmwork goes; the activity is often undertaken by very young children with such plants as watercress of the always-loved Chia Pet toys. Hydroponics is also a standard technique in Biology research and techical teaching and a popular hobby.\

Due to its arid climate, Israel has developed advanced hydroponic technology. The largest commercial hydroponics facility in the world is Eurofresh Farms situated in Wilcox, Arizona, USA. Eurofresh has 256 acres under glass and represents about a third of the commercial hydroponic greenhouse area in the U.S.A. Eurofresh does not consider their tomatoes organic, but they are pesticide-free. They are grown in rockwool using drip irrigation.

Some commercial installations use no pesticides or herbicides, preferring integrated pest management techniques. There is often a price premium willingly paid by eager consumers for fresh produce which is legally labeled "organic". Some states in the USA require soil as an essential to obtain organic certification. There are also overlapping and somewhat contradictory rules established by the US Federal Government, so some food grown with hydroponics can be certified organic. In fact, they are the cleanest plants possible because there is no environment variable and the dirt in the food supply is extremely limited. Hydroponics also saves an great amount of freshwater; it uses as little as 1/20 the amount as a regular farm to produce the same amount of food. The water-table can be adversely impacted by the freshwater use as well as the contamination of runoff water with chemicals used by farmers. Hydroponics may assist in minimizing such impacts, as well as having the advantage that water use and water returns are easier to measure. This can save the commercial farmer money expenditures by allowing reduced freshwater use and the ability to measure consequences to the unused land surrounding farms.

The environment in a hydroponics glasshouse is strictly controlled for maximum efficiency,and this new industrial mindset is called "Soil-less/Controlled Environment Agriculture" (S/CEA). With this, farmers can make ultra-premium foods just about anywhere in the world, regardless of temperature and growing seasons. Growers monitor the temperature, humidity, and PH level constantly. Hydroponics have been used to enhance vegetable nutrition.

Advantages, disadvantages, and misconceptions:

- While removing soil-grown crops from the ground effectively kills them, hydroponically grown crops such can be packaged and sold while still alive, greatly increasing the length of freshness at purchase time.

- Solution-culture hydroponics does not require disposal of a solid medium or sterilization and reuse of a solid medium.
- Solution-culture hydroponics allows greater control over the root zone environment of plants than does common soil culture.
- Over- and under-watering is prevented
- Hydroponics is often the best crop production method in remote areas that lack suitable soil, such as the Arctic, Antarctica, orbitingstations in outer space, future space colonies, or on Earthly ocean atolls such as Wake Island.
- In solution-culture hydroponics, plant roots are visible to the farmer.
- Soil-borne plant diseases are virtually absent.
- Weeds are virtually absent.
- Fewer pesticides may be required because of the above two reasons.
- Edible crops are not contaminated with soil.
- Freshwater use can be substantially less than with outdoor irrigation of soil-grown crops.
- Hydroponics may cost more than 20% less than common cultivation techniqus.
- Many hydroponic systems give the plants more nutrition while at the same time using less energy and space.
- Hydroponics allow for easier fertilization since it is possible to use an automatic timer-run machine to fertilize plants.
- It provides the growing green plant with a fully balanced nutrition because the essential nutrients are dissolved into the water-soluble nutrient-solution.
- If timers or electric pumps fail, or the system clogs or springs a leak, plants can die very quickly in many kinds of existing hydroponic systems.
- Hydroponics usually requires a greater technical knowledge and technical mastery than geoponics (soil-grown crops).
- For the previous two reasons, and the fact that most hydroponic crops are grown in greenhouses or controlled environment agriculture, hydroponic crops are usually more expensive than geoponic crops.
- Solution culture hydroponics requires that the plants be supported because the roots have no anchorage without a solid medium.
- The plants will die if not frequently monitored while soil plants do not require such close attention.

There are many widely held misconceptions regarding hydroponics, as noted by the following facts:

- Hydroponics will not always produce greater crop yields than good quality soil.
- Hydroponic plants cannot always be spaced closer together than soil-grown crops under the same overall environmental conditions.
- Hydroponic produce will not necessarily be more nutritious or better tasting than geoponics.

Hydroponics will, however, grow 30% faster and cost less markedly less. They are also proven to be healthier and more productive foods.

With pest problems reduced, and nutrients constantly fed to the plant roots, productivity in hydroponics is high, plant growth being limited by the low levels of carbon dioxide gas in the atmosphere, or limited light. To increase plant yield further, some sealed greenhouses inject carbon dioxide into their environment to stimulate growth (CO_2-enrichment), or add artificial lights to lengthen the crop growing day and control vegetative growth.

Hydroponic technology allows for growing green plants where no one has grown before, be it underground, or above, in outer space or in the ocean, this technology allows humanity to live where humanity chooses. If used for our own survival or our colonisation of extra-Earth places, hydroponics is and will be a major part of humankind's future.

3. Control of Local Weather

Governments spend billions of dollars merely studying the weather. The many big government scientific research organizations, and perhaps a hundred thousand scientists, have been studying Earth's weather for more than a hundred years. There are innumerable published and compiled scientific reports on weather control. Most reports are of little practical value, some are magnificently impractical proposals to study or control our world's weather. We cannot exactly predict weather at long-range, to avert a rain storm, strong wind, tornado, or hurricane. We cannot control the movement of the clouds, temperature and humidity of the atmosphere. We cannot yet make much more tolerable a harsh winter or very hot summer. We cannot yet convert a naturally cold Earth region to subtropics, a desolate wilderness to an economically prosperous region. We can only observe the cyclonic storms and hurricanes, and approximately predict their direction of movement. We can only advise people on where they ought not to be present. Every year damaging storms, hurricanes, strong winds and rains and inundations destroy thousands of houses, kill thousands of people.

2. Description and Innovations

Our idea is a closed dome covering a local region by a thin film with controlled heat conductivity and optionally-controlled clarity (reflectivity, albedo, carrying capacity of solar spectrum). The film is located at an altitude of ~50 – 300 m. The film is supported at this altitude by a small additional air pressure produced by ventilators sitting on the ground. The film is connected to Earth's surface by tethering cables. The cover may require double-layer film. We can control the heat conductivity of the dome cover by pumping in air between two layers of the dome film cover and change the solar heating due to sunlight heating by control of the cover's clarity. That allows selecting for different conditions (solar heating) in the covered area and by pumping air into dome. Envisioned is a cheap film having liquid crystal and conducting layers. The clarity is controlled by application of selected electric voltage. These layers, by selective control, can pass or blockade the available sunlight (or parts of solar spectrum) and pass or blockade the Earth's radiation. The incoming and outgoing radiations have different wavelengths. That makes control of them separately feasible and, therefore, possible to manage the heating or cooling of the Earth's surface under this film. In conventional conditions about 50% of the solar energy reaches the Earth surface. Much is reflected back to outer space by white clouds that shade approximately 65% of the Earth's land/water surface. In our closed water system the rain (or at least condensation) will occur at night when the temperature is low. In open atmosphere, the Sun heats the ground; the ground must heat the whole troposphere (4 – 5 km) before stable temperature rises happen. In our case, the ground heats ONLY the air in the dome (as in a hotbed). We have, then, a literal greenhouse effect. That means that many cold regions (Alaska, Siberia, northern Canada) may absorb more solar energy and became a temperate climate or sub-tropic climate (under the dome, as far as plants are concerned). That also means the Sahara and other deserts can be a prosperous regions with a fine growing and living climate and with a closed-loop water cycle.

The building of a film dome is easy. We spread out the film over Earth's surface, turn on the pumping propellers and the film is raised by air over-pressure to the needed altitude limited by the support cables. Damage to the film is not a major trouble because the additional air pressure is very small (0.0001- 0.01 atm) and air leakage is compensated for by spinning propeller pumps. Unlike in an outer space colony or extra-Earth planetary colony, the outside air is friendly and, at worst, we might lose some heat (or cold) and water vapor.

The first main innovation of our dome, and its main difference from a conventional hotbed, or glasshouse, is the inflatable HIGH span of the closed cover (up to 50 – 300 m). The elevated height of the enclosed volume aids organizing of a CLOSED LOOP water cycle - accepting of water vaporized by plants and returning this water in the nighttime when the air temperature decreases. That allows us to perform irrigation in the vast area of Earth's land that does not have enough freshwater for agriculture. We can convert the desert and desolate wildernesses into Eden-like gardens without expensive delivery of remotely obtained and transported freshwater. The initial amount of freshwater for water cycle may be collected from atmospheric precipitation in some period or delivered. Prime soil is not a necessity because hydroponics allows us to achieve record harvests on any soil.

The second important innovation is using a cheap controlled heat conductivity, double-layer cover (controlled clarity is optionally needed for some regions). This innovation allows to conserve solar heat (in cold regions), to control temperature (in hot climates). That allows two to three rich crops annually in the Earth's middle latitudes and to conversion of the cold zones (Siberia, northern Canada, Alaska) to good single-crop regions.

The third innovation is control of the cover height, which allows adapting to local climatic seasons.

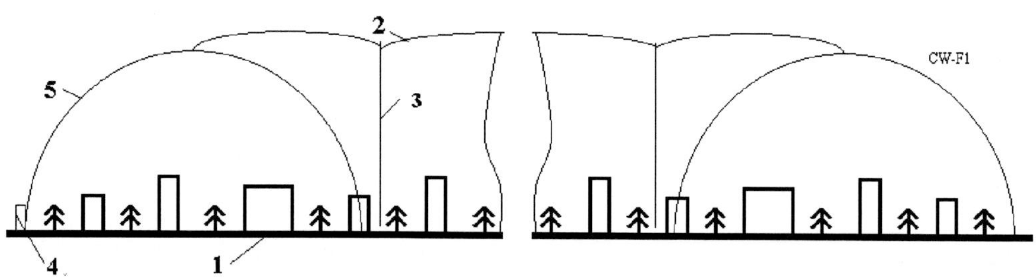

Figure 3. Film dome over agriculture region or a city. *Notations*: 1 - area, 2 - thin film cover with a control heat conductivity and clarity, 3 – control support cable and tubes for rain water (height is 50 – 300 m), 4 - exits and ventilators, 5 - semi-cylindrical border section.

The fourth innovation is the use of cheap, thin film as the high altitude cover. This innovation decreases the construction cost by thousands of times in comparison with the conventional very expensive glass-concrete domes offered by some for city use.

Lest it be objected that such domes would take impractical amounts of plastic, consider that the world's plastic production is today on the order of 100 million metric tons. If, with the expectation of future economic growth, this amount doubles over the next generation, and the increase is used for doming over territory at 500 tons a square kilometer, about 200,000 square kilometers could be roofed over annually. While small in comparison to the approximately 150 million square kilometers of land area, consider that 200,000 1 kilometer sites scattered over the face of the Earth made newly inhabitable could revitalize vast swaths of land surrounding them—one square kilometer could grow local vegetables for a city sited in the desert, one over there could grow bio-fuel, enabling a desolate South Atlantic island to become independent of costly fuel imports; at first, easily a billion people a year could be taken out of sweltering heat, biting cold and drenching rains, saving money that purchase, installation and operation of HVAC equipment—heating, ventilation, air-conditioning—would require.

Our dome design is presented in Figure 3 includes the thin inflated film dome. The innovations are listed here: (1) the construction is air-inflatable; (2) each dome is fabricated with very thin, transparent film (thickness is 0.1 to 0.3 mm) having controlled clarity and controlled heat conductivity without rigid supports; (3) the enclosing film has two conductivity layers plus a liquid crystal layer between them which changes its clarity, color and reflectivity under an electric voltage (Figure 4); (4) the bounded section of the dome proposed that has a hemispheric shape (#5, Figure 3). The air pressure is greater in these sections, and they protect the central sections from wind outside.

Figure 3 illustrates the thin transparent control dome cover we envision. The inflated textile shell—technical "textiles" can be woven or films—embodies the innovations listed: (1) the film is very thin, approximately 0.1 to 0.3 mm., implying under 500 tons per square kilometer. A film this thin has never before been used in a major building; (2) the film has two strong nets, with a mesh of about 0.1 × 0.1 m and $a = 1 \times 1$ m, the threads are about 0.5 mm for a small mesh and about 1 mm for a big mesh. The net prevents the watertight and airtight film covering from being damaged by vibration; (3) the film incorporates a tiny electrically conductive wire net with a mesh of about 0.1 x 0.1 m and a line width of about 100 μ and a thickness near 10 μ. The wire net is electric (voltage) control conductor. It can inform the dome maintenance engineers concerning the place and size of film damage (tears, rips); (4) the film may be twin-layered with the gap — $c = 1$ m and $b = 2$ m—between film layers for heat insulation. In Polar (and hot Tropic) regions this multi-layered covering is the main means for heat isolation and puncture of one of the layers wont cause a loss of shape because the second film layer is unaffected by holing; (5) the airspace in the dome's covering can be partitioned, either hermetically or not; and (6) part of the covering can have a very thin shiny aluminum coating that is about 1μ (micron) for reflection of unneeded sunlight in the equatorial region, or collect additional solar radiation in the polar regions [1].

The authors offer a method for moving off the accumulated snow and ice from the film in polar regions. After snowfall we decrease the heat cover protection, heating the snow (or ice) by warm air flowing into channels 5 (Figure 4) (between cover layers), and water runs down into tubes 3 (Figure 3).

The town cover may be used as a screen for projecting of pictures, films and advertising on the cover at nighttime.

Brief Data on Cover Film

Our dome filmic cover has 5 layers (Figure 4c): transparant dielectric layer, conducting layer (about 1 - 3 μ), liquid crystal layer (about 10 - 100 μ), conducting layer (for example, SnO_2), and transparant dielectric layer. Common thickness is 0.1 - 0.5 mm. Control voltage is 5 - 10 V. This film may be produced by industry relatively cheaply.

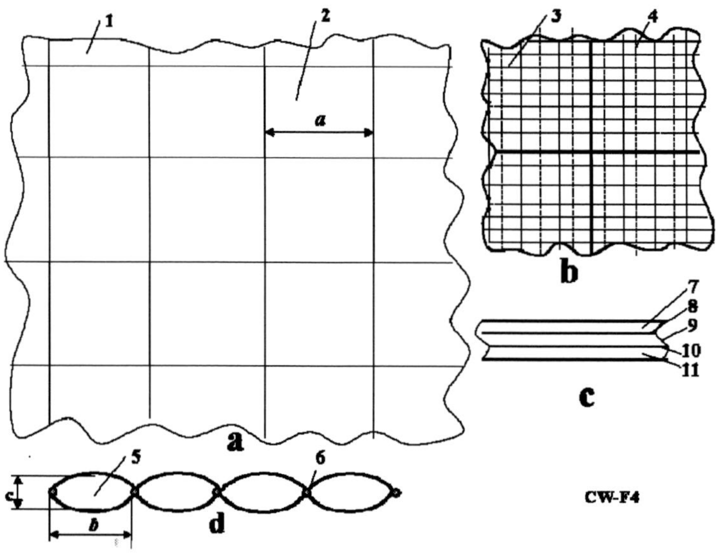

Figure 4. Design of membrane covering. *Notations*: (a) Big fragment of cover with control clarity (reflectivity, carrying capacity) and heat conductivity; (b) Small fragment of cover; (c) Cross-section of cover (film) having 5 layers; (d) Longitudinal cross-section of cover for cold and hot regions; 1 - cover; 2 -mesh; 3 - small mesh; 4 - thin electric net; 5 - cell of cover; 6 - tubes;: 7 - transparant dielectric layer, 8 - conducting layer (about 1 - 3 μ), 9 - liquid crystal layer (about 10 - 100 μ), 10 - conducting layer, and 11 - transparant dielectric layer. Common thickness is 0.1 - 0.5 mm. Control voltage is 5 - 10 V.

1. Liquid Crystals (LC)

Liquid crystals (LC) are substances that exhibit a phase of matter that has properties between those of a conventional liquid, and those of a solid crystal.

Liquid crystals find general employment in liquid crystal displays (LCD), which rely on the optical properties of certain liquid crystalline molecules in the presence or absence of an electric field. On command, the electric field can be used to make a pixel switch between clear or dark. Color LCD systems use the same technique, with color filters used to generate red, green, and blue pixels. Similar principles can be used to make other liquid crystal-based optical devices. Liquid crystal in fluid form is used to detect electrically generated hotspots for failure analysis in the semiconductor industry. Liquid crystal memory units with extensive capacity were used in the USA's Space Shuttle navigation equipment. Worth noting also is the fact that many common fluids are, in fact, liquid crystals. Soap, for instance, is a liquid crystal, and forms a variety of LC phases depending on its concentration in water.

The conventional control clarity (transparancy) film reflected all superfluous energy to outer space. If the film has solar cells then it may convert the once superfluous solar energy into harnessed electricity.

2. Transparency

In optics, transparency is the material property of passing natural and artificial light through any material. Though transparency usually refers to visible light in common usage, it may correctly be used to refer to any type of radiation. Examples of transparent materials are air and some other gases, liquids such as water, most non-tinted glasses, and plastics such as Perspex and Pyrex. The degree of material transparency varies according to the wavelength of the light. From electrodynamics it results that only a vacuum is really transparent in the strictist meaning, any matter has a certain absorption for electromagnetic waves. There are transparent glass walls that can be made opaque by the application of an electric charge, a technology known as electrochromics.Certain crystals are transparent because there are straight-lines through the crystal structure. Light passes almost unobstructed along these lines. There exists a very complicated scientific theory "predicting" (calculating) absorption and its spectral dependence of different materials.

3. Electrochromism

Electrochromism is the phenomenon displayed by some chemical species of reversibly changing color when a burst of electric charge is applied.

One good example of an electrochromic material is polyaniline which can be formed either by the electrochemical or chemical oxidation of aniline. If an electrode is immersed in hydrochloric acid which contains a small concentration of aniline, then a film of polyaniline can be grown on the electrode. Depending on the redox state, polyaniline can either be pale yellow or dark green/black. Other electrochromic materials that have found technological application include the viologens and polyoxotungstates. Other electrochromic materials include tungsten oxide (WO_3), which is the main chemical used in the production of electrochromic windows or smart windows.

As the color change is persistent and energy need only be applied to effect a change, electrochromic materials are used to control the amount of light and heat allowed to pass through windows ("smart windows"), and has also been applied in the automobile industry to automatically tint rear-view mirrors in various lighting conditions. Viologen is used in conjunction with titanium dioxide (TiO_2) in the creation of small digital displays. It is hoped that these will replace LCDs as the viologen (which is typically dark blue) has a high contrast to the bright color of the titanium white, therefore providing a high visibility of the display.

3. Theory and Computations of the AB Dome

1. General Theory

As wind flows over and around a fully exposed, nearly completely sealed inflated dome, the weather affecting the external film on the windward side must endure positive air pressures as the wind stagnates. Simultaneously, low air pressure eddies will be present on the leeward side of the dome. In other words, air pressure gradients caused by air density differences on different parts of the dome's envelope is characterized as the "buoyancy effect". The buoyancy effect will be greatest during the coldest weather when the dome is heated and the temperature difference between its interior and exterior are greatest. In extremely cold climates such as the Arctic and Antarctic the buoyancy effect tends to dominate dome air pressurization.

Current rigid dome.

Our basic computed equations, below, are derived from a Russian-language textbook [13]. Solar radiation impinging the orbiting Earth is approximately 1400 W/m^2. The average Earth reflection by white-colored clouds and the sub-aerial surfaces (such as water, ice and land) is about 0.3. The Earth-atmosphere absorbs about 0.2 of the Sun's radiation. That means about q_0 = 700 W/m^{2}s of solar energy (heat) reaches our planet's surface in cloudy weather at the Equator. That means we can absorb about 30 - 80% of solar energy. It is enough for normal plant growth in wintertime (up to 40-50° latitude) and in circumpolar regions with a special variant of the dome design.

The solar spectrum is graphically portrayed in Figure 5.

The visible part of the Sun's spectrum is only λ = 400 – 800 nm (0.4 to 0.8 μ.). Any warm body emits radiation. The emission wavelength depends on the body's temperature.

The wavelength of the maximum intensity (see Figure 5) is governed by the black-body law originated by Max Planck (1858-1947):

$$\lambda_m = \frac{2.9}{T}, \quad [mm], \tag{1}$$

where T is body temperature, ^{o}K. For example, if a body has an ideal temperature 20 ^{o}C (T = 293 ^{o}K), the wavelength is $\lambda_m = 9.9$ μ.

The energy emitted by a body may be computed by employment of the Josef Stefan-Ludwig Boltzmann law:

$$E = \varepsilon\sigma_S T^4, \ [\mathrm{W/m^2}], \tag{2}$$

where ε is coefficient of body blackness (ε =0.03 ÷ 0.99 for real bodies), $\sigma_s = 5.67\times10^{-8}$ [W/m^2·K] Stefan-Boltzmann constant. For example, the absolute black-body (ε = 1) emits (at T = 293 ^{0}K) the energy E = 418 W/m^2.

2. Cold Regions

Amount of the maximum solar heat flow on one square meter per second of Earth surface is

$$q = q_o \cos(\varphi \pm \theta) \ [\mathrm{W/m^2}], \tag{3}$$

where $\varphi < 90^{o}$ is Earth longevity, $\theta < 23.5^{o}$ is angle between projection of Earth polar axis to the plate which is perpendicular to the ecliptic plate and contains the line Sun-Earth and the perpendicular to ecliptic plate. The sign "+" signifies Summer and the "-" signifies Winter, $q_o \approx 700$ W/m^2 is the annual average solar heat flow to Earth at equator corrected for Earth reflectance. For our case this magnitude can reach $q_o \approx 1000$ - 1100 W/m^2 or in clouded sky the magnitude decreases up 100 W/m^2.

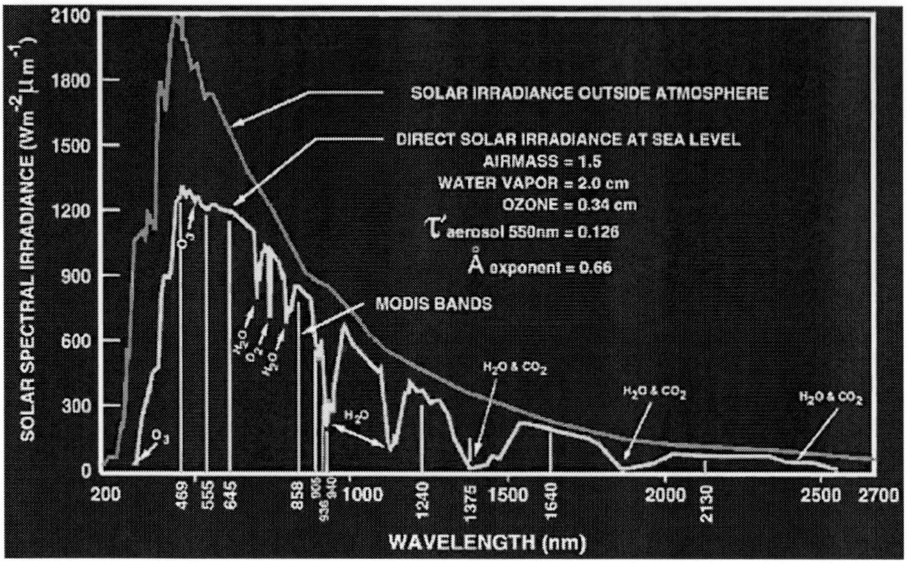

Figure 5. Spectrum of solar radiation. Visible light is 400 - 800 nm.

This angle is changed during the passage of a year and may be estimated for Earth's North Polar region hemisphere by the following the first approximation equation:

$$\theta = \theta_m \cos\omega, \quad \text{where} \quad \omega = 2\pi\frac{N}{364}, \tag{4}$$

where θ_m is maximum θ, $|\theta_m| = 23.5^{\circ} = 0.41$ radian; N is number of day in a year. The computations for summer and winter are presented in Figure 6.

The heat flow for a hemisphere having reflector [1] at noon may be computed by the equation:

$$q = c_1 q_0 [\cos(\varphi - \theta) + S\sin(\varphi + \theta)], \tag{5}$$

where S is fraction (relative) area of reflector to service area of "Evergreen" dome [1]. For reflector of Figure 1 [1] $S = 0.5$; c_1 is film transparency coefficient ($c_1 \approx 0.8$ - 0.95).

The daily average solar irradiation (energy) is calculated by equation

$$Q = 86400\, c\, qt, \quad \text{where} \quad t = 0.5(1 + \tan\varphi\tan\theta), \quad |\tan\varphi\tan\theta| \le 1, \tag{6}$$

where c is daily average heat flow coefficient, $c \approx 0.5$ without dome, $c \approx 0.75$ with dome; t is relative daily illuminated time, $86400 = 24 \times 60 \times 60$ is the number of seconds in a day.

The computation for relative daily light time is presented in Figure 7.

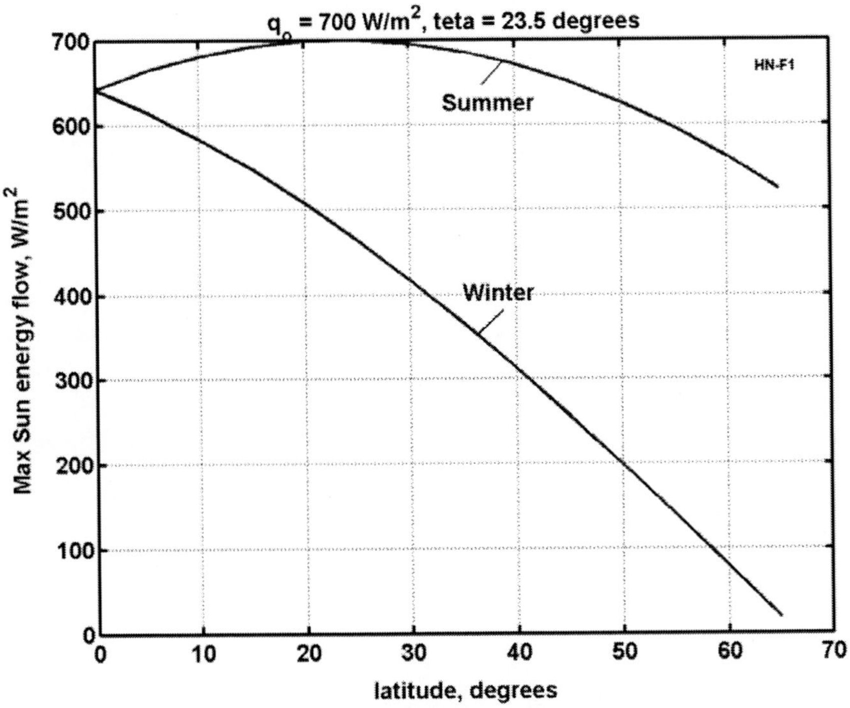

Figure 6. Maximum Sun radiation flow at Earth surface via Earth latitude and season without dome.

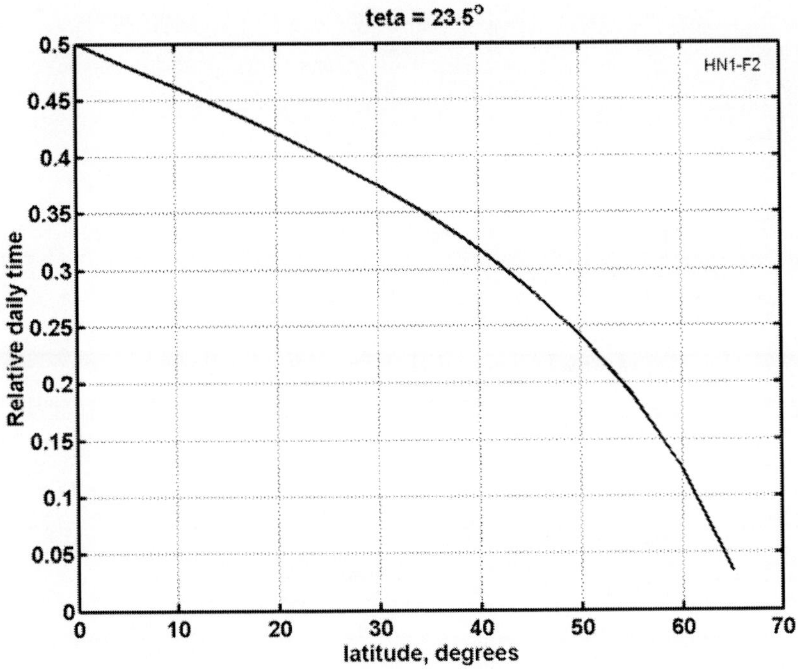

Figure 7. Relative daily light-time via Earth latitude.

The convective (conductivity) heat flow per square meter of dome film cover by convection and heat conduction is (see [2]):

$$q = k(t_1 - t_2), \quad \text{where} \quad k = \frac{1}{1/\alpha_1 + \sum_i \delta_i / \lambda_i + 1/\alpha_2}, \tag{7}$$

where k is heat transfer coefficient, W/m^2·K; $t_{1,2}$ are temperatures of the inter and outer multi-layers of the heat insulators, $^\circ$C; $\alpha_{1,2}$ are convention coefficients of the inter and outer multi-layers of heat insulators ($\alpha = 30 \div 100$), W/m^{2}K; δ_i are thickness of insulator layers; λ_i are coefficients of heat transfer of insulator layers (see Table 1), m; $t_{1,2}$ are temperatures of initial and final layers $^\circ$C.

The radiation heat flow per square meter of the service area computed by equations (2):

$$q = C_r\left[\left(\frac{T_1}{100}\right)^4 - \left(\frac{T_2}{100}\right)^4\right],$$

$$\text{where} \quad C_r = \frac{c_s}{1/\varepsilon_1 + 1/\varepsilon_2 - 1}, \quad c_s = 5.67, \tag{8}$$

where C_r is general radiation coefficient [W/m^{2}K^4], ε are black body rate (emittance) of plates (see Table 2); T is temperatures of plates (surface), ^{0}K.

The radiation flow across a set of the heat reflector plates is computed by the equation:

$$q = 0.5\frac{C'_r}{C_r}q_r, \tag{9}$$

where C'_r is computed by equation (8) between plate and reflector.

The data of some construction materials is found in Table 1 Attn. or [13], p.331, 2 [13], p. 465.

Table 1. [13], p.331. Heat Transfer Data

Material	Density, kg/m^3	Thermal conductivity, λ, W/m·°C	Heat capacity, kJ/kg.°C
Concrete	2300	1.279	1.13
Baked brick	1800	0.758	0.879
Ice	920	2.25	2.26
Snow	560	0.465	2.09
Glass	2500	0.744	0.67
Steel	7900	45	0.461
Air	1.225	0.0244	1

Table 2. [13], p. 465. Emittance, ε (Emissivity)

Material	Temperature, T °C	Emittance, ε
Bright Aluminum	50 ÷ 500 °C	0.04 - 0.06
Bright copper	20 ÷ 350 °C	0.02
Steel	50 °C	0.56
Asbestos board	20 °C	0.96
Glass	20 ÷ 100 °C	0.91 - 0.94
Baked brick	20 °C	0.88 - 0.93
Tree	20 °C	0.8 - 0.9
Black vanish	40 ÷ 100 °C	0.96 – 0.98
Tin	20 °C	0.28

As the reader will see, the air layer is the best heat insulator. We do not limit its thickness δ.

As the reader will notice, the shiny aluminum louver coating is an excellent means of retention against radiation losses from the dome.

The general radiation heat Q computes by equation (6). Equations (1) – (9) allow computation of the heat balance and comparison of incoming heat (gain) and outgoing heat (loss).

The computations of heat balance of a dome (with reflector mirror [1]) of any size in the coldest wintertime of the Polar Regions are presented in Figure 8.

The thickness of the dome envelope, its sheltering shell of film, is computed by formulas (from equation for tensile strength):

$$\delta_1 = \frac{Rp}{2\sigma}, \quad \delta_2 = \frac{Rp}{\sigma}, \tag{10}$$

where δ_1 is the film thickness for a spherical dome, m; δ_2 is the film thickness for a cylindrical dome, m; R is radius of dome, m; p is additional pressure into the dome (10÷1000), N/m^2; σ is safety tensile stress of film (up to 2×10^9), N/m^2.

The dynamic pressure from wind is

$$p_w = \frac{\rho V^2}{2}, \tag{11}$$

where $\rho = 1.225$ kg/m^3 is air density; V is wind speed, m/s.

For example, a storm wind with speed $V = 20$ m/s, standard air density is $\rho = 1.225$ kg/m^3. Then dynamic pressure is $p_w = 245$ N/m^2. That is four time less when internal pressure $p = 1000$ N/m^2. When the need arises, sometimes the internal pressure can be voluntarily decreased, bled off.

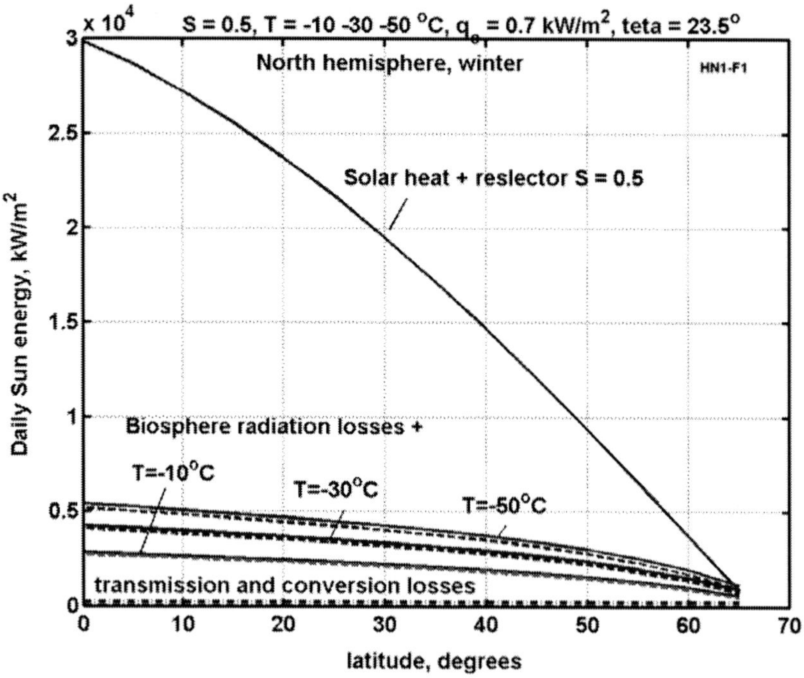

Figure 8. Daily heat balance through 1 m^2 of dome with mirror during coldest winter day versus Earth's latitude (North hemisphere example). Data used for computations (see Eq. (1) - (9)): temperature inside of dome is t_1= +20 °C, outside are t_2 = -10, -30, -50 °C; reflectivity coefficient of mirror is c_2= 0.9; coefficient transparency of film is c_1 = 0.9; convectively coefficients are $\alpha_1 = \alpha_2 = 30$; thickness of film layers are $\delta_1 = \delta_2 = 0.0001$ m; thickness of air layer is $\delta = 1$ m; coefficient of film heat transfer is $\lambda_1 = \lambda_3 = 0.75$, for air $\lambda_2 = 0.0244$; ratio of cover blackness $\varepsilon_1 = \varepsilon_3 = 0.9$, for louvers $\varepsilon_2 = 0.05$.

In Figure 8 the alert reader has noticed: *the daily heat loss is about the solar heat in the very coldest winter day when a dome located above 60^0 North or South Latitude and the outside air temperature is –50 ^{0}C.*

In [1] we show the heat loss of the dome in Polar region is less than 14 times the heat of the buildings inside unprotected by an inflated dome.

We consider a two-layer dome film and one heat screen. If needed, better protection can further reduce the heat losses as we can utilize inflated dome covers with more layers and more heat screens. One heat screen decreases heat losses by 2, two screens can decrease heat flow by 3 times, three by 4 times, and so on. If the Polar Region domes have a mesh structure, the heat transfer decreases proportional to the summary thickness of its' enveloping film layers.

The dome shelter innovations outlined here can be practically applied to many climatic regimes (from Polar to Tropical). The North and South Poles may, during the 21st Century, become places of cargo and passenger congregation since the a Cable Space Transportation System, installed on Antarctica's ice-cap and on a floating artificial ice island has been proposed that would transfer people and cargoes to and from the Moon [3].

Table 3. Maximum partial pressure of water vapor in atmosphere for given air temperature

t, C	-10	0	10	20	30	40	50	60	70	80	90	100
p,kPa	0.287	0.611	1.22	2.33	4.27	7.33	12.3	19.9	30.9	49.7	70.1	101

3. Irrigation without Water. Closed-loop Water Cycle

A reader can derive the equations below from well-known physical laws [12]. Therefore, no detailed explanations of these are furnished here.

1. *Amount of water in atmosphere.* Amount of water in Earth's atmosphere depends entirely upon temperature and humidity. For relative humidity 100%, the maximum partial pressure of water vapor is shown in Table 3.

The amount of water in 1 m^3 of air may be computed by equation

$$m_W = 0.00625\ [p(t_2)h - p(t_1)], \tag{12}$$

where m_W is mass of water, kg in 1 m^3 of air; *p(t)* is vapor (steam) pressure from Table 3, relative $h = 0 \div 1$ is relative humidity. The computation of equation (12) is presented in Figure 9. Typical relative humidity of atmosphere air is 0.5 - 1.

Computation of Closed-loop Water Cycle

Assume the maximum safe temperature is achieved in the daytime. When dome reaches the maximum (or given) temperature, then the control system fills the space 5 with air (Figure 4) between double–layers of the film cover. That protects the inside part of the dome from further heating by hot air outside the dome. The control system decreases also the solar radiation input, increasing reflectivity of the liquid crystal layer of the film cover. In short, we can then support a constant temperature inside our imagined filmic dome.

The *heating* of the dome in the daytime may be computed by equations:

$$q(t) = q_0 \sin(\pi t / t_d), \quad dQ = q(t)dt, \quad Q = \int_0^{t_d} dQ, \quad Q(0) = 0, \quad M_w = \int_0^{t_d} a dT,$$

$$dT = \frac{dQ}{C_{p1}\rho_1\delta_1 + C_{p2}\rho_2 H + rHa}, \quad a = 10^{-5}(5{,}28T + 2), \quad T = \int_0^{t_d} dT, \quad T(0) = T_{\min}, \tag{13}$$

where q is heat flow, J/m^2 s; q_o is maximal Sun heat flow in noon daily time, $q_o \approx 100 \div 1000$, J/ m^2s; t is time, s; t_d is daily (Sun) time, s; Q is heat, J; T is temperature in dome (air, soil), °C; C_{pl} is heat capacity of soil, $C_{pl} \approx 1000$ J/kg; $C_{p2} \approx 1000$ J/kg is heat capacity of air; $\delta_1 \approx 0.1$ m is thickness of heating soil; $\rho_1 \approx 1000$ kg/m^3 is density of the soil; $\rho_2 \approx 1.225$ kg/m^3 is density of the air; H is thickness of air (height of cover), $H \approx 50 \div 300$ m; $r = 2{,}260{,}000$ J/kg is evaporation heat, a is coefficient of evaporation; M_w is mass of evaporation water, kg/m^3; T_{min} is minimal temperature into dome after night, °C.

The convective (conductive) *cooling* of dome at night time may be computed as below

$$q_t = k(T_{\min} - T(t)), \quad \text{where} \quad k = \frac{1}{1/\alpha_1 + \sum_i \delta_i / \lambda_i + 1/\alpha_2}, \tag{14}$$

where q_t is heat flow through the dome cover by convective heat transfer, J/m^{2}s or W/m^2; see the other notation in Eq. (7). We take $\delta = 0$ in nighttime (through active control of the film).

The radiation heat flow (from dome to night sky, radiation cooling) may be estimated by equations:

$$q_r = C_r\left[\left(\frac{T_{\min}}{100}\right)^4 - \left(\frac{T(t)}{100}\right)^4\right], \quad \text{where} \quad C_r = \frac{c_s}{1/\varepsilon_1 + 1/\varepsilon_2 - 1}, \quad c_s = 5.67, \tag{15}$$

where q_r is heat flow through dome cover by radiation heat transfer [W/m^2/K^4], J/m^{2}s or W/m^2; see the other notation in Eq. (8). We take $\varepsilon = 1$ in night time (through active control of the film).

The other equations are same (13)

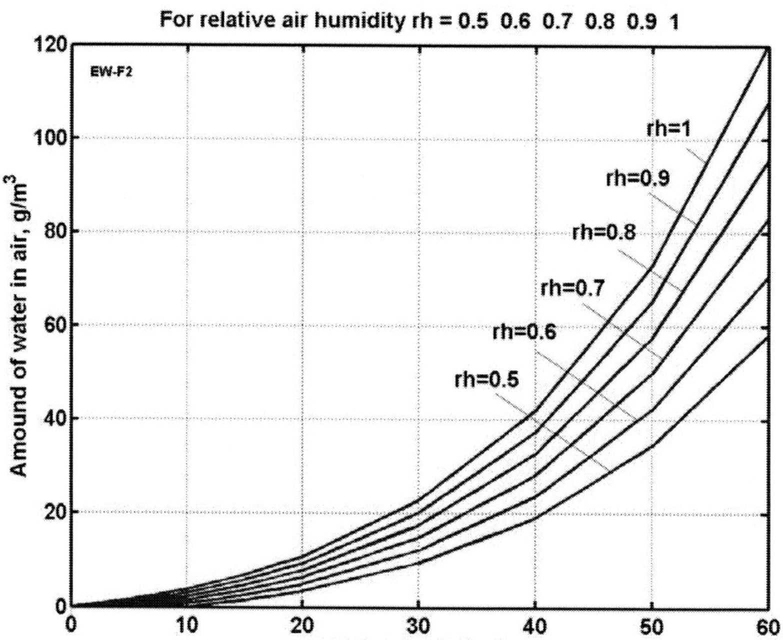

Figure 10. Amount of water in 1 m^3 of air versus air temperature and relative humidity (rh). $t_l = 0$ °C.

$$dQ = [q_t(t) + q_r(t)]dt, \quad Q = \int_0^{t_d} dQ, \quad Q(0) = 0, \quad M_w = \int_0^{t_d} a dT,$$

$$dT = \frac{dQ}{C_{p1}\rho_1\delta_1 + C_{p2}\rho_2 H + rHa}, \quad a = 10^{-5}(5{,}28T + 2), \quad T = \int_0^{t_d} dT, \quad T(0) = T_{\min}, \tag{16}$$

Let us take the following parameters: H = 135 m, α =70, δ = 1 m between cover layers (see #5 in Figure 4), λ = 0.0244 for air. Result of computation for given parameter are presented in Figures 10 – 13.

For dome cover height H = 135 m the night precipitation (maximum) is 0.027×135 = 3.67 kg (liter) or 3.67 mm/day (Figure 11). The annual precipitation is 1336.6 mm (maximum). If it is not enough, we can increase the height of dome cover. The globally-averaged annual precipitation is about 1000 mm.

As you see, we can support the same needed temperature in a wide range of latitudes at summertime and wintertime. That means the covered regions are not hostage to their geographical location on the Earth's surface (up to latitude 40° -50°), nor Earth's seasons, or the vagaries of sometime inclement exterior weather. Our dome design, most assuredly, is not optimal but rather it is selected for realistic parameters.

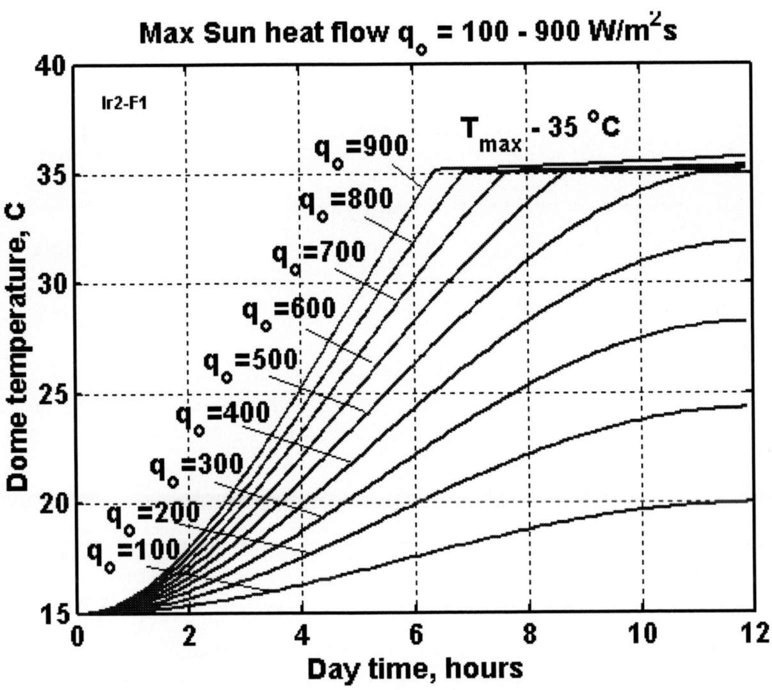

Figure 10. Heating of the dome by solar radiation from the night temperature of 15° C to 35° C via daily maximal solar radiation (W/m^2) for varying daily time. Height of dome film cover equals H = 135 m. The control temperature system limits the maximum internal dome temperature to 35° C. Compare with Figure 6. *The AB Dome can support the average daily temperature of 18° C in wintertime up to planetary latitude of 50°.*

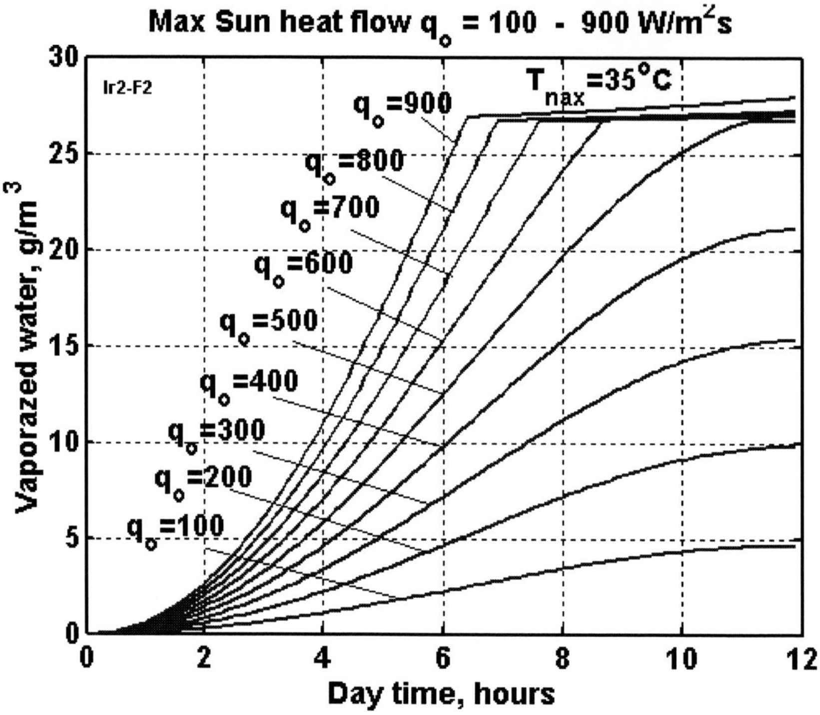

Figure 11. Water vaporization for 100% humidity of the air for different maximal solar radiation (W/m^2) levels delivered over varying daily time. Height of dome film cover equals H = 135 m. The temperature control system limits the maximum internal dome temperature to 35^0 C.

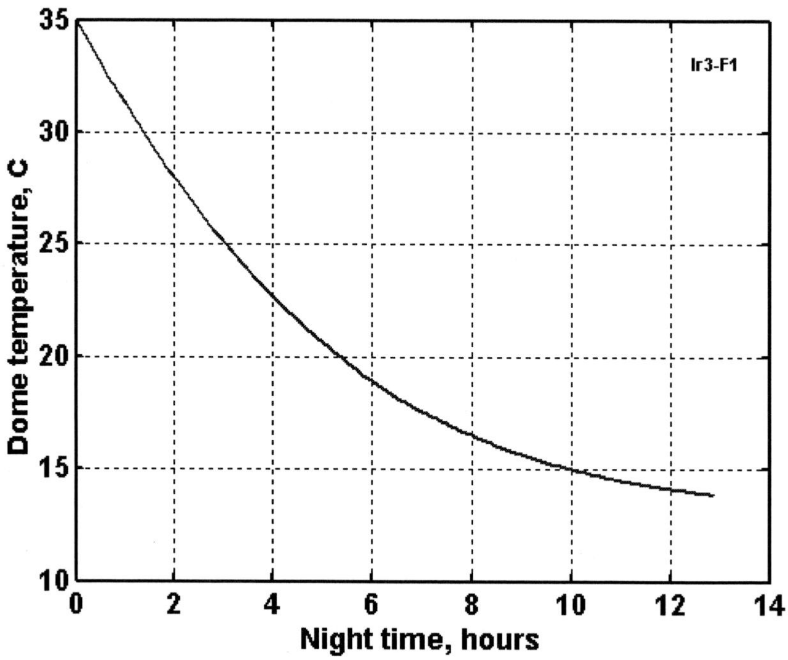

Figure 12. Cooling of the Dome via nighttime for initial daily temperature 35° C and the night outer temperature 13° C.

4. Discussion

As with any innovative macro-project proposal, the reader will naturally have many questions. We offer brief answers to the four most obvious questions our readers are likely to ponder.

1. *How can snow and ice be removed from the dome?*
   If water appears over film (rain), it flows downwards through a special evacuation tube. If snow (ice) appears atop the film, the control system passes the warm air between two cover layers. The warm air melts the snow (ice) and water flows down. The film cover is flexible and has a lift force of about 10 -100 kg/m^2.
2. *Storm wind.*
   The storm wind can only be on the bounding (outside) sections of dome. They are special semi-cylindrical form (Figure 3) and stronger than central sections.
3. *Cover damage.*
   The envelope contains a rip-stop cable mesh so that the film cannot be damaged greatly. Electronic signals alert supervising personnel of any rupture problems. The needed part of cover may be reeled down by control cable and repaired.
4. *What is the design life of the film covering?*
   Depending on the kind of materials used, it may be as much a decade or, if existing under lucky circumstances, more. In all or in part, the cover can be replaced periodically.

5. Conclusion

One half of Earth's human population is chronically malnourished. *The majority of Earth's surface area is not suitable for unshielded human life*. The increasing of agriculture area, crop capacity, carrying capacity by means of converting the deserts, desolate wildernesses, taiga, tundra permafrost into gardens are an important escape-hatch from some of humanity's most pressing macro-problems. The offered cheapest ($0.1÷ 0.3/m^2) AB method may dramatically increase the potentially realizable sown area, crop capacity; indeed the range of territory suitable for human living. In theory, converting all Earth land such as Alaska, northern Canada, Siberia, or the Sahara or Gobi deserts into prosperous gardens would be the equivalent of colonizing another Solar System planet. The suggested method is very cheap (cost of covering 1 m^2 is about 10 - 30 USA cents) and may be utilized immediately. We can start from small regions, such as towns in bad regions and, gradually, extend the practice over a large region—and what is as important, earning monetary profits most of the time.

Filmic domes can foster the fuller economic development of dry, hot, and cold regions such as the Earth's Arctic and Antarctic, the Sahara and, thus, increase the effective area of territory dominated by 21st Century humans. Normal human health can be maintained by ingestion of locally grown fresh vegetables and healthful "outdoor" exercise. The domes can also be used in the Tropics and Temperate Zone. Eventually, technical adaptations may find application on the Moon or Mars since a vertical variant, inflatable towers to outer space, are soon to become available for launching spacecraft inexpensively into Earth-orbit or interplanetary flights [12].

The related problems are researched in references [1]-[12].

Let us shortly summarize some advantages of this offered AB Dome method of climate moderation:

(1) Method does not need large amounts of constant input freshwater for irrigation;
(2) Low cost of inflatable filmic Dome per area reclaimed: (10 - 30 cents/m^2);
(3) Control of inside temperature is total;
(4) Usable in very hot and cool climate regions;
(5) Covered region is not at risk from exterior weather;
(6) Possibility of flourishing crops even with a sterile hydroponics soil;
(7) 2 – 3 harvests each year; without farmers' extreme normal risks.
(8) Rich harvests, at that.
(9) Converting deserts, desolate wildernesses, taiga, tundra, permafrost terrain, and the ocean into gardens;
(10) Covering towns and cities with low-cost, even picturesque domes;
(11) Using the dome cover for income-generating neighborhood illumination, picture displays, movies and videos as well as paid advertising.

We can concoct generally agreeable local weather and settle new territory for living with an agreeable climate (without daily rain, wind and low temperatures) for agriculture. By utilizing thin film, gigantic territorial expanses of dry and cold regions can be covered. Countries having big territory (but also much bad land) may be able to use domes to increase their population and became powerful states during the 21st Century.

The offered method may be used to conserve a vanishing sea such as the Aral Sea or the Dead Sea. A closed-loop water cycle could save these two seas for a future generation of people, instead of bequeathing a salty dustbowl.

A.A. Bolonkin has further developed the same method for the ocean. By controlling the dynamics and climate there, oceanic colonies may increase the Earth's humanly useful region another three times concurrent with or after the doubling of useful land outlined above. Our outlined method would allow the Earth's human population to increase by 5 – 10 times, without the starvation.

The offered method can solve the problem of apparent problem of global warming because the AB domes will be able to confine until use much carbon dioxide gas which appreciably increases a harvest. This carbon dioxide gas will show up in yet more productive crop harvests! The dome lift-force reaches up to 300 kg/m^2. The telephone, TV, electric, water and other communications can be suspended from the dome cover.

The offered method can also help to defend cities (or an entire given region) from rockets, nuclear warheads, and military aviation. Details are offered in a later chapter.

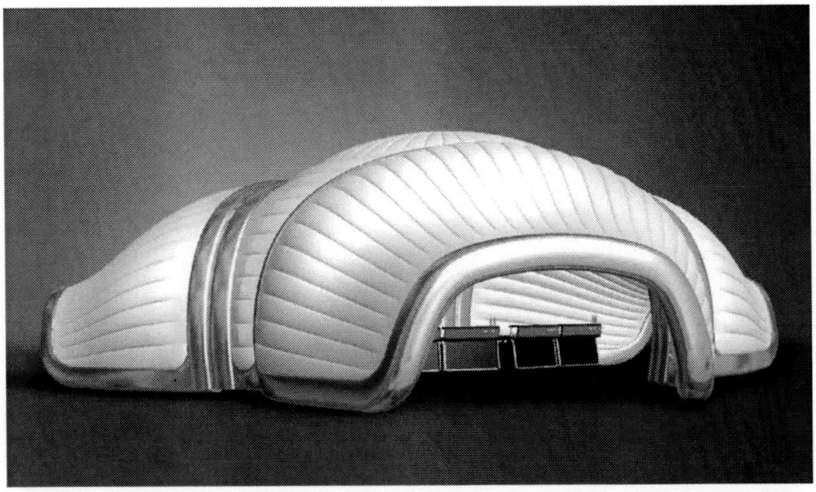

Inflatable dome.

(Continued)

Current rigid domes.

(Continued)

Inflatable Domes.

REFERENCES

(The reader may find some of these articles at the author's web page: *http://arxiv.org* seach "Bolonkin", *http://Bolonkin.narod.ru/p65.htm* and in two published books: "*Non-Rocket Space Launch and Flight*" (Elsevier, London, 2006) and in "*New Concepts, Ideas, Innovations in Aerospace, Technology and Human Science*" (NOVA, 2008).

[1] Bolonkin, A.A. and Cathcart, R.B., Inflatable 'Evergreen' Dome Settlements for Earth's Polar Regions. *Clean Technologies and Environmental Policy.* DOI 10.1007/s10098.006-0073.4 (2006).

[2] Bolonkin A.A., *Control of Regional and Global Weather*. 2006. *http://arxiv.org*.

[3] Bolonkin, A.A. and R.B. Cathcart, "A Cable Space Transportation System at the Earth's Poles to Support Exploitation of the Moon", *Journal of the British Interplanetary Society* 59: 375-380, 2006.

[4] Bolonkin, A.A., *Cheap Textile Dam Protection of Seaport Cities against Hurricane Storm Surge Waves, Tsunamis, and Other Weather-Related Floods,* 2006. http://arxiv.org.

[5] Bolonkin, A.A. and R.B. Cathcart, *Antarctica: A Southern Hemisphere Wind-power Station?,* 2007. *http://arxiv.org*.

[6] Cathcart, R.B. and Bolonkin, A.A., *Ocean Terracing,* 2006. http://arxiv.org.

[7] Bolonkin, A.A. and R.B. Cathcart, *The Java-Sumatra Aerial Mega-Tramway,* 2006. *http://arxiv.org*.

[8] Bolonkin, A.A., "Optimal Inflatable Space Towers with 3-100 km Height", *Journal of the British Interplanetary Society* Vol. 56, pp. 87 - 97, 2003.

[9] Bolonkin, A.A., Non-Rocket Space Launch and Flight, *Elsevier*, London, 2006, 488 pages.

[10] Badescu, V., R.B. Cathcart and R.D. Schuiling, Macro-Engineering: A Challenge for the Future. Springer, 2006. 318 pages.

[11] Bolonkin, A.A., Cathcart R.B., *Inflatable 'Evergreen' Polar Zone Dome* (EPZD) Settlements, 2006. *http://arxiv.org*.

[12] Bolonkin, A.A., Inflatable Dome for Moon, Mars, Asteroids and Satellites, *Presented as paper AIAA-2007-6262 by AIAA Conference "Space-2007",* 18-20 September 2007, Long Beach. CA, USA.

[13] Naschekin, V.V., Technical thermodynamic and heat transmission. Public House High University, Moscow, USSR. 1969 [in Russian].

[14] Wikipedia. *Some background material in this article is gathered from Wikipedia under the Creative Commons license.*

Chapter 2

INFLATABLE DOME FOR MOON, MARS, ASTEROIDS, AND SATELLITES*

ABSTRACT

On a planet without atmosphere, sustaining human life is very difficult and dangerous, especially during short sunlit period when low temperature prevails. To counter these environmental stresses, the author offer an innovative artificial "Evergreen" dome, an inflated hemisphere with interiors continuously providing a climate like that of Florida, Italy and Spain. The "Evergreen" dome theory is developed, substantiated by computations that show it is possible for current technology to construct and heat large enclosed volumes inexpensively. Specifically, a satisfactory result is reached by using high altitude magnetically supported sunlight reflectors and a special double thin film as an enclosing skin, which concentrates solar energy inside the dome while, at the same time, markedly decreasing the heat loss to exterior space. Offered design may be employed for settlements on the Moon, Mars, asteroids and satellites.

Keywords: artificial biosphere, inflatable film building, Moon and Mars settlements, Space evergreen, magnetic solar energy reflector

1. INTRODUCTION

The real development of outer space (permanent human life in space) requires two conditions: all-sufficient space settlement and artificial life conditions close to those prevailing currently on the Earth. (Such a goal extends what is already being attempted in the Earth-biosphere—for example at the 1st Advanced Architecture Contest, "Self-Sufficient Housing", sponsored by the Institute for Advanced Architecture of Catalonia, Spain, during 2006.) The first condition demands production of all main components needed for human life: food, oxidizer, and energy within the outer space and Solar System body colony. The second requisite condition is a large surface settlement having useful plants, attractive flowers, splashing water pools, walking and sport areas, etc. All these conditions may be realized within large 'greenhouses' [1] that will produce food, oxidizer and "the good life" conditions.

Human life in outer space and on other planetary or planet-like places will be more comfortable if it uses A.A. Bolonkin's macro-project proposal - staying in outer space without special spacesuit [2], p.335 (mass of current spacesuit reaches 180 kg). The idea of this paper may be used also for control of Earth's regional and global weather and for converting our Earth's desert and cold polar zone regions into edenic subtropical gardens [3]-[4].

The current conditions in Moon, Mars and Space are far from comfortable. For example, the Moon does not have any useful atmosphere, the day and night continues for 14 Earth days each, there are deadly space radiation and meteor bombardments, etc.

Especially during wintertime, Mars could provide only a meager and uncomfortable life-style for humans, offering low temperatures, strong winds. The distance north or south from that planet's equator is amongst the most significant measured environmental variables underlying the physical differences of the planet. In other words, future humans living in the Moon and Mars must be more comfortable for humans to explore and properly exploit these distant and dangerous places.

Moon base.

* Presented as paper AIAA-2007-6262 to AIAA Conference "Space-2007", 18-20 [illegible] 2007, Long Beach. CA, USA. Published in http://arxiv.org.

2. 'Evergreen' Inflated Domes

Possibly the first true architectural attempt at constructing effective artificial life-support systems on the climatically harsh Moon will be the building of greenhouses. Greenhouses are maintained nearly automatically by heating, cooling, irrigation, nutrition and plant disease management equipment. Humans share commonalities in their responses to natural environmental stresses that are stimulated by night cold, day heat, absent atmosphere, so on. Darkness everywhere inflicts the same personal visual discomfort and disorientation as cosmonauts/astronauts experience during their space-walks—that of being adrift in featureless space! With special clothing and shelters, humans can adapt successfully to the well-landmarked planet Mars, for example. Incontrovertibly, living on the Moon, beneath Mars' low-density atmosphere is difficult, even when tempered by strong conventional protective buildings.

Our macro-engineering concept of inexpensive-to-construct-and-operate "Evergreen" inflated surface domes is supported by computations, making our macro-project speculation more than a daydream. Innovations are needed, and wanted, to realize such structures upon the Moon of our unique but continuously changing life.

3. Description and Innovations

Dome

Our basic design for the Moon-Mars people-housing "Evergreen" dome is presented in Figure 1, which includes the thin inflated double film dome. The innovations are listed here: (1) the construction is air-inflatable; (2) each dome is fabricated with very thin, transparent film (thickness is 0.2 to 0.4 mm) without rigid supports; (3) the enclosing film is a two-layered structural element with air between the layers to provide insulation; (4) the construction form is that of a hemisphere, or in the instance of a roadway/railway a half-tube, and part of the film has control transparency and a thin aluminum layer about 1 μ or less that functions as the gigantic collector of incident solar radiation (heat). Surplus heat collected may be used to generate electricity or furnish mechanical energy; and (5) the dome is equipped with sunlight controlling louvers [also known as, "jalousie", a blind or shutter having adjustable slats to regulate the passage of air and sunlight] with one side thinly coated with reflective polished aluminum of about 1 μ thickness. Real-time control of the sunlight's entrance into the dome and nighttime heat's exit is governed by the shingle-like louvers or a controlled transparency of the dome film.

Variant 1 of artificial inflatable Dome for Moon and Mars is shown in Figure 1. Dome has top thin double film 4 covered given area and single under ground layer 6. The space between layers 4 - 6 is about 3 meters and it is filled by air. The support cables 5 connect the top and underground layers and Dome looks as a big air-inflated beach sunbathing or swimming mattress. The Dome includes hermetic sections connected by corridors 2 and hermetic lock chambers 3. Topmost film controls the dome's transparency (and reflectivity). That allows people to closely control temperature affecting those inside the dome. Topmost film also is of a double-thickness. When a meteorite pushes hole in the topmost double film, the lowermost layer closes the hole and puts temporary obstacles in the way of the escaping air. Dome has a fruitful soil layer, irrigation system, and cooling system 9 for supporting a selected given humidity. That is, a closed-biosphere with a closed life-cycle that regularly produces an oxidizer as well as sufficient food for people and their pets, even including some species of farm animals. Simultaneously, it is the beautiful and restful Earth-like place of abode. The offered design has a minimum specific mass, about 7-12 kg/m^2 (air - 3 kg, film - 1 kg, soil - 3 - 8 kg). Mass of an example area of 10×10 m is about 1 metric ton (oftentimes spelt "tonnes").

Figure 2 illustrates the second thin transparent dome cover we envision. The Dome has double film: semispherical layer (low pressure about 0.01 - 0.1 atmosphere, atm.) and lower layer (high 1 atm. pressure). The hemispherical inflated textile shell—technical "textiles" can be woven (weaving is an interlacement of warp and weft) or non-woven (homogenous films)—embodies the innovations listed: (1) the film is very thin, approximately 0.1 to 0.3 mm. A film this thin has never before been used in a major building; (2) the film has two strong nets, with a mesh of about 0.1 × 0.1 m and a = 1 × 1 m, the threads are about 0.3 mm for a small mesh and about 1 mm for a big mesh.

The net prevents the watertight and airtight film covering from being damaged by micrometeorites; the film incorporates a tiny electrically-conductive wire net with a mesh of about 0.001 x 0.001 m and a line width of about 100 μ and a thickness near 1μ. The wire net can inform the "Evergreen" dome supervisors (human or automated equipment) concerning the place and size of film damage (tears, rips, punctures, gashes); the film is twin-layered with the gap — c = 1 m and b = 2 m—between the layer covering. This multi-layered covering is the main means for heat insulation and anti-puncture safety of a single layer because piercing won't cause a loss of shape since the film's second layer is unaffected by holing; the airspace in the dome's twin-layer covering can be partitioned, either hermetically or not; and part of the covering may have a very thin shiny aluminum coating that is about 1μ for reflection of non-useful or undesirable impinging solar radiation.

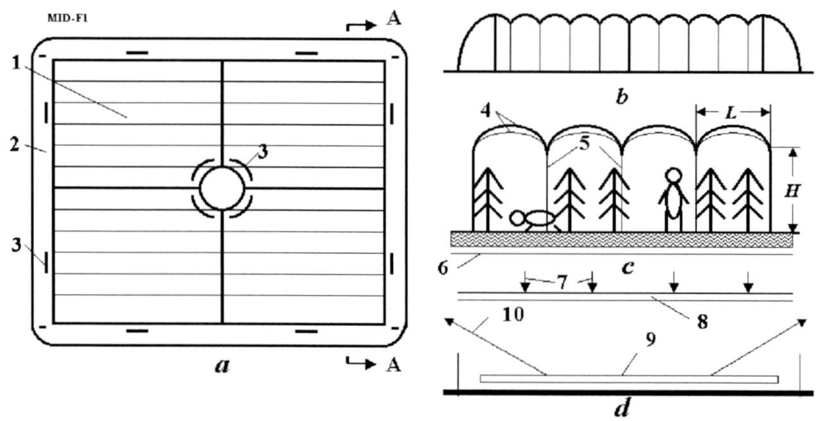

Figure 1. Variant 1 of artificial inflatable Dome for Moon and Mars. (a) top view of dome; (b) cross-section AA area of dome; (c) inside of the Dome; (d) Cooling system. Notations: 1 - internal section of Dome; 2 - passages; 3 - doors; 4 - transparence thin double film ("textiles") with control transparency; 5 - support cables; 6 - lower underground film; 7 - solar light; 8 - protection film; 9 - cooling tubes; 10 - radiation of cooling tubes.

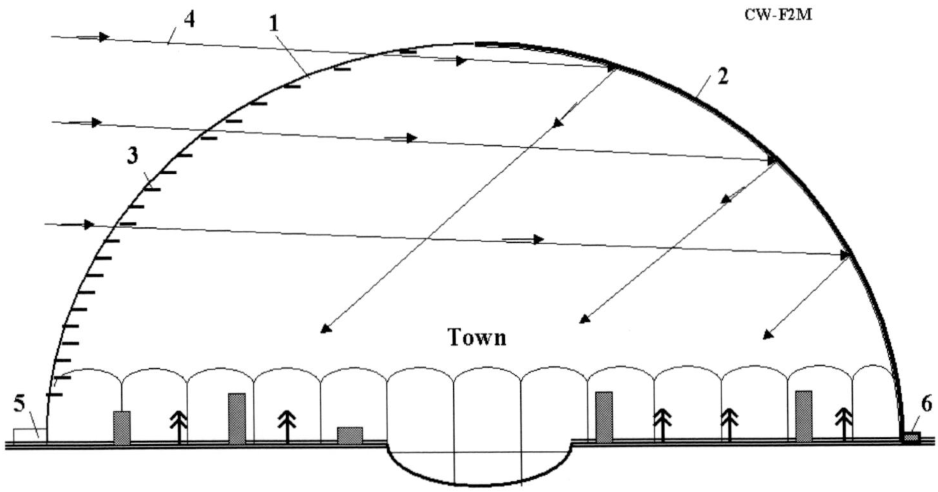

Figure 2. Variant 2 of artificial inflatable Dome for Moon and Mars. Notations: 1 - transparent thin double film ("textiles"); 2 - reflected cover of hemisphere; 3 - control louvers (jalousie); 4 - solar beams (light); 5 - enter (dock chamber); 6 - water extractor from air. The lower section has air pressure about 1 atm. The top section has air pressure of 0.01 - 0.1 atm.

Offered inflatable Dome can cover a big region (town) and create beautiful Earth-like conditions on an outer space solid body (Figure 4a). In future, the "Evergreen" dome can cover a full planetary surface (Moon, Mars, asteroid) (Figure 4b). Same type of domes can cover the Earth's lands, converting them (desert, cool regions) into beautiful gardens with controlled weather and closed material life cycles.

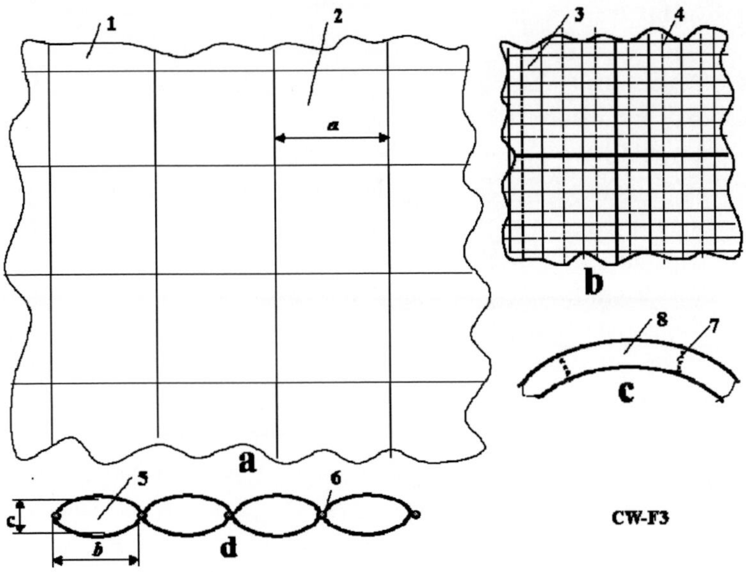

Figure 3. Design of "Evergreen" cover. Notations: (a) Big fragment of cover; (b) Small fragment of cover; (c) Cross-section of cover; (d) Longitudinal cross-section of cover; 1 - cover; 2 -mesh; 3 - small mesh; 4 - thin electric net; 5 - sell of cover; 6 - tubes; 7 - film partition (non hermetic); 8 - perpendicular cross-section area.

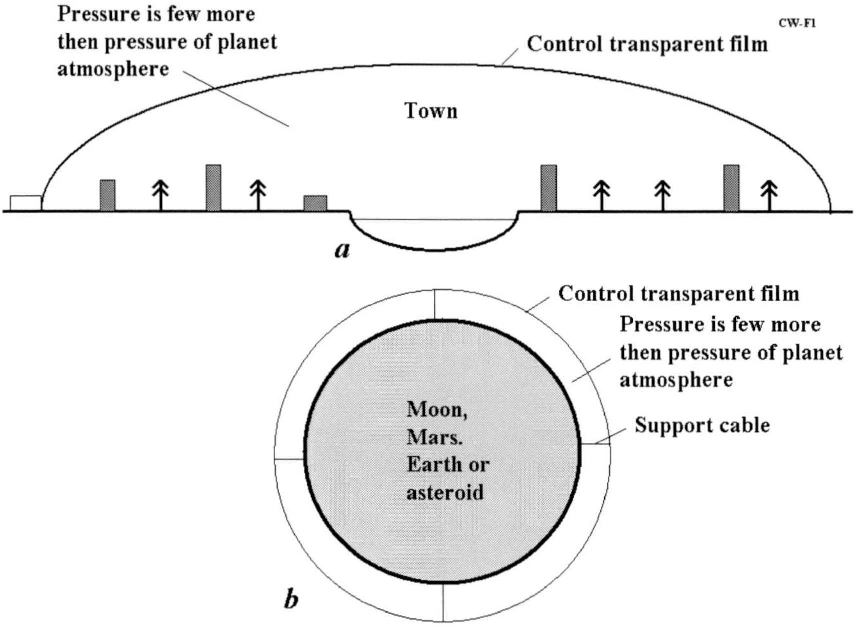

Figure 4. (*a*) Inflatable film dome over a single town; (*b*) Inflatable film dome covering a planet (Moon, Mars) and asteroid. Same type of domes can cover the Earth's extreme climate regions and convert them (desert, cool regions) into beautiful gardens with controlled weather and closed life cycles.

Location, Illumination and Defending Human Settlements from Solar Wind and Space Radiation

The Moon makes one revolution in about 29 Earth days. If we want to have conventional Earth artificial day and natural solar lighting, the settlement must locate near one or both of the Moon's poles and have a magnetic control mirror suspended at high altitude in given (stationary) place (Figure 5). For building this mirror (reflector) may use idea and theory of magnetic levitation developed by A.A. Bolonkin in [5]. If reflector is made with variable focus, as in [3] p. 306, Figure 16.3, then it may well be employed as a concentrator of sunlight and be harnessed for energy during "night" (Earth-time).

The second important feature of the offered installation is defense of the settlement from solar wind and all cosmic radiation. It is known that the Earth's magnetic field is a natural defense for living animals, plants and humans against high-energy particles, such as protons, of the solar wind. The artificial magnetic field near Moon settlement is hundreds of times stronger than the Earth's magnetic field. It will help to defend delicate humans. The polar location of the planned settlement also decreases the intensity of the solar wind. Location of human settlement in polar zone(s) Moon craters also decreases the solar wind radiation. People can move to an underground cosmic radiation protective shelter, a dugout or bunker, during periods of high Sun activity (solar flashes, coronal mass ejections).

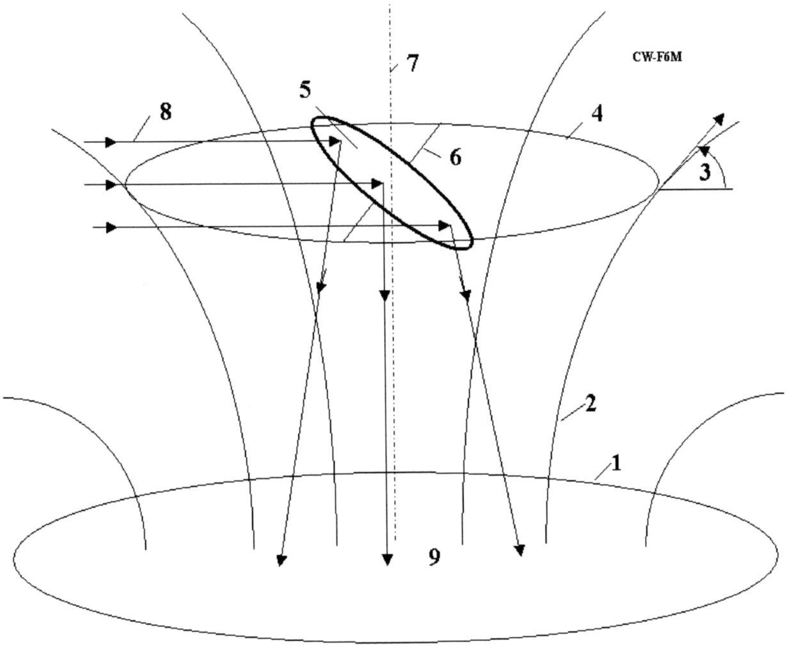

Figure 5. Magnetic control mirror is suspended at high altitude over human Moon settlement. Notations: 1 -superconductivity ground ring; 2 - magnetic lines of ground superconductivity ring; 3 - angle (α) between magnetic line of the superconductivity ground ring and horizontal plate (see Eq. (6)); top superconductivity ring for supporting the mirror (reflector) 5; 6 - axis of control reflector (which allows turning of mirror); 7 - vertical axis of the top superconductivity ring; 8 - solar light; 9 – human settlement.

The theory and computation of this installation is in theoretical section, below. The mass of the full reflector (rings, mirror, head screens is about 70 - 80 kg; if the reflector is used also as powerful energy source, then the mass can reach 100 - 120 kg. Note: for lifting, the reflector does not need a rocket. The magnetic force increases near ground (see Eq. (3)). This force lifts the reflector to the altitude that is required by its usage. The reflector also will be structurally stable because it is located in magnetic hole of a more powerful ground ring magnet.

The artificial magnetic field may be used, too, for free flying of men and vehicles, as it is described in [4]-[5]. If planet does not have enough gravity, then electrostatic artificial gravity may be used [3], Ch. 15.

The magnetic force lifts the reflector to needed altitude.

Figure 6 illustrates a light-weight, possibly portable house, using the same basic construction materials as the dwelling/workplace.

Inflatable Space Hotel

We live during the 21st Century when Earthly polar tourism is just becoming a scheduled pastime and the world public anticipates outer space tourism. The offered inflatable outer space (satellite) hotel for tourists is shown in Figure 7. That has the common walking area (garden) covered by a film having the controlled transparency (reflectivity), internal sections (living rooms, offices, restaurants, concert hall, storage areas, etc.). Hotel has electrostatic artificial gravity [3], and magnetic field. The electrostatic artificial gravity creates usual Earth environment, the magnetic field allows people to easily fly near the outer space hotel and still be effectively defended from the dangerous, and sometimes even lethal, solar wind.

Hotel has electrostatic artificial gravity and magnetic field that will permit people to freely fly safely near the hotel even when radiation in outer space is closely present and intense.

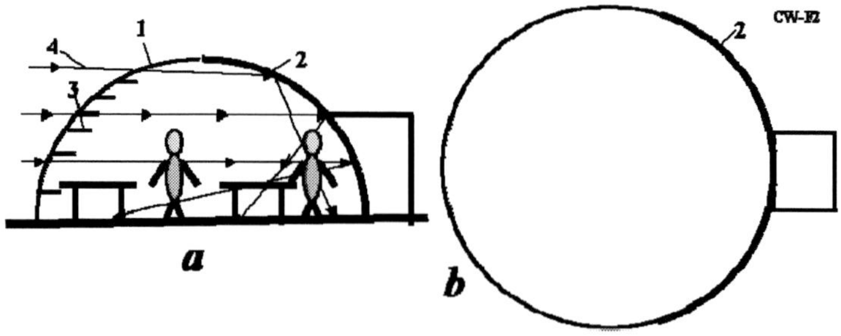

Figure 6. Inflatable film house for planet. Notation: (a) Cross-section area; (b) Top view. The other notations are same with Figure 2.

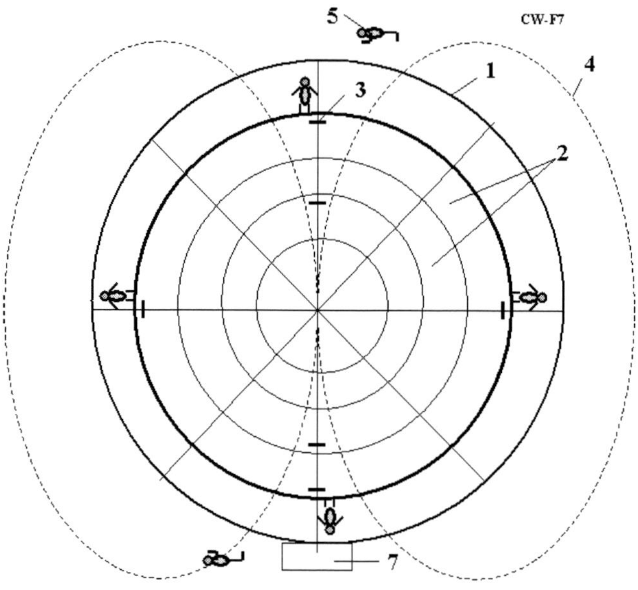

Figure 7. Inflatable space (satellite) hotel. Notations: 1 - inflatable hotel (control transparency cover film); 2 - internal sections of hotel (living rooms, offices, café, music hall, storage, etc.); 3 - door and windows in internal sections; 4 - magnetic line; 5 – outer space flying person (within hotel's magnetic field, [5]); 6 - common walking area (garden). 7 - docking chamber.

Visit Outer Space without a Spacesuit

Current spacesuit designs are very complex and expensive "machines for living". They must, at minimum, unfailingly support human life for some period of time. However, the spacesuit makes a cosmonaut/astronaut barely mobile, slow moving, prevents exertive hard work, creates bodily discomfort such as pain or irritations, disallows meals in outer space, has no toilet, etc. Mass of current spacesuits is about 180 kg. Cosmonauts/Astronauts—these should be combined into "Spationauts" as the 20th Century descriptions were derived from Cold War superpower competition—must have spaceship or special outer space home habitat located not far from where they can undress for eating, toilet, and sleep as well as rest.

Why do humans need the special spacesuit in outer space, or on atmosphere-less bodies of the Solar System? There is only one reason – we need an oxygen atmosphere for breathing, respiration. Human evolution in the Earth-biosphere has created lungs that aerate our blood with oxygen and delete the carbonic acid. However, in a particularly harsh environment, we can do it more easily by artificial apparatus. For example, surgeons when they perform surgery on heart or lungs connect the patient to the apparatus "Heart-lung machine", temporarily stopping the patient's respiration and hear-beat. In [3] at page 335, it is suggested that a method exists by donating some human blood, with the use of painless suture needles, is possible and that the blood can then be passed through artificial "lungs", just as is done in hospitals today.

We can design a small device that will aerate people's blood with oxygen infusion and delete the carbonic acid. To make offshoots from main lungs arteries to this device, we would turn on/off the artificial breathing at anytime and to be in vacuum (asteroid or planet without atmosphere) or bad or poisonous atmosphere, underwater a long time. In outer space we can be in conventional spacesuit defending the wearer from harmful solar light. Some type of girdle-like total body wrapping is required to keep persons in outer space from expanding explosively.

This idea may be checked with animal experiments in the Earth. We use the current "Heart-Lung" medical apparatus and put an animal under bell glass and remove the air inside the bell jar.

We can add into the blood all appropriate nutrition and, thusly, be without normal eating food for a long period of time; it is widely known that many humans in comas have lived fairly comfortably for many years entirely with artificial nourishment provided by drip injection.

The life in outer space without spacesuit will be easier, comfortable and entirely safe.

4. Theory, Estimation and Computations of Inflatable Outer Space Dome

1. Specific mass of inflatable Dome. The mass (and relative mass) of film is summary of top double layer and support cable (Figure 1) is

$$M = \frac{pS\gamma}{\sigma}H + \frac{\pi pS\gamma}{2\sigma}L \quad or \quad \overline{M} = \frac{M}{S} = \frac{p\gamma}{\sigma}(H + \frac{\pi}{2}L), \tag{1}$$

where M is film and cable mass, kg; p is air pressure, N/m^2; S is cover area, m^2; γ is specific mass of film and support cables, kg/m^3; σ is safety tensile stress of film and the support cable, N/m^2; H is height of Dome, m; L is distance between support cable, m.

The needed thickness of film δ is

$$\delta = \frac{\pi pL}{2\sigma}. \quad (2)$$

Example: Let us take $p = 10^5$ N/m^2; $\sigma = 10^9$ N/m^2 = 100 kgf/mm^2; $\gamma = 1800$ kg/m^3; $H = 3$ m; $L = 2$ m. Then $\overline{M} = 1.1 \quad kg/m^2, \quad \delta = 0.314 \quad mm.$

2. *Magnetic stationary solar space reflector.* Magnetic intensity from ground ring

$$B \approx \mu_0 \frac{iS}{2\pi H^3}, \quad S = \pi R^2, \quad (3)$$

where B is magnetic intensity, T; $\mu_0 = 4\pi \times 10^{-7}$ is magnetic constant; i is electric currency, A; S is area of ground ring, m^2; R is ground ring radius, m; $H >> R$ is altitude of reflector, m.

Example: for $R = 1000$m, $H = 1000$ m; $i = 10^5$ A, magnetic intensity is $B = 6.3 \times 10^{-5}$ T.

The mass of electronically superconducting wire is

$$M_R = 2\pi R s \gamma_w, \quad (4)$$

where s is cross-section area of wire, m^2; γ_w is specific mass of wire, kg/m^3.

For density of electric currency $j = 10^5$ A/mm^2 and $\gamma_w = 8000$ kg/m^3 the mass density of ground wire is about 50 kg. The mass of thin film heat screens, which defend the wire from solar and Moon heat radiation, is about 20 kg [5].

The mass of solar thin film reflector is

$$m_r = k_1 \pi r^2 \delta_r \gamma_r, \quad (5)$$

where r is reflector radius, m; δ_r is thickness of reflector film, m; γ_r is mass density of reflector, k_1 is coefficient of reflector mass increasing from additional support parts (for example, from inflatable ring). For $r = 20$ m, $\delta_r = 5$ μ, $\gamma_r = 1800$ kg/m^3, $k_1 = 1.2$ the reflector mass is 13.6 kg.

The mass of top ring is

$$m = \frac{m_r}{(B j_r \cos\alpha / g_m \gamma_w) - k_2}, \quad (6)$$

where j_r is density of electric currency, A/m^2; g_m is gravity of planet, m/s^2 (for Moon g_m=1.62 m/s^2); α is angle between magnetic line and planet surface (Figure 5); $k_2 > 1$ is coefficient of top ring mass increasing from heat radiation screens. The mass of top ring is remarkably quite small (less 0.5 kg).

The energy emitted by a body may be computed by employment of the Josef Stefan-Ludwig Boltzmann law.

$$E = \varepsilon\sigma_S T^4, \text{ [W/m}^2\text{]}, \tag{7}$$

where ε is coefficient of body blackness (ε =0.03 ÷ 0.99 for real bodies), σ_S = 5.67×10 $^{-8}$ Stefan-Boltzmann constant. For example, the absolute blackbody (ε = 1) emits (at T = 293 ^{0}C) the energy E = 418 W/m^2.

The common daily average Sun-furnished heat (energy) at Earth's orbit is calculated by equation

$$Q = 86400\, c\, qt, \tag{8}$$

where c is daily average heat flow coefficient, $c \approx 0.5$; t is relative daily light time, 86400 = 24 × 60 × 60 is number of seconds of an Earth-day, q = 1400 W/m^{2}s is heat flow at Earth's orbit.

The heat loss flow per 1 m^2 of dome film cover by convection and heat conduction is (see [6]):

$$q = k(t_1 - t_2), \quad \text{where} \quad k = \frac{1}{1/\alpha_1 + \sum_i \delta_i/\lambda_i + 1/\alpha_2}, \tag{9}$$

where k is heat transfer coefficient, $t_{1,2}$ are temperatures of the initial and final multi-layers of the heat insulators, $\alpha_{1,2}$ are convention coefficients of the initial and final multi-layers of heat insulators (α = 30 ÷ 100), δ_i are thickness of insulator layers; λ_i are coefficients of heat transfer of insulator layers (see Table 1), $t_{1,2}$ are temperatures of initial and final layers, $^{\circ}$C.

The radiation heat flow per 1 m^{2}s of the service area computed by equations:

$$q = C_r\left[\left(\frac{T_1}{100}\right)^4 - \left(\frac{T_2}{100}\right)^4\right], \quad \text{where} \quad C_r = \frac{c_s}{1/\varepsilon_1 + 1/\varepsilon_2 - 1}, \quad c_s = 5.67, \tag{10}$$

where C_r is general radiation coefficient, ε are black body rate of plates (see Table 2); T is temperatures of plates, $^{\circ}$K.

The radiation flow across a set of the heat screens is computed by equation

$$q = 0.5\frac{C'_r}{C_r}q_r, \tag{11}$$

where C'_r is computed by equation (10) between plate and reflector.

The data of some construction materials is found in Table 1 Attn. or [6], p.331, and Table 2 below.

As the reader will see, the air layer is the very best heat insulator. We do not limit the used material's thickness δ.

Table 2. [6], p. 465. Blackness

Material,	Blackness, ε	Material	Blackness, ε	Material	Blackness, ε
Bright Aluminum	0.04 - 0.06	Baked brick	0.88 - 0.93	Glass	0.91 - 0.94
t = 50 ÷ 500 °C		t = 20 °C		t = 20 ÷ 100 °C	

As the astute reader will notice, the shiny aluminum louver coating is excellent mean jalousie offsetting radiation losses from the dome.

The general radiation heat Q computes by equation [10]. Equations [7] – [11] allow computation of the heat balance and comparison of incoming heat (gain) and outgoing heat (loss).

The heat from combusted fuel is found by equation

$$Q = c_t m/\eta, \tag{12}$$

where c_t is heat rate of fuel [J/kg]; c_t = 40 MJ/kg for liquid oil fuel; m is fuel mass, kg; η is efficiency of heater, η = 0.5 - 0.8.

The thickness of the dome envelope, its sheltering shell of film, is computed by formulas (from equation for tensile strength):

$$\delta_1 = \frac{Rp}{2\sigma}, \quad \delta_2 = \frac{Rp}{\sigma}, \tag{13}$$

where δ_1 is the film thickness for a spherical dome, m; δ_2 is the film thickness for a cylindrical dome, m; R is radius of dome, m; p is additional pressure into the dome, N/m^2; σ is safety tensile stress of film, N/m^2.

For example, compute the film thickness for dome having radius R =100 m, additional air pressure p = 0.01 atm at top section in Figure 2 (p = 1000 N/m^2), safety tensile stress σ = 50 kg/mm^2 ($\sigma = 5\times10^8$ N/m^2), cylindrical dome.

$$\delta = \frac{100\times1000}{5\times10^8} = 0.0002\, m = 0.2\, mm \tag{14}$$

The dynamic pressure from Mars surface wind is

$$p_w = \frac{\rho V^2}{2}, \tag{15}$$

where ρ is atmospheric density, kg/m^3; V is wind speed, m/s.

If a planet has a long nighttime, the heat loss protection can reduce the head losses as we can utilize inflated dome covers with additional layers and additional heat screens. One heat screen decreases heat losses by 2, two screens can decrease heat flow by 3 times, three by 4 times, and so on. If the inflatable domes have a multi-layer structure, the heat transfer decreases proportional to the total thickness of its enveloping film layers.

5. Macro-Projects

The dome shelter innovations outlined here can be practically applied to many cases and climatic regimes. We suggest initial macro-projects could be small (10 m diameter) houses (Figure 6) followed by an "Evergreen" dome covering a land area 200 m × 1000 m, with irrigated vegetation, homes, open-air swimming pools, playground, "under the stars style" concert hall.

The house and "Evergreen" dome have several innovations: magnetic suspended Sun reflector, double transparent insulating film, controllable jalousies coated with reflective aluminum (or film with transparency control properties and/or structures) and an electronic cable mesh inherent to the film for dome safety/integrity monitoring purposes. By undertaking to construct a half-sphere house, we can acquire experience in such constructions and explore more complex constructions. By computation, a 10 m diameter home has a useful floor area of 78.5 m^2, airy interior volume of 262 m^3 covered by an envelope with an exterior area of 157 m^2. Its film enclosure material would have a thickness of 0.0003 m with a total mass of about 100 kg.

A city-enclosing "Evergreen" dome of 200 m × 1000 m (Figure 2, with spherical end caps) could have calculated characteristics: useful area = 2.3×10^5 m^2, useful volume 17.8×10^6 m^3, exterior dome area of 3.75×10^5 m^2, comprised of a film of 0.0003 m thickness and about 200 tonnes. If the "Evergreen" dome were formed with concrete 0.25 m thick, the mass of the city-size envelope would be 200×10^3 tonnes, which is a thousand times heavier. Also, just for comparison, if we made a gigantic "Evergreen" dome with stiff glass, thousands of tonnes of steel, glass would be necessary and such materials would be very costly to transport hundreds or thousands of kilometers into outer space to the planet where they would be assembled by highly-paid but risk-taking construction workers. Our film is flexible and plastically deformable. It can be relatively cheap in terms of manufacturing cost. The single greatest boon to "Evergreen" dome construction, whether on the Moon, in Mars or elsewhere, is the protected cultivation of plants under a protective dome that efficiently captures energy from the available and technically harnessed sunlight.

6. Discussion

As with any innovative macro-project proposal, the reader will naturally have many questions. We offer brief answers to the two most obvious questions our readers are likely to ponder.

(1) *Cover damage.*

The envelope contains a rip-stopping cable mesh so that the film cannot be damaged greatly. Its structured cross-section of double layering governs the escape of air inside the living realm. Electronic signals alert supervising personnel of all ruptures and permit a speedy repair effort by well-trained responsive emergency personnel. The topmost cover has a strong double film.

(2) *What is the design-life of the dome film covering?*

Depending on the kinds of materials used, it may be as much a decade (up 30 years). In all or in part, the durable cover can be replaced periodically as a precautionary measure by its owners.

7. Conclusion

Utilization of "Evergreen" domes can foster the fuller economic development of the Moon, Mars and Earth itself - thus, increasing the effective area of territory dominated by humans on, at least, three celestial bodies. Normal human health can be maintained by ingestion of locally grown fresh vegetables and healthful "outdoor" exercise. "Evergreen" domes can also be used in the Earth's Tropics and Temperate Zone. Eventually, "Evergreen" domes may find application on the Moon or Mars since a vertical variant, inflatable space towers [2], are soon to become available for launching spacecraft inexpensively into Earth orbit or on to long-duration interplanetary outer spaceflights.

Moon base.

Mars base.

REFERENCES

(Parts of these articles can be found at A.A. Bolonkin's website: *http://Bolonkin.narod.ru/p65.htm* and in his two most recent books: "*Non-Rocket Space Launch and Flight*" (Elsevier, London, 2006) and in "*New Concepts, Ideas and Innovations in Aerospace and Technology*" (NOVA, Hauppauge NY, 2008.)

[1] Bolonkin A.A., Cathcart R.B., "Inflatable 'Evergreen' dome settlements for Earth's Polar Regions", *Clean Technologies and Environmental Policy,* Vol. 9, No. 2, May 2007, pp.125-132.

[2] Bolonkin, A.A., "Non-Rocket Space Launch and Flight", Elsevier, London, 2006, 488 pages.

[3] Bolonkin, A.A., *Control of Regional and Global Weather,* 2006.

[4] Bolonkin, A.A., "New Concepts, Ideas, and Innovations in Aerospace, Technology and Human Life" (NOVA, 2007).

[5] Bolonkin A.A., AB Levitation and Energy Storage, *This work presented as paper AIAA-2007-4613 to 38th AIAA Plasma dynamics and Lasers Conference in conjunction with the16th International Conference on MHD Energy Conversion* on 25-27 June 2007, Miami, USA.

[6] Naschekin, V.V., Technical thermodynamic and heat transmission. Public House High University, Moscow, 1969 [in Russian].

Chapter 3

PASSENGER LIFE-SAVING IN A BADLY DAMAGED AIRCRAFT SCENARIO*

ABSTRACT

Offered is an entirely new method for saving passenger lives in any catastrophic situation, including total failure of aircraft control, extreme damage and loss of an aircraft's wings, tail, the shutdown of all propelling engines, etc. It is shown that previous works which have proposed using only parachutes are useless because their proposers failed to consider the likely overload of the parachute jerk stress (at the moment of parachute release and blossom) and the sudden-stop impact of crashing aircraft on Earth's surface. Such jerk and impact can destroy aircraft and, therefore, kill passengers and destroy cargo.

Offered is a connected series of inter-related technical innovations which overcome these obvious difficulties and allow for a soft, near zero-speed landing in any topographically and vegetationally suitable stopping place on land, allowing potential to save aircraft. This method may be applied to all known types of existing airplanes and increases their weight by only about 1.5 to 2.5%. Also, the method may be used for vertically landing extant, already built aircraft as, for example, when any paved airport runway is damaged or could become overloaded.

Keywords: aircraft safety, saving of air passengers, air catastrophe

INTRODUCTION

Relative deaths of air passengers are sometimes more than the death rate due to railroads or ships. Especially, the probability of death in an aircraft catastrophe has increased since the start of the War on Terror subsequent to the 9/11/01 attacks at the tip of New York City's famous Manhattan Island. However, people cannot abandon 21st Century aviation. Quite simply, there is no substitute! Aviation transport is the most comfortable, quickest and

---

* Presented as paper AIAA-2007-5844 to 43 Joint Propulsion Conference, 8-11 July 2007, Cincinnati, OH, USA.

sometimes only means to reach many distant Earth regions. So, macro-engineers must find the best method of saving passenger lives when aircraft are damaged and even when a full loss of aerial vehicle control happens.

A.A. Bolonkin offers some unique innovations and shows that his offered devices, installed on any current aircraft, can save passenger lives in any airplane damage scenario including wing failure, tail, engine, control and steeply diving from the stratosphere. The relative weight of offered devices is only 1.5 to 2.5% of aircraft's current weight.

Aircraft catastrophes are very different from sea or railway catastrophes [1]. In any ship catastrophe, except for rogue wave sinkings, there is time to call land bases or other nearby ships for help and rescue. Ships have lifeboats. In any railway catastrophe, for the most part, passengers (especially those riding in the train's cars farthest to the rear of the locomotive) are, generally speaking, safer that passengers towards the front of the train that stops suddenly or is derailed while moving. In marked contrast, aviation catastrophes usually result in all passengers being killed as the aircraft plummets from the sky and violently hits the ground or, sometimes, water surface of the Earth. About 40 passengers out of 1 million are fatalities in every flight (It is without terrorist attack. The terrorist attack increases it by 4 times). American Airlines Flight 587 crashed in Queens, NY, killing 256 people. Aircraft crashes dissuade people from flying as passengers by inspiring trepidation and latent fear. September 11, 2001 demonstrated that big, fully-fueled commercial aircraft are potential super-bombs that can kill not only all 200 - 300 passengers but thousands of people concentrated in buildings immovably affixed to the soil.

Common passenger aircraft Boeing 737.

1. Aircraft Accidents

Approximately 80% of all aviation accidents occur shortly before, during or after takeoff or landing, and are typically the result of human, also known as "pilot error", and/or ignored technical problems within an aircraft; mid-flight catastrophes such as aircraft collisions and sabotage or bombings are exceedingly rare. Among other things, the latter have been caused by smuggled bombs as in the 1988 Lockerbie, Scotland incident, mid-air collisions such as the 2002 Überlingen crash or in cases of (purportedly) mistaken identity where civilian aircraft were shot down by military (Korean Air Flight 007 in Asia).

An accident survey of 2145 aircraft accidents occurring during the period 1950 - 2004 were determined by accident investigators to be caused by:

- 45%: Pilot error—almost a catch-all category, especially if they are dead
- 33%: Undetermined or unrecorded
- 13%: Mechanical failure of the air vehicle
- 7%: Weather
- 5%: Sabotage (obvious bombings, known hijackings, suspected shoot-downs)
- 4%: Other human error (air traffic controller mistakes, improper cargo loading of aircraft, improper aircraft maintenance, fuel contamination, language miscommunication amongst assorted workers, etc.)
- 1%: Other cause, undefined and unknown.

2. Aircraft Accident Statistics

Aircraft accident statistics [2] provide a valuable source of data to set the proper priorities, as well as to monitor progress made, by the world's aviation industry. They are calculated for many kinds of aircraft. The accident statistics presented in this section apply to worldwide commercial jet-propelled airplanes that are heavier than ~27 tonnes maximum gross weight. However, airplanes manufactured in the former Soviet Union (CIS) are not included.

Aircraft accident statistics provide a valuable source of information in order to develop the correct priorities, as well as to monitor progress made, by the world's aviation industry. They are calculated for many types of aircraft.

Globally, air traffic has increased since the early 20th Century. But, after the World Trade Center attack on 9/11/01, many commercial airlines were in put into financial difficulty, and worldwide air traffic decreased. Two years later economic growth of the airline sector recommenced.

In 2004, airplanes flew 34.1 million flight hours. This number must be kept in mind for statistical analysis and failure rates interpretation: if an event has a probability to occur of 1 out of 1 million (which is rather low), it will occur several times a year.

The following graph shows the total number of certified commercial jets that are heavier than ~27 tonnes maximum gross weight (it does not include any airplanes manufactured in the former Soviet Union).

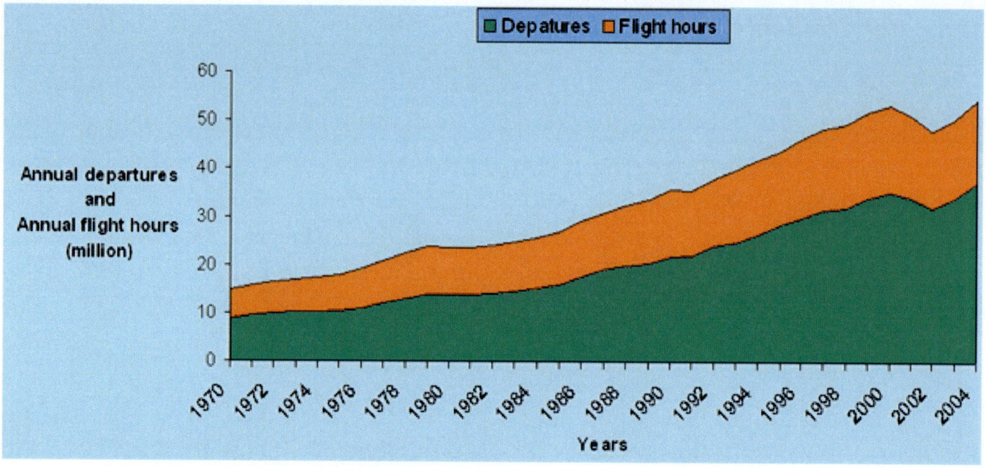

Figure 1. Annual departures and flight hours (millions).

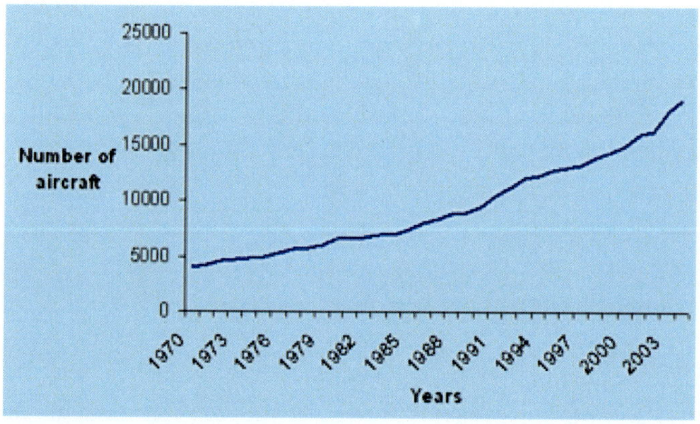

Figure 2. Number of aircraft in the World.

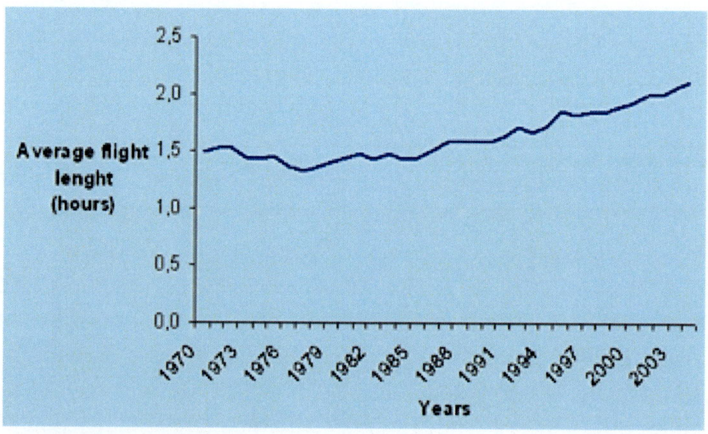

Figure 3. Duration of an average aircraft's flight (in hours).

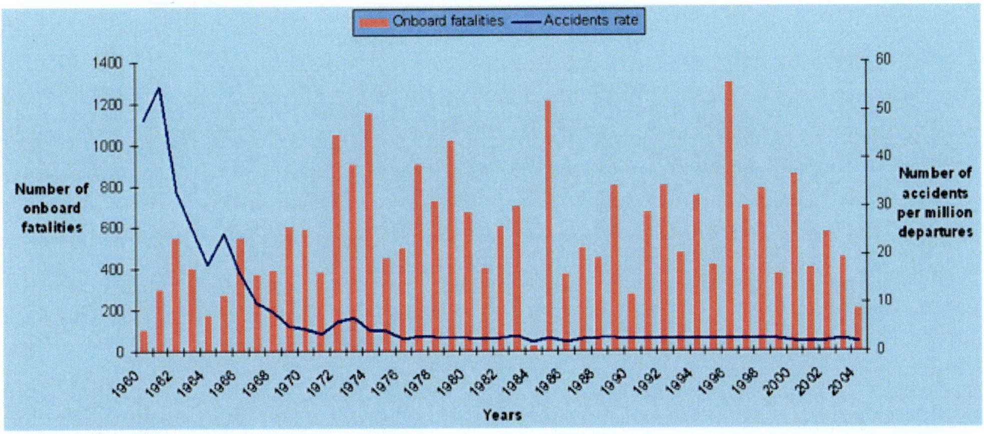

Figure 4. The Number of on-board human fatalities and number of aircraft accidents per million air terminal departures.

The number of aircraft ceaselessly increases to satisfy the world public's growing transportation demands. Fortunately, even if the busiest airspaces are congested, the risk of collision is very low owing to the new aviation technologies that make possible accurate absolute geographic position and atmospheric altitude measurements above land and water, helping both flight crews and official ground Air Traffic Controllers.

For example, in order to accommodate all the aircraft traffic over some well traveled ocean routes, specific airspaces where aircraft are vertically separated by only 1000 feet were created; such airspaces are labeled "RVSM" airspaces, which signifies Reduced Vertical Separation Minima.

The informative graph above illustrates that the average flight-time also increases with the passage of years. Some state-of-the-art aircraft can perform 22-hour, non-stop long-distance flights.

3. Aircraft Accident Rates

The number of accidents per flight decreases with time's elapse. But the number of human fatalities per year is virtually unchanging (no significant fatality decrease).

Nowadays, aircraft accidents are less likely than several decades ago. Nevertheless, the growing number of flying aircraft, and their increasing passenger/cargo capacity, cannot result in a reduction of on-board fatalities.

Table 1.

	1959-2004	1995-2004
Number of accidents	1 402	376
Number of onboard fatalities	25 664	5 612

4. Scheduled Passenger Operations Are less Likely to Have an Accident than any other Type of Flight-related Operation

As newspapers always report, you are less likely to have an accident if you fly a regular flight than if you get on other types of flights (unscheduled passenger and charter, cargo, ferry, test, training, and demonstration). This is a truth of life which you may appreciate by examining the following graph. This graph also illustrates that scheduled passenger operations are 5 to 6 times more numerous.

It is quite exceptional for an accident to be related to a single cause. Almost every accident is the final consequential outcome of a causal chain of events. Accident reports usually make a distinction between the main, or sufficient cause, and a number of contributing factors. The graph below illustrates the percentage pie-chart of all of the main causes identified for early-21st Century aircraft crashes.

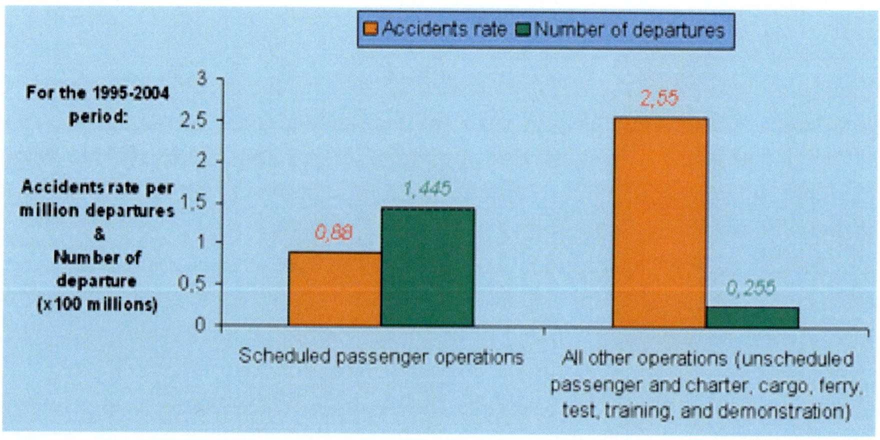

Figure 5. Aircraft accident causes.

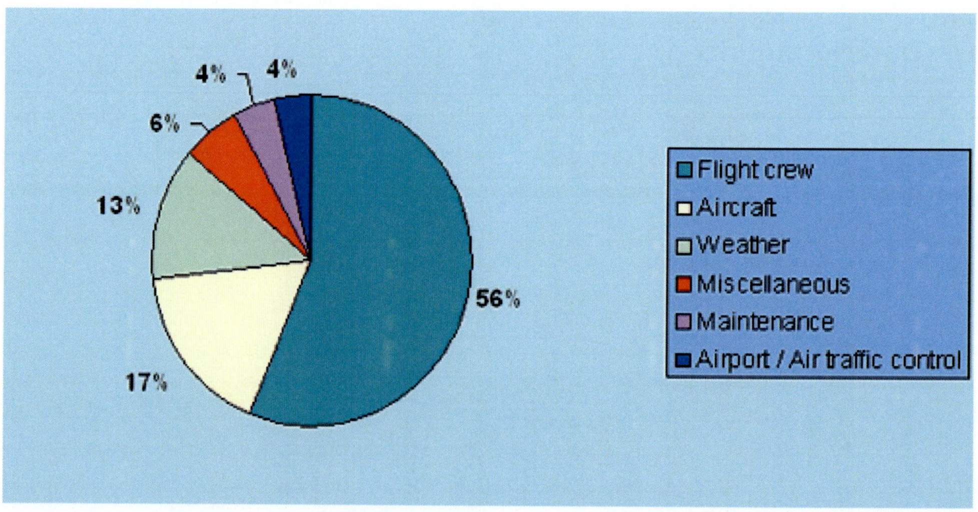

Figure 6. Percentage of different aircraft loss causes.

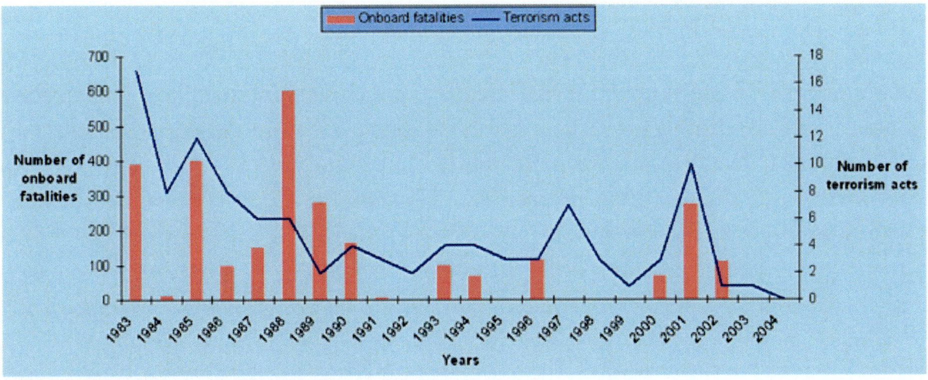

Figure 7. Number of onboard fatalities and number of terrorism acts.

The main cause is the infamous "human factors". In order to prevent this source of accidents, aircraft crews are legally required to periodically re-train in their imperative skills set. Next comes the aircraft mechanical and structural failures, but this cause is much less likely with thoroughly engineered modern aircraft than with early-20th Century (experimental) passenger aircraft!

5. Aviation Terrorism

Only airport security services and alert people such as airport/airline staffs and observant passengers or airport guests can prevent aviation terrorism such as bombing and hijacking. Aircraft are exposed to such attacks, because they are vulnerable while waiting on the airport's tarmac and subject to various kinds of shoot-downs while still in the air (landing and take-off), and because an aircraft crash is newsworthy. Nowadays, cargo bay containers are bomb-resistant, but not aircraft since such over-building would make them too heavy to fly!

Fortunately, and thanks to the numerous ideas that have been imagined to prevent such attacks, the number of terrorism acts has been reduced with years, especially since the commencement of the War on Terror after 9/11/01. But terrorism fatalities have no obvious trend. New aircraft that may carry many persons (+600 passengers) may become a special terrorism target, which could lead to extremely deadly attacks in the coming years.

About 40 passengers out of 1 million become actual flight fatalities. We must decrease the flight death rate and total killed and give more hope to the traveling public: in other words, whatever the flight occurrence, the passengers must have a high probability to survive the aircraft terrorism event.

INNOVATIONS

In case of failure, or full damage of main aircraft system, airframe, wing, tail and controls (for example, by terrorist surface-to-air or air-to-air missile) for saving of passenger lives the pilot activates and uses the system described herein. The offered installation has the following peculiarities and devices:

1. Small Braking Parachute

The parachute is a well-known device used by parachutists for air jumps and airborne re-supply aviation for air dropping of materials to persons on the ground during emergency and wartime conditions. But, usually, a parachute is not suitable for saving conventional air passenger and other freight-hauling aircraft. The conventional aircraft parachute must be very heavy and ordinarily gives a high landing speed. The aircraft has a touchdown speed of about 0.3 to 1 m/s and the undercarriage brake distance is 0.5 to 0.8 m. That way, the passenger feels only a small blow during touchdown moment. The parachutist is a trained man or woman who is landing on his spring-like flexible legs. His parachute has an air drag only about 10 N/m^2. However, the parachutist has a drop speed of 5 to 7m/s and endures 5 to 7 "g" overload during such physically taxing ground landings. Regular aviation passengers are untrained persons (men and women plus frail children) who are usually sitting in a passenger aircraft's well-upholstered, padded cabin-floor secured chair. Passengers do not always have their seat safety harness locked. In contrast, the parachutist lands on his legs and has a brake distance of 0.6 - 0.8 m. The regular passenger seats in aircraft have a brake distance of 5-8 cm. That means, obviously, that the passenger overload is 50 to 70 "g". That is a deadly overload for any person, because the trained person can just barely endure the maximum overload of 16 "g" for only a very short time. This stress will fully destroy an aircraft because the safe aircraft overload is 2 - 3 "g". We can save our regular passengers and aircraft if the touchdown speed will be close to zero! There were a lot of R&D schemes for saving small aircraft by parachutes [3], but most of them were useless because even an enormous diameter (and materially heavy) parachute cannot fully ensure slow touchdown speed in order to save passengers and the aircraft.

As noted, we use a small chute, which decreases the speed only to 25 to 40 m/s. The parachute drag depends upon the speed as V^2. That means our parachute will be 25 to 100 times less (and less weighty) than conventional parachute used in other instances. (How we decrease speed from 20 - 40 m/s to 0 m/s, you will see soon.) Clearly, we can use the inflatable wing parachute having ratio lift/drag equals 6 - 7 and glide a long distance.

The other difficulty is aircraft overload in the moment of canopy opening. If aircraft speed is high, the overload is very large. We use a special parachute (or double) which canopy opens slowly with permanent overload (2 to 3g) to successful save the aircraft and its passengers/cargo.

The next problem is aircraft control. The pilot must have a choice in a landing place away from buildings, high-voltage electric power lines, tall and mature trees, etc. We use the controllable 'square' ram-air parachute (airfoil, parafoil) that gives a limited, but at least some, control in the pilot's choice of a disstressed aircraft landing place.

At the present time, there are available strong, light artificial material fibers which can be used for the proposed aircraft parachute [4] p. 33 (or see Table 3 in Attn.).
The fibrous material PRD-49 had the following properties in 1972 [5]:

Tensile ult. 312 kgf/mm^2; specific gravity 1500 kg/m^3; filament diameter 0.0003 in.; minimum available yarn size (called the "denier") is 200.

Summary of the aircraft parachute innovative features:

a) We use small, light parachutes which decrease aircraft speed only up V = 20 to 40 m/s.

b) Our stowed aircraft parachute, when deployed, opens with the constant overload which is safest for passengers and the aircraft.
c) Our parachute is ram-air and has limited control.
d) It is made from currently available commercial artificial fibers that are strong.
e) Our parachute is an outer-stored deployable device and can be joined to (or disconnected from) any existing aircraft to mainframe or special round type (Figure 8).

2. Rocket Retard Booster

For decreasing (braking) the touchdown speed to zero solid fuel rockets are used. The change of speed (20 - 40 m/s) is small and rocket booster is also small. Its weight is 1 - 2% of aircraft weight. The small solid rocket engine is simplest, very reliable, inexpensive. Its combustion camera (body) may be made from current composed materials and has a small weight (10 - 15% from fuel weight). The rocket booster is packed with parachute and is taken out together simultaneously. It is located under open parachute and works automatically only when the aircraft elevation above the ground is appropriate (see Figure 8).

3. Aircraft Underbelly Inflatable Pillows

For soft landing and passenger rescue in landing on sea or lake, our system may to have inflate pillows located below of fuselage in a special low-drag container. They are filled with air before the aircraft's touchs down and, in any case, saves the aircraft when one lands on its fuselage (not its normal rubber-wheeled landing gear), or in sufficiently deep water such as a lake or ocean. Their weight is small and inconsequential.

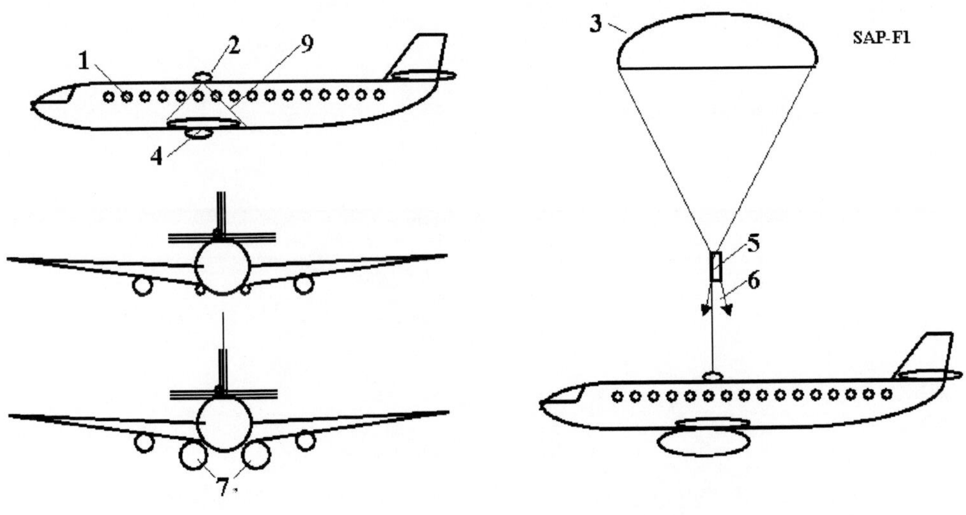

Figure 8. Method for saving passenger lives in any damaged aircraft. Notations: 1 - aircraft, 2 - parachute booster container, 3 - parachute, 4 - inflatable pillow-float container, 5 - rocket booster, 6 - booster jet, 7 – underbelly pillow-float in swollen form, 9 - support type.

The offered method has some very important advantages: in some emergency situations, the aircraft might be able to make successfully an unusual vertical landing virtually anyplace without great damage or loss of human life!

COMPUTATIONS AND ESTIMATIONS

1. Parachute Drag

Relative parachute drag computed by equation

$$\overline{D} = \frac{D}{S} = C_D \frac{\rho V^2}{2}, \tag{1}$$

where D is air drag, N/m^2; S is parachute area, m^2; $C_D = 1 \div 1.6$ is drag coefficient; $\rho = 1.225$ is air density, kg/m^3; V is parachute speed, m/s. Computations are presented in Figure 9 graph.

2. Parachute Parameters

The equation below allows the computation of the mass of circular parachute

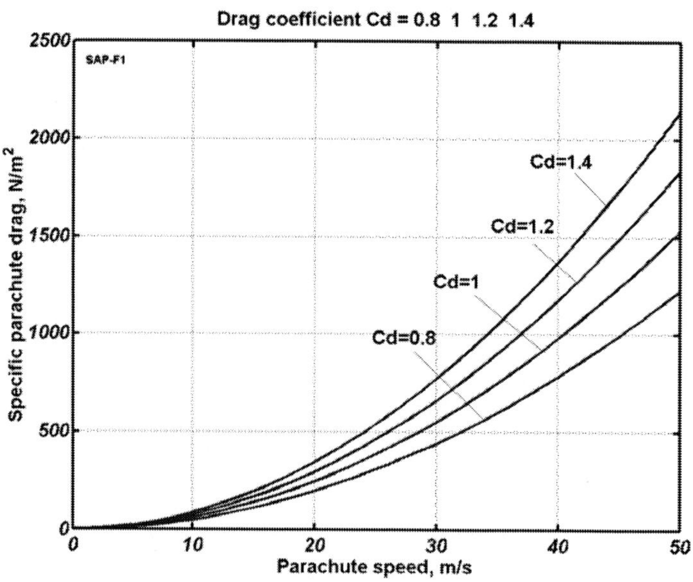

Figure 9. Specific parachute drag (N/m^2) via parachute speeds for different parachute drag coefficient.

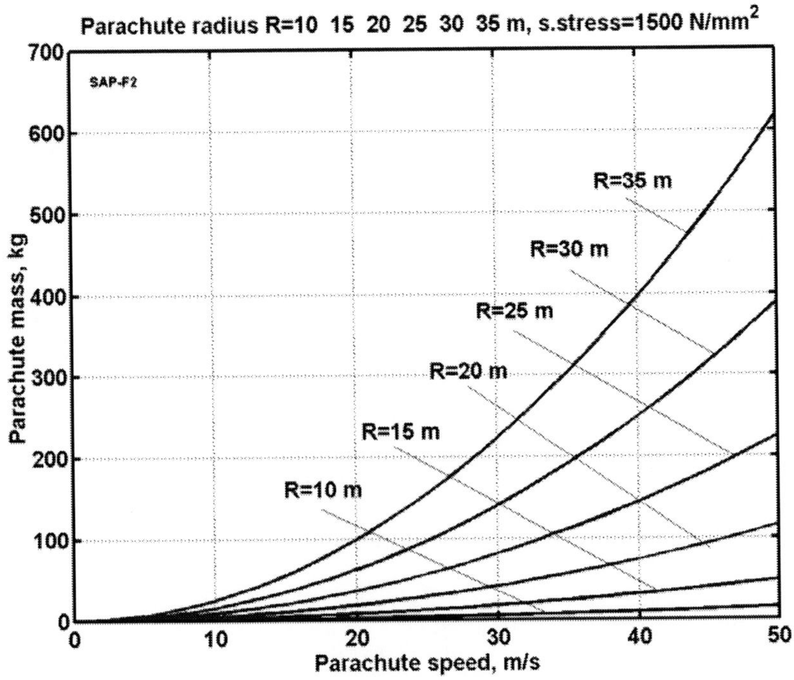

Figure 10. Parachute mass via parachute speed for different parachute radius. $C_D = 1$.

$$S = \pi R^2, \quad S_c = 2S, \quad \delta = \frac{\overline{D}R}{2\sigma}, \quad M = 2\pi R^3 \overline{D} \frac{\gamma}{\sigma}, \qquad (2)$$
$$for \; C_D = 1 \; M = 2\pi R^3 \frac{\rho V^2}{2} \frac{\gamma}{\sigma}$$

where R is parachute radius, m; δ is parachute thickness; $\sigma \approx 150\times10^7$ is safety stress of parachute material, N/m^2; M is parachute mass, kg; $\gamma \approx 1500$ kg/m^3 is density of parachute material, kg/m^3.

Coefficient 2 means that the weight of the parachute cord approximately equals the weight of the parachute canopy. The computations are shown in Figure 10.

3. Mass of Rocket Booster

The fuel mass of the solid rocket booster may be estimaded by equation

$$\overline{M}_b = \frac{V}{v}, \qquad (3)$$

where $v \approx 1800 \div 2600$ m/s is specific impulse (exhaust jet speed) of firing rocket booster. Results of computation is presented in Figure 11. The body mass of booster rocket is about 10 $\div$ 15% for current composite material.

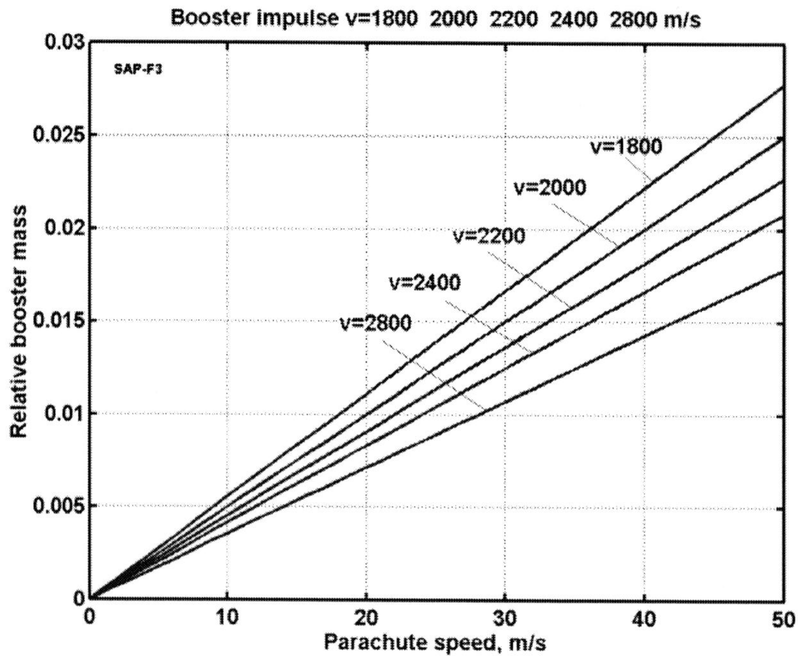

Figure 11. Relative booster mass via parachute speed for different booster fuel.

MACRO-PROJECT

As example, we calculate the parachute parameters for aircraft having weight W = 100 tonnes (2×10^6 N/m^2). Take as initial data: the parachute speed V = 30 m/s, drag coefficient C_D = 1.2, specific impulse of booster v = 2200 m/s. The required area of parachute is

$$S = \frac{2W}{C_D \rho V^2} = \frac{2\times10^6}{1.2\cdot1.225\cdot30^2} = 1500 \ \text{m}^2, \quad R = \sqrt{\frac{S}{\pi}} = 21.8 \ \text{m}^2.$$

Canopy thickness and total mass of parachute for σ = 150×10^7 N/m^2, γ = 1500 kg/m^2 are

$$q = \frac{\rho V^2}{2} = 551, \quad \delta = C_D q \frac{R}{2\sigma} = \frac{551\cdot21.8}{2\cdot1.5\cdot10^9} = 4\cdot10^{-6} \ \text{m},$$
$$M = 4S\delta\gamma = 4\cdot1500\cdot4\cdot10^{-6}1500 = 36 \ kg$$

Mass of rocket booster for v = 2200 m/s is

$$\overline{M}_b = 1.15\frac{V}{v} = 1.15\frac{30}{2200} = 0.0157,$$

where weight of booster body is 15%.

Mass of air pillows, together with air balloon, is ~14 kg, support types 20 kg.

The total installation mass is 1640 kg. We assume 1700 kg. That is 1.7% of usual total mass of commercial aircraft. That is small price for saving passenger lives in any catastrophic situation!

CONCLUSION

Offered is an entirely new method of saving passenger lives in almost any imaginable catastrophic potential crash landing situation, including full failure of aircraft controls, damage and loss of the aircraft's wings, tail, breakdown and/or shutdown of all propelling engines, etc. It shown that previous unworkable schemes, proposing the use of only the parachute, are utter failures because the "g" overload during the parachute jerk (occurring during first release of parachute) and the blow (touchdown) of aircraft on the Earth's surface were not fully considered. Instead, here a series of innovations, which overcome these great difficalties, and allow a soft landing in any suitable place, permits the preservation of the passengers' lives and, possibily, a repairable commercial airliner. This technique may be applied to all existing airplanes and increases their effective operational flying weight only by1.5 to 2.5%. This invention may be also used for vertically landing endangered aircraft when, for example, the airport's runway is damaged, overloaded or simply out-of-commission, temporarily or otherwise.

See also the works [6]-[7].

REFERENCES

(Part of this chapter can found at the main author's website: *http://arxiv.org*. search "Bolonkin", *http://Bolonkin.narod.ru/p65.htm* and in the printed books: "*Non-Rocket Space Launch and Flight*" (Elsevier, 2006) and "*New Concepts, Ideas and Innovations in Aerospace and Technology*" (Nova, NY, 2008.)

[1] Bolonkin A.A., A N*ew Method of Atmospheric reentry for Space Shuttles.* 2006.

[2] Go to: *http://www.1001crash.com/*

[3] Eilertson W.H., Aircraft recovery using an inflatable wing, *AIAA, Paper No.73-470.*

[4] Bolonkin A.A., Non-Rocket Space Launch and Flight, Elsevier, 2006, 488 pages.

[5] Harris J.T., Advanced material and assembly methods for inflatable structures, *AIAA, Paper No. 73-448.*

[6] Nguyen N. and et al., *Dynamics and Adaptive Control for Stability Recovery of Damaged Asymmetric Aircraft,* AIAA, 2006.

[7] S. Dellicker (U.S. Army, Yuma Proving Ground, AZ), R. Benney, S. Patel (U.S. Army, Natick, MA), T. Williams, C. Hewgley, O. Yakimenko, R. Howard, and I. Kaminer (U.S. Naval Postgraduate School, Monterey, CA), Performance, control, and simulation of the Affordable Guided Airdrop System, AIAA-2000-4309, *AIAA Modeling and Simulation Technologies Conference,* Denver, CO, Aug. 14-17, 2000.

[8] *Wikipedia*. Some background material in this article is gathered from Wikipedia under the Creative Commons license.

Chapter 4

ELECTROSTATIC CLIMBER FOR SPACE ELEVATOR AND LAUNCHER*

ABSTRACT

Here, the main author details laboratory and library research on the new, and intrinsically prospective, Electrostatic Space Elevator Climber. Based on a new electrostatic linear engine previously offered at the 42nd Joint Propulsion Conference (AIAA-2006-5229) and published in "AEAT", Vol.78, No.6, 2006, pp. 502-508, the electrostatic climber described below can have any speed (and braking), the energy for climber movement is delivered by a light-weight high-voltage line into a Space Elevator-holding cable from Earth-based electricity generator. This electric line can be used for delivery electric energy to a Geosynchronous Space Station. At present, the best solution of the climber problem, announced by NASA, is very problematic.

Shown also, the linear electrostatic engine may be used nowadays as a realistic power space launcher. Two macro-projects illustrate the efficacy of these new devices.

Keywords: Space elevator, Electrostatic climber for space elevator, Electrostatic space launcher, Electrostatic accelerator

INTRODUCTION

General Statement

Our world's functioning aviation, space, and energy industries need and seek truly revolutionary ideas which will significantly improve the employment capability of all future ground, air and outer space vehicles. A.A. Bolonkin has, over the past few years, offered a series of new ideas [1-73] contained in (a) numerous patent applications [3 -17], and in (b) presentations and manuscripts that have been given at the World Space Congress (WSC)-

* This work is presented as paper AIAA-2007-5838 for 43 Joint Propulsion Conference, Cincinnati, Ohio, USA, 9 - 11 July, 2007.

1992, 1994 [19 -22], the WSC-2002 [23 -31], as well as at several Propulsion Conferences [32 -39], and (c) other additional articles [40 -73].

In this chapter, a revolutionary method, and doable means of proper implementation, for future manned and unmanned spaceflights and ground systems are revealed. The method exposed here uses highly charged cylindrical bodies. The proposed space launch system creates tens of tonnes of thrust and can accelerate various space-worthy scientific and commercial apparatus to extremely high traveling velocities.

History

In early published works and public-record patent applications (1965 - 1991), during World Space Congress-2002 and at other scientific forums, A.A. Bolonkin suggested a series new cable vehicle launchers, outer space transport systems, space elevator, anti-gravitator, kinetic space tower, and other unique space systems, which decrease the cost of space launch by thousands times, or radically increase the use possibilities of major related ground system infrastructures. All such beneficial space systems need a linear engine. In particular, there are: Cable Space Launcher [23-25, 40], Earth-Moon Transport system [29,39], Earth-Mars Transport System [30], Circle Space Launcher [31], Hypersonic tube gas launcher [32], Air Cable aircraft [41, 42], Non-Rocket Transport System for Space Elevator (Elevator climber)[36], Centrifugal Keeper [38], Asteroid Propulsion System [27, 40], Kinetic Space Towers [43], Long Transfer of Mechanical Energy [45], High Speed Catapult Aviation [52], Kinetic Anti-Gravitator [55], Electrostatic Levitation [59], AB Levitator [67] and so on [1]-[73] (Figure 1). Part of these works is outlined rather fully in several printed books [60] and [73].

Particulars of an electrostatic engine [66], which can be used for every noted installation as a linear driver, are broadcast. Specifically, in this chapter of our book, much more essential detail about the likely future applications of the previously offered electrostatic linear engine is related directly to the space elevator climber and to the Earth-based accelerator used as a low-cost launcher of robotic and manned spaceships and interplanetary as well as interstellar space probes.

Description of Electrostatic Linear Climber and Launcher

The linear electrostatic engine [66], climber, for Space Elevator includes the following main parts (Figure 2): plate (type) stator 1 (special cable of Space elevator), cylinders 3 inside having conducting layer (or net) (cylinder may be vacuum or inflatable film), conducting layer insulator, chargers (switches 6) of cable cylinders, high-voltage electric current line 6, linear rotor 7. Linear rotor has permanent charged cylinder 4. As additional devices, the engine can have a gas-pressurizing capability and a vacuum pump [66].

The cable (stator) has a strong cover 2 (it keeps tensile stress - thrust/braking) and variable cylindrical charges contained dielectric cover (insulator). The conducting layer is very thin and we neglect its weight. Cylinders of film are also very light-weight. The charges can be connected to high-voltage electric lines 6 that are linked to a high-voltage device (electric generator) located on the ground.

The electrostatic engine works in the following mode. The rotor has a stationary positive charge. The cable has the variable positive and negative charges. These charges can be received by connection to the positive or negative high-voltage electric line located in cable (in stator). When positive rotor charge is located over given stator cylinder this cylinder connected by switch to positive electric lines and cylinder is charged positive charge but simultaneously the next stator cylinder is charged by negative charge. As result the permanent positive rotor charge repels from given positive stator charge and attracts to the next negative stator charge. This force moves linear rotor (driver). When positive rotor charge reaches a position over the negative stator charge, that charge re-charges to positive charge and next cylinder is connected to the negative electric line and then the whole cycle is repeated. To increase its efficiency, the positive and negative stator charges, before the next cycle, can run down through a special device, and their energy is returned to the electric line. It is noteworthy that the linear electrostatic engine can have very high efficiency!

Earth-constant potential generator creates a running single wave of charges along the stationary stator. This wave (charges) attracts (repel) the opposed (same) charges in rotor (linear driver) and moves (thrust or brake use) climber.

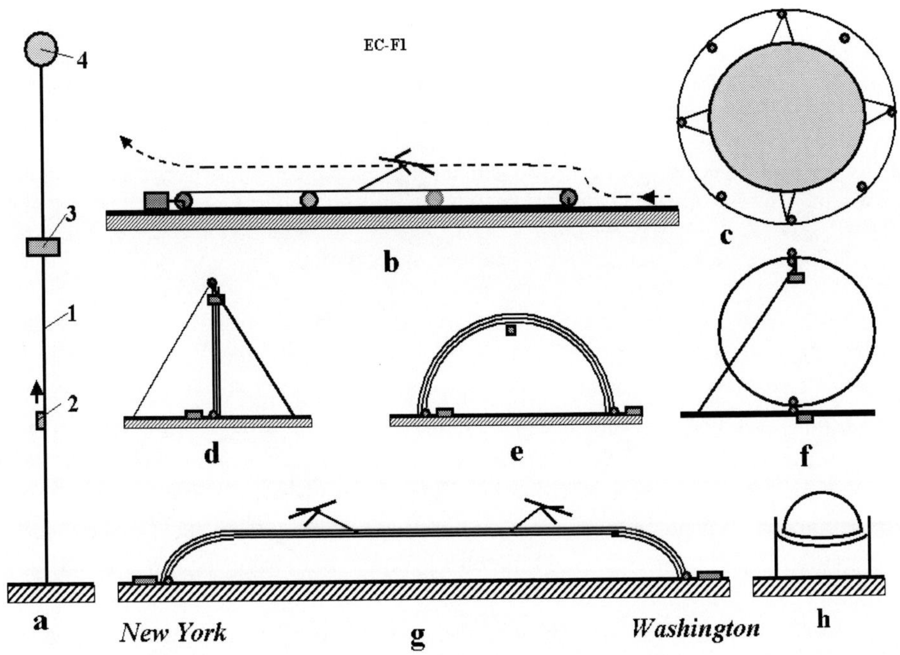

Figure 1. Installations needing the linear electrostatic engine. (a) Space elevator [36]. Notation: 1 - space elevator, 2 - climber, 3 - geosynchronous space station, 4 - balancer of space elevator. (b) Space cable launcher [23-25, 40]; (b) Circle launcher [31]; (c) Earth round cable space keeper [38]; (d) Kinetic space tower [43]; (e - f) Space keeper [38]; (g) (f) Cable aviation [41,42]; (g) Levitation train [59].

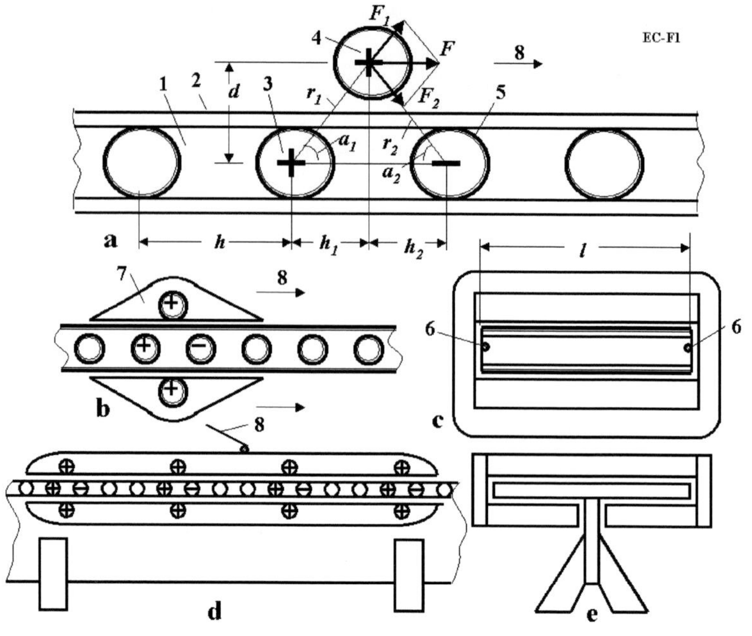

Figure 2. Electrostatic linear engine (accelerator) for Space Elevator and Space Launcher [23 -25, 40]. (a) Explanation of force in electrostatic engine; (b) Two cylindrical electrostatic engines for Space Elevator (here, a side view); (c) Two cylindrical electrostatic engine for Space Elevator (forward view); (d) Eight cylindrical electrostatic engine for Space Launcher (side view); (e) Eight cylindrical electrostatic engine for Space Launcher (forward view).

Notations: 1 - plate cable of Space Elevator with inserted variable cylindrical charges; 2 - part of cable-bearing tensile stress; 3 - insulated variable charging cylinder of stator; 4 - insulated permanent charging cylinder of rotor; 6 – high-voltage wires connected with Earth's generator and switch; 7 - mobile part of electrostatic linear engine; 8 - cable to space aircraft.

The space launcher works same (Figure 2d, 2e). That has a stationary stator and mobile rotor (driver). The stationary stator (monorail) located upon the Earth's surface below the atmosphere. Driver is connected to space-aircraft and accelerates the aircraft to a needed speed (8 km/s and more) [23-25]. For increasing a thrust, the driver of the space launcher can have some charges (Figure 2d) separated by enough neutral non-charged stator cylinders.

Bottom and topmost parts of cable (or stator) have small different charge values. This difference creates a vertical electric field which supports the driver in its suspended position about the stator, non-contact bearing and zero friction. The driver position about the stator is controlled by electronic devices.

Charges have cylindrical form (row of cylinders), and are located within a good dielectric having high disruptive voltage. The cylinders have conducting layer which allows changing the charges with high frequency and produces a running high-voltage wave of charges. The engine creates a large thrust (see computation below), reaches a very high (practically speaking, virtually unlimited) variable speed of driver (km/s), to change the moving of driver in opposed direction, to fix a driver in selected given position. The electrostatic engine can also operate as a high-voltage electric generator when the climber (a cabin-style spaceship) is braking or is moved by some controlled mechanical force. The Space Elevator climber (and many other mobile apparatuses) has a constant charge; the cable (stator) has a running charge. The weight of electric wires is small, almost insignificant, because the voltage is very high.

ACCELERATING ELECTROSTATIC ENGINE THEORY

1. Estimation of Thrust

Let us here consider a single charged cylinder 3, 4, in cable (Figure 2a). The rotor charge 4 attracts to the opposed charges and repels from the same charges 3 located into stator. Let us compute the sum force acting to this single charge.

$$E_i = k\frac{2\tau}{\varepsilon r_i}, \quad F_i = \sum_i qE_i, \quad i = 1, 2, \quad q = \tau l, \tag{1}$$

where E_i is electric intensity [V/m], $k = 9\times10^9$ is electric coefficient [N·m^2/C^2], τ is linear charge [C/m], ε is dielectric constant, r_i is distance between centers of charges [m], F_i is force [N], q is charge [C], l is length of linear charge [m] (Figure 2a).

The sum of two forces is

$$F = q(E_1\cos\alpha_1 + E_2\cos\alpha_2) = \frac{2k\tau^2 l}{\varepsilon}\left(\frac{h_1}{r_1^2} + \frac{h_2}{r_2^2}\right), \quad \text{where} \quad \cos\alpha_i = \frac{h_i}{r_i}, \quad \tau = \frac{\varepsilon a E}{2k}, \tag{2}$$

h_i is horizontal distances between charges (Figure 2a), m; a is radius of charged cylinders, m.

Define the relative ratios of distances (Figure 2a)

$$\bar{h}_i = \frac{h_i}{a}, \quad \bar{r}_i = \frac{r_i}{a}, \quad i = 1, 2, \quad \bar{h} = \frac{h}{a}, \quad \bar{d} = \frac{d}{a}, \tag{3}$$

where d is distance between center of charged cylinders of stator and rotor, m; h is distance between two nearest stator charges, m.

Substituting all of these values to (2) we receive

$$F = b\frac{\varepsilon a E^2 l}{k}, \quad where \quad b = \frac{1}{2}\left(\frac{\bar{h}_1}{\bar{r}_1^2} + \frac{\bar{h}_2}{\bar{r}_2^2}\right), \quad \bar{r}_i^2 = \bar{d}^2 + \bar{h}_i^2, \quad i = 1, 2, \quad \bar{h} = \bar{h}_1 + \bar{h}_2 \tag{4}$$

where b is trust/brake coefficient.

If we take distance between charged rows of stator and cable $d = h = 3a$ (Figure 2a), where a is radius of charged cylinders, the value b is:

a) On ends of segment h the coefficient $b_e = 1/12 = 0.08333$.
b) In the middle of the segment h the coefficient is maximum $b_m = 2/15 = 0.1333$.

c) The average we accept and take $b = 0.5(b_e + b_m) = 0.108$.

For other ratio *d, h* the coefficient *b* is following: when $d = 3a$, $h = 4a$, the $b = 0.117$; when $d = h = 4a$, the $b = 0.08125$.

When we take the *N* symmetric pair stator charges the maximum force may be estimated by equation [66]

$$F_{\max} = b_m \frac{eaE^2 l}{k}, \quad \text{where} \quad b_m = \left[\sum_{i=1}^{N} (-1)^{i+1} \frac{\bar{h}(i-0.5)}{\bar{d}^2 + (\bar{h})^2 (i-0.5)^2} \right] \tag{5}$$

For $N = i = 1$ we received equation (4). That pair has maximum force. For $N=100$, $d = 3a$, $h = 4a$, the $b_m = 0.074$, $b_e \approx 0$. $b = 0.037$,

The trust of linear electric engine can be tons from square meter of charged areas.

2. Charge of One Cylinder is

$$q = \frac{ea^2 El}{k}. \tag{6}$$

3. Needed Maximum Voltage is

$$U \approx Ed. \tag{7}$$

4. Needed Power is

$$P = FV, \tag{8}$$

where *V* is speed apparatus, m/s.

5. Requested Average Electric Currency (*I*, A) is

$$I = P/U, \tag{9}$$

6. Voltage Loss (*ΔU*, V) and Coefficient Efficiency of Electric Line is

$$\Delta U = IR, \quad \eta = \frac{U - \Delta U}{U}, \quad \text{where} \quad R = \rho \frac{L}{s}, \tag{10}$$

R is electric resistance, Ω; η is coefficient efficiency; ρ is specific electric resistance, Ω·m (take from table); L is length of wire, m; s is cross-section area of wire, m^2.

Note: For elevator climber the requested cross-section area is small (<1 mm^2) and the coefficient efficiency of electric line is good ($\eta > 0.5$) because the voltage is very large ($U \approx 15\times10^6$ V).

7. The Main Switching Number is

$$\nu = 2V / h. \tag{11}$$

8. Dielectric Strength of Current Materials

In our computation we used electric intensity over the electric strength of air $E_s \approx 3\times10^6$ V/m. That means, obviously, the air located inside of the engine, between stator and cable, can be ionized. That is unimportant primarily because the amount of air is so small. We can extract (pump out) the air from engine, or fill up with an appropriate dielectric liquid, causing a vacant or liquid-filled volume to exist located between the stator and cable unit. Also we can cover the electrode units with a thin dielectric layer having a high-voltage dielectric strength. In the hard vacuum of outer space, this problem won't be relevant.

The data for computations are in Table 4 in Attn.

Note: Dielectric constant ε can reach 4.5 - 7.5 for mica (E is up 200 MV/m), 6 -10 for glasses (E = 40 MV/m), and 900 -3000 for special ceramics (marks are CM-1, T-900)[74], p. 321, (E = 13 - 28 MV/m). Ferroelectrics have ε up to 10^4 - 10^5. Dielectric strength appreciably depends from surface roughness, thickness, purity, temperature and other conditions of materials. Very clean material without admixture (for example, quartz) can have electric strength up 1000 MV/m. It is necessary to find some really good isolative (insulation) materials and to research environmental conditions which increase the dielectric strength.

9. The Half-life of the Charge

Let us estimate of lifetime of charged driver.

(a) *Charge in spherical ball.* Let us take a very complex condition; where the unlike charges are separated only by an insulator (charged spherical condenser):

$$Ri - U = 0, \quad U = \delta E, \quad E = \frac{kq}{\delta^2}, \quad R = \rho\frac{\delta}{4\pi a^2},$$
$$U = \frac{q}{C}, \quad R\frac{dq}{dt} + \frac{a}{C} = 0, \quad \frac{dq}{q} = \frac{dt}{RC}, \quad C = \frac{\varepsilon\, a}{k},$$
$$q = q_0 \exp\left(-\frac{4\pi\varepsilon a k}{\rho\delta}t\right), \quad \frac{q}{q_0} = \frac{1}{2}, \quad -\frac{4\pi\varepsilon a k}{\rho\delta}t_h = \ln\frac{1}{2} =$$
$$-0.693 \approx -0.7, \quad \text{final} \quad t_h = 0.693\frac{\rho\delta}{4\pi\varepsilon k a}, \tag{12-13}$$

where: t_h – half-life time, [sec]; R – insulator resistance, [Ohm]; i – electric current, [A]; U – voltage, [V]; δ – thickness of insulator, [m]; E – electrical intensity, [V/m]; q – charge, [C]; t - time, [seconds]; ρ – specific resistance of insulator, [Ohm-meter, Ω·m]; a – internal radius of the ball, [m]; C – capacity of the ball, [C]; $k = 9\times10^9$ [N·m^2/C^2, m/F]. Last equation is result.

Example: Let us take typical data: $\rho = 10^{19}$ Ω-m, $k = 9\times10^9$, $\delta/a = 0.2$, then $t_h = 1.24\times10^7$ seconds = 144 days.

(b) *Half-life of cylindrical tube*. The computation is same as for tubes (1 m charged cylindrical condenser):

$$q = q_0 \exp\left(-\frac{1}{RC}t\right), \quad C = \frac{\varepsilon}{k\ln(1+\delta/a)}, \quad R = \frac{\rho\delta}{2\pi a}, \quad -0.693 = -\frac{1}{RC}t_h,$$
$$t_h = \frac{0.693\rho\delta\varepsilon}{2\pi k a\ln(1+\delta/a)}, \quad \textit{for} \quad \delta \to 0, \quad \textit{final} \quad t_h \approx 0.7\frac{\rho\varepsilon}{2\pi k}. \tag{14}$$

Example: Let us take typical data (polystyrene) : $\rho = 10^{18}$ Ω·m, $k = 9\times10^9$, $\varepsilon = 2$, then $t_h = 2.5\times10^7$ seconds = 290 days.

10. Condenser as Accumulator of Launch Energy

Space launcher needs much energy in short timeframe. Most investigators of the electromagnetic launcher concept offer condensers for energy storage. Let us estimate the maximum energy which can be accumulated by1 kg plate electric condenser.

$$W_M = \frac{1}{2}Q_M U = \frac{\varepsilon E_s^2}{8\pi k\gamma}, \quad \textit{where} \quad Q_M = \frac{Q}{M} = \frac{\varepsilon E_s}{4\pi k\gamma d}, \quad C_M = \frac{Q_M}{U} = \frac{\varepsilon}{4\pi k\gamma d^2}, \tag{15}$$

where W_M is energy [J/kg], Q_M is electric charge [C/kg], U is voltage [V], C_M is value of capacitor [C/kg], γ is specific density of dielectric [kg/m^3], d is distance between plate (layers) in plate condenser [m].

For $\varepsilon = 3$, $E_s = 3\times10^8$ V/m, $k = 9\times10^9$ we have $W_M = 660$ J/kg. The industry capacities have energy density 0.02 ÷ 0.08 Wh/kg, the ultra-capacity has 3 ÷ 5 Wh/kg (1Wh = 3600 J).

That is very small value. The energy of a battery is 30 ÷ 40 Wh/kg, a gunpowder is about 3 MJ/kg, the energy of a rocket fuel is 9 MJ/kg ($C + O_2 = CO_2$). In previous works (see, for example, [25],[40], and [60]) A.A. Bolonkin offered to use an unremarkable and low-cost flywheel as an energy accumulator. The flywheel energy storage is

$$W_M = \frac{1}{2}\frac{\sigma}{\gamma}, \tag{16}$$

where σ is safety tensile stress [N/m^2] of fly-wheel material. For $\sigma = 300$ kg/mm^2, $\gamma = 1800$ kg/m^3 (it is current composite matter from artificial fibers) we have $W_M = 0.83\times10^5$ J/kg. When composite matter composed of very fine whiskers and carbon nanotubes become widely used then that critical value will increase by many times.

The other method is to obtain a high electric energy output from an impulse magneto-dynamic electric generator.

Macro-Projects

Below the reader will find the brief estimation of two macro-projects: Climber for space elevator and Earth space AB launcher. The taken parameters are not optimal. Our aim is to merely illustrate the exciting foreseeable possibilities of systems offered here as well as the method for useful computation.

1. Space Elevator Climber

Let us take the following data: $a = 0.05$ m, $E = 10^8$ V/m, $l = 0.3$ m, $h = d = 3a$, $\varepsilon = 3$. $k = 9\times10^9$, $b = 0.109$.

Then the trust of two cylinder electrostatic engine (Figure 2a,b,c) is [Eq. (4)]

$$F = 2b\frac{\varepsilon E^2 al}{k} = 10.8\times10^3\ N\,.$$

The charge of one cylinder is [Eq. (6)]

$$q = \frac{\varepsilon a^2 El}{k} = 2.5\times10^{-5}\ C\,.$$

Requisite voltage of electric line is $U = Ed = 10^8\times0.15 = 15\times10^6$ V, [Eq. (7)].
Requisite maximum power for $V = 0.5$ km/s is $P = FV = 5.4\times10^7$ W, [Eq. (8)].
Requisite currency flowing in the electric line is $I = P/U = 3.6$ A, [Eq. (9)].

Maximum loss of voltage in electric line for double-braid aluminum wire having length L = 36,000 km (Geosynchronous Earth Orbit), cross-section areas s = 1 mm^2 ($\rho = 2.8\times10^{-6}$ Ω·cm) and coefficient of efficiency of Earth-based electric line [Eq. (10)].

$$R = \rho\frac{L}{s} = 2.02\times10^6 \ \Omega, \quad \Delta U = IR = 7.27\times10^6 \ V,$$

$$\eta = \frac{U-\Delta U}{U} = \frac{15\times10^6 - 7.27\times10^6}{15\times10^6} = 0.52$$

Clearly, the wire's cross-section may be small and coefficient of efficiency is good for this super-long electric power-line. We take a maximum length to geosynchronous orbit (GEO). That is efficient delivery of energy to GEO receiving space station, where the weight of the climber is zero. (The mass will remain the same, of course.) In reality, if we take an average distance and the average climber weight, the loss of energy decreases by 5 times and an efficiency is reached of 0.8 ÷ 0.9.

The maximum number of main switching is $\upsilon \approx 1.06\times10^5$ 1/s [Eq. (11)].

2. Earth electrostatic AB Space Launcher

Let us take the following data inputs: a = 0.1 m, $E = 10^8$ V/m, l =1 m, $h = d = 3a$, ε=3, N = 4 (N is number of pair driver cylinders), b = 0.109.

Then the thrust of the eight cylinder electrostatic engine (Figure 2d,e) is [Eq. (4)]

$$F = 2\times4\times b\frac{\varepsilon E^2 al}{k} = 29\times10^4 \ N.$$

The charge of one cylinder is [Eq. (6)]

$$q = \frac{\varepsilon a^2 El}{k} = 3.3\times10^{-2} \ C.$$

Required voltage of the electricity line is $U = Ed = 10^8\times0.3 = 30\times10^6$ V, [Eq. (7)].

Needed maximum power for V = 8 km/s is $P = FV = 2.32\times10^9$ W, [Eq. (8)].

Requisite currency in the hanging electric line is $I = P/U$ = 77.3 A, [Eq.(9)]. Note that the thrust F = 29 tonnes is enough to launch a 10 tonne spaceship (acceleration is a = 3 g) with conventional people (100 to 150 tourists) if the monorail's length is ~1100 km. Trained people (Spationauts) can survive an overload of 6 g. The needed length of track is 530 km. The payload which can keep overload 300 g needs only the 11 km-long track [60].

In future, by using the electrostatic linear engine, a cheap and highly productive manned (or unmanned) outer space catapult can be put together at the present time! This catapult system decreases launch costs by as much as 2 - 4 $/kg and, thusly, allowing humans to launch thousands tonnes each year into outer space.

That will be simpler than A.A. Bolonkin's cable catapult offered in [23]-[25],[40],[60] since nanotubes, mobile cable and 109 drive stations are not needed. There is only electrostatic motionless monorail-stator which produces a running electrostatic wave of charges and the linear permanent charged rotor (driver) connected by cable to a spaceship. The monorail-stator (cable) is suspended on columns (or in air as described and illustrated in [41] - [42]).

Installation may also be used as high-speed conventional tramway.

3. Other Useful Applications with Estimations

1. *Electrostatic Interplanetary Space Launcher or Spaceship Propulsion.* Assume we want to launch mass M = 2 kg interplanetary probe by L = 100 m electrostatic accelerator (launcher). We use a thrust F = 120 tonnes. Then the acceleration will be $a = F/M = 1.2\times10^6/2 = 0.6\times10^6$ m/s^2, final speed $V = (2\times10^2\times 0.6\times10^6)^{0.5}$ =11 km/s. If we launch from spaceship, the space ship receives the momentum MV in opposed direction.

3. *Transport cable systems (space climber) for Earth to Moon, Earth to Mars.* In [25], [29], [30], [36], [39], [60] A.A. Bolonkin researched and reported the mechanical cable transport systems for Space Elevator and for Earth-Moon, Earth-Mars vehicle trips. All these systems need a high speed engine for moving of spacecraft. One cable version of the suggested transport system is noted in the above-cited works. However, the system offered in any given article may be used in many structures. The cable is stator, the vehicle has linear rotor. The cable delivers the energy in the form of a running charge wave. The vehicle (climber) follows this running wave. The speed of the running wave (and, consequently, the vehicle too) can be very great. The voltage is extremely large and the weight of the electric wires is small.

Let us to make the simplest estimation. Assume, the climber weighs W = 1 ton =10,000 N and has speed V= 1 km/s = 1000 m/s. The power is $P = WV = 10^4\times10^3 = 10^7$ W. For voltage $U = 10^8$ V, the electric currency is $i = 10^7/10^8 = 0.1$ A. For safety currency 20 A/mm^2, the needed wire diameter is about 0.1 mm^2.

4. *Suspended satellite system.* In [25], [36] A.A. Bolonkin tactfully suggested a single cable-ring, rotating around the Earth, with "motionless" satellites suspended within the cable-ring (Figure 1e). The linear engine can be used as engine for compensation of the cable air friction and as a non-contact bearing for a suspended system.

5. *Electrostatic levitation train and linear engine.* In [59] the author suggested the electrostatic levitation train (Figure 1h). The offered linear engine can be used as propulsion engine for this train. In braking the energy of acceleration will be returned in electric line.

6. *Electrostatic rotary engine.* At present time industry uses conventional low voltage electric engine. When we have a high voltage electric line it may be easier to use high voltage electrostatic rotary engine.

7. *Electrostatic levitation bearing.* Some technical installations need low friction bearings. The mono-electrets can be used as non-contact bearing having zero mechanical friction.

8. *Electrostatic Gun System.* Cannonry fires high-speed shells. However, the shell speed is limited by gas speed inside cannon barrel. The suggested linear electrostatic engine can be

used as the high efficiency shell ejector in armor-piercing cannonry (having very high initial shell speed) because the initial shell speed of linear engine does not have the speed limit (see application 1 above). That means the electrostatic gun can shoot projectiles thousands of kilometers!

Conclusion

Presently, the suggested space climber is the single immediately buildable high-efficiency transport system for a space elevator. The electromagnetic beam transfer energy is very complex, expensive and has very low efficiency, especially at a long-distance from divergence of electromagnetic beam. The laser has similar operational disadvantages. The conventional electric line, equipped with conventional electric motor, is very heavy and decidedly unacceptable for outer space.

The offered electrostatic engine could find wide application in many fields of technology. It can drastically decrease the monetary costs of launch by hundreds to thousands of times. The electrostatic engine needs a very high-voltage but this voltage, however, is located in a small area inside of installations, and is not particularly dangerous to person living or working nearby. Currently used technology does not have any other way for reaching a high speed except by the use of rockets. But crewed or un-crewed rockets, and rocket launches for expensive-to-maintain-and-operate Earth bases, are very expensive and space businesses do not know ways to cut the cost of rocket launch by hundreds to thousands of times.

Figure. (Continued)

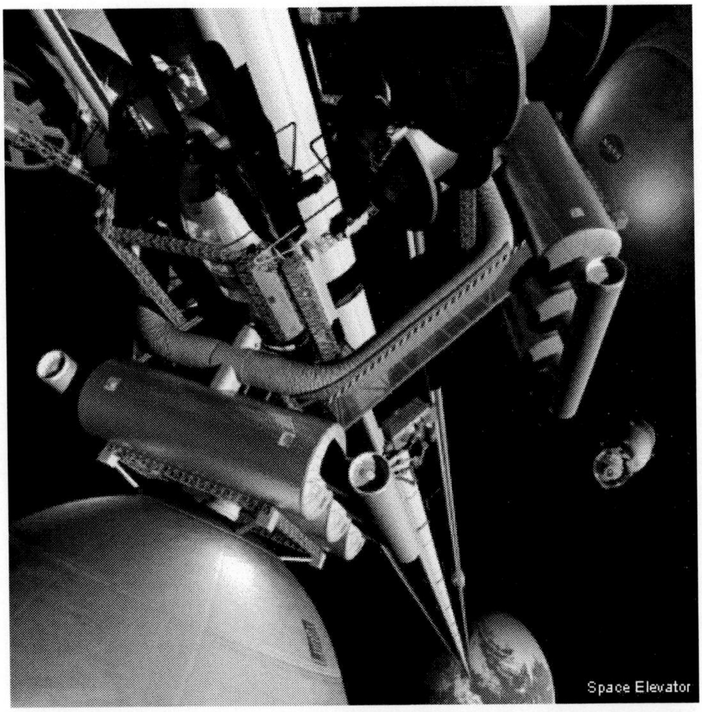

Space Elevator.

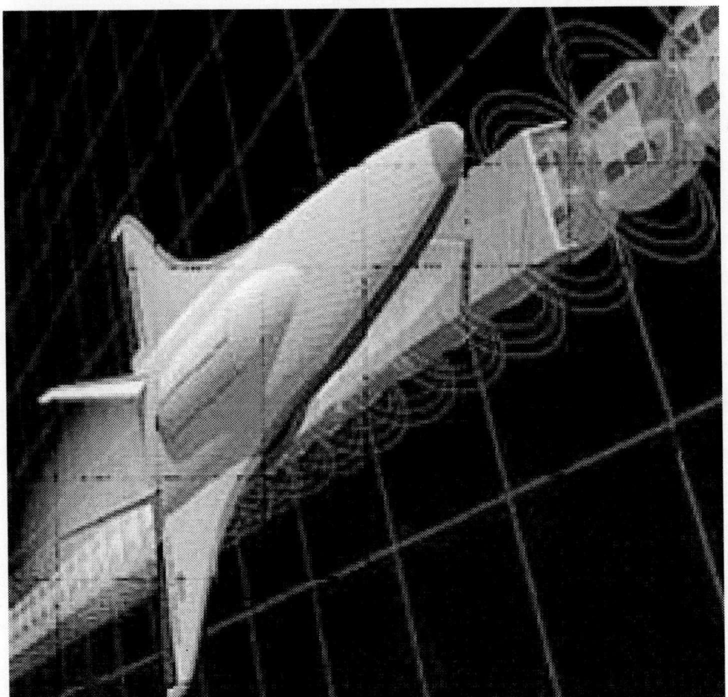

Figure. (Continued)

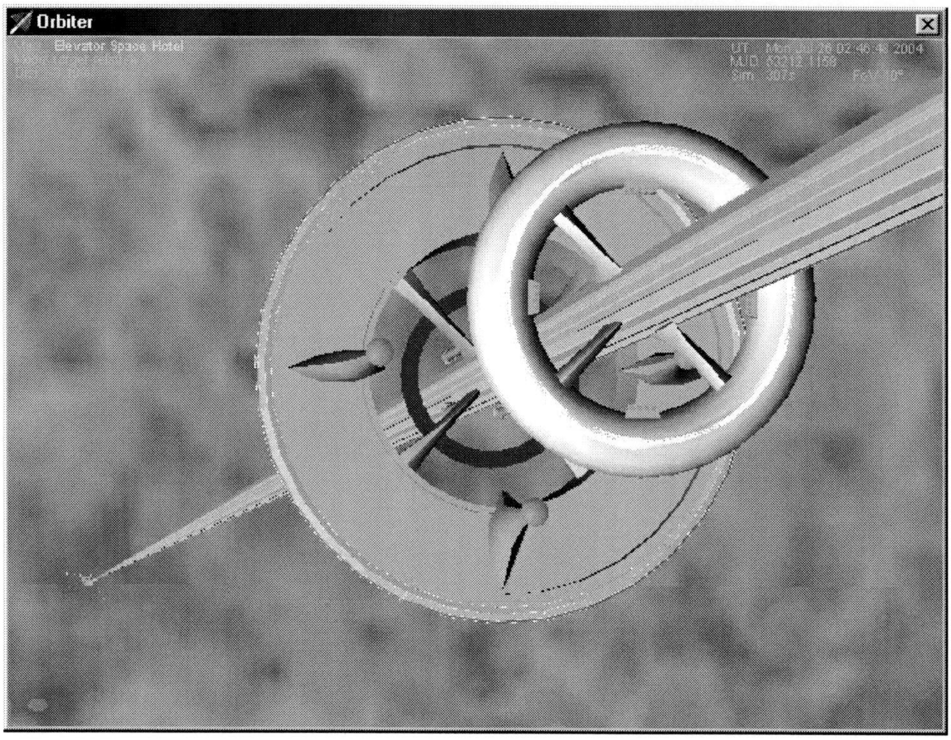

Proposed system of Space Elevator

REFERENCES

(Part of these articles the reader can be found at A.A. Bolonkin's website: *http://arxiv.org*. search "Bolonkin", *http://Bolonkin.narod.ru/p65.htm* and in several published books: "*Non-Rocket Space Launch and Flight*" (Elsevier, London, 2006) and "*New Concepts, Ideas and Innovations in Aerospace and Technology*" (Nova, 2007.)

[1] Bolonkin, A.A., (1965a), "Theory of Flight Vehicles with Control Radial Force". Collection Researches of Flight Dynamics, Mashinostroenie Publisher, Moscow, pp. 79 -118, 1965, (in Russian). Intern. Aerospace Abstract A66-23338# [English].

[2] Bolonkin, A.A., (1965c), Optimization of Trajectories of Multistage Rockets. Collection Researches of Flight Dynamics. Mashinostroenie Publisher, Moscow, 1965, p. 20 -78 (in Russian). International Aerospace Abstract A66-23337# [English].

[3] Bolonkin, A.A., (1982a), Installation for Open Electrostatic Field, *Russian patent application* #3467270/21 116676, 9 July, 1982 [in Russian], RF PTO.

[4] Bolonkin, A.A., (1982b), Radioisotope Propulsion. *Russian patent application* #3467762/25 116952, 9 July 1982 [in Russian], RF PTO.

[5] Bolonkin, A.A., (1982c), Radioisotope Electric Generator. *Russian patent application* #3469511/25 116927. 9 July 1982 [in Russian], Russian PTO.

[6] Bolonkin, A.A., (1983a), Space Propulsion Using Solar Wing and Installation for It, *Russian patent application* #3635955/23 126453, 19 August, 1983 [in Russian], RF PTO.

[7] Bolonkin, A.A., (1983b), Getting of Electric Energy from Space and Installation for It, *Russian patent application* #3638699/25 126303, 19 August, 1983 [in Russian], RF PTO.

[8] Bolonkin, A.A., (1983c), Protection from Charged Particles in Space and Installation for It, *Russian patent application* #3644168 136270, 23 September 1983, [in Russian], RF PTO.

[9] Bolonkin, A. A., (1983d), Method of Transformation of Plasma Energy in Electric Current and Installation for It. *Russian patent application* #3647344 136681 of 27 July 1983 [in Russian], RF PTO.

[10] Bolonkin, A. A., (1983e), Method of Propulsion using Radioisotope Energy and Installation for It. of Plasma Energy in Electric Current and Installation for it. *Russian patent application* #3601164/25 086973 of 6 June, 1983 (in Russian), RF PTO.

[11] Bolonkin, A. A., (1983f), Transformation of Energy of Rarefaction Plasma in Electric Current and Installation for it. *Russian patent application* #3663911/25 159775, 23 November 1983 [in Russian], RF PTO.

[12] Bolonkin, A. A., (1983g), Method of a Keeping of a Neutral Plasma and Installation for it. *Russian patent application* #3600272/25 086993, 6 June 1983 [in Russian], RF PTO.

[13] Bolonkin, A.A., (1983h), Radioisotope Electric Generator. *Russian patent application* #3620051/25 108943, 13 July 1983 [in Russian], RF PTO.

[14] Bolonkin, A.A., (1983i), Method of Energy Transformation of Radioisotope Matter in Electricity and Installation for it. *RF patent application* #3647343/25 136692, 27 July 1983 [in Russian], RF PTO.

[15] Bolonkin, A.A., (1983j), Method of stretching of thin film. *Russian patent application* #3646689/10 138085, 28 September 1983 [in Russian], RF PTO.

[16] Bolonkin, A.A., (1987), "New Way of Thrust and Generation of Electrical Energy in Space". *Report ESTI,* 1987, (USSR Classified Projects).

[17] Bolonkin, A.A., (1990), "Aviation, Motor and Space Designs", Collection *Emerging Technology in the Soviet Union*, 1990, Delphic Ass., Inc., pp.32-80 [in English].

[18] Bolonkin, A.A., (1991), *The Development of Soviet Rocket Engines*, 1991, Delphic Ass.Inc.,122 pages. Washington, [in English].

[19] Bolonkin, A.A., (1992a), "A Space Motor Using Solar Wind Energy (Magnetic Particle Sail)". *The World Space Congress,* Washington, DC, USA, 28 Aug. - 5 Sept., 1992, IAF-0615.

[20] Bolonkin, A.A., (1992b), "Space Electric Generator, run by Solar Wing". *The World Space Congress*, Washington, DC, USA, 28 Aug. -5 Sept. 1992, IAF-92-0604.

[21] Bolonkin, A.A., (1992c), "Simple Space Nuclear Reactor Motors and Electric Generators Running on Radioactive Substances", *The World Space Congress,* Washington, DC, USA, 28 Aug. - 5 Sept., 1992, IAF-92-0573.

[22] Bolonkin, A.A. (1994), "The Simplest Space Electric Generator and Motor with Control Energy and Thrust", *45th International Astronautical Congress*, Jerusalem, Israel, 9-14 Oct., 1994, IAF-94-R.1.368.

[23] Bolonkin, A.A., (2002a), "Non-Rocket Space Rope Launcher for People", IAC-02-V.P.06, 53rd International Astronautical Congress, *The World Space Congress* - 2002, 10-19 Oct 2002, Houston, Texas, USA.

[24] Bolonkin, A.A,(2002b), "Non-Rocket Missile Rope Launcher", IAC-02-IAA.S.P.14, 53rd International Astronautical Congress, *The World Space Congress* - 2002, 10-19 Oct 2002, Houston, Texas, USA.

[25] Bolonkin, A.A., (2002c), "Inexpensive Cable Space Launcher of High Capability", IAC-02-V.P.07, 53rd International Astronautical Congress, *The World Space Congress* - 2002, 10-19 Oct 2002, Houston, Texas, USA.

[26] Bolonkin, A.A.,(2002d), "Hypersonic Launch System of Capability up to 500 tons per day and Delivery Cost $1 per Lb". IAC-02-S.P.15, *53rd International Astronautical Congress, The World Space Congress* - 2002, 10-19 Oct 2002, Houston, Texas, USA.

[27] Bolonkin, A.A.,(2002e), "Employment Asteroids for Movement of Space Ship and Probes". IAC-02-S.6.04, *53rd International Astronautical Congress, The World Space Congress* - 2002, 10-19 Oct 2002, Houston, Texas, USA.

[28] Bolonkin, A.A., (2002f), "Optimal Inflatable Space Towers of High Height". COSPAR-02 C1.1-0035-02, *34th Scientific Assembly of the Committee on Space Research (COSPAR), The World Space Congress* - 2002, 10-19 Oct 2002, Houston, Texas, USA.

[29] Bolonkin, A.A., (2002g), "Non-Rocket Earth-Moon Transport System", COSPAR-02 B0.3-F3.3-0032-02, 02-A-02226, *34th Scientific Assembly of the Committee on Space Research (COSPAR), The World Space Congress* - 2002, 10-19 Oct 2002, Houston, Texas, USA.

[30] Bolonkin, A. A.,(2002h) "Non-Rocket Earth-Mars Transport System", COSPAR-02 B0.4-C3.4-0036-02, *34th Scientific Assembly of the Committee on Space Research (COSPAR), The World Space Congress* - 2002, 10-19 Oct 2002, Houston, Texas, USA.

[31] Bolonkin, A.A.,(2002i). "Transport System for Delivery Tourists at Altitude 140 km". IAC-02-IAA.1.3.03, *53rd International Astronautical Congress, The World Space Congress* - 2002, 10-19 Oct. 2002, Houston, Texas, USA.

[32] Bolonkin, A.A., (2002j), "Hypersonic Gas-Rocket Launch System." AIAA-2002-3927, 38th AIAA/ASME/SAE/ASEE *Joint Propulsion Conference and Exhibit,* 7-10 July 2002. Indianapolis, IN, USA.

[33] Bolonkin, A.A., (2003a), "Air Cable Transport", *Journal of Aircraft*, Vol. 40, No. 2, March-April 2003.

[34] Bolonkin, A.A., (2003b), "Optimal Inflatable Space Towers with 3-100 km Height", *JBIS*, Vol. 56, No 3/4, pp. 87-97, 2003.

[35] Bolonkin, A.A.,(2003c), "Asteroids as Propulsion Systems of Space Ships", *JBIS*, Vol. 56, No 3/4, pp. 97-107, 2003.

[36] Bolonkin A.A., (2003d), "Non-Rocket Transportation System for Space Travel", *JBIS*, Vol. 56, No 7/8, pp. 231-249, 2003.

[37] Bolonkin, A.A., (2003e), "Hypersonic Space Launcher of High Capability", *Actual problems of aviation and aerospace systems*, Kazan, No. 1(15), Vol. 8, 2003, pp. 45-58.

[38] Bolonkin, A.A., (2003f), "Centrifugal Keeper for Space Stations and Satellites", *JBIS*, Vol. 56, No 9/10, pp. 314-327, 2003.

[39] Bolonkin A.A., (2003g), "Non-Rocket Earth-Moon Transport System", *Advances in Space Research*, Vol. 31/11, pp. 2485-2490, 2003.

[40] Bolonkin, A.A., (2003h), "Earth Accelerator for Space Ships and Missiles". *Journal of the British Interplanetary Society*, Vol. 56, No. 11/12, 2003, pp. 394-404.

[41] Bolonkin, A.A., (2003i), "Air Cable Transport and Bridges", TN 7567, International Air & Space Symposium - The Next 100 Years, 14-17 July 2003, Dayton, Ohio, USA.

[42] Bolonkin, A.A., (2003j), "Air Cable Transport System", *Journal of Aircraft*, Vol. 40, No. 2, March-April 2003, pp. 265-269.

[43] Bolonkin, A.A., (2004a), "Kinetic Space Towers and Launchers ', *JBIS*, Vol. 57, No 1/2, pp. 33-39, 2004.

[44] Bolonkin, A.A., (2004b), "Optimal trajectory of air vehicles", *Aircraft Engineering and Space Technology*, Vol. 76, No. 2, 2004, pp. 193-214.

[45] Bolonkin, A.A., (2004c), "Long Distance Transfer of Mechanical Energy", International Energy Conversion Engineering Conference at Providence RI, Aug. 16-19, 2004, AIAA-2004-5660.

[46] Bolonkin, A.A., (2004d), "Light Multi-Reflex Engine", *JBIS*, Vol. 57, No 9/10, pp. 353-359, 2004.

[47] Bolonkin, A.A., (2004e), "Kinetic Space Towers and Launchers", *JBIS*, Vol. 57, No 1/2, pp. 33-39, 2004.

[48] Bolonkin, A.A., (2004f), "Optimal trajectory of air and space vehicles", *AEAT*, No. 2, pp. 193-214, 2004.

[49] Bolonkin, A.A., (2004g), "Hypersonic Gas-Rocket Launcher of High Capacity", *JBIS*, Vol. 57, No 5/6, pp. 167-172, 2004.

[50] Bolonkin, A.A., (2004h), "High Efficiency Transfer of Mechanical Energy". *International Energy Conversion Engineering Conference at Providence,* RI, USA. 16-19 August, 2004, AIAA-2004-5660.

[51] Bolonkin, A.A., (2004i), "Multi-Reflex Propulsion System for Space and Air Vehicles", *JBIS*, Vol. 57, No 11/12, 2004, pp. 379-390.

[52] Bolonkin, A.A.,(2005a) "High Speed Catapult Aviation", AIAA-2005-6221, *Atmospheric Flight Mechanic Conference* - 2005, 15-18 August, 2005, USA.

[53] Bolonkin, A.A., (2005a), Electrostatic Solar Wind Propulsion System, AIAA-2005-3653. *41st Propulsion Conference, 1*0-12 July, 2005, Tucson, Arizona, USA.

[54] Bolonkin, A.A., (2005b), Electrostatic Utilization of Asteroids for Space Flight, AIAA-2005-4032. *41st Propulsion Conference,* 10-12 July, 2005, Tucson, Arizona, USA.

[55] Bolonkin, A.A., (2005c), Kinetic Anti-Gravitator, AIAA-2005-4504. *41st Propulsion Conference,* 10-12 July, 2005, Tucson, Arizona, USA.

[56] Bolonkin, A.A., (2005d), Sling Rotary Space Launcher, AIAA-2005-4035. *41st Propulsion Conference,* 10-12 July, 2005, Tucson, Arizona, USA.

[57] Bolonkin, A.A., (2005e), Radioisotope Space Sail and Electric Generator, AIAA-2005-4225. *41st Propulsion Conference,* 10-12 July, 2005, Tucson, Arizona, USA.

[58] Bolonkin, A.A., (2005f), Guided Solar Sail and Electric Generator, AIAA-2005-3857. *41-st Propulsion Conference,* 10-12 July, 2005, Tucson, Arizona, USA.

[59] Bolonkin A.A., (2005g), Problems of Electrostatic Levitation and Artificial Gravity, AIAA-2005-4465. *41 Propulsion Conference,* 10-12 July, 2005, Tucson, Arizona, USA.

[60] Bolonkin, A.A., (2006a), "*Non-Rocket Space Launch and Flight*", Elsevier, London, 2006, 488 pgs.

[61] Bolonkin, A.A., (2006b), *Electrostatic AB-Ramjet Space Propulsion,* AIAA-2006-6173, AEAT, Vol.79, No. 1, 2007, pp. 3 - 16.

[62] Bolonkin, A.A., (2006c), *Beam Space Propulsion,* AIAA-2006-7492, (published in *http://arxiv.org* search "Bolonkin").

[63] Bolonkin, A.A., (2006d), *High Speed Solar Sail,* AIAA-2006-4806, (published in *http://arxiv.org* search "Bolonkin").

[64] Bolonkin, A.A., (2006e), *Suspended Air Surveillance System,* AIAA-2006-6511, (published in *http://arxiv.org* search "Bolonkin").

[65] Bolonkin, A.A., (2006f), *Optimal Solid Space Tower* (Mast), (published in *http://arxiv.org* search "Bolonkin").

[66] Bolonkin, A.A., (2006g), Electrostatic Linear Engine, presented as paper AIAA-2006-5229 to *42nd Propulsion Conference,* USA and published in AEAT, Vol.78, No.6, 2006, pp. 502-508.

[67] Bolonkin, A.A., (2006h). *AB Levitator and Electricity Storage.* Published in *http://arxiv.org* search "Bolonkin".

[68] Bolonkin, A.A., (2006i), Theory of Space Magnetic Sail Some Common Mistakes and Electrostatic MagSail. *Presented as paper AIAA-2006-8148 to 14-th Space Planes and Hypersonic System Conference,* 6-9 November 2006, Australia. *http://arxiv.org* search "Bolonkin".

[69] Bolonkin, A.A., (2006j) Micro - Thermonuclear AB-Reactors for Aerospace. *Presented as paper AIAA-2006-8104 in 14th Space Plane and Hypersonic Systems Conference,* 6-8 November, 2006, USA. Published in *http://arxiv.org* search "Bolonkin".

[70] Bolonkin, A.A. (2006k). *Simplest AB-Thermonuclear Space Propulsion and Electric Generator.* Published in *http://arxiv.org* search "Bolonkin".

[71] Bolonkin, A.A., (2006*l*). *Wireless Transfer of Electricity in Outer Space.* AIAA-2007-0590. Published in *http://arxiv.org* search "Bolonkin".

[72] Bolonkin, A.A., New Concepts, Ideas and Innovations in Aerospace and Technology, Nova, 2007. See also: *Macro-Engineering - A challenge for the future*, Edited by V. Badescu, R.B Cathcart and R.D. Schuiling, Springer, 2006. (Collection contains two Bolonkin's articles: Space Towers; Cable Anti-Gravitator, Electrostatic Levitation and Artificial Gravity).

[73] Kikoin, I.K., (Ed.), *Tables of Physical Values.* Atomuzdat, Moscow, 1976 (in Russian).

Chapter 5

PROTECTION OF NEW YORK CITY'S URBAN FABRIC WITH LOW-COST TEXTILE STORM SURGE BARRIERS*

ABSTRACT

Textile storm surge barriers, sited at multiple locations, are literally extensions of the city's world famous urban fabric—another manifestation of the dominance of the man-made city over local Nature. Textile Storm Surge Barriers (TSSB) are intended to preserve the City from North Atlantic Ocean hurricanes that cause sea waves impacting the densely populated and high-value real estate, instigating catastrophic, and possibly long-term, infrastructure and monetary losses. Complicating TSSB installation macro-project planning is the presence of the Hudson and other rivers, several small tidal straits, future climate change and other factors. We conclude that TSSB installations made of homogeneous construction materials are worthwhile investigating because they may be less expensive to build, and more easily replaced following any failure, than concrete and steel storm surge barriers, which are also made of homogeneous materials. We suppose the best macro-project outcome will develop in the perfect Macro-engineering planning way and at the optimum time-of-need during the early-21st Century by, among other groups, the Port Authority of New York and New Jersey. TSSB technology is a practical advance over wartime harbor anti-submarine/anti-torpedo steel nets and rocky Churchill Barriers used in the UK.

INTRODUCTION

For our purposes, a storm surge is an above normal rise in sea level accompanying a hurricane passing near or over New York City. New York City has endured North Atlantic Ocean storm surges caused by hurricanes—for example, lower Manhattan Island was flooded on 3 September 1821 by a Category 3 storm that made landfall near today's JFK Airport. The earliest calendar date of a hurricane affecting the State of New York's coast is 4 June, while

* Presented in http://arxiv.org in 2007.

the latest is 13 November. September is the month of greatest frequency for storms of tropical origin, although the storms of greatest intensity tend to arrive later in the hurricane season. 21st Century North Atlantic Ocean hurricane activity will probably not be significantly different from that of the 20th Century since all future Atlantic Ocean tropical storm activity will critically stem directly from the warming of the tropical Atlantic Ocean relative to that of the combined Indian Ocean-Pacific region. Indeed, recent climate studies suggest that North Atlantic hurricane activity is greater during La Nina years and suppressed during El Nino year due primarily to increased vertical wind shear in strong El Nino years hindering hurricane development. (An excellent summary of current scientific knowledge about tropical storms is available in Patrick J. Fitzpatrick's HURRICANES: A REFERENCE HANDBOOK, 2nd Edition, 2005, 412 pages.)

Some previous cases of weak storm surge inundations of New York City are illustrated below (Figures 1 -10A – 10D).

A. B.

C. D.

Source. New York City Office of Emergency Management.

Figure 1. A. Lower East Side, Manhattan. November 24, 1950. B: La Guardia Airport, Queens. November 25, 1950. C: Hoboken PATH Station, New Jersey. 1992 Nor'easter. D: FDR Drive Northbound at 80th Street, East River, Manhattan. 1992 Nor'easter.

North Atlantic Ocean hurricanes are rated according to wind speed by the Saffir-Simpson Scale first offered in 1969 by a civil engineer born in New York City, Herbert Saffir (1917-2007), and by a meteorologist, Robert Simpson. Weather forecasters and infrastructure emergency managers in the USA utilize the Saffir-Simpson Scale to anticipate macro-problems for urban fabrics, which even extends into the countryside for the purpose of estimating the potential damage and flooding that might be stimulated by a hurricane's landfall. Managers and other decision makers desire to know with some certainty whether either an evacuation or a shelter-in-place public announcement is required and what boroughs of New York City are likely to be most impacted as well as for the storm impact period of duration—in other words, New York City's exposure and its weather sensitivity economically. Hurricanes normally weaken after landfall and the weakening is sometimes evidenced by potentially destructive short-term phenomena (tornadoes)! Coastal hills and a city's skyline—Manhattan Island, covered as it is with tall buildings and other structures, including landfills, is almost the equivalent of a rocky geological formation—can influence the dynamics of such storms through blocking and resultant disruption of the storm wind circulation, causing torrential rain in the urbanized and rural coastal zone. The near-surface winds are decelerated by increased surface friction with natural and made landforms as well as artificial structures and buildings. Construction mistakes of a landmark Manhattan skyscraper, the 278 m-high Citigroup Center completed in 1977, necessitated costly special reinforcement after its occupancy to successfully resist future impacting hurricane winds.

New York City's meteorological vulnerability is a subject still being vigorously debated by all stakeholders. It is possible that abrupt global climate change could cause a noticeable rise in local sea levels, thereby aggravating the situation as regards hurricane impacts. All existing and planned infrastructure embodies the extant Macro-engineering design criteria codifying today's technology and science. Even so, there are few successful macro-projects. Any macro-project to protect New York City from hurricane storm surges must be indivisible: Thus, for a macro-project to proceed it must be in the best interests of all key participants, *i.e.*, it must be and *remain* superior to any other practical courses open to any of these key parties, including doing absolutely nothing whatsoever! All protective macro-projects must have the capability to endure not only hurricanes but earthquakes and aviation terrorist attacks such as that which occurred on 11 September 2001 or mobile truck bombings. A focus of planning in the metropolitan region since its inception in 1921, the Port Authority of New York and New Jersey which oversees seaport operations, and associated operations, must play a leading role in the planning and construction of the several required hurricane storm surge barriers needed to shield New York City and adjacent urban fabrics. We assume that impermeable barriers like the Canso Causeway across the Strait of Canso, Nova Scotia, linking Cape Breton Island with Canada's mainland or the Churchill Barriers in the Orkney isles built to protect the UK's Navy's Scapa Flow naval anchorage are impractical.

NYC Storm Surge Barrier Purpose

Storm surges, the flooding induced by surface wind stresses and the barometric pressure reduction associated directly with hurricanes and northeasters, are a threat to New York City's infrastructure as well as other famous cities. The Thames Barrier in the UK and the

Delta Project in The Netherlands were designed with heights to exceed the surge elevations of certain macro-project design storms. However, these barriers will assuredly become more ineffective as global and regional sea level rises during the 21st Century. Computer modeling of storm surge flooding of any urban fabric is rife with uncertainties. Our TSSB (Textile Storm Surge Barrier) macro-project is meant to exclude only storm surges predicted to affect New York City's infrastructure. Unlike the Thames Barrier and the Delta Project, the TSSB can be easily elevated (through refurbishment) to accommodate future regional sea level rise. Here, we examine the potential effectiveness of hypothetical TSSB installations emplaced at The Narrows, the upper East River near College Point (near the Bronx-Whitestone Bridge opened to vehicle traffic in 1939) and at the mouth of Arthur Kill near Perth Amboy, New Jersey; in other words, we hope to prevent damaging hurricane and northeaster storm surges from entering the New York and New Jersey metropolitan region by way of Lower New York Bay, Raritan Bay and Long Island Sound. Much of the land facing these bodies—estimated at ~260 km^2—of seawater are <3 m ASL. The TSSB sites we have selected are identical to those suggested by the Stony Brook Storm Surge Research Group currently led by Professor Malcolm J. Bowman. (GOTO: http://stormy.msrc.sunysb.edu/sbss-Banner.htm.) Emplacement of three TSSB facilities, which would remain open to shipping until closed prior to expected hurricane and northeaster storm surge events, means that, like other concretized blockages, the TSSB will retain ~133 km^2 of water surface.

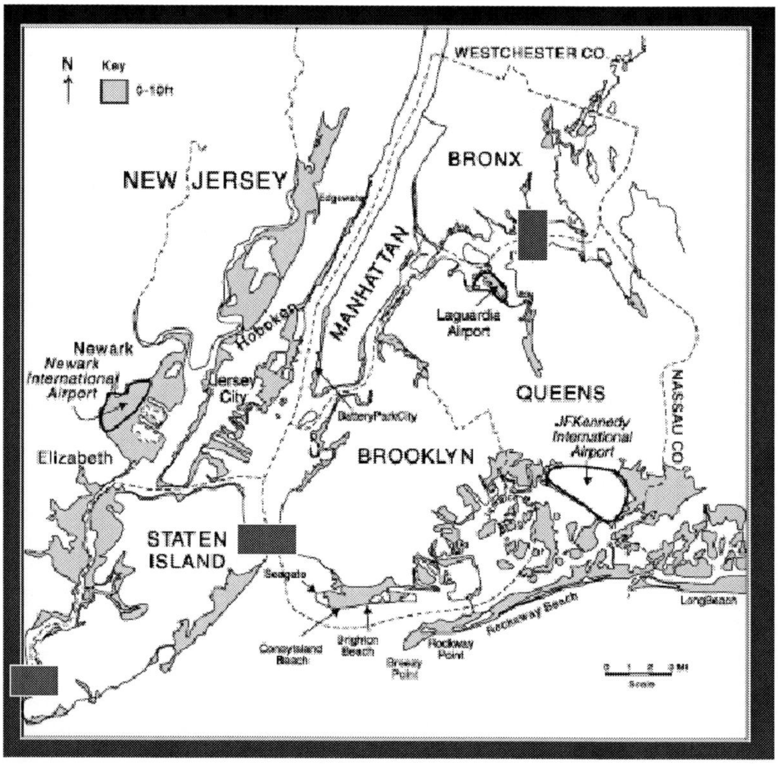

Figure 2. The Dams are offered by M. Bowman.

During the 1971-72 of investigations preparatory to a final builder-owner decision on the Thames Barrier's design, an ingenious fabric dam was considered after its submission by Andrew Noel Schofield. At page 55 of Stuart Gilbert and Ray Horner's THE THAMES BARRIER (1984, Thomas Telford Ltd) Schofield's dam is described: "A vast fabric sheet was to rest on the bed of the river (it could be removed for maintenance during the summer non-surge season). The upriver edge of the sheet would be linked through running gear to a heavy hawser or cable, both sheet and cable spanning the river. The sheet could be raised by winches on the river banks hauling in the cable. Other cables attached to strong anchors in the bed of the river downriver would be secured to the main cable across the river. By under-draining below the sheet and pumping, the lower edge of the sheet would be gripped between the water pressure and the river bed. As the surge came in the depth of water and holding pressure would increase. The under-drainage would be formed by perforated plastic pipes inserted in layers of clean gravel dropped into a trench dredged across the river bed. Any silt settling on the sheet could be tipped off it by raising the sheet on the outgoing tide. This was an unusual solution and had its attractions…". The environmental impact of the use of fabric and textiles is a little-examined field; the short and long-term environmental impact of textile production has been studied. There is only one published report available that deals with something quite similar environmentally to Schofield's 30+ year-old macro-project proposal. [Here, it is necessary to note that fiber-reinforced polymer (FRP) employment is increasing worldwide. For example, the 114 m-long cable-stayed Aberfeldy Footbridge over the Tay River in Scotland, finished in 1990, has a pedestrian deck of FRP composite panels and a supporting pylon composed of FRP and the cable stays transmitting the deck's dead and live loads to the pylon's top are comprised of woven Kevlar.]

At certain work sites, macro-engineers will have to span nearly 1.6 km of tidal strait water! (The Thames Barrier protecting London is only 523 m-long.) At the Narrows, it may be possible to form a TSSB "add-on" to the 1964 Verrazano-Narrows Bridge. The so-called East River, which is actually a tidal strait, is the site of the USA's first major tidal-power macro-project; ultimately, ~100 underwater turbines, each of ~35 kw generating capacity, will be installed in the East River. Such a facility will be affected adversely by closure of the planned TSSB! No seawater flow, no electricity output! Closure of the TSSB will not damage the facility whatsoever. However, what is the point of worrying about a temporary power shutdown when the urban fabric is at risk of flooding and extreme, long-lasting physical damage? Kate Ascher's ANATOMY OF A CITY (2005, The Penguin Press, 228 pages) well illustrates that complex NYC urban fabric, but without any cautionary mention of hurricane storm surge threats (or suggested Macro-engineering cures)! Reference #7 cited below shows vividly the progress that has been made in modeling cities—see, for example, old-fashion physical models of New York City as it looked by 1992.

The Narrows separates the New City boroughs of Staten Island and Brooklyn. It is the main channel through which the post-Ice Age Hudson River drains. The Hudson estuary has an extremely low stability ratio (0.4) which is defined as the ratio of the volume of an estuary (in km^3) to the mean flow of freshwater in m^3/s into the system. This implies that it has a very rapid response to storms and hurricanes, with the salt front moving long distances very rapidly. Episodic hurricanes deliver sediment to the Hudson River estuary since the estuary's geologic creation less than 13,000 years ago. The possible worksite for our TSSB beneath or very near the Verrazano-Narrows Bridge offers a Quaternary alluvium geology typified by the "Harbor Hill Terminal Moraine" formation. (The welcoming suspension bridge spanning

2040 m of seawater, built during 1959-64, then cost approximately 60 million USA dollars to construct.) We presume a TSSB located thereabouts won't exceed the dollar cost incurred nearly fifty years ago! And, we certainly reject an alleged estimate of the cost for total metropolitan anti-storm surge protection as tens of billions of dollars. As we mentioned previously, there must be three macro-projects to effectively guard the metropolitan region (See Figure 2) from infrequent storm surge events because there is the potential for an infrastructure loss of utility amounting to an estimated $300 billions! The Stony Brook Storm Surge Research Group's *circa* 2002 recommendation for three barriers is controversial because a three-dam macro-project's basic economics, cost/benefit and political feasibility and long-term environmental impacts are as yet entirely unknown. It would be a capital-intensive 'structural' solution, a hardening of the shoreline as it were.

Following is a brief presentation of the physics of the Textile Storm Surge Barrier as exemplified by a single installation at the Narrows.

DESCRIPTION OF INNOVATION [25]-[30]

Current coast-protection dams are built from solid material (heaped stones, concrete, piled soil). They are expensive to emplace and, sometimes, are ugly. Such dams require detailed on-site research of the surface and sub-surface environment, costly construction and high-quality building efforts over a long period of time (years). Naturally, the coast city inhabitants lose the beautiful sea view and ship passengers are unable to admire the city panorama of the partly hidden city (New Orleans, which is below sea level). The sea coastal usually has a complex geomorphic configuration that greatly increases the length and cost of dam protections.

We offer to protect seaport cities against hurricane storm surge waves, tsunami, and other weather related-coastal and river inundations by new special design of the water and land textile dams.

The offered dam is shown in Figure 3a below. One contains the floats 4, textile (thin film) 3 and support cables 5. The textile (film) is connected a top edge to the floats, the lower edge to a sea bottom. In calm weather the floats are located on the sea surface (Figure 3a) or at the sea bottom (Figure 3b). In stormy weather, hurricane, predicted tsunami the floats automatically rise to top of wave and defend the city from any rapid increase of seawater level (Figure 3c).

This textile-based dam's cost-to-build is thousands of times cheaper than a massive, and possibly unsightly, concrete dam, and a textile infrastructure may be assembled in few months instead of years! They may be installed on ground surface around vital or important infrastructure objects (entries to subway tunnels, electricity power plants, civic airport, and so on) or around a high-value part of the City (example, Manhattan Island) if inundation poses a threat to the City (Figure 4). These textile protections are mobile and can be relocated and installed in few days if hurricane is predictably moving to given city. They can defend the noted object or city from stormy weather inundation, tsunami, and large waves of a height up to 10 m and probably even more.

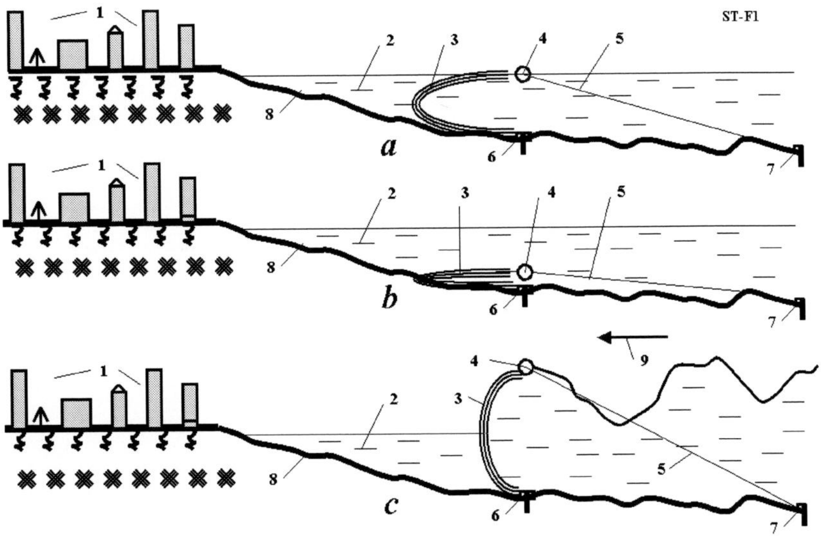

Figure 3. Protection of coastal city against hurricane storm surge waves, tsunami, and other weather related-coastal inundations with textile (film) membrane located in seawater. (*a*) - position of membrane on ocean surface during a calm weather; (*b*) - position of membrane on a sea-bottom in a calm weather; (*c*) - position of membrane in hurricane storm surge waves, tsunamis, and other weather related-coastal inundations. Notations: 1 - city, 2 - ocean, 3 - membrane, 4 - float, 5 - support cable, 6 - connection of membrane to a seabed, 7 - connection of support cable to a seafloor, 8 – sea-bottom, 9 - wind.

The offered tension textile dam may be also used as a big predictable source of electricity because such barriers can be built as dams in rivers and it is used as water dams for the electric station (Figure 5).

They also can be used as dams for an ebbing sea water flow electric station (Figure 6).

The membranes must be made from artificial fiber or a very stout film. The many current artificial fibers available are cheap, have very high safety tensile stress (some times more the steel!) and chemical stability. They can work safely as dams for tens years of service. They are easily repaired.

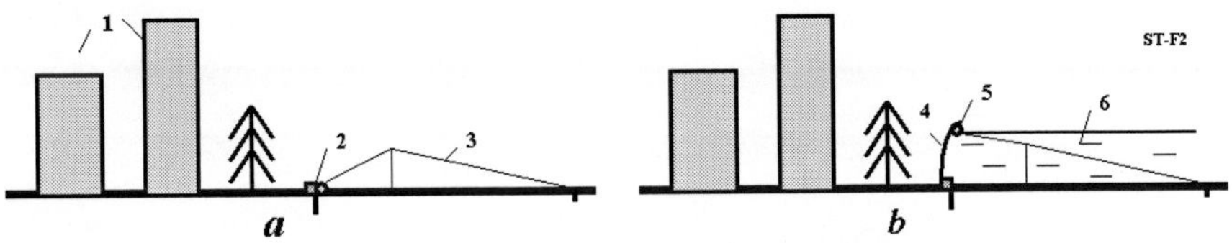

Figure 4. Protection for a city against hurricane storm surge waves, tsunamis, and other weather related-coastal inundations by textile (film) membrane located on ground surface. (*a*) - position of membrane on a ground surface in a compact form in a calm weather; (*b*) - position of membrane in hurricane storm surge waves, tsunamis, and other weather related-coastal inundations. Notations: 1 - city, 2 - membrane in the compact form, 3 - support cable, 4 - membrane, 5 - float, 6 -water, 7 - connection of support cable to a sea-bottom.

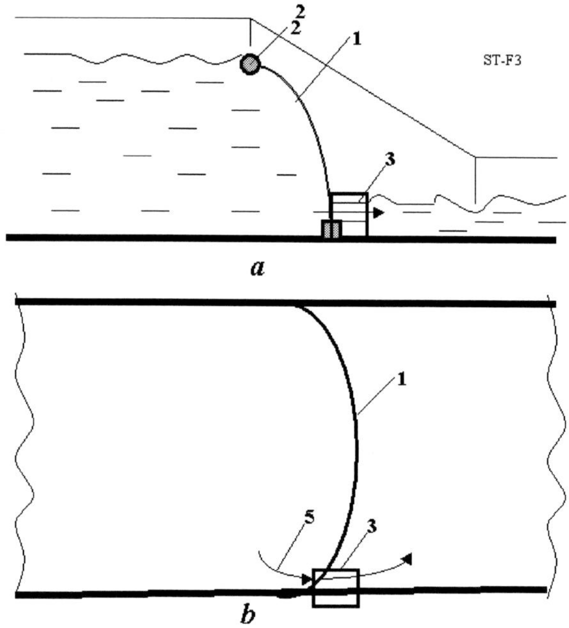

Figure 5. Textile dam and electric station in a river. (*a*) side view; (*b*) - top view. Notations: 1 - textile dam, 2 - float, 3 - electric station, 4 - water flow.

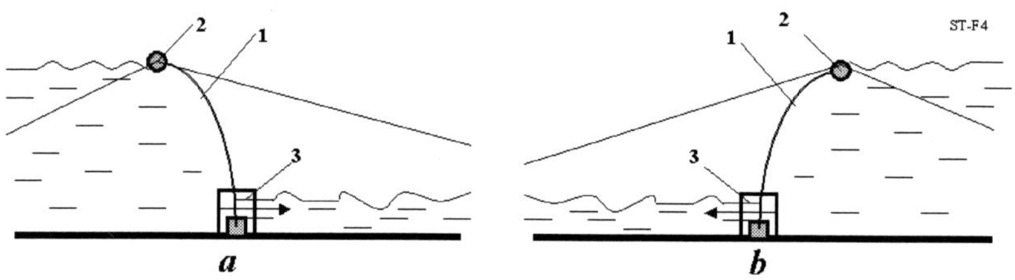

Figure 6

. Tidal (ebb and flow) electric station with textile dam. (*a*) - ebb; (*b*) - flow. Notations: 1 -textile membrane, 2 - float, 3 - electric station.

THEORY AND COMPUTATION

1. Force P [N/m^2] for 1 m^2 of dam is

$$P = g\gamma h, \qquad (1)$$

where g = 9.81 m/s2 is the Earth gravity; γ is water density, γ=1000 kg/m3; h is difference between top and lower levels of water surfaces, m (see computation in Figure 10).

2. Waterpower N [W] is

$$N = \eta g m h, \quad m = \gamma v S, \quad v = \sqrt{2gh}, \atop N = \eta g \gamma h S \sqrt{2gh}, \quad N/S \approx 43.453 \eta h^{1.5}, \quad [\text{kW/m}^2] \qquad (2)$$

where m is mass flow across one meter of width expressed in kg/m; v is water speed, m/s; S is turbine area, m^2; η is coefficient efficiency of the water turbine, N/S is specific power of water turbine, kW/m^2.

Computation is presented in Figure 11.

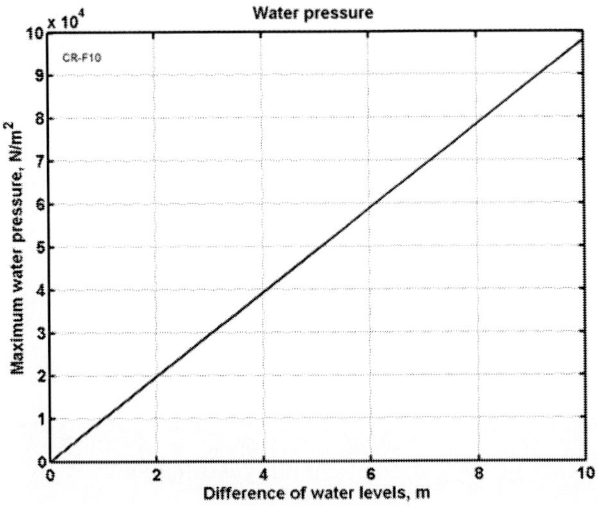

Figure 10. Water pressure via difference of water levels.

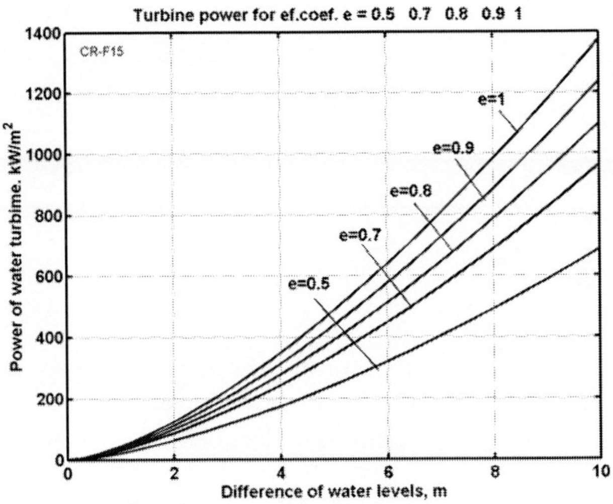

Figure 11. Specific power of a water turbine via difference of water levels and turbine efficiency coefficient.

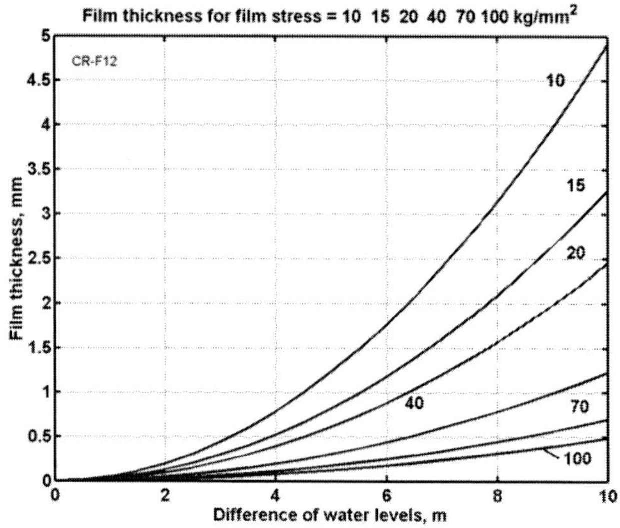

Figure 12. Film (textile) thickness via difference of water levels safety film (textile) tensile stress.

3. *Film thickness* is

$$\delta = \frac{g\gamma h^2}{2\sigma}, \tag{3}$$

where σ is safety film tensile stress, N/m^2. Results of computation are in Figure 12. The fibrous material (Fiber B, PRD-49) has $\sigma = 312$ kg/mm^2 and specific gravity $\gamma = 1.5$ g/cm^3.

4. *The film weight* of 1m width is

$$W_f = 1.2\delta\gamma H \tag{4}$$

Computation is shown in Figure 13. If our dam has length *L* in m, we must multiple this results by *L*.

5. *The diameter d* of the support cable is

$$T = \frac{Pl_2}{2}, \quad S = \frac{T}{\sigma}, \quad d = \sqrt{\frac{4S}{\pi}}, \tag{5}$$

where *T* is cable force, N; l_2 is distance between cable, m; *S* is cross-section area, m^2.

Computation is presented in Figure 14. The total weight of supporting cable is

$$W_c \approx 2\gamma_c HSL / l_2, \quad W_a = \gamma_c SL, \tag{6}$$

where γ_c is cable density, kg/m^3; L is length of dam, m; W_a is additional (connection of banks) cable, m. The cheap current fiber has $\sigma = 620$ kg/mm^2 and specific gravity $\gamma = 1.8$ g/cm^3.

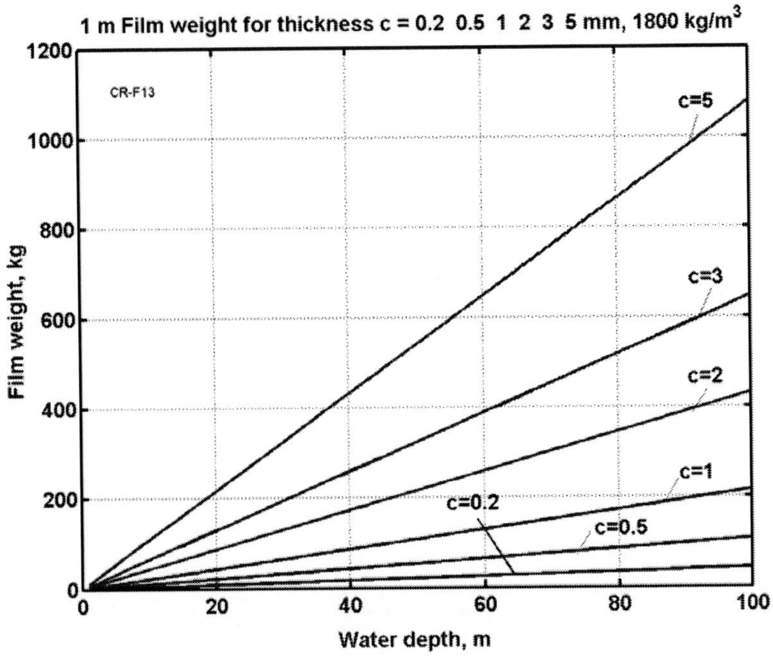

Figure 13. One meter of film weight via the depth of dam and requisite film thickness c, density 1800 kg/m^3.

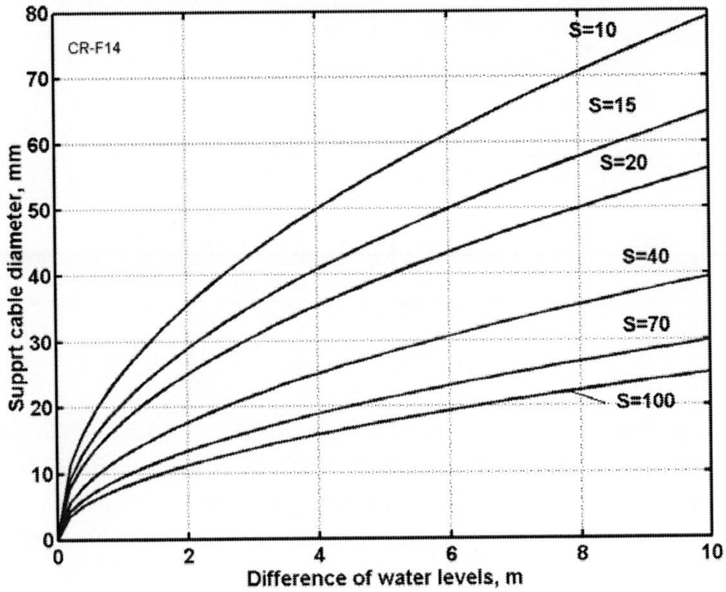

Figure 14. Diameter of the support cable via difference of a water levels and the safety tensile stress for every 10 m textile dam.

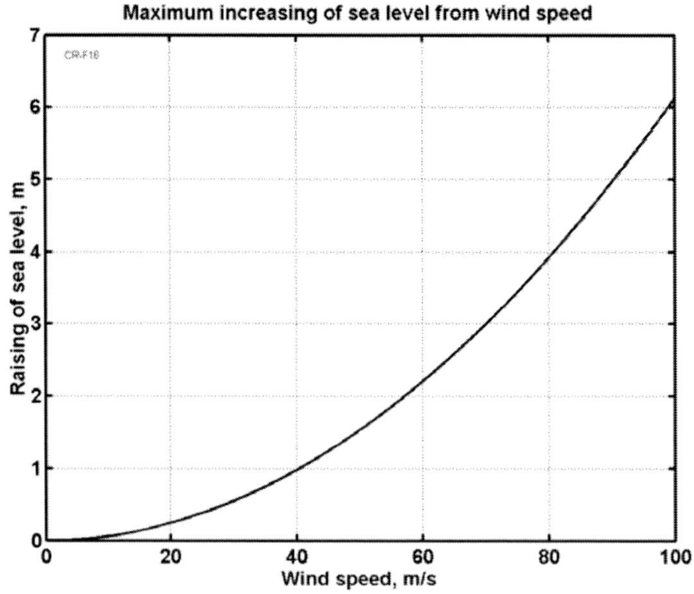

Figure 15. Rising sea level via wind speed.

6. Maximum rise of seawater caused by a hurricane versus wind speed is

$$h = \frac{\rho V^2}{2\gamma}, \tag{7}$$

where h is water raising, m; $\rho = 1.225$ kg/m^3; V is wind speed, m/s; $\gamma = 1000$ kg/m^3 is water density.

Computation is presented in Figure 15.

Wind speed is the main magnitude which influences in the seawater rise. The direction of wind, rain, general atmospheric pressure, depth and relief of seafloor, and the Moon's phase also influence to the seawater rise and can decrease or increase the local sea level computed by Equation (7). For example, in hurricane "eye" the wind is absent, but atmospheric pressure is very low and affected sea level rises consequently.

SUMMARY

We have researched a new method, and cheap design, for land-based and sea-anchored tension textile and film dams. The offered method of seaport and city protection against destructive hurricane storm surge waves, tsunami, and other weather-related inundations is the cheapest so far proposed by anyone and has the very perspective application for defense from natural weather disasters. That is also a new method for producing a useful amount of renewable cheap energy, getting new land for some countries. However, there are important details not considered yet in our research.

Application

Using the graphs above, we can estimate the relevant physical parameters NYC dams and of many interesting macro-projects suitable at other places [25] - [30].

Reflections

The closest approximation of a TSSB at the Narrows is to be found in our Chapter 3, "Preservation of the Mediterranean Sea during Global Sea Level Rise with a Gibraltar Strait Textile Barrage (GSTB)" and in our Chapter 1 of Part B, "The Golden Gate Textile Barrier: Preserving California's Bay of San Francisco from a Rising North Pacific Ocean". Here we examined similar proposed installation that seems obviously far more economical than what has been discussed in the world's popular media as far as we are aware. Finally, we would like others to re-examine a Macro-engineering project proposed by Nigel Chattey—the *Feasibility Study of the Icconn-Erie Incorporated* (1981). Chattey suggested, among other macro-projects, the creation of a large area artificial island offshore of New York City. The presence of such an industrialized island, especially shaped to serve also as breakwater, may have very beneficial oceanographic effects on the Narrows approach to New York City!

References

[1] Vermette, Stephen, "*Storms of tropical origin: a climatology for New York State,* USA (1851-2005)", *Natural Hazards* 42: 91-103 (July 2007).

[2] Bolonkin, A.A., *Cheap Textile Dam Protection of Seaport Cities against Hurricane Storm Surge Waves,* Tsunamis, and Other Weather-Related Floods. February 2007, (published in http://arxiv.org search "Bolonkin").

[3] Bolonkin, A.A. and Cathcart, R.B., *Fabric Aerial Dams/Wind Turbines on Antarctica,* (published in *http://arxiv.org* search "Bolonkin").

[4] Cathcart, R.B. and Bolonkin A.A., Ocean terracing, (published in http://arxiv.org search "Bolonkin").

[5] Bolonkin, A.A. and Cathcart, R.B., *Inflatable 'Evergreen' Dome Settlements,* (published in *http://arxiv.org* search "Bolonkin").

[6] Cathcart, R.B. "Sethusamudram ship channel macro-project: Anti-tsunami and storm surge textile arrestors protecting Palk Bay (India and Sri Lanka)", *Current Science* 91: 1474-1476 (2006).

[7] Bolonkin, A.A. and Cathcart, R.B., *The Java-Sumatra Aerial Mega-Tramway,* (published in *http://arxiv.org* search "Bolonkin").

[8] Macro-Engineering: *A Challenge for the Future.* Edited by V. Badescu, R.B. Cathcart and R.D. Schuiling, Springer, 2006.

Chapter 6

CHEAP AB-SHIELD FOR A NUCLEAR WARHEAD ASSAULTED CITY*

ABSTRACT

In this chapter it is researched a cheap, closed AB-Dome which protects the densely populated cities from nuclear, chemical, biological weapon (bombs) delivered by warheads, strategic missiles, rockets, and various incarnations of aviation technology. The offered AB-Dome is also very useful in peacetime because it shields a city from exterior weather and creates a fine climate within the AB-Dome. The hemispherical AB-Dome is the inflatable, thin transparent film, located at altitude up to as much as 15 km, which converts the city into a closed-loop system. The film may be armored the stones which destroy the rockets and nuclear warhead. AB-Dome protects the city in case the World nuclear war and total poisoning the Earth's atmosphere by radioactive fallout (gases and dust). Construction of the AB-Dome is easy; the enclosure's film is spread upon the ground, the air pump is turned on, and the cover rises to its planned altitude and supported by a small air over-pressure. The offered method is cheaper by thousand times than protection of city by current anti-rocket systems. The AB-Dome may be also used (height up to 15 and more kilometers) for TV, communication, telescope, long distance location, tourism, high placed windmills (energy), illumination and entertainments. The author developed theory of AB-Dome, made estimation, computation and computed a typical project. Discussion and results are in the end of article.

Keywords: Protection from nuclear weapon, protection from chemical, biological weapon, inflatable structures, control of a local weather

* Presented to http://arxiv.org on Jan.10, 2008.

INTRODUCTION

1. Effects of Nuclear Explosion

The dominant effects of a nuclear weapon where people are likely to be directly affected (blast and thermal radiation) are identical physical damage mechanisms to conventional explosives. However the energy produced by a nuclear explosive is millions of times more powerful per gram and the temperatures reached are, briefly, Sun-like. (See: G.I. Brown's *The Big Bang: A History of Explosives,* 1998).

Energy from a nuclear explosive is initially released in several forms of penetrating radiation. When there is a surrounding material such as air, rock, or water, this radiation interacts with and rapidly heats it to an equilibrium temperature. This causes vaporization of surrounding material resulting in its rapid expansion. Kinetic energy created by this expansion contributes to the formation of a shockwave. When a nuclear detonation occurs in air near sea level, much of the released energy interacts with the atmosphere and creates a shockwave which expands spherically from the hypocenter. Intense thermal radiation at the hypocenter forms a fireball and if the burst is low enough, it is often associated an aerial mushroom cloud. In a burst at high altitudes, where the air density is low, more energy is released as ionizing gamma radiation and x-rays than an atmosphere displacing shockwave.

A low-yield USA fission nuclear test in Nevada.

Shockwave. The high temperatures and pressures cause gas to move outward radially in a thin, dense shell called "the hydrodynamic front." The front acts like a piston that pushes against and compresses the surrounding medium to make a spherically expanding shock wave. At first, this shock wave is inside the surface of the developing fireball, which is created in a volume of air by the X-rays. However, within a fraction of a second the dense shock front obscures the fireball, making the characteristic double pulse of light seen from a nuclear detonation. For air bursts at or near sea-level between 50-60% of the explosion's energy goes into the blast wave, depending on the size and the yield-to-weight ratio of the bomb. As a general rule, the blast fraction is higher for low yield and/or high bomb mass. Furthermore, it decreases at high altitudes because there is less air mass to absorb radiation energy and convert it into blast. This effect is most important for altitudes above 30 km, corresponding to <1 per cent of sea-level air density.

Much of the destruction caused by a nuclear explosion is due to blast effects. Most buildings, except reinforced or blast-resistant structures, will suffer moderate to severe damage when subjected to overpressures of only 35.5 kilopascals (kPa).

The blast wind may exceed several hundred km/h. The range for blast effects increases with the explosive yield of the weapon and also depends on the burst altitude. Contrary to what one might expect from geometry the blast range is not maximal for surface or low altitude blasts but increases with altitude up to an "optimum burst altitude" and then decreases rapidly for higher altitudes. This is due to the nonlinear behaviour of shock waves. If the blast wave reaches the ground it is reflected. Below a certain reflection angle the reflected wave and the direct wave merge and form a reinforced horizontal wave, the so-called Mach stem (named after Ernst Mach). For each goal overpressure there is a certain optimum burst height at which the blast range is maximized. In a typical air burst, where the blast range is maximized for 35 to 140 kPa, these values of overpressure and wind velocity noted above will prevail at a range of 0.7 km for 1 kiloton (kt) of TNT yield; 3.2 km for 100 kt; and 15.0 km for 10 megatons (Mt) of TNT.

Two distinct, simultaneous phenomena are associated with the blast wave in air:

- *Static overpressure,* i.e., the sharp increase in pressure exerted by the shock wave. The overpressure at any given point is directly proportional to the density of the air in the wave.
- *Dynamic pressures,* i.e., drag exerted by the blast winds required to form the blast wave. These winds push, tumble and tear objects.

Most of the material damage caused by a nuclear air burst is caused by a combination of the high static overpressures and the blast winds. The long compression of the blast wave weakens structures, which are then torn apart by the blast winds. The compression, vacuum and drag phases together may last several seconds or longer, and exert forces many times greater than the strongest hurricane.

Acting directly on the human body, the shock waves cause pressure waves through the tissues. These waves mostly damage junctions between tissues of different densities (bone and muscle) or the interface between tissue and air. Lungs and the abdominal cavity, which contain air, are particularly injured. The damage causes severe haemorrhaging or air embolisms, either of which can be rapidly fatal. The overpressure estimated to damage lungs is about 70 kPa. Some eardrums would probably rupture around 22 kPa (0.2 atm) and half would rupture between 90 and 130 kPa (0.9 to 1.2 atm).

Blast Winds: The drag energies of the blast winds are proportional to the cubes of their velocities multiplied by the durations. These winds may reach several hundred kilometers per hour.

Thermal Radiation

Nuclear weapons emit large amounts of electromagnetic radiation as visible, infrared, and ultraviolet light. The chief hazards are burns and eye injuries. On clear days, these injuries can occur well beyond blast ranges. The light is so powerful that it can start fires that spread rapidly in the debris left by a blast. The range of thermal effects increases markedly with weapon yield. Thermal radiation accounts for between 35-45% of the energy released in the explosion, depending on the yield of the device.

There are two types of eye injuries from the thermal radiation of a weapon:

Flash blindness is caused by the initial brilliant flash of light produced by the nuclear detonation. More light energy is received on the retina than can be tolerated, but less than is required for irreversible injury. The retina is particularity susceptible to visible and short wavelength infrared light, since this part of the electromagnetic spectrum is focused by the lens on the retina. The result is bleaching of the visual pigments and temporary blindness for up to 40 minutes.

The mushroom cloud from the first "true" Soviet hydrogen bomb test in 1955.

The desolation after the atomic bombing of Hiroshima in 1945.

When thermal radiation strikes an object, part will be reflected, part transmitted, and the rest absorbed. The fraction that is absorbed depends on the nature and color of the material. A thin material may transmit a lot. A light colored object may reflect much of the incident radiation and thus escape damage. The absorbed thermal radiation raises the temperature of the surface and results in scorching, charring, and burning of wood, paper, fabrics, etc. If the material is a poor thermal conductor, the heat is confined to the surface of the material.

Actual ignition of materials depends on how long the thermal pulse lasts and the thickness and moisture content of the target. Near ground zero where the energy flux exceeds 125 J/cm², what can burn, will. Farther away, only the most easily ignited materials will flame. Incendiary effects are compounded by secondary fires started by the blast wave effects such as from upset stoves and furnaces.

In Hiroshima, a tremendous fire storm developed within 20 minutes after detonation and destroyed many more buildings and homes. A fire storm has gale force winds blowing in towards the center of the fire from all points of the compass. It is not, however, a phenomenon peculiar to nuclear explosions, having been observed frequently in large forest fires and following incendiary raids during World War II.

Because thermal radiation travels more or less in a straight line from the fireball (unless scattered) any opaque object will produce a protective shadow. If fog or haze scatters the light, it will heat things from all directions and shielding will be less effective, but fog or haze would also diminish the range of these effects.

Indirect Effect

Ionizing Radiation

About 5% of the energy released in a nuclear air burst is in the form of ionizing radiation: neutrons, gamma rays, alpha particles, and electrons moving at incredible speeds, but with different speeds that can be still far away from the speed of light (beta particles). The neutrons result almost exclusively from the fission and fusion reactions, while the initial gamma radiation includes that arising from these reactions as well as that resulting from the decay of short-lived fission products.

The intensity of initial nuclear radiation decreases rapidly with distance from the point of burst because the radiation spreads over a larger area as it travels away from the explosion. It is also reduced by atmospheric absorption and scattering.

The character of the radiation received at a given location also varies with distance from the explosion. Near the point of the explosion, the neutron intensity is greater than the gamma intensity, but with increasing distance the neutron-gamma ratio decreases. Ultimately, the neutron component of initial radiation becomes negligible in comparison with the gamma component. The range for significant levels of initial radiation does not increase markedly with weapon yield and, as a result, the initial radiation becomes less of a hazard with increasing yield. With larger weapons, above fifty kt (200 TJ), blast and thermal effects are so much greater in importance that prompt radiation effects can be ignored.

The neutron radiation serves to transmute the surrounding matter, often rendering it radioactive. When added to the dust of radioactive material released by the bomb itself, a large amount of radioactive material is released into the environment. This form of radioactive contamination is known as nuclear fallout and poses the primary risk of exposure to ionizing radiation for a large nuclear weapon.

Electromagnetic Pulse

Gamma rays from a nuclear explosion produce high energy electrons through Compton scattering. These electrons are captured in the earth's magnetic field, at altitudes between twenty and forty kilometers, where they resonate. The oscillating electric current produces a coherent electromagnetic pulse (EMP) which lasts about one millisecond. Secondary effects may last for more than a second.

The pulse is powerful enough to cause long metal objects (such as cables) to act as antennae and generate high voltages when the pulse passes. These voltages, and the associated high currents, can destroy unshielded electronics and even many wires. There are no known biological effects of EMP. The ionized air also disrupts radio traffic that would normally bounce off the ionosphere.

One can shield electronics by wrapping them completely in conductive mesh, or any other form of Faraday cage. Of course radios cannot operate when shielded, because broadcast radio waves can't reach them.

Summary of the Effects

The following table summarizes the most important effects of nuclear explosions under certain conditions.

Survivability. This is highly dependent on factors such as proximity to the blast and the direction of the wind carrying fallout.

Table 1. Nuclear bomb effect

Effects	Explosive yield / Height of Burst			
	1 kT / 200 m	20 kT / 540 m	1 MT / 2.0 km	20 MT / 5.4 km
Blast—effective ground range *GR* / km				
Urban areas almost completely leveled (20 PSI)	0.2	0.6	2.4	6.4
Destruction of most civil buildings (5 PSI)	0.6	1.7	6.2	17
Moderate damage to civil buildings (1 PSI)	1.7	4.7	17	47
Thermal radiation—effective ground range *GR* / km				
Conflagration	0.5	2.0	10	30
Third degree burns	0.6	2.5	12	38
Second degree burns	0.8	3.2	15	44
First degree burns	1.1	4.2	19	53
Effects of instant nuclear radiation—effective slant range[1] *SR* / km				
Lethal[2] total dose (neutrons and gamma rays)	0.8	1.4	2.3	4.7
Total dose for acute radiation syndrome[2]	1.2	1.8	2.9	5.4

[1] For the direct radiation effects the slant range instead of the ground range is shown here, because some effects are not given even at ground zero for some burst heights. If the effect occurs at ground zero the ground range can simply be derived from slant range and burst altitude.

[2] "Acute radiation syndrome" corresponds here to a total dose of one gray, "lethal" to ten grays. Note that this is only a rough estimate since biological conditions are neglected here.

2. Possible Types of Nuclear Wars

The possibility of using nuclear weapons in war is usually divided into two subgroups, each with different effects and potentially fought with different types of nuclear armaments.

The first, a *limited nuclear war* (sometimes *attack* or *exchange*), refers to a small scale use of nuclear weapons by one or more parties. A "limited nuclear war" would most likely consist of a limited exchange between two nuclear superpowers targeting each other's military facilities, either as an attempt to pre-emptively cripple the enemy's ability to attack as a defensive measure or as a prelude to an invasion by conventional forces as an offensive measure. It will also refer to a nuclear war between minor nuclear powers, who lack the ability to deliver a decisive strike. This term would apply to any limited use of nuclear weapons, which may involve either military or civilian targets.

The second, a *full-scale nuclear war*, consists of large numbers of weapons used in an attack aimed at an entire country, including both military and civilian targets. Such an attack would seek to destroy the entire economic, social, and military infrastructure of a nation by means of an overwhelming nuclear attack. Such a nuclear war would be unlikely to remain contained between only the two countries involved, especially if either of the nuclear superpowers were involved.

Some Cold War strategists argued that a limited nuclear war could be possible between two heavily armed superpowers (such as the United States and the Soviet Union) and if so several predicted that a limited war could "escalate" into an all-out war. Others have called limited nuclear war "global nuclear holocaust in slow motion" arguing that once such a war took place others would be sure to follow over a period of decades, effectively rendering the planet uninhabitable in the same way that a "full-scale nuclear war" between superpowers would, only taking a much longer and more agonizing path to achieve the same result.

Even the happiest thoughts about the effects of a major nuclear exchange predict the death of millions of civilians within a very short amount of time; more pessimistic predictions argue that a full-scale nuclear war could bring about the extinction of the human race or its near extinction with a handful of survivors (mainly in remote areas) reduced to a pre-medieval quality of life and life expectancy for centuries after and cause permanent damage to most complex life on the planet, Earth's ecosystems, and the global climate. It is in this latter mode that nuclear warfare is usually alluded to as a doomsday scenario.

A third category, not usually included with the above two, is *accidental nuclear war*, in which a nuclear war is triggered unintentionally. Possible scenarios for this have included malfunctioning early warning devices and targeting computers, deliberate malfeasance by rogue military commanders, accidental straying of planes into enemy airspace, reactions to unannounced missile tests during tense diplomatic periods, reactions to military exercises, mistranslated or miscommunicated messages, and so forth. A number of these scenarios did actually occur during the Cold War, though none resulted in a nuclear exchange. Many such scenarios have been depicted in popular culture, such as in the 1962 novel *Fail-Safe* and the 1964 film *Dr. Strangelove or: How I Learned to Stop Worrying and Love the Bomb*.

Post-Cold War

Although the collapse of the Soviet Union ended the Cold War and, subsequently, greatly reduced tensions between the United States of America and the Russian Federation, both nations remained in a "nuclear stand-off" due to the continuing presence of a significant number of fabricated warheads in both nations. Additionally, the end of the Cold War led the USA to become increasingly concerned with the development of nuclear technology by other nations outside of the former USSR. In 1995, a branch of the U.S. Strategic Command produced an outline of forward-thinking strategies in the document "Essentials of Post-Cold War Deterrence".

The United Nations Organization's Disarmament Committee has announced that there are probably more than 16,000 strategic and tactical nuclear weapons ready for deployment and another 14,000 in storage. The U.S.A. has nearly 7,000 ready for action and 3,000 in storage and Russia (RF) has about 8,500 on hand and 11,000 in storage. China has, possibly, 400 nuclear weapons, Britain 400, France 350, India 95, and Pakistan 50. North Korea is confirmed as having nuclear weapons, though it is not known how many (a common estimate is between 1 and 10). Also, despite denials, Israel is also widely believed to have nuclear weapons, possibily as many as 200. NATO has stationed 480 U.S.A-built nuclear weapons in Belgium, the Netherlands, Italy, Germany, and Turkey, with several other countries in pursuit of an arsenal of their own.

A key development in nuclear warfare in the 21st Century has been the proliferation of nuclear weapons to the developing world, with Pakistan and India both publicly testing nuclear devices and North Korea conducting an underground nuclear test on October 9, 2006. The U.S. Geological Survey measured a 4.2 magnitude earthquake in the area where some form of testing occurred. It may have been a dud. Iran, meanwhile, has embarked on a nuclear program which, while officially for civilian purposes, has come under scrutiny by the United Nations and individual states.

Israel has been involved in wars with its neighbours on numerous occasions, and its small geographic size would mean that in the event of future wars the Israeli military might have very little time to react to a future invasion or other major threat; the situation could escalate to nuclear warfare very quickly in some scenarios. In addition, the fact that Iran appears to many observers to be in the process of developing a nuclear weapon has heightened fears of a nuclear conflict in the Middle East, either with Israel or with Iran's Sunni neighbours.

Potential Consequences of a Regional Nuclear War

A study presented at the annual meeting of the American Geophysical Union in December 2006 asserted that even a small-scale, regional nuclear war could produce as many direct fatalities as all of World War II and disrupt the global climate for a decade or more. In a regional nuclear conflict scenario where two opposing nations in the subtropics would each use 50 Hiroshima-sized nuclear weapons (ca. 15 kiloton each) on major populated centers, the researchers estimated fatalities from 2.6 million to 16.7 million per country. Also, as much as five million tons of soot would be released, which would produce a cooling of several degrees Centrigrade over large areas of North America and Eurasia, including most of the grain-growing regions. The cooling would last for years and could be "catastrophic" according to the researchers. This event-process is, sometimes, dubbed "Nuclear Winter".

Sub-strategic Use

The above examples envisage nuclear warfare at a strategic level, i.e., total war. However, many nuclear powers are believed to have the ability to launch more limited engagements. The United Kingdom has reserved the possibility of launching a sub-strategic nuclear strike against an enemy, described by its Parliamentary Defence Select Committee as "the launch of one or a limited number of missiles against an adversary as a means of conveying a political message, warning or demonstration of resolve". This would see the deployment of strategic nuclear weapons in a very limited role rather than the battlefield exchanges of tactical nuclear weapons.

British Trident SSBN submarines are believed to carry some missiles for this purpose, potentially allowing a strike as low as one kiloton against a single target. Former Defence Secretary Malcolm Rifkind argued that this capacity offset the reduced credibility of fullscale strategic nuclear attack following the end of the Cold War *circa* 1990.

Commodore Tim Hare, former Director of Nuclear Policy at the UK's Ministry of Defence, has described it as offering the Government "an extra option in the escalatory process before it goes for an all-out strategic strike which would deliver unacceptable damage".

However, this sub-strategic capacity has been criticised as potentially increasing the acceptability of using nuclear weapons. The related consideration of new generations of limited yield battlefield nuclear weapons by the United States has also alarmed anti-nuclear groups, who believe it will make the use of nuclear weapons more acceptable.

Nuclear Terrorism

Early in the 21st century, concerns began that "rogue states" such as North Korea and Iran could acquire or manufacture nuclear weapons and use them to attack a foe indirectly through terrorism. Nuclear terrorism by non-state organizations may be more likely, as states possessing nuclear weapons are susceptible to retaliation in kind. Geographically dispersed and mobile terrorist organizations are not so easy to discourage by the threat of retaliation. Furthermore, while the collapse of the Soviet Union ended the Cold War, it greatly increased the risk that former Soviet nuclear weapons might become available on the black market. Indeed, it has been alleged that several suitcase-size nuclear fission bombs might have been available. Using such a weapon as a foundation, a terrorist might even create a salted bomb capable of dispersing radioactive contamination over a large area, killing a greater number of people than the explosion itself. Also, "Dirty Bombs" can be made with common explosives used to spread radioactive particles, poisoning and sickening people directly affected by the fallout rather than the actual bomb shockwave.

3. Current Methods of Protection from the Nuclear Warheads and their Disadvantages

The *Strategic Defense Initiative* (SDI) was a proposal by U.S. President Ronald Reagan on March 23, 1983 to use ground-based and space-based systems to protect the United States from attack by strategic nuclear ballistic missiles. The initiative focused on strategic defense rather than the prior strategic offense doctrine of mutual assured destruction (MAD).

Though it was never fully developed or deployed, the research and technologies of SDI paved the way for some anti-ballistic missile systems of today. The Strategic Defense Initiative Organization (SDIO) was set up in 1984 within the United States Department of Defense to oversee the Strategic Defense Initiative. (Amongst its negative critics, it gained the popular name *Star Wars* after the 1977 Hollywood-made movie by George Lucas.) Under the administration of President Bill Clinton in 1993, its name was changed to the Ballistic Missile Defense Organization (BMDO) and its emphasis was shifted from national missile defense to theater missile defense; from global to regional coverage. BMDO was later renamed to the Missile Defense Agency. This article covers defense efforts under the SDIO.

SDI was not the first U.S.A. defensive system against nuclear ballistic missiles. In the 1960s, The Sentinel Program was designed and developed to provide a limited defensive capability, but was never deployed. Sentinel technology was later used in the Safeguard Program, briefly deployed to defend one U.S. location. In the 1970s the Soviet Union deployed a missile defense system, still operational today, which defends Moscow and nearby missile sites.

SDI is unique from the earlier U.S.A. and Soviet missile defense efforts. It envisioned using space-oriented basing of defensive systems vs solely ground-launched interceptors. It also initially had the ambitious goal of a near total defense against a massive sophisticated ICBM attack, vs previous systems which were limited in defensive capacity and geographic coverage.

In 1984, the Strategic Defense Initiative Organization (SDIO) was established to oversee the program, which was headed by Lt. General James Alan Abrahamson, USAF, a past Director of the NASA Space Shuttle program. Research and development initiated by the SDIO created significant technological advances in computer systems, component miniaturization, sensors and missile systems that form the basis for current systems.

Initially, the program focused on large scale systems designed to defeat a Soviet offensive strike. However, as the threat diminished, the program shifted towards smaller systems designed to defeat limited or accidental launches.

By 1987, the SDIO developed a national missile defense concept called the Strategic Defense System Phase I Architecture. This concept consisted of ground and space based sensors and weapons, as well as a central battle management system. The ground-based systems operational today trace their roots back to this concept.

In his 1991 State of the Union Address George H. W. Bush shifted the focus of SDI from defense of North America against large scale strikes to a system focusing on theater missile defense called Global Protection Against Limited Strikes (GPALS).

In 1993, the Clinton administration, further shifted the focus to ground-based interceptor missiles and theater scale systems, forming the Ballistic Missile Defense Organization (BMDO) and closing the SDIO. Ballistic missile defense has been revived by the George W. Bush administration as the National Missile Defense and Ground-based Midcourse Defense.

SDI included the following Programs:

- Ground-based programs
  1 Extended Range Interceptor (ERINT)
  2 Homing Overlay Experiment (HOE)
  3 Exo-atmospheric Reentry-vehicle Interception System (ERIS)
- Directed-energy weapon (DEW) programs
  1 X-ray laser
  2 Chemical laser
  3 Neutral Particle Beam
  4 Laser and mirror experiments
  5 Hypervelocity Rail Gun (CHECMATE)
- Space-based programs
  1 Space-Based Interceptor (SBI)
  2 Brilliant Pebbles
- Sensor programs
  1 Boost Surveillance and Tracking System (BSTS)
  2 Space Surveillance and Tracking System (SSTS)
  3 Brilliant Eyes
  4 Other sensor experiments.

There was a lot of criticism of non efficiency SDI. Another criticism of SDI was that it would not be effective against non-space faring weapons, namely cruise missiles, bombers, and non-conventional delivery methods such as delivery via commercial naval vessels. This latter method in particular would be attractive to terrorists and rogue states as it would be inexpensive, difficult to trace, and technologically undemanding.

The USA Government spent hundreds billions of dollars but the USA does not have SDI now after 24 years R&D.

Anti-ballistic Missile

An *anti-ballistic missile* (ABM) is a missile designed to counter ballistic missiles. A ballistic missile is used to deliver nuclear, chemical, biological or conventional warheads in a ballistic flight trajectory. The term "anti-ballistic missile" describes any antimissile system designed to counter ballistic missiles. However the term is more commonly used for ABM systems designed to counter long range, nuclear-armed Intercontinental ballistic missiles (ICBMs).

Only two ABM systems have previously been operational against ICBMs, the U.S. Safeguard system, which utilized the LIM-49A Spartan and Sprint missiles, and the Russian A-35 anti-ballistic missile system which used the Galosh interceptor, each with a nuclear warhead themselves. Safeguard was only briefly operational; the Russian system has been improved and is still active, now called A-135 and using two missile types, Gorgon and Gazelle, both with conventional warheads. However the U.S. Ground-Based Midcourse Defense (GMD, previously called NMD) system has recently reached initial operational capability. It does not have an explosive charge, but launches a kinetic projectile.

Three shorter range tactical ABM systems are currently operational: the U.S. Army Patriot, U.S. Navy Aegis combat system/Standard SM-3, and the Israeli Arrow. The longer-range U.S. Terminal High Altitude Area Defense (THAAD) system is scheduled for deployment in 2011. In general short-range tactical ABMs cannot intercept ICBMs, even if within range. The tactical ABM radar and performance characteristics do not allow it, as an incoming ICBM warhead moves much faster than a tactical missile warhead. However it is possible the higher performance THAAD missile could be upgraded to intercept ICBMs.

Latest versions of the U.S. Hawk missile have a limited capability against tactical ballistic missiles, but is usually not described as an ABM. Similar claims have been made about the Russia's long-range surface-to-air S-300 and S-400 series.

Israel ABM

The Arrow research effort was first co-funded by the U.S.A. and Israel on May 6, 1986.

The Arrow ABM system was designed and constructed in Israel with financial support by the United States in a multi-billion dollar development program called "Minhelet Homa" with the participation of companies such as the Israel Military Industries, Tadiran and Israel Aerospace Industries.

In 1998 the Israeli military conducted a successful test of their Arrow ABM. Designed to intercept incoming missiles travelling at up to 2 mile/s (3 km/s), the Arrow is expected to perform much better than the Patriot did in the Gulf War. On July 29, 2004 Israel and the United States carried out joint experiment in the USA, in which the Arrow was launched against a real Scud missile. The experiment was a success, as the Arrow destroyed the Scud with a direct hit. In December 2005 the system was successfully deployed in a test against a replicated Shahab-3 missile. This feat was repeated on February 11, 2007.

Anti-satellite Weapons (ASATs)

These are space weapons designed to destroy satellites for strategic military purposes. Currently, only the USA, the former USSR and the People's Republic of China are known to have developed these weapons, with India claiming the technical capability to develop such weapons. On January 11, 2007, China destroyed an old orbiting weather satellite, the world's first test since the 1980s.

ASAT in the Era of Strategic Defense

The era of the Strategic Defense Initiative (proposed in 1983) focussed primarily on the development of systems to defend against nuclear warheads, however, some of the technologies developed may be useful also for antisatellite use.

After the Soviet Union collapsed, there were proposals to use this aircraft as a launch platform for lofting commercial and science packages into orbit. Recent political developments (see below) may have seen the reactivation of the Russian Air-Launched ASAT program, although there is no confirmation of this as yet.

The Strategic Defense Initiative gave the U.S.A. and Russian ASAT programs a major boost; ASAT projects were adapted for ABM use and the reverse was also true. The initial US plan was to use the already developed MHV as the basis for a space based constellation of about 40 platforms deploying up to 1,500 kinetic interceptors. By 1988 the US project had evolved into an extended four stage development. The initial stage would consist of the Brilliant Pebbles defense system, a satellite constellation of 4,600 kinetic interceptors (KE ASAT), of 100 lb (45 kg) each, in Low Earth orbit, and their associated tracking system. The next stage would deploy the larger platforms and the following phases would include the laser and charged particle beam weapons that would be developed by that time from existing projects such as MIRACL. The first stage was intended to be completed by 2000 at a cost of around $125 billion.

Research in the USA and Russia was proving that the requirements, at least for orbital based energy weapon systems, were, with available technology, close to impossible. Nonetheless, the strategic implications of a possible unforeseen breakthrough in technology forced the USSR to initiate massive spending on research in the 12th Five Year Plan, drawing all the various parts of the project together under the control of GUKOS and matching the USA-proposed deployment date of 2000.

Both countries began to reduce expenditure from 1989 and the Russian Federation unilaterally discontinued all SDI research in 1992. Research and Development (both of ASAT systems and other space based/deployed weapons) has, however reported to have be been resumed under the government of Vladimir Putin as a counter to renewed US Strategic Defense efforts post Anti-Ballistic Missile Treaty. However the status of these efforts, or indeed how they are being funded through National Reconnaissance Office projects of record, remains unclear. The U.S. has begun working on a number of programs which could be foundational for a space-based ASAT. These programs include the Experimental Spacecraft System (XSS 11), the Near-Field Infrared Experiment (NFIRE), and the space-based interceptor (SBI).

Recent Developments

On 14 October 2002, a ground based interceptor launched from the Ronald Reagan Ballistic Missile Defense Site destroyed a mock warhead 225 km above the Pacific. The test included three decoy balloons.

On 16 December 2002 President George W. Bush signed National Security Presidential Directive 23 which outlined a plan to begin deployment of operational ballistic missile defense systems by 2004. The following day the U.S.A. formally requested from the UK and Denmark use of facilities in Fylingdales, England, and Thule, Greenland, respectively, as a part of the NMD program. The projected cost of the program for the years 2004 to 2009 will be $53 billion, making it the largest single expenditure in America's defence department budget.

Since 2002, the US has been in talks with Poland and other European countries over the possibility of setting up a European base to intercept long-range missiles. A site similar to the US base in Alaska would help protect the US and Europe from missiles fired from the Middle East or North Africa. Poland's prime minister Kazimierz Marcinkiewicz said in November 2005 he wanted to open up the public debate on whether Poland should host such a base.

In 2002, NMD was changed to Ground-Based Midcourse Defense (GMD), to differentiate it from other missile defense programs, such as space-based, sea-based, and defense targeting the boost phase and the reentry phase (see flight phases).

On 22 July 2004, the first ground-based interceptor was deployed at Ft. Greely, Alaska (63.954° N 145.735° W). By the end of 2004, a total of six had been deployed at Ft. Greely and another two at Vandenberg Air Force Base, California. Two additional were installed at Ft. Greely in 2005. The system will provide "rudimentary" protection.

On 15 December 2004, an interceptor test in the Marshall Islands failed when the launch was aborted due to an "unknown anomaly" in the interceptor, 16 minutes after launch of the target from Kodiak Island, Alaska.

"I don't think that the goal was ever that we would declare it was operational. I think the goal was that there would be an operational capability by the end of 2004," Larry DiRita said on 13 January 2005 at a Pentagon press conference. However, the problem is and was funding "There has been some expectation that there will be some point at which it is operational and not something else these expectations are not unknown, if Congress pours more attention and funding to this system, it can be operational relatively quick."

On 18 January 2005, the Commander, United States Strategic Command issued direction to establish the Joint Functional Component Command for Integrated Missile Defense. JFCC IMD, once activated, will develop desired characteristics and capabilities for global missile defense operations and support for missile defense.

On 14 February 2005, another interceptor test failed due to a malfunction with the ground support equipment at the test range on Kwajalein Island, not with the interceptor missile itself.

On 24 February 2005, the Missile Defense Agency, testing the Aegis Ballistic Missile Defense System, successfully intercepted a mock enemy missile. This was the first test of an operationally configured Standard missile 3 interceptor and the fifth successful test intercept using this system. On 10 November 2005, the USS *Lake Erie* detected, tracked, and destroyed a mock two-stage ballistic missile within two minutes of the ballistic missile launch.

On 1 September 2006, the Ground-Based Midcourse Defense System was successfully tested. An interceptor was launched from Vandenberg Air Force Base to hit a target missile launched from Alaska, with ground support provided by a crew at Colorado Springs. This test was described by Missile Defense Agency director Lieutenant General Trey Obering as "about as close as we can come to an end-to-end test of our long-range missile defense system."

Deployment of the Sea-based X-band Radar system is presently underway.

On 24 February 2007, The Economist, a magazine published in the UK, reported that the United States ambassador to NATO, Victoria Nuland, had written to her fellow envoys to advise them regarding the various options for missile-defence sites in Europe. She also confirmed that "The United States has also been discussing with the UK further potential contributions to the system."

In February 2007 US started formal negiotiations with Poland and Czech Republic concerning construction of missile shield installations in those countries for a Ground-Based Midcourse Defense System. According to press reports Czech republic agreed to host a missile defence radar on its territory while a base of missile interceptors is supposed to be built in Poland. The objective is reportedly to protect most of Europe from long-range missile strikes from Iran.

The Ustka-Wicko base of Polish Army is mentioned as a possible site of US missile interceptors.

Criticism. Report of the American Physical Society

There has been controversy among experts about whether it is technically feasible to build an effective missile defense system. One technical criticism came from U.S.A. physicists and culminated in the publication of a critical study on the subject by the American Physical Society (APS).

This study focused on the feasibility of intercepting missiles in the boost phase, which the current NMD system does not attempt. The study found it might be possible to develop a limited system capable of destroying a liquid-fuel propelled ICBM during the boost phase. This system could also possibly destroy some solid-propellant missiles from Iran, but not those from North Korea, because of differences in the boost time and range to target. However, there is a trend toward using solid-fueled ICBMs which are harder to intercept during boost phase.

Using orbital launchers to provide a reliable boost-phase defense against solid fuel missiles from Iran or North Korea was found to require at least 1,600 interceptors in orbit. Intercepting liquid-fueled missiles would require 700 interceptors. Using two or more interceptors per target would require many more orbital launchers.

The only boost phase system the U.S.A. contemplates for near term use is the Airborne laser (ABL). The study found the ABL possibly capable of intercepting missiles if within 300 km for solid fuel missiles or 600 km for liquid fuel missiles, however solid fuel missiles are more resistant to damage.

While the APS report did not address the current U.S. mid-course NMD system, it concluded that were the U.S.A. in the future to develop a boost-phase ABM defense, there could be significant technical problems limiting effectiveness.

Summary

The current defense systems are very complex, requiring a very high technology and are VERY EXPENSIVE. Only highly industrialized and very rich countries can develop or buy these systems. In many instances such defense systems cost multiple times more than the attack systems (strategic missiles and nuclear warheads), expecially the defense system from small rockets (as, for example, the infamous and widely-used home-made Qassam) or mortar shells. We recognize that future fuzes on assault warheads may exploit the urban dome defense system outlined here, but there are also defeating counter-measures under development to negate such anticipated weapon fuzing penetration strategies.

Acknowledgement: Some data in this work is garnered from Wikipedia under the Creative Commons License.

Description of AB-Dome and Innovations

The author here offers a new kind of city-wide protection system against incoming warheads carrying a nuclear weapon. The AB-Dome is described in the author's works [1]-[4].

His idea is a thin dome covering a city with that is a very transparent film 2 (Figure 1). The film has thickness 0.05 – 0.3 mm. One is located at high altitude (5 - 20 km). The film is supported at this altitude by a small additional air pressure produced by ground ventilators. That is connected to Earth's ground by managed cables 3. The film may have a controlled transparency option. The system can have the second lower film 6 with controlled reflectivity, a further option.

The small additional pressure creats a signufically lift force. For example, the additional pressure only p = 0.01 atm produces the lift force about 100 kg/m^2. At altitute H = 7 km this force is more 40 kg/m^2; at altitude H =10 km, the force is 26 kg/m^2; at altitude H =15 km, the force is 12 kg/m^2. The support cable has a weigth about 5-10 kg/m^2, the 1 m^2 of film weights less 0.05 - 0.5 kg (for example, Kevlar film of thickness 0.2 mm has the weight 0.3 kg/m^2). That means every square meter of dome can support a useful load 10 - 30 kg. At high altitude the useful load decreses, but if it is needed, it can be increased by increasing the interior air pressure.

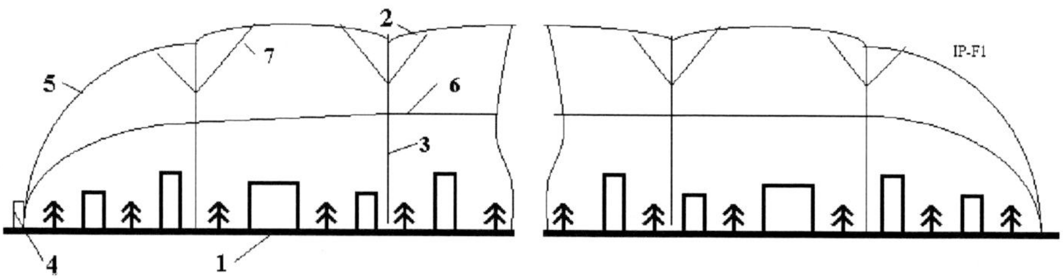

Figure 1. Film AB-Dome for big city. *Notations*: 1 - area, 2 - thin film cover with a control clarity (option), 3 – control support cable height is 10 – 15 km), 4 - exits and ventilators, 5 - border section, 6 – the second (lower) control reflectivity film cover (option), 7 – additional support cables.

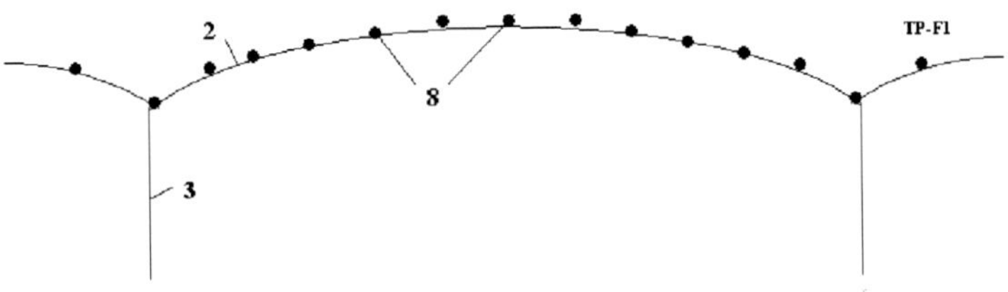

Figure 2. The top film is armored by strong stones (pebbles) 8.

The top film may have the cheap pebbles (stones) 8 (Figure 2) on the upper surface. If the pebble has mass 0.5 kg and a step 0.5 m, the total stone weight will be 2 kg/m^2.

Wartime

The offered protection defends in the following way. The smallest space warhead has a minimum cross-section area 1 m^2 and a huge speed 3 – 5 km/s. The warhead gets a blow and overload from film (mass about 0.5 kg). This overload is 500 – 1500 g and destroys the warhead (see computation below). Warhead also gets an overpowering blow from 2 -5 (every mass is 0.5 - 1 kg) of the strong stones. Relative (about warhead) kinetic energy of every stone is about 8 millions Joules! (It is in 2-3 more than energy of 1 kg explosive!). The film destroys the high speed warhead (aircraft, bomber, wing missile) especially if the film will be armored by stone as it is described in Figure 2, above.

Note: a nuclear bomb is very different from a conventional explosive bomb. Small inaccuracy in the connection of nuclear fuel parts (in the location or times) makes the bomb a pieces of junk or decreases it efficiency by hundreds times! In most cases, the destroyed nuclear warhead (bomb) will fall to Earth.

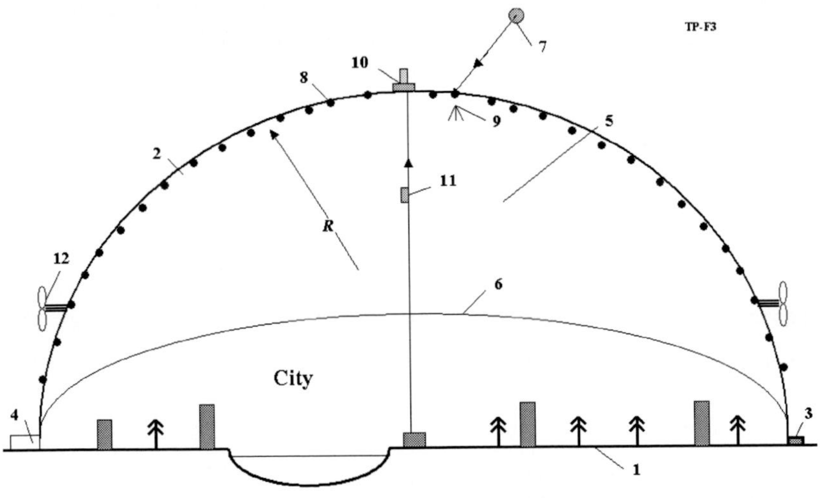

Figure 3. Spherical AB-Dome for big city against thermonuclear warheads, strategic rockets, missiles, aviation, chemical and biological weapons. *Notations*: 1 – protected area; 2 – thin film; 3 – ventilator (air pump); 4 – exit; 5 –spherical thin film AB-Dome; 6 – control reflectivity thin film (optional); 7 – nuclear warhead, rocket, missile, bomb or aviation; 8 – strong stones; 9 – fragments of destroyed warhead, rocket, missile or aircraft; 10 – TV, communication, telescope, locator, tourists; 11 – elevator; 12 – windmills.

The nuclear fuel can be taken from it (as part of any widespread decontamination effort) and then, possibly, used for making the defender's own nuclear bomb since the city and nation affected will still have operating weapons construction infrastructure! The basic design for a nuclear bomb is well known, and the main problem is to get the nuclear fuel (explosive). That means the defending country receives the nuclear weapon for the deserved punishment or attack enemy.

The optimal (for physical damage) height of nuclear explosion is 50-500 m (it depends entirely on the nuclear bomb's explosive yield). The greater the bomb's yield, accordingly, the height ought to be greater. Low detonation height results in the strong air blast wave and energetic neutrons to reach a ground surface and, therefore, produce a lot of radioactive isotopes (fallout dust). If the nuclear bomb explodes at altitude 7 -15 km the bomb efficiency decreases by hundreds and thousands of times (see theoretical, and computation sections below). Why? The nuclear bomb has two main effects: air blast and thermal radiation. Both effects decrease in third order from blast site distance. That means if distance increases in 10 times, the efficiency decreases in 10^3 =1000 times! For example, the bomb has efficiency 0.5 km. That is exploded at altitude 10 km. Their efficiency decreases in 20^3 = 8000 times! Add to this that a relative air density at altitude 10 km is 4 times less than the air density at sea level, the air blast then decreases additionally by 4 times (total decrease will be 32,000 times!).

The same situation occurs with post-detonation thermal radiation. In above example, it decreases by 8000 times. The lower film layer will automatically increase their reflectivity and one can decrease the thermal radiation in additionally about 10 times (total decreasing will be by a factor of 80,000 times!). Internal AB-Dome white-colored water vapor clouds (located at altitude 2-4 km) additionally decreases the thermal radiation present.

The lower film automatically increases the light reflectivity and additionally decreases the thermal radiation in 10 - 20 times. The bomb exploded at high altitude (in rare atmosphere) has significantly less air blast and small radioisotope contamination (the neutrons don't reach the ground).

The city saves the resident life and gets a minimal damage.

In *peacetime,* the offered AB-Dome produces a fine warm climate (weather) in a covered city. The films can have a controlled transparency and reflectivity. That allows provision of different solar heating conditions within the shielded city. These gigantic protective covers are composed of a cheap film having liquid crystal and conducting layers. The clarity of them is controlled by electric voltage. They can pass or block the incidental sunlight (or parts of the solar spectrum) and pass or blockade the Earth radiation. The outer and inside radiations have different wavelengths. That makes it possible to control of them separately and to control heating into (and re-radiation from) the Earth's surface. In conventional conditions about 50% of the solar energy reaches the Earth surface. The most part is reflected back to outer space by the white clouds. In our closed system the clouds (and rain, or at least condensation based dripping) will occur at night when the temperature is low. That means the many cold regions (Alaska, Siberia) may absorb more net solar energy and became, within the bubbles, lands with a temperate or sub-tropic climate. That also means the Sahara desert can be a prosperous area with a fine climate and with closed-loop water cycle.

The building of a film dome is very easy. Don't think of a present science fiction dome city made of impervious thick crystal geodesic panels. We simply spread out the film over Earth surface, turn on the pumping fans, and the film is gradually and controllably raised by air over-pressure inside to needed altitudes, limited by the support cables. Damage to the film is not a major trouble because the additional air pressure is very small and propeller pumps compensate for any air leakage. Unlike in an outer space colony or thin-atmosphere planetary colony (for example, Mars), the outside air is friendly and at worst we lose some heat (or cold) and water vapor.

The other advantages of the suggested method include the possibility to paint pictures on the sky (dome), to show films on the artificial sky by projector, to suspend illuminations, decorations, and air tramways and any other utilities and conveniences (and engineering works) from this new type of roofing.

Long distance aircraft fly at altitude between 8 - 11 km and our dome (less 7- 10 km) does not trouble them unless the dome is built on the edge of a glide path (inbound) or outbound departure path to an airport! The restraining cables will have safety illumination lights (red, flashing, in a string) and internal helicopters will take normal precautions in avoiding contact with them.

More detail the offered AB-Dome is described in [1]-[4] and [13]. Additional information is repeated below.

Our design for the AB-Dome is presented in Figures 1-3, which include the thin inflated film dome. The *innovations* are listed here: (1) the construction is air-inflatable; (2) each dome is fabricated with very thin, transparent film (thickness is 0.05 to 0.3 mm, implying under 150-500 tons a square kilometer) having the control clarity quality without rigid supports; (3) the enclosing film can have (option) two conductivity layers plus a liquid crystal layer between them which changes its clarity, color and reflectivity under an electric voltage; (4) the bound section of dome has a hemisphere form. The air pressure is more in these sections and they protect the central sections from outer wind.

Figures 1-3 illustrate the thin transparent control dome cover we envision. The inflated textile shell—technical "textiles" can be woven or non-woven (films)—embodies the innovations listed: (1) the film is very thin, approximately 0.1 to 0.3 mm. A film this thin has never before been used in a major building; (2) the film has two strong nets, with a mesh of about 0.1 × 0.1 m and $a = 1 \times 1$ m, the threads are about 0.5 mm for a small mesh and about 1 mm for a big mesh. The net prevents the watertight and airtight film covering from being damaged by vibration; (3) the film incorporates a tiny electrically conductive wire net with a mesh about 0.1 x 0.1 m and a line width of about 100 μ and a thickness near 10 μ. The wire net is electric (voltage) control conductor. It can inform the dome supervisors concerning the place and size of film damage (tears, rips, etc.); (4) the film may be twin-layered with the gap — $c = 1$ m and $b = 2$ m—between covering's layers for heat saving. In polar regions this multi-layered low height covering is the main means for heat insulation and puncture of one of the layers won't cause a loss of shape because the film's second layer is unaffected by holing; (5) the airspace in the dome's covering can be partitioned, either hermetically or not; and (6) part of the covering can have a very thin shiny aluminum coating that is about 1μ for reflection of unnecessary solar radiation in equatorial or polar regions [1]-[4].

THEORY AND COMPUTATION (ESTIMATION) OF AB-DOME

a) General Information

Our dome cover (film) has 2 layers (Figures 1, 3): top transparant layer 2, located at a maximum altitude (up 5 -20 km), and lower transparant layer 4 having control reflectivity, located at altitude of 1-3 km (option). Upper transparant cover has thickness about 0.05 – 0.3 mm and supports the protection strong stones (rebbles) 8. The stones have a mass 0.2 – 1 kg and locate the step about 0.5 m.

Brief Information about Cover Film

If we want to control temperature in city, the top film must have some layers: transparant dielectric layer, conducting layer (about 1 - 3 μ), liquid crystal layer (about 10 - 100 μ), conducting layer (for example, SnO_2), and transparant dielectric layer. Common thickness is 0.05 - 0.5 mm. Control voltage is 5 - 10 V. This film may be produced by industry relatively cheaply.

1. Liquid crystals (LC)(Ch.1, Figure 4) are substances that exhibit a phase of matter that has properties between those of a conventional liquid, and those of a solid crystal.

Liquid crystals find wide use in liquid crystal displays (LCD), which rely on the optical properties of certain liquid crystalline molecules in the presence or absence of an electric field. The electric field can be used to make a pixel switch between clear or dark on command. Color LCD systems use the same technique, with color filters used to generate red, green, and blue pixels. Similar principles can be used to make other liquid crystal based optical devices. Liquid crystal in fluid form is used to detect electrically generated hot spots for failure analysis in the semiconductor industry. Liquid crystal memory units with extensive capacity were used in Space Shuttle navigation equipment. It is also worth noting that many common fluids are in fact liquid crystals. Soap, for instance, is a liquid crystal, and forms a variety of LC phases depending on its concentration in water.

The conventional control clarity (transparancy) film reflected a superfluos energy back to space. If film has solar cells that converts the superfluos solar energy into electricity.

2. Transparency. In optics, transparency is the material property of allowing light to pass through. Though transparency usually refers to visible light in common usage, it may correctly be used to refer to any type of radiation. Examples of transparent materials are air and some other gases, liquids such as water, most glasses, and plastics such as Perspex and Pyrex. Where the degree of transparency varies according to the wavelength of the light. From electrodynamics it results that only a vacuum is really transparent in the strict meaning, any matter has a certain absorption for electromagnetic waves. There are transparent glass walls that can be made opaque by the application of an electric charge, a technology known as electrochromics.Certain crystals are transparent because there are straight lines through the crystal structure. Light passes unobstructed along these lines. There is a complicated theory "predicting" (calculating) absorption and its spectral dependence of different materials.

3. Electrochromism is the phenomenon displayed by some chemical species of reversibly changing color when a burst of charge is applied.

One good example of an electrochromic material is polyaniline which can be formed either by the electrochemical or chemical oxidation of aniline. If an electrode is immersed in hydrochloric acid which contains a small concentration of aniline, than a film of polyaniline can be grown on the electrode. Depending on the redox state, polyaniline can either be pale yellow or dark green/black. Other electrochromic materials that have found technological application include the viologens and polyoxotungstates. Other electrochromic materials include tungsten oxide (WO_3), which is the main chemical used in the production of electrochromic windows or smart windows.

As the color change is persistent and energy need only be applied to effect a change, electrochromic materials are used to control the amount of light and heat allowed to pass through windows ("smart windows"), and has also been applied in the automobile industry to automatically tint rear-view mirrors in various lighting conditions. Viologen is used in conjunction with titanium dioxide (TiO_2) in the creation of small digital displays. It is hoped that these will replace LCDs as the viologen (which is typically dark blue) has a high contrast to the bright color of the titanium white, therefore providing a high visibility of the display.

4. Film and cable properties [16]-[19]. Artificial fibers are currently being manufactured, which have tensile strengths of 3-5 times more than steel and densities 4-5 times less than steel. See section Nanotubes in Attachment of this book.

For example, whiskers of Carbon nanotube (CNT) material have a tensile strength of 200 Giga-Pascals and a Young's modulus over 1 Tera Pascals (1999). The theory predicts 1 Tera Pascals and a Young's modulus of 1-5 Tera Pascals. The hollow structure of nanotubes makes them very light (the specific density varies from 0.8 g/cc for SWNT's (Single Wall Nano Tubes) up to 1.8 g/cc for MWNT's, compared to 2.26 g/cc for graphite or 7.8 g/cc for steel). Tensile strength of MWNT's nanotubes may reach 150 GPa.

Specific strength (strength/density) is important in the design of the systems presented in this paper; nanotubes have values at least 2 orders of magnitude greater than steel. Traditional carbon fibers have a specific strength 40 times that of steel. Since nanotubes are made of graphitic carbon, they have good resistance to chemical attack and have high thermal stability. Oxidation studies have shown that the onset of oxidation shifts by about 100^0 C or higher in nanotubes compared to high modulus graphite fibers. In a vacuum, or reducing atmosphere, nanotube structures will be stable to any practical service temperature (in vacuum up 2800 °C. in air up 750°C).

Nanotubes are produced about 60 tons/years now (2007). Price is about $100 - 50,000/kg. Experts predict production of nanotubes 6000 tons/years and price $1 – 100/kg to 2012.

The artificial fibers are cheap and widely used in tires and everywhere. The author has found only old information about textile fiber for inflatable structures (Harris J.T., Advanced Material and Assembly Methods for Inflatable Structures, AIAA, Paper No. 73-448, 1973). This refers to DuPont textile Fiber B and Fiber PRD-49 for tire cord. They are 6 times strong as steel (psi is 400,000 or 312 kg/mm^2) with a specific gravity only 1.5. Minimum available yarn size (denier) is 200, tensile module is 8.8×10^6 (B) and 20×10^6 (PRD-49), and ultimate elongation (percent) is 4 (B) and 1.9 (PRD-49). Some data are in Table 5 Attn.

Industrial fibers have $\sigma = 500$ - 600 kg/mm^2, $\gamma = 1800$ kg/m^3, and $\sigma/\gamma = 2{,}78\text{x}10^6$. But we are used in all our projects the cheapest films and cables (safety $\sigma = 50$ - 100 kg/mm^2).

5. Wind effect. As wind flows over and around a fully exposed, nearly completely sealed inflated dome, the weather affecting the external film on the windward side must endure positive air pressures as the wind stagnates. Simultaneously, low air pressure eddies will be present on the leeward side of the dome. In other words, air pressure gradients caused by air density differences on different parts of the dome's envelope is characterized as the "buoyancy effect". The buoyancy effect will be greatest during the coldest weather when the dome is heated and the temperature difference between its interior and exterior are greatest. In extremely cold climates such as the Arctic and Antarctic Regions the buoyancy effect tends to dominate dome pressurization.

6. Solar radiation. Solar radiation impinging the orbiting Earth is approximately 1400 W/m^2. The average Earth reflection by clouds and the sub-aerial surfaces (water, ice and land) is about 0.3. The Earth-atmosphere absorbs about 0.2 of the Sun's radiation. That means about q_0 = 700 W/m^{2}s of solar energy (heat) reaches our planet's surface in cloudy weather at the Equator. That means we can absorb about 30 - 80% of solar energy. It is enough for normal plant growth in wintertime (up to 40-50° latitude) and in circumpolar regions with a special variant of the dome design.

The solar spectrum is graphically portrayed in Figure 4.

The visible part of the Sun's spectrum is only λ = 400 – 800 nm (0.4 to 0.8 μ.). Any warm body emits radiation. The emission wavelength depends on the body's temperature. The wavelength of the maximum intensity (see Figure 5) is governed by the black-body law originated by Max Planck (1858-1947):

$$\lambda_m = \frac{2.9}{T}, \quad [mm], \tag{1}$$

where T is body temperature, ^{o}K. For example, if a body has an ideal temperature 20 ^{o}C (T = 293 ^{o}K), the wavelength is λ_m = 9.9 μ.

The energy emitted by a body may be computed by employment of the Josef Stefan-Ludwig Boltzmann law:

$$E = \varepsilon\sigma_S T^4, \text{ [W/m}^2\text{]}, \tag{2}$$

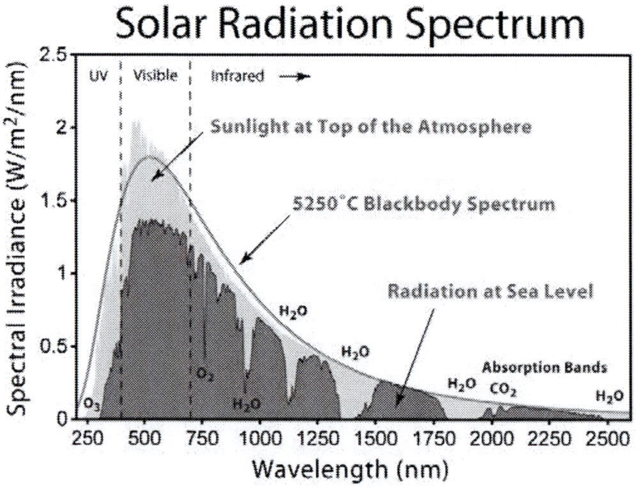

Figure 4. Spectrum of solar irradiance outside atmosphere and at sea level with absorption of electromagnetic waves by atmospheric gases. Visible light is 0.4 - 0.8 μ (400 – 800 nm).

Table 2. Standard Earth atmosphere

H km	0	1	2	3	4	5	6	7
$\bar{p} = p_h/p_o$	1	0.887	0.784	0.692	0.609	0.533	0.466	0.406
H km	8	9	10	11	12	13	14	15
$\bar{p} = p_h/p_o$	0.362	0.304	0.261	0.224	0.191	0.164	0.14	0.12

Table 3. Effects of nuclear explosion

Yield (megatons)	0.001	0.01	0.1	1	10
Thermal radiation radius (3rd degree burns)	687 m	1.8 km	4.5 km	11.7 km	30 km
Air blast radius (4.6 psi, widespread destruction)	739 m	1.6 km	3.4 km	7.2 km	15.4 km
Air blast radius (20 psi, near-total fatalities)	280 m	599 m	1.3 km	2.7 km	5.9 km
Ionizing radiation radius (500 rem)	840 m	1.3 km	2 km	3.1 km	4.8 km
Fireball duration	0.2 s	0.6 s	1,6 s	4.5 s	12.7 s
Fireball radius (minimum)	30 m	70 m	170 m	430 m	1.1 km
Fireball radius (airburst)	30 m	80 m	210 m	530 m	1.3 km
Fireball radius (ground-contact airburst)	40 m	110 m	280 m	700 m	1.8 km

- All figures assume optimum burst height.
- Thermal radiation is non-ionizing electromagnetic radiation which has a significant heating effect. Air is virtually transparent to thermal radiation. At the destructive radius, the thermal radiation intensity is sufficient to cause lethal burns.
- The first air blast is 4.6psi overpressure, which is sufficient to collapse most residential and industrial structures. Note that exposed humans can actually survive such a blast, about 1/3 bar above standard. However, that much pressure exerted against the face of a building exerts very high force (a 40 foot tall, 50 foot wide structure would be hit with more than 600 tons-force).
- The second air blast category is 20 psi over-pressure, which is sufficient to destroy virtually any large above-ground structure and cause nearly 100% fatalities.
- Ionizing radiation is electromagnetic radiation of sufficient frequency (and hence energy) to literally "knock off" electrons from atoms, thus ionizing them. Ionizing radiation is extremely dangerous but it is also strongly absorbed by air, unlike thermal radiation. At the 500rem dosage, mortality is between 50% and 90%, although this can be mitigated with prompt and sophisticated medical care (which may not be available in the aftermath of a nuclear attack).
- Fireball duration is based on emission intensity reduction to 10% of peak.
- Fireball radius is based on a scaling law from "The Effects of Nuclear Weapons" (1977), Chapter IIc, from excerpts reprinted at EnviroWeb. According to that source, fireball radius scales with (Y^0.4), where Y is yield. Also note that a ground-contact airburst creates a larger fireball because some of the energy is reflected back up from the surface.

where ε is coefficient of body blackness ($\varepsilon = 0.03 \div 0.99$ for real bodies), $\sigma_s = 5.67\times10^{-8}$ [W/m^2.K] Stefan-Boltzmann constant. For *example*, the absolute black-body ($\varepsilon = 1$) emits (at $T = 293$ ^{0}K) the energy $E = 418$ W/m^2.

7. *Earth's atmosphere.* The property of Earth's atmosphere needed for computations are presented in Table 2 below.

b) General Theory

1. Effect of Nuclear Explosion

Effect of nuclear explosion depends from many factors. For example, from atmosphere conditions. The results of computation for typical conditions [15] are presented in Table 3. CONVERSION NOTE: 1 psi = 0.45359 kilogram per square inch.

The data used in HYDESim are based on information found in *"The Effects of Nuclear Weapons", 3rd Edition,* by Samuel Glasstone and Philip J. Dolan [15].

Over-pressure Key (1 psi = 70.3 g/cm^2)

15 psi	Complete destruction of reinforced concrete structures, such as skyscrapers, will occur within this ring. Between 7 psi and 15 psi, there will be severe to total damage to these types of structures.
5 psi	Complete destruction of ordinary houses, and moderate to severe damage to reinforced concrete structures, will occur within this ring.
2 psi	Severe damage to ordinary houses, and light to moderate damage to reinforced concrete structures, will occur within this ring.
1 psi	Light damage to all structures, and light to moderate damage to ordinary houses, will occur within this ring.
0.25 psi	Most glass surfaces, such as windows, will shatter within this ring, some with enough force to cause injury.

Results of air blast and the thermal radiation computation via bomb yield are presented in Figure 5.

2. Decrease of Bomb Effects with Increased Detonation Altitude

The decreasing of bomb efficiency from height of explosion may be estimated by equations:

$$T_r = \left(\frac{R}{H}\right)^3, \quad A_b = \left(\frac{p_h}{p_0}\right)\left(\frac{A}{H}\right)^3, \qquad (3)$$

where T_r is relative thermal radiation; R is thermal radiation (3rd degree burns), km; A_b is Air blast radius (4.6psi,widespread destruction), km; H is altitude of explosion, km.

Result of computation is presented in Figure 6.

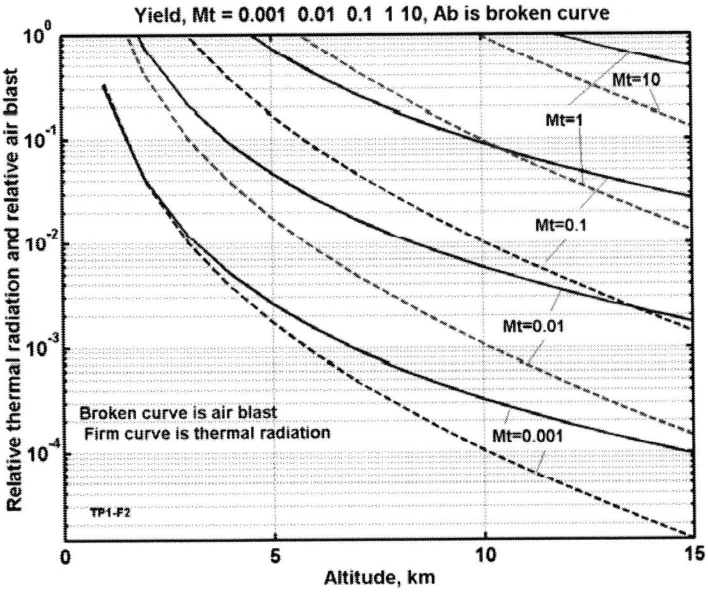

Figure 5. Thermal radiation radius (3rd degree burns) and Air blast radius (4.6 psi, wide-spread destruction) via bomb yield. Reminder: the 1945 Hiroshima bomb had yield about 0.015 Mt, the hydrogen bomb has 0.1 (and more) Mt.

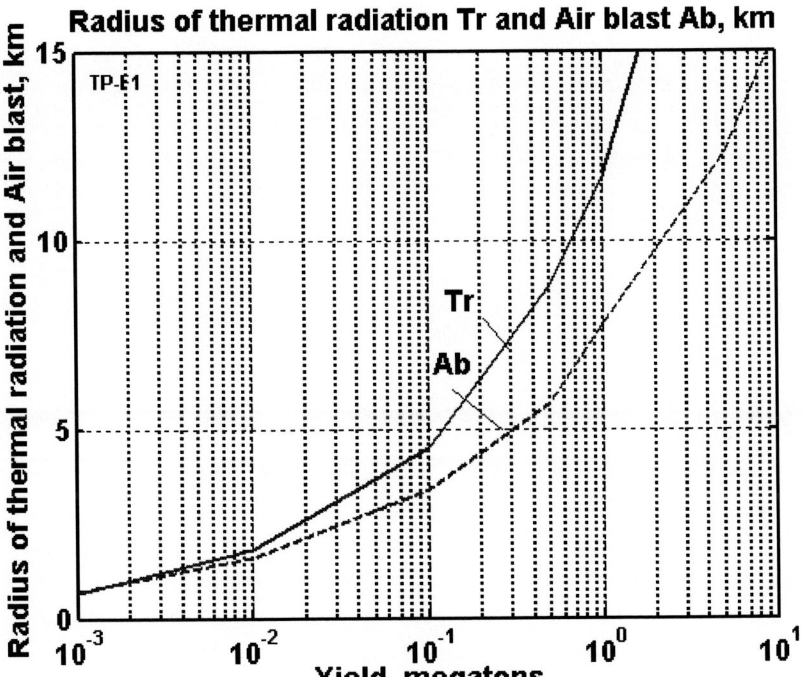

Figure 6. Decreasing of bomb effect versus altitude for different bomb yield (Mt is Megatons). The broken curve is the thermal effect; the firm line is the air blast effect.

As you see the 10 kt bomb exploded at altitude 10 km decreases the air blast effect about in 1000 times and thermal radiation effect without the second cover film in 500 times, with the second reflected film about 5000 times. The hydrogen 100kt bomb exploded at altitude 10 km decreases the air blast effect about in 10 times and thermal radiation effect without the second cover film in 20 times, with the second reflected film about 200 times.

Only power 1000 kt thermonuclear (hydrogen) bomb can damage city. But this damage will be in 10 times less from air blast and in 10 times less from thermal radiation. If the film located at altitude 15 km, the damage will be in 85 times less from the air blast and in 65 times less from the thermal radiation.

For protection from super thermonuclear (hydrogen) bomb we need in higher dome altitudes (20-30 km and more). We can cover by AB-Dome the important large region and full country.

3. Warhead Overload when Film Break

When the impacting warhead penetrates the city-shielding film, the warhead breaks the film. The film break is made at a very high speed (3-6 km/s) and, therefore, the penetrating warhead has instantaneous overload. This overload appears from the film tensile stress and from part of the film which get instantaneous acceleration.

The overload from film break may be estimated by equation

$$a = F / gM, \quad F \approx \sigma_{\max} \delta L, \tag{4}$$

where a is overload in g; g = 9.81 m/s^2 is Earth's acceleration; M is warhead mass, kg; F is break force, N; σ_{max} is maximum tensile stress of film, N/m^2; δ is film thickness, m; L is length of break, m.

For *example*, if the warhead makes a single hole with only a 1 m diameter ($L \approx 3$ m), σ_{max} = 250 kg/mm^2 = 2.5×10^9 N/m^2, δ = 0.25 mm = 0.00025 m, Warhead mass M = 100 kg. We receive the $F = 19\times10^5$ N =190 metric tons, a = 1900 g. That is a gigantic material overload! Only special (high mass) design of warhead can possibly successfully resist this instantaneous pressure overload. The film mass $m \approx 0.4$ kg/m^2, adding an additional destructive overload to the incoming warhead (as 1 kg of explosive!).

Who doubt, doubts this, one can take a conventional customer plastic shopping bag, roll it into a tube shape and break it. Note, the polyethylene bag has thickness only δ = 7 -12 μm, $L \approx 0.5$ and not a high σ.

The film produces a dispersed stress. If the film is armored by strong stones, the 1 – 5 stones (in example over) full destroyed the any bomb, missile, rocket, and aircraft because the relative speed of flight apparatus is closed to that of artillery projectiles. In high-speed blow, the mass is a more important factor than the strength of stone studding the AB-Dome. The author saw the result of blow test when a small (some grams) stone struck with speed 7 km/s into an alumina armor of 10 cm thickness. One made a hole of 10 cm diameter!

The results of computation of equations (4) are shown in Figure 7.

4. The Thickness and Weight of the AB-Dome

The thickness and weight of the AB-Dome, its sheltering shell of film, is computed by formulas (from equation for tensile strength):

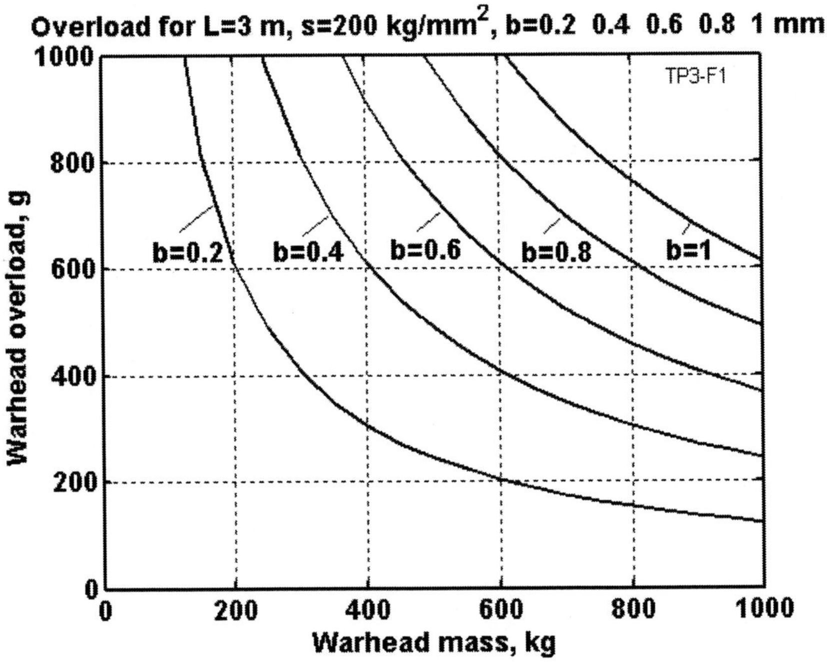

Figure 7. Warhead percussion overload when warhead strike in cover dome film via warhead mass for different the film thickness b = δ, maximal film tensile stress s = σ = 200 kg/mm^2, and length of rupture L = 3 m.

$$\delta_1 = \frac{Rp}{2\sigma}, \quad \delta_2 = \frac{Rp}{\sigma}, \tag{5}$$

where δ_1 is the film thickness for a spherical dome, m; δ_2 is the film thickness for a cylindrical dome, m; R is radius of dome or radius of cover cell between cable (it may be half of distance between top cable), m; p is additional pressure into the dome, N/m^2, (p depends from altitude); σ is safety tensile stress of film, N/m^2.

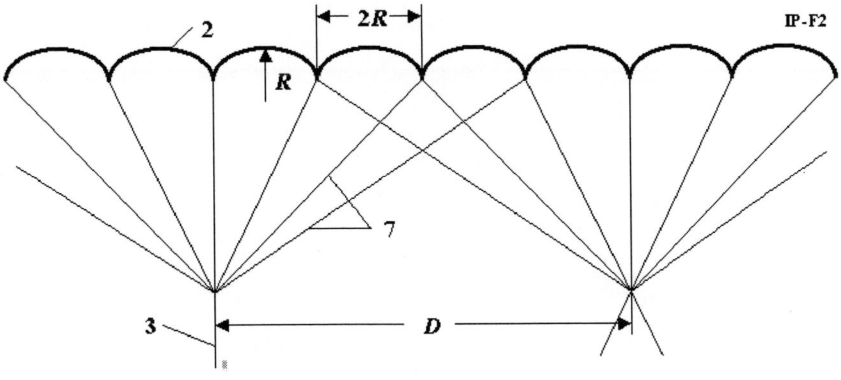

Figure 8. Cable-retaining system. Radius, R, spherical cell of dome cover and distance D between main cable.

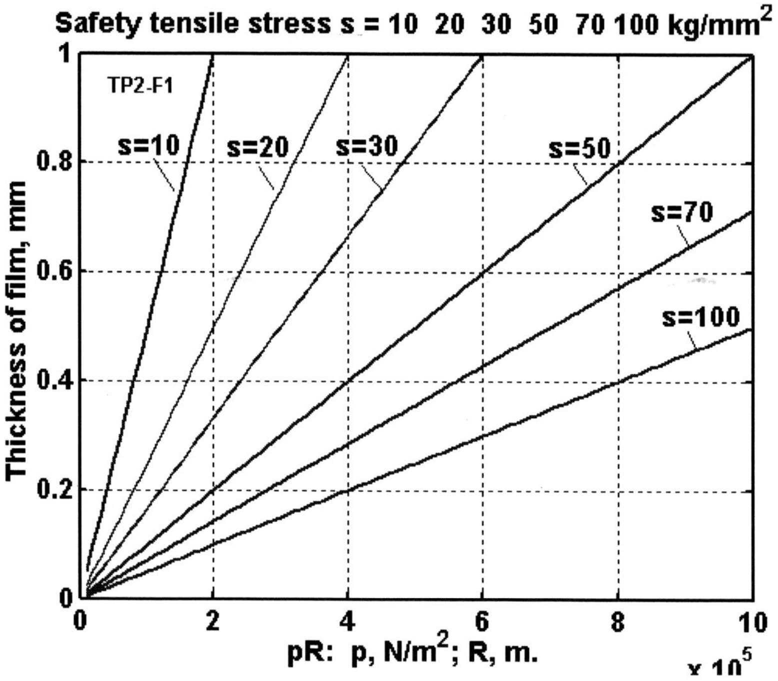

Figure 9. The thickness of top cover via the production of overpressure and radius spherical dome cell (distance between top cables for different safety film tensile stress.

For *example*, compute the film thickness for dome having radius R = 100 m (distance between top cable 7 is 400 m), additional air pressure p = 0.01 atm (p = 1000 N/m^2), safety tensile stress σ = 50 kg/mm^2 (σ = 5×10^8 N/m^2), hemi-spherical dome. We receive δ_1 = 0.1 mm. Distance between main cable 3 is D = 0.8 km (Figure 9).

The computation for others case are presented in Figure 9 below.

The cover weight (mass) of 1 m^2 is computed by equation

$$m = \gamma\delta\,, \tag{6}$$

where m is 1 m^2 film mass, kg/m^2; γ is cover density, m. For *example*, if the cover thickness is δ = 0.2 mm = 0.0002 m and γ = 1500 kg/m^3, the m = 0.3 kg/m^2.

Area S_c of semi-sphere diameter R, film cover mass M_f and cost C of AB-Dome plastic envelope are

$$S_c = 2\pi R^2, \quad M_f = m_f S_c, \quad C = cS_c, \quad C = c_m M_f \quad , \tag{7}$$

where R is radius of hemisphere, m; m_f is average cover area of 1 m^2 ; c is cost of 1 m^2, \$US/m^2; c_m is cover cost of 1 kg, \$US/m^2; C is cost of total cover, \$US.

Example. Let us take the semi-sphere Dome. If m_f = 0.3 kg/m^2, film cost c = \$0.1 /m^2. The film mass covered of 1 km^2 of ground area is M_1 = 2×10^6 m_c = 600 tons/km^2 and film cost is \$60,000/km^2.

The area of big city diameter 20 km is 314 km^2. Area of semi-spherical dome is 628 km^2. The cost of Dome cover is 62.8 millions \$US. We can take less the overpressure (p = 0.001atm) and decrease the cover cost in 5 – 7 times.

The total cost of installation is about 30-90 million \$US.

That is less in hundred times, than the cost of anti-rocket system (tens of billions \$U.S.A.). The anti-rocket system is useless in peacetime and it may be useless soon in wartime because the weapon is permanently improved. The offered AB-Dome is very useful in peacetime (control weather, temperature inside!), The AB-Dome defense also against any biological, chemical, radioactivity-emitting weapons. AB-Dome is CLOSED-LOOP system. All Earth can be poisoned by radioactive precipitations, poison-gas, harmful microbes but a big city (country) can exist under AB-Dome. (That is the idea and use of the "Doomsday Weapon".)

5. The Mass of Holding Cable for 1 m^2 Projection of AB-Dome

The mass of the support cable for every projection 1 m^2 of dome cover may be computed by equation:

$$m_c \approx \gamma_c \frac{\bar{p}p}{\sigma} H, \quad M_c = \sum_S m_c, \quad M_c \approx \gamma_c \frac{p_a}{\sigma} S H_a, \tag{8}$$

where m_c is cable mass supported the projection one m^2 of the dome cover, kg/m^2; M_c is total mass of cable, kg; $S=\pi R^2$ is projection of cover on ground, m^2; γ_c is density of cable, kg/m^3; p is overpressure, N/m^2; σ is safety cable tensile stress, N/m^2; H is height of cable, m; $\bar{p}$ is relative air pressure at given altitude (Table 2); p_a is average over pressure, N/m^2; H_a is average height of the support cable, m.

Example, for cable having the $\sigma = 3\times10^9$ N/m^2, γ_c = 1500 kg/m^3 (Table 5. Attn.), H = 10 km = 10^4 m, ($\bar{p}$ = 0.261), p = 0.01 atm = 1000 N/m^2; the mass of cable is m_c = 1.3 kg/m^2. If we take p = 0.001 atm = 100 N/m^2, R = 10 km, H_a = 5 km, average overpressure will be p_a = 63 N/m^2, $\sigma = 10^9$ N/m^2, than the mass of support cable is about 1500 tons.

We can design the dome cover without the support cable. In this case we compute the thickness of dome cover for radius R of full Dome (see Eq.(5)). The cover thickness will be more. That is better for defense from warhead. If we wand to compute more exactly and spend less cover mass, we must compute the variable overload (from altitude). In this case we accepted the variable thickness of dome cover.

6. Lift Force of 1 m^2 Projection of Dome Cover

The lift force of 1 m^2 projection of dome cover is computed by equation:

$$L_1 = \bar{p}p, \tag{9}$$

where L_1 is lift force of 1 m^2 projection of dome cover, N/m^2; p is overpressure, N/m^2; $\bar{p}$ is relative air pressure at given altitude (Table 2).

Example, for p = 0.01 atm = 1000 N/m^2, H = 10 km ($\bar{p}$ = 0.261, Table 2) we get L_1 = 261 N = 26 kg.

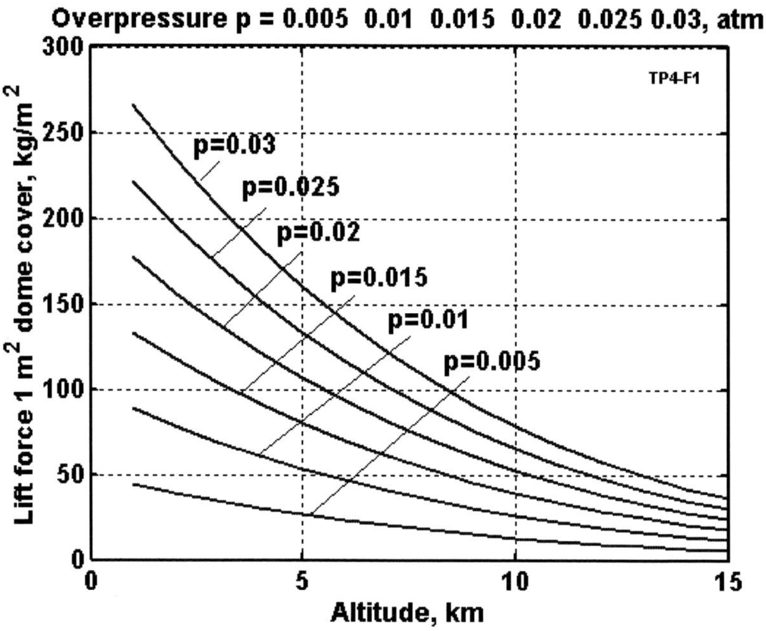

Figure 10. Lift force 1 m^2 of vertical projection the AB-Dome versus altitude for different overpressures.

Result of computation for different p is shown in Figure 10.

1. Leakage of air through a hole. The leakage of air through hole, requested power of ventilator, and time of sinking of Dome cover (in case large of hole) may be estimates by equation:

$$V = \sqrt{\frac{2p}{\rho}}, \quad M_a = \rho V S_h, \quad N = \frac{pVS_h}{\eta}, \tag{10}$$

where V is speed of air leakage, m/s; p is overpressure, N/m^2; ρ is air density at given altitude, ρ = 1,225 kg/m^3 at H = 0; S_h is area of hole, m; N is motor power, W; η is coefficient efficiency of motor.

Example. The area of hole equals $S_h = 10^2$ m^2 (10×10 m) at H = 0 m, p = 0.001 atm = 100 N/m^2, η = 0.8. Computation gives the V = 12.8 m/s, M_a = 1568 kg/s, N =161 kW.

Let us to estimate the time of Dome cover sinking if no supercharge (pumping). Take the sphere radius R = 500 m. The volume of semi-sphere is $v = 262\times10^6$ m^3, the air rate is $q = VS_h$ =12.8×100 = 1280 m^3. The time of the full sinking is $t = v/q = 204\cdot10^3$ s = 57 hours.

Note, the overpressured air will flow outward from the dome. That means the radioactive bomb pollution cannot penetrate into the dome. The outer air pumping in dome can be filtered from radioactive dust.

The repair of dome hole is easy. The support cable of the need part of dome is reeled on and film patch closes the hole.

9. Protection Studding Stones

For increasing protection the AB-Dome cover may be armored with selected strong stones (for example, pebbles, rock flakes, pieces of concrete, and other cheap and available materials), having mass 0.1 – 1 kg. The current anti-rocket systems widely used the kinetic method for destroying the strategic missiles and warhead. The anti-rocket scatters the small balls in the trajectory of rocket (warhead) and when the rocket (warhead) strike ball with very high speed (a lot of times more than conventional projectile, bullet, shell, explpsove) the missile is destroyed. It is not necessary to apply explosive.

The number of the stones for 1 km^2 square area may be estimated by following way. Assume the step of stones is 1 m. Then the number of stones in 1 km^2 is $N_s \approx L^2$, where length L = 1000 m is side of square. If the average mass of a stone is 0.2 kg, the total mass of stones in 1 km^2 will be 200 tonnes per square kilometer. If we then put an additional, similar stone net in this same region, then the stone mass increases by two times and will be 400 tons/km^2, but stone step will be equal about 0.5 m. That is more than enough for destroying an incoming warhead. If warhead has cross-section area 1 m^2 then 1-5 stones will destroy it.

10. The wind Dynamic Pressure is:

$$p_w = \frac{\rho V^2}{2}, \tag{11}$$

where ρ = 1.225 kg/m^3 is air density; V is wind speed, m/s.

For example, a storm wind with speed V = 20 m/s, standard air density is ρ = 1.225 kg/m^3. Than dynamic pressure is p_w = 245 N/m^2. That is four time less when internal pressure p = 1000 N/m^2 = 0.01 atm. When the need arises, sometimes the internal pressure can be voluntarily decreased, bled off.

MACRO-PROJECT

Let us to consider the protection of typical big city having diameter 20 km by semi-spherical AB-Dome, having radius (altitude) R = 10 km. The covered region is $S = \pi R^2 = 314$ km^2, the semi-spherical area is $S_s = 2S = 628$ km^2. The world's biggest city (Mumbai, India) has an area of 484 km^2 and a population of about 14.3 million persons. One can be protected by spherical AB-Dome having radius R = 12 – 15 km.

The many computations for this Dome are made as examples in theoretical section above. We summarize these data in one macro-project below.

Let us take the overpressure 0.01 atm at sea level. Note, a man (deep diver) can keep tolerate overpressure of some atmospheres. The Earth's atmosphere changes the pressure routinely by some percents and we do not feel it. Take a film having safety tensile stress σ = 50 kg/mm^2, local R = 100 m (distance between main cable is 0.8 km, Figure 9), then the film has a thickness δ_1 = 0.1 mm at sea level H = 0 and δ_1 = 0.026 mm at altitude H = 10 km [Eq. (5)]. The average δ_1 = 0.063 mm.

Area of semi-sphere is $S_s = 2\pi R^2 = 628$ km^2. If density of film equals γ =1500 kg/m^3, the total mass of cover film is $M = \gamma\delta S_s$ = 60,000 tons. If cost of film is \$1/kg, the total cost of top cover is C = \$60 millions. Take the thickness of lower film δ = 0.01 mm and density γ = 1500 kg/m^3, then the mass of lower film is $M = \gamma\delta S$ = 4,700 tons and cost \$4.7 millions.

If the stone step is s = 0.5 m and stone mass equals m_s = 0.2 kg each, we early calculated 400 tons/km^2 of stones, the needed total mass of stone is $M_s = m_s S_s$ = 251,200 tons. If stones price is U.S.A. \$10/ton, the stone costs \$2.5 millions.

The average overpressure is p = 0.0063 atm = 630 N/m^2. The total cover lift force is $L = pS$ = 20 millions tons. If support cables has safety tensile stress σ = 100 kg/mm^2, γ =1500 kg/m^3, price \$0.5/kg and average height 5 km, the total mass of support cable is M_c = 150,000 tons, the cost of cables is \$75 millions.

The total mass of construction is 466,000 tons and cost of construction material \$142 millions. The total cost of AB-Dome cover big city is about \$170 millions.

Our dome lift force (20,000,000 metric tons) in 43 times is over then a total weight of the dome construction (466,000 tonnes). If we decrease the overpressure in 10 times (p = 0.001 atm), the need mass of construction material decreases approximately in 10 times. The AB-Dome will be cost about \$30 millions.

As you see the most mass and cost have the support cables. If we use the more thickness or strong film (σ = 100 kg/mm^2) we can make the AB-Dome without support cable. Let us to make estimation of this case for overpressure p = 0.001 atm. The average thickness of top film is 0.63 mm, the cover mass is 600,000 tonnes. The cost of construction material is \$75 millions (for the cost of top cover \$0.5/kg). The total cost of Dome is about \$90 millions. The total dome lift force is about L = 2,000,000 tons. The non-cable Dome is more expensive about in 3 times (in comparison \$30 millions). However, the AB-Dome without internal cables is more comfortable for internal helicopter flight, but one can be less comfortable for repair. The non-cable AB-Dome requests a more cover thickness (by about 3-4 times), which increases the overload of the warhead and may make the stones unneeded.

Our Dome is far from optimal. The average lift force (6.3 kg/m^2) is over by a factor of 4.7 times the weight of cover (0.94 kg/m^2) plus the weight of stones (0.4 kg/m^2). We can increase the mass (and number) of stones by 3-5 times.

The author is prepared to discuss the problems with organizations which are interested in research and development related projects.

DISCUSSION

As you see the 10 kt bomb exploded at altitude 10 km decreases the air blast effect about in 1000 times and thermal radiation effect without the second cover film in 500 times, with the second reflected film about 5000 times. The hydrogen 100kt bomb exploded at altitude 10 km decreases the air blast effect about in 100 times and thermal radiation effect without the second cover film in 10 times, with the second reflected film about 100 times.

Only power 1000kt (1 Megaton) thermonuclear (hydrogen) bomb can damage a city. But this damage will be in 10 times less from air blast and in 10 times less from thermal radiation. If the film located at altitude 15 km, the damage will be in 85 times less from the air blast and in 65 times less from the thermal radiation. The white clouds (they are located at altitude 2 - 4 km inside the AB-Dome) additionally decreases the thermal radiation effect. For more security the cover must be located at altitude up 20 km.

When warhead penetrates the film, the warhead breaks the film. The break is made at extremely high speed (3-6 km/s) and warhead has instantaneous overload. This over-load appears from the film tensile stress and from part of the film which get instantaneous acceleration. The over-load reaches 1000g (and more) and destroys any warhead, missile, rocket, and aircraft. The stones have energy (relative warhead) in 2 – 3 times more them explosive of same mass.

The area of a big city with a diameter of 20 km is about 314 km^2. The total cost of the AB-Dome installation is about 50 - 100 million $U.S.A.

That is less by hundred times, then the cost of anti-rocket system (tens of billions $U.S.A.). The anti-rocket system is useless in peacetime and it may be useless soon in wartime because the weapon is permanently improved. The offered AB-Dome is very useful in peacetime because it can be used to control climate, weather and the air temperature inside. The AB-Dome defense also works against any biological, chemical, radioactive outer weapons. AB-Dome is CLOSED-LOOP system. All Earth can be poisoned by radioactive precipitations, poison-gas, harmful microbe, but a big city (country) can exist under AB-Dome.

Note, the over-pressure air will flow out from dome. That means the radioactive bomb pollution cannot penetrate to dome in case of AB-Dome (cover) damage. The expelling air pumping in dome can be filtered from radioactive dust.

Constructing the AB-Dome is rather easy, as building macro-projects go. The covering spreads on the ground, air pump turned on and the air-supported AB-Dome then rises.

The repair of an AB-Dome hole is not difficult. The support cable of the need part of dome is reeled on and film patch closes the hole. If the dome does not have the support cable, one can have suspended cables which allow reaching any part of AB-Dome. The non-cable AB-Dome requests a more cover thickness (about in 3 - 4 times), which increases the overload of the warhead and may make the stones unneeded.

For increasing protection, the dome cover may be armored with many strong stones (for example, pebbles, rock flakes, pieces of concrete, basalt and other selected cheap materials), having mass 0.1 – 1 kg. The current anti-rocket systems widely used the kinetic method for destroying the strategic missiles and warhead. The anti-rocket scatters the small balls in the trajectory of rocket (warhead) and when the rocket (warhead) strikes a ball with very high speed (a lot of times more then conventional projectile, bullet, shell) the missile is destroyed. It is not necessary to apply explosive.

The control of Earth's regional and global weather is an important problem for humanity. That ability dramatically increases the territory suitable for people to live in, the sown area, and crop capacity. In the long term, it allows us to convert all Earth land such as Alaska, North Canada, Siberia and deserts such as the Sahara or the Gobi into a prosperous garden. The suggested method is very cheap (cost of covering 1 m^2 is about 5 - 25 U.S.A. cents) and may be utilized at the present time. We can start from a small area, from small towns in bad climactic regions and extended to a large area.

Film domes can foster the fuller economic development of cold regions such as the Earth's Arctic and Antarctic and, thus, increase the effective area of territory dominated by humans. Normal human health can be maintained by ingestion of locally grown fresh vegetables and healthful "outdoor" exercise. The domes can also be used in the Tropics and

Temperate Zone. Eventually, they may find application on the Moon or Mars since a vertical variant, inflatable towers to outer space, are soon to become available for launching spacecraft inexpensively into Earth-orbit or interplanetary flights. AB-Dome can keep at high altitude the load up 300 kg/sq. m.

Lest it be objected that such AB-Domes would take impractical amounts of plastic, consider that the world's plastic production is today on the order of 100 million tons. If, with economic growth, this amount doubles over the next generation and the increase is used for doming over territory, at 300 - 500 tons a square kilometer 200,000 square kilometers could be roofed over annually. While small in comparison to the approximately 150 million square kilometers of land area, consider that 200,000 one kilometer sites scattered over the face of the Earth newly made productive and more habitable could revitalize vast swaths of land surrounding them—one square kilometer here could grow local vegetables for a city in the desert, one over there could grow biofuel, enabling a desolate South Atlantic island to become fuel independent; at first, easily a billion people a year could be taken out of sweltering heat, biting cold and slashing rains, saving the money buying and running heating and air conditioning equipment would require. Additionally, clean rain water could flow directly to cisterns, away from the pollution of the storm sewers.

In effect, by doming over inhospitable land as specified, in exchange we get new territory for living with a wonderful climate.

The associated problems are researched in references [1]-[12].

Results

Author offers the cheap AB-Dome which protects the big cities from nuclear, chemical, biological weapon (bombs) delivered by warheads, strategic missiles, rockets, and aviation. The offered AB-Dome is also very useful in peacetime because that protests the city from outer weather and creates a fine climate into Dome.

Main advantages of offered AB-Dome:

1. AB-Dome is cheaper in hundreds times then current anti-rocket systems.
2. AB-Dome does not need in high technology and can build by poor country.
3. It is easy for building.
4. Dome is used in peacetime; it creates the fine climate (weather) into Dome.
5. AB-Dome protects from nuclear, chemical, biological weapon.
6. Dome produces the autonomous existence of the city population after total World nuclear war and total confinement (infection) all planet and its atmosphere.
7. Dome may be used for high region TV, for communication, for long distance locator, for astronomy (telescope).
8. Dome may be used for high altitude tourism.
9. Dome may be used for the high altitude windmills (getting of cheap renewable wind energy).
10. Dome may be used for a night illumination and entertainment.
11. The additional applications of offered AB-Dome the reader finds in [1]-[12].

The Worlds Largest Air Supported Structure.

August, 2006. Southern Inflatables announced the completion of the installation of the largest span air supported structure ever constructed in the world, 215m x 215m x 45m high (705' x 705' x 150' high). The structure was installed over a waste disposal site, 45m deep (150') in South Korea.

New technologies allowed the Generations Sports Complex to cover an area 2 football fields in length by almost a football field wide without support columns to get in the way.

REFERENCES

(The reader may find some of these articles at the author's web page *http://Bolonkin.narod.ru/p65.htm*, *http://arxiv.org* search "Bolonkin", and in the book "*Non-Rocket Space Launch and Flight*", Elsevier, London, 2006, 488 pages and in his book "*New Concepts, Ideas, Innovations in Aerospace, Technology and Human Science*", NOVA, NY, 2008, 430 pages.)

[1] Bolonkin, A.A., Cathcart R.B., *Inflatable 'Evergreen' Polar Zone Dome* (EPZD) Settlements, 2006. *http://arxiv.org* search "Bolonkin".

[2] Bolonkin, A.A. and R.B. Cathcart, Inflatable 'Evergreen' Dome Settlements for Earth's Polar Regions. *Clean Technologies and Environmental Policy.* DOI 10.1007/s10098.006-0073.4.

[3] Bolonkin, A.A., *Control of Regional and Global Weather.* 2006. *http://arxiv.org* search for "Bolonkin".

[4] Bolonkin, A.A., *AB Method of Irrigation without Water* (Closed-loop water cycle). 2007. *http://arxiv.org* search for "Bolonkin".

[5] 5. Bolonkin, A.A., *Cheap Textile Dam Protection of Seaport Cities against Hurricane Storm Surge Waves, Tsunamis, and Other Weather-Related Floods,* 2006. http://arxiv.org.

[6] Bolonkin, A.A. and R.B. Cathcart, *Antarctica: A Southern Hemisphere Windpower Station? http://arxiv.org*, 2007.

[7] Cathcart, R.B. and Bolonkin, A.A. *Ocean Terracing,* 2006. http://arxiv.org.

[8] Bolonkin, A.A. and R.B. Cathcart, *The Java-Sumatra Aerial Mega-Tramway*, 2006. http://arxiv.org.

[9] Bolonkin, A.A., "Optimal Inflatable Space Towers with 3-100 km Height", *Journal of the British Interplanetary Society* Vol. 56, pp. 87 - 97, 2003.

[10] Bolonkin A.A., Inflatable Dome for Moon, Mars, Asteroids and Satellites, Presented as paper AIAA-2007-6262 by *AIAA Conference* "Space-2007", 18-20 September 2007, Long Beach. CA, USA

[11] Bolonkin, A.A., Electrostatic Space Towers, 2007, *http://arxiv.org*, search "Bolonkin".

[12] Bolonkin, A.A., *Non-Rocket Space Launch and Flight, Elsevier*, London, 2006, 488 pages.

[13] Bolonkin, A.A., "*New Concepts, Ideas, Innovations in Aerospace, Technology and Human Science*", NOVA, 2008, 430 pgs.

[14] Macro-Engineering: *A Challenge for the Future.* Springer, 2006. 318 pages. Collection articles.

[15] *"General Principles of Nuclear Explosions",* Chapter 1 in Samuel Glasstone and Phillip Dolan, eds., The Effects of Nuclear Weapons, 3rd Edition, (Washington D.C.: U.S. Department of Defense/U.S. Energy Research and Development Administration, 1977); provides information about the relationship of nuclear yields to other effects (radiation, damage, etc.).

[16] Galasso, F.S., *Advanced Fibers and Composite,* Gordon and Branch Scientific Publisher, 1989.

[17] Carbon and High Performance Fibers Directory and Data Book, London-New. York: Chapman & Hall, 1995, 6th ed., 385 page.
[18] Concise Encyclopedia of Polymer Science and Engineering, Ed. J.I. Kroschwitz, N. Y., Wiley, 1990, 1341 pages.
[19] Dresselhaus, M.S., Carbon Nanotubes, by, Springer, 2000.

Chapter 7

SIMPLEST AB-THERMONUCLEAR SPACE PROPULSION AND ELECTRIC GENERATOR*

ABSTRACT

The author applies, develops and researches mini-sized Micro- AB Thermonuclear Reactors for space propulsion and space power systems. These small engines directly convert the high speed charged particles produced in the thermonuclear reactor into vehicle thrust or vehicle electricity with maximum efficiency. The simplest AB-thermonuclear propulsion offered allows spaceships to reach speeds of 20,000 - 50,000 km/s (1/6 of light speed) for fuel ratio 0.1 and produces a huge amount of useful electric energy. Offered propulsion system permits flight to any planet of our Solar system in short time and to the nearest non-Sun stars by E-being or intellectual robots during a single human life period.

Keywords: AB-propulsion, thermonuclear propulsion, space propulsion, thermonuclear power system

INTRODUCTION

At present, both solid and liquid chemical fueled rockets are used for launch to and flights in interplanetary outer space. They have been intensively developed since War II when German engineer Wernher von Braun (1912-1977) successfully designed the first long distance rocket FAU-2. In the subsequent years, liquid and solid rockets reached their developmental peak. Their main shortcomings are (1) very high space launch cost of $20,000 – 50,000/kg; (2) large fuel consumption; (3) liquid fuel storage problems because oxidizer and fuel (for example; oxygen and hydrogen) require cryogenic temperatures, or they are poisonous substances (for example; nitric acid, N_2O_3).

---

* This work presented as paper AIAA-2007-4613 to 38th AIAA Plasmadynamics and Lasers Conference in conjunction with the16th International Conference on MHD Energy Conversion on 25-27 June 2007, Miami, USA. See also http://arxiv.org search "Bolonkin".

In past years, the author and other scientists have published series of new methods that promise to revolutionize space launch and space flight [1-14]. These include cable accelerator, circle launcher and space keeper, space elevator transport system, space towers, kinetic towers, the gas-tube method, sling rotary method, electromagnetic and electrostatic accelerators, tether system, Earth-Moon or Earth-Mars non-rocket cable transport system. There include new propulsion and power systems such as solar and magnetic sails, Solar wind sail, radioisotope sail, electrostatic space sail, laser beam, kinetic anti-gravitator, multi-reflective beam propulsion system, asteroid employment electrostatic levitation, etc. (Too, there are new ideas in aviation that can be useful for flights in the atmosphere.)

Some of these have the potential to decrease space launch costs thousands of times, others allow changing the speed and direction of space apparatus without expending fuel.

The thermonuclear propulsion and power method is very perspective -- though not speculative -- because it promises high vehicular apparatus speed up 50,000 km/s. This method needs a small, special thermonuclear reactor that will allow the direct and efficient utilization of the kinetic energy of nuclear particles – the AB Thermonuclear Reactor –first offered by author [15].

Description of Innovations

The AB thermonuclear propulsion and electric generator are presented in Figure 1. As it is shown in [15] the minimized, or micro-thermonuclear reactor 1 generates high-speed charged particles 2 and neutrons that leave the reactor. The emitted charged particles may be reflected by electrostatic reflector, 4, or absorbed by a semi-spherical screen 3; the neutrons may only be absorbed by screen 3.

In *screen* of the AB-thermonuclear reactor (Figure 1*a*) the forward semi-spherical screen 3 adsorbed particles that move forward. The particles, 2, of the back semi-sphere move freely and produce the vehicle's thrust. The forwarded particles may to warm one side of the screen (the other side is heat protected) and emit photons that then create additional thrust for the apparatus. That is the *photon* AB-thermonuclear thruster.

In *reflector* AB-thermonuclear reactor (Figure 1*b*) the neutrons fly to space, the charged particles 5 are reflected the electrostatic reflector 4 to the side opposed an apparatus moving and create thrust.

The *screen-reflector* AB-thermonuclear reactor (Figure 1*c*) has the screen and reflector

The *spherical* AB-propulsion-generator (Figure 1*d*) has two nets which stop the charged particles and produced electricity same as in [14] Chapter 17. Any part 8 of the sphere may be cut-off from voltage and particles 9 can leave the sphere through this section and, thusly, create the thrust. We can change direction of thrust without turning the whole apparatus.

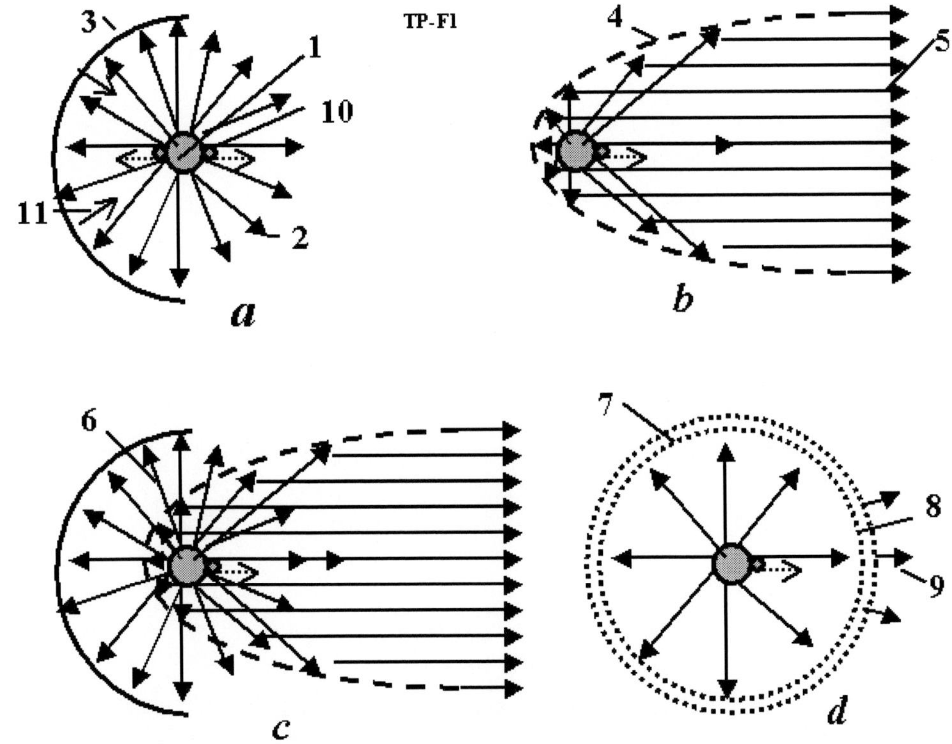

Figure 1. Types of the suggested propulsion and power system. (*a*) *screen* AB-thermonuclear propulsion and *photon* AB-thermonuclear propulsion ; (*b*) (electrostatic) *reflector* AB-thermonuclear propulsion; (*c*) *screen-reflector* AB-thermonuclear propulsion; (*d*) *spherical* AB-propulsion-generator. Notations: 1 - micro (mini) AB-thermonuclear reactor [15], 2 - particles (charged particles and neutrons), 3 - screen for particles, 4 - electrostatic reflector; 5 - charged particles, 6 - neutrons, 7 - spherical net of electric generator, 8 - transparency (for charged particles) part of spherical net, 9 - charged particle are producing the thrust, 10 - electron discharger, 11 - photon radiation.

THEORY OF THE THERMONUCLEAR REACTOR, PROPULSION AND POWER

List of Main Equations

Below are the main equations for the proper estimation of benefits from the offered innovations.

1. Energy needed to overcome the Coulomb barrier

$$F = k\frac{Q_1 Q_2}{r^2}, \quad E = \int_{r_0}^{\infty} F dr, \quad E = \frac{kZ_1 Z_2 e^2}{r_0}, \qquad , \tag{1}$$

$$r_0 = (1.2 \div 1.5)\cdot 10^{15}\sqrt[3]{A}, \; r_0 = r_1 + r_2, \quad A = Z + N$$

where $k = 9\times10^9$; Z_1, Z_2 are charge state of 1 and 2 particles respectively; $e = 1{,}6\times10^{-19}$ C is charge of electron; r_o is radius of nuclear force, m; A is number of element; F is force, N; E is energy, J; Q is charge of particles.

For example, for reaction H+H (hydrogen, $Z_1 = Z_2 = 1$, $r_o \approx 2\times10^{-15}$ m) this energy is ≈ 0.7 MeV or 0.35 MeV for every particle. The real energy is about 30 times less because some particles have more than average speed and there is a tunnel effect.

2. Energy needed for ignition. Figure 8 [15] shows a magnitude $n\tau$ (analog of Lawson criterion) required for ignition.

Present-day industry produces powerful lasers:

- Carbon dioxide lasers emit up to 100 kW at 9.6 μm and 10.6 μm, and are used in industry for cutting and welding.
- Carbon monoxide lasers must be cooled, but can produce up to 500 kW.

Special laser and ICF reactors:

- NOVA (1999, USA). Laser 100 kJ (wavelenght $\lambda=1054\times10^{-9}$ m) and 40 kJ (wavelenght $\lambda=351\times10^{-9}$ m), power few tens of terawatts (1 TW = 10^{12} W), time of impulse $(2 \div 4)\times10^{-9}$ s, 10-beams, matter is Nd:class.
- OMERA (1995, USA). 60-beam, neodyminm class laser, 30 kJ, power 60 TW.
- Z-machine (USA, under construction), power is up 350 TW. It can create currency impulses up to 20×10^6 A.
- NIF (USA). By 2005, the National Ignition Facility is worked on a system that, when complete, will contain 192-beam, 1.8-megajoule, 700-terawatt laser system adjoining a 10-meter-diameter target chamber.
- 1.25 PW - world's most powerful laser (claimed on 23 May 1996 by Lawrence Livermore Laboratory).

3. Radiation energy from hot solid black body is (Stefan-Boltzmann Law):

$$E = \sigma T^4, \tag{2}$$

where E is emitted energy, W/m^2; $\sigma = 5.67\times10^{-8}$ - Stefan-Boltzmann constant, W/m^2 ^{o}K^4; T is temperature in ^{o}K.

4. Wavelength corresponded of maximum energy density (Wien's Law) is

$$\lambda_0 = \frac{b}{T}, \quad \omega = \frac{2\pi}{\lambda_0}, \tag{3}$$

where $b = 2.8978\times10^{-3}$ is constant, m ^{o}K; T is temperature, ^{o}K; ω is angle frequency of wave, rad/s.

5. Pressure for one full reflection is

$$F = 2E/c, \tag{4}$$

where F - pressure, N/m^2; $c = 3\times10^8$ is light speed, m/s, E is radiation power, W/m^2. If plasma does not reflect radiation the pressure equals

$$F = E/c. \tag{5}$$

6. Pressure for plasma multi-reflection [8, 14] is

$$F = \frac{2E}{c}\left(\frac{2}{1-q}\right), \tag{6}$$

where q is plasma reflection coefficient. For example, if $q = 0.98$ the radiation pressure increases by 100 times. We neglect losses of prism reflection.

7. The Bremsstrahlung (brake) loss energy of plasma by radiation is ($T > 10^6$ °K)

$$P_{Br} = 5.34\cdot10^{-37} n_e^2 T^{0.5} Z_{eff}, \quad \text{where} \quad Z_{eff} = \sum(Z^2 n_z)/n_e, \tag{7}$$

where P_{Br} is power of Bremsstrahlung radiation, W/m^2; n_e is number of particles in m^3; T is a plasma temperature, KeV; Z is charge state; Z_{eff} is cross-section coefficient for multi-charges ions. For reactions H+D, D+T the Z_{eff} equals 1.

Losses may be very high. For some reactions, they are more then useful nuclear energy and fusion nuclear reaction may be stopped. The Bremsstrahlung emission has continuous spectra.

8. Electron frequency in plasma is

$$\omega_{pe} = \left(\frac{4\pi n_e e^2}{m_e}\right)^{1/4}, \quad \text{or} \quad \omega_{pe} = 5.64\times10^4 (n_e)^{1/4}, \tag{8}$$

in "cgs" units, or $\omega_{pe} = 56.4(n)^{1/4}$ in CI units

where ω_{pe} is electron frequency, rad/s; n_{e} is electron density, [1/cm^3]; n is electron density, [1/m^3]; $m_{\text{e}} = 9.11\times10^{-28}$ is mass of electron, g; $e = 1.6\times10^{-19}$ is electron charge, C.

The plasma is reflected an electromagnet radiation if frequency of electromagnet radiation is less then electron frequency in plasma, $\omega < \omega_{\text{pe}}$. That reflectivity is high. For $T > 15\times10^6$ °K it is more than silver and increases with plasma temperature as $T^{3/2}$. The frequency of laser beam and Bremsstrahlung emission are less then electron frequency in plasma.

9. The deep of penetration of outer radiation into plasma is

$$d_p = \frac{c}{\omega_{pe}} = 5.31\cdot10^5 n_e^{-1/2}. \text{ [cm]} \tag{9}$$

For plasma density $n_e = 10^{22}$ 1/cm^3 $d_p = 5.31\times10^{-6}$ cm.

10. The gas (plasma) dynamic pressure, p_k, is

$$p_k = nk(T_e + T_i) \quad \text{if} \quad T_e = T_k \quad \text{then} \quad p_k = 2nkT\,, \tag{10}$$

where $k = 1.38\times10^{-23}$ is Boltzmann constant; T_e is temperature of electrons, ^{o}K; T_i is temperature of ions, ^{o}K. These temperatures may be different; n is plasma density, 1/m^3; p_k is plasma pressure, N/m^2.

11. The gas (plasma) ion pressure, p, is

$$p = \frac{2}{3} nkT\,, \tag{11}$$

Here n is plasma density in 1/m^3.

12. The magnetic p_m and electrostatic pressure, p_s, are

$$p_m = \frac{B^2}{2\mu_0}, \quad p_s = \frac{1}{2}\varepsilon_0 E_S^2\,, \tag{12}$$

where B is electromagnetic induction, Tesla; $\mu_0 = 4\pi\times10^{-7}$ electromagnetic constant; $\varepsilon_0 = 8.85\times10^{-12}$, F/m, is electrostatic constant; E_S is electrostatic intensity, V/m.

13. Ion thermal velocity is

$$v_{Ti} = \left(\frac{kT_i}{m_i}\right)^{1/2} = 9.79\times10^5 \mu^{-1/2} T_i^{1/2} \quad \text{cm/s}\,, \tag{13}$$

where $\mu = m_i/m_p$, m_i is mass of ion, kg; $m_p = 1.67\times10^{-27}$ is mass of proton, kg.

14. Transverse Spitzer plasma resistance

$$\eta_\perp = 1.03\times10^{-2} Z \ln\Lambda T^{-3/2}, \quad \Omega\ \text{cm} \quad \text{or} \quad \rho \approx \frac{0.1Z}{T^{3/2}} \quad \Omega\ \text{cm}\,, \tag{14}$$

where $\ln\Lambda = 5 \div 15 \approx 10$ is Coulomb logarithm, Z is charge state.

15. Reaction rates <σv> (in cm^3 s^{-1}) averaged over Maxwellian distributions for low energy (T<25 keV) may be represent by

$$\begin{aligned} (\overline{\sigma v})_{DD} &= 2.33\times10^{-14} T^{-2/3} \exp(-18.76 T^{-1/3}) \quad \text{cm}^3\text{s}^{-1}, \\ (\overline{\sigma v})_{DT} &= 3.68\times10^{-12} T^{-2/3} \exp(-19.94 T^{-1/3}) \quad \text{cm}^3\text{s}^{-1}, \end{aligned} \tag{15}$$

where T is measured in keV.

16. The power density released in the form of charged particles is

$$P_{DD} = 3.3\times10^{-13} n_D^2 (\overline{\sigma v})_{DD}, \quad \text{W cm}^{-3}$$
$$P_{DT} = 5.6\times10^{-13} n_D n_T (\overline{\sigma v})_{DT}, \quad \text{W cm}^{-3} \qquad (16)$$
$$P_{DHe^3} = 2.9\times10^{-12} n_D n_{He^3} (\overline{\sigma v})_{DHe^3}, \quad \text{W cm}^{-3}$$

Here in P_{DD} equation it is included D + T reaction.

RESULTS OF COMPUTATION

1. Some thermonuclear reactions. The primary nuclear reaction is D-D reaction that takes place when two nuclei of deuterium collide. Deuterium can be obtained from seawater, its abundance being about 0.0148% that of hydrogen, and used as a fuel resource, this amount can be regarded as almost inexhaustible.

The D-D reaction consists of the following two reactions:

$$\text{D+D} \rightarrow {}^3\text{He} + \text{n} + 3.27 \text{ MeV, } 50\% \qquad (17)$$

$$\text{D+D} \rightarrow \text{T} + \text{H} + 4.03 \text{ MeV. } 50\% \qquad (18)$$

In reaction (17) an isotope of helium (^{3}He) and neutron (n) are produced by the collision of two deuterium nuclei (D). In reaction (2), a tritium (T) and a proton (H) are produced. The numbers on the right-side denote the kinetic energy released by the reaction, which can be calculated us follows: If we denote the mass defect of each particle in the unit of MeV (10^6 eV), we have D: 13.1359 Mev, He: 14.9313 MeV, and n: 8.0714 Mev ([16], p. 1295), so that the energy released by reaction (17) is

$$2 \times 13.1359 - (14.9313 + 8.0714) = 3.2691 = 3.27 \text{ MeV.}$$

For reaction (18), we can use for T: 14.9500 Mev and for H: 7.289 MeV.

The partition of the released energy from the reaction products can be estimated from energy and momentum conservation. Kinetic energy of D before collision is very small compared to the energy released by the reaction. We can ignore the initial kinetic energy and treat the deuterium nuclei as being at rest. Denoting the mass and speed of helium and n by $_1$ and $_2$ respectively, we have for reaction (17)

$$0.5 m_1 v_1^2 + 0.5 m_2 V_2^2 = E = 3.27 \quad \text{MeV}, \quad m_1 v_1 = m_2 v_2, \qquad (19)$$

where in the second formula we assumed that He and n fly out in opposite directions. From these relations, we find

$$E_1 = \frac{1}{2} m_1 v_1^2 = \frac{E}{1 + m_1 / m_2} = 0.82 \quad \text{MeV}, \quad E_2 = \frac{1}{2} m_2 v_2^2 = \frac{E}{1 + m_2 / m_1} = 2.45 \quad \text{MeV}. (20)$$

Obviously, the lighter particle acquires more energy than the heavier particle.

Current nuclear fusion research is focused on the D + T thermonuclear fusion reaction

$$D + T \rightarrow {}^4He(3.5\text{ MeV}) + n\ (14.1\text{ MeV}), \tag{21}$$

Reaction (21) can occur in high-temperature deuterium-tritium plasma. Most energy released by the reaction is converted to the kinetic energy of the neutron. Since the neutron is not confined or reflected by a magnetic or electrostatic field it leaves, going outwards to surrounding space or hits the screen or vessel wall (or blanker) immediately after reaction. In last instance, the neutron kinetic energy is converted to heat. The heat is taken away from the screen by direct radiation or and indirect circulating coolant and can be used to run an electric generator. If we add ^{6}Li inside the blanket, then tritium can be produced by reaction

$$n + {}^6Li \rightarrow {}^4He\ (2.1\text{ MeV}) + T\ (2.7\text{ MeV}) \tag{22}$$

and then used as the fuel. Another reaction product is the alpha particles ^{4}He carrying 3.5 MeV which can be directed or confined by electro-magnetic field.

The reaction that produces only charged particles are best for the proposed propulsion system and generator. Unfortunately, these reactions are not great (see Table 1).

However, since the nuclei are positively charged, they must have enough energy to overcome the Coulomb repulsion between them, in order for a few of them to be able to combine. The required energy can be estimated by equation (1).

2. *Rocket Impulse.* When we know the energy of the thermonuclear particles, the particle speed can be calculated by equation

$$V_i = 1.384\times10^4 (E/N_i)^{0.5}, \tag{23}$$

where E is particle energy in MeV, N is number of nucleons in particle (in mass units, for example, $N = 1$ for proton, neutron, $N = 2$ for deuterium, $N = 3$ for tritium, $N = 4$ for helium).

Conventionally, we have two components on the equation's right-side having different mass and speed. The average efficiency particle speed V (impulse) of thermonuclear reaction may be estimated by equation

$$V = \eta_1 \frac{N_1}{N} V_1 + \eta_2 \frac{N_2}{N} V_2. \tag{24}$$

where lower index "$_i$" is number of particle, η is coefficient utilization of kinetic particle energy, N is total number of nucleons in single reaction. For particles adsorbed by screen (Figure 1*a*) $\eta = 0.25$, for particles reflected by reflector (Figure 1*b*) $\eta = 1$.

When we use the radiation (photon) energy of one hot side of screen, the efficiency particle speed (24) has additional member

$$\Delta V = \eta_3 \frac{E_J}{m_p c N}, \tag{25}$$

where m_p =1.67495×10^{-27} kg is mass of neutron, c = 3×10^8 m/s is the light speed, E_J=1.6×10^{-19}E energy of particle in J, η_3 = 0.25 is coefficient utilization of heat energy.

The apparatus speed is

$$V_m = -V \ln \frac{M_f}{M_0}, \tag{26}$$

where V_m is maximum speed of space apparatus, m/s; M_f /M_0 is ratio of a final mass (apparatus without thermonuclear fuel) to the initial apparatus mass.

Results of computation are presented in Table 1.

The *first* column shows the thermonuclear reaction. In left side from pointer it is shown the components of thermonuclear reactor fuel. In the right-side it is shown the particles which appear in the reaction and kinetic energy every particle in MeV.

The *second* column shows the efficiency impulse (in m/s) computed by equation (18) for *AB-screen* engine.

The *third* column shows the maximum speed which apparatus (equipped with an engine screen) reaches a fuel mass ratio equal to 0.1 (equation (20)).

The *fourth* and *fifth* column shows the efficiency impulse and maximum apparatus speed for *AB-reflector* engine.

Table 1. Reaction for thermonuclear propulsion

Type of propulsion→ Thermonuclear reaction, MeV ↓	AB-screen, Impulse from charged particles ×10^6 m/s	AB-screen. Max speed of apparatus for M_f /M_0=0.1 Speed×10^{6}m/s	AB-reflector. Impulse from charged particles ×10^6 m/s	AB-reflector. Max speed of apparatus for M_f /M_0=0.1 Speed×10^{6}m/s	Mass ratio M_f /M_0 for fuel=0	+ Photon. Add speed ×10^6 m/s
1	2	3	4	5	6	7
D+T→^{4}He(3.5)+n(14.1)	5.18	8.13	10.3	23.8	0.19	0.23
T+T→^{4}He(3.77)+2n(7.53)	4.48	7.03	8.96	20.6	0.25	0.15
D+^{3}He→^{4}He(3.6)+p(14.7)	5.28	8.29	21.1	48.5	0.1	-
D+^{6}Li→2^{4}He(22.4)	5.8	9.11	23.6	54.3	0.1	-
^{3}He+^{3}He→^{4}He(4.3)+2p(8.6)	5	7.85	20	46	0.1	-
^{3}He+^{6}Li→2^{4}He(1.9)+p(16.9)	3	4.47	12	27.6	0.1	-
p+^{11}B→3^{4}He(8.7)	2.95	4.63	11.78	27.1	0.1	-

Here are: D - deuterium, T - tritium, He - helium, Li - lithium, B - boron, n - neutron, p - proton.

The *sixth* column shows the final mass of the space apparatus when the mass of initial fuel equals zero. In the first two reactions this ratio is not 0.1 because the neutrons are adsorbed by the screen. That decreases the apparatus speed, but the neutrons can be harnessed to get additional fuel and to sustain other useful and valuable thermonuclear reaction. All computation in Table 1 is made for mass ratio 0.1.

The last (*seventh*) column shows the additional speed from hot one-sided screen emitting photons. This additional speed is small.

The thrust of the spherical thruster-generator (Figure 1*d*) for small angles may be computed as for AB-screen engine. Thrust is proportion the ratio of the open area to the full sphere surface.

The power of electric generator may be estimated by equation

$$W = 0.5 \times 10^{14} \eta \frac{N_p}{N} m_f E , \qquad (27)$$

where W is power, W; N_p is number of the charges (protons) nucleons; N is total number of nucleons; E is energy of charged particles, MeV; m_f is fuel consumption, kg/s. η is coefficient efficiency.

The trust propulsion is

$$T = m_f V , \qquad (28)$$

where T is trust, N.

The relative mass of fuel converted to energy and thrust is

$$\Delta \overline{m} = \frac{E}{938N} , \qquad (29)$$

where E is reaction energy in MeV.

For example, let us take the reaction D+T and fuel consumption $m_f = 10^{-5}$ kg/s. Then N_p $/N$ = 4/5, <u>E</u> = 3.5 MeV and $W \approx 1.34\times10^6$ kW; thrust $T \approx 100$ N; the relative mass converted into energy is 3.75×10^{-3}. If the fuel consumption is $m_f = 10^{-2}$ kg/s, the thrust is $T \approx 10^5$ N. The energy is gigantic $W \approx 1.34\times10^9$ kW.

Table 1 shows that the offered thermonuclear AB-propulsions can accelerate the space apparatus up the speed $(20 \div 50)\times10^6$ m/s (or up 1/6 of light speed) with a fuel ratio of M_f/M_0 = 0.1. The AB-propulsion is the most efficient of all thermonuclear propulsions, capable of reaching the theoretic maximum impulse from currently known thermonuclear propulsions and known thermonuclear reactions.

Please note that the reaction that produces only charged particles is more efficient than reactions producing neutrons and charged particles (two in the first lines of Table 1). The neutrons accept a lot of a common energy, but this energy does not produce any thrust. Converting this energy into photons (column 7 of table 1) is also ineffective. The neutrons do not leave the space apparatus, increasing its final launch and travel weight (column 6) and decreasing the final apparatus speed. They may be used for next reaction (see (22) and below), but technical realization of such reaction is decidedly complex and presently unproductive. See some of these reactions below. Unfortunately, most neutron reactions are exoteric.

3. The thermonuclear reactions used the slow neutrons:

Table 2.

Reaction	Energy, MeV	Cross section, burns
$^{3}He + n \rightarrow {}^{3}H + p$	+0.764	5400
$^{6}Li + n \rightarrow {}^{3}H + \alpha$	+4.785	945
$^{7}Be + n \rightarrow {}^{7}Li + p$	+1.65	51000
$^{10}B + n \rightarrow {}^{7}Li + \alpha$	+2.791	3837
$^{14}N + n \rightarrow {}^{14}C + p$	+0.626	1.75
$^{17}O + n \rightarrow {}^{14}C + \alpha$	+1.72	0.5
$^{33}S + n \rightarrow {}^{33}P + p$	+0.75	0.002
$^{35}Cl + n \rightarrow {}^{35}H + p$	+0.62	0.3

The spherical AB-engine can produce much electrical energy, but conversion of this energy into vehicular thrust by common electric (ion) propulsion is inefficient in comparison with the offered AB-engine.

DISCUSSION

The potential space traveling apparatus speed 1/6 of light speed is maximum velocity predicted by thermonuclear AB-propulsion. That speed allows Mars to be a destination in minutes (or some days when apparatus has limited acceleration); that very high speed allows short period trips throughout our Solar system. However, it is not sufficient for easy interstellar space trips. The nearest star system is located at a distance of 3 - 5 light-years. That means the trip requires a minimum of 40 - 60 years. But required fuel ratio M_f/M_0 is very high: acceleration and braking of moving apparatus needs 4-stage rocket having the ratio for every stage $M_f/M_0 = 0.1$. The total weight fuel ratio will be $M_f/M_0 = 10^{-4}$. If useful weight is 10 tonnes, the starting rocket mass is $M_0/M_f = 10^4$ tonnes. The relative mass of thermonuclear reaction converted into energy (and thrust) is only 0.3 ÷ 0.4% of total fuel mass. The author's research so far shows that the magnet cannot absorb the big amount of interstellar matter in the high apparatus speed mode; consequently, the envisioned apparatus must take fuel for the entire trip.

Human interstellar flight is very expensive and complex. We can develop long-distance communication system and send, instead, E-men [18, 19] or artificial intelligent robot.

However, only an annihilation reaction can efficiently solve the interstellar trip macro-problem. Otherwise, new physics discoveries that allow such trips are required.

Conclusion

The authors suggests the simplest maximally efficient thermonuclear AB-propulsion (and electric generators) based in the early offered size-minimized Micro-AB-thermonuclear reactor [15]. These engines directly convert high-speed charged particles produced in thermonuclear reactor into vehicular thrust or onboard vehicle electricity resource. Offered propulsion system allows travel to any of our Solar System's planets in a short time as well as trips to the nearest stars by E-being or intellectual robot in during a single human life [18]-[20].

References

(See some of the below articles in *http://Bolonkin.narod.ru/p65.htm* and *http://arxiv.org* search "Bolonkin", in books "*Non-Rocket Space Launch and Flight*". Elsevier, 2006. "*New Concepts. Ideas, Innovations in Technology and Human Science*", NOVA, 2008).

[1] Bolonkin A.A., Beam Space Propulsion, *Presented as paper AIAA-2006-7492 to Conference "Space-2006",* 19-21 September, 2006, San-Jose, CA, USA. Published in *http://Arxiv.org*.

[2] Bolonkin A.A., Electrostatic AB-Ramjet Space Propulsion, *Presented as paper AIAA-2006-6173 to AIAA/AAS Astrodynamics Specialist Conference,* 21-24 August 2006, USA. Published in *http://Arxiv.org*. Search: Bolonkin.

[3] Bolonkin A.A., Theory of Space Magnetic Sail Some Common Mistakes and Electrostatic MagSail, *Presented as paper AIAA-2006-8148 to 14th AIAA/AHI Space Planes and Hypersonic Systems and Technologies Conference,* 6 - 9 Nov 2006 National Convention Centre, Canberra, Australia. Published in *http://Arxiv.org*.

[4] Bolonkin A.A., High Speed AB-Solar Sail, *This work is presented as paper AIAA-2006-4806 for 42 Joint Propulsion Conference*, Sacramento, USA, 9-12 July, 2006. Published in *http://Arxiv.org*.

[5] Bolonkin A.A., Electrostatic Utilization of Asteroids as Propulsion System. *Presented as paper AIAA-2005-4032 at the 41 Propulsion Conference,* 10–12 July 2005, Tucson, Arizona, USA.

[6] Bolonkin A.A., Electrostatic Solar Sail. Book "Non-Rocket Space Launch and Flight", Elsevier, 2006, Ch.18, pp. 317-326.

[7] Bolonkin A.A., Electronic Solar Sail. Book "Non-Rocket Space Launch and Flight", Elsevier, 2006, Ch.19, pp. 334-335.

[8] Bolonkin A.A., Multi-reflex Propulsion System for Space and Air Vehicles and Energy Transfer for Long Distance. Book "Non-Rocket Space Launch and Flight", Elsevier, 2006, Ch.12, pp. 223-244. See also: JBIS, 2004, vol.57, No. 11/12, pp.379-390.

[9] Bolonkin A.A., Electrostatic Solar Wind Propulsion. Presented as paper AIAA-2005-3857 at the *41st Propulsion Conference*, 10-12 July 2005, Tucson, AZ, USA. See also book "Non-Rocket Space Launch and Flight", Elsevier, 2006, Ch.13, pp. 245-270.

[10] Bolonkin A.A., Electrostatic Utilization of asteroids for space flight. Presented as paper AIAA-2005-4032 at the *41st Propulsion Conference*, 10-12 July 2005, Tucson, AZ,

USA. See also book "Non-Rocket Space Launch and Flight", Elsevier, 2006, Ch.14, pp. 271-280.

[11] Bolonkin A.A., Guided solar sail and energy generator. Presented as paper AIAA-2005-3857 at the *41st Propulsion Conference*, 10-12 July 2005, Tucson, AZ, USA. See also book "*Non-Rocket Space Launch and Flight*", Elsevier, 2006, Ch.15, pp. 302-308.

[12] Bolonkin A.A., Radioisotope Space Sail and Electr0-Generator. Presented as paper AIAA-2005-4225 at the *41st Propulsion Conference*, 10-12 July 2005, Tucson, AZ, USA. See also book "Non-Rocket Space Launch and Flight", Elsevier, 2006, Ch.17, pp. 309-316.

[13] Bolonkin A.A., Wireless Transfer of Electricity in Outer Space. *Presented as paper AIAA-2007-0590 to 45th AIAA Aerospace Science Meeting,* 8 - 11 January 2007, Reno, Nevada, USA. Published in *http://Arxiv.org.*

[14] Bolonkin A.A., *Non-Rocket Space Launch and flight, Elsevier,* 2006, *http://Bolonkin.narod.ru/p65.htm.*

[15] Bolonkin A.A., Micro-Thermonuclear AB-Reactors for Aerospace. *Presented as paper AIAA-2006-8104 in 14th Space Plane and Hypersonic Systems Conference,* 6-8 November, 2006, USA; Published in *http://arxiv.org.*

[16] Nishikawa K., Wakatani M., Plasma Physics, Spring, 2000.

[17] *Handbook of Physical Quantities,* Ed. Igor S. Grigoriev, 1997, CRC Press, USA.

[18] Bolonkin A.A., Breakthrough to Immortality, 2004, *http://Bolonkin.narod.ru.*

[19] Bolonkin A.A., *Humanity and Space Civilization,* 2000, *http://Bolonkin.narod.ru/p101.htm.*

[20] Bolonkin A.A., New concepts, ideas and innovations in aerospace and technology, Nova, 2008.

[21] Bolonkin A.A., *Human Immortality and Electronic Civilization,* Lulu, 2006. 3-rd Edition. *http://lulu.com* serach “Bolonkin”.

[22] *Wikipedia.* Some background material in this article is gathered from Wikipedia under the Creative Commons license.

Chapter 8

PROTECTION OF CITIES FROM SMALL ROCKETS, MISSILES, PROJECTILES, AND MORTAR SHELLS*

ABSTRACT

The authors suggest a low cost closed AB-Dome, which may protect small cities such as Sederot from rockets, mortar shells, chemical and biological weapons. The offered AB-Dome is also very useful in peacetime because it protects the city from outside weather (violent storms, hail) and creates a fine climate within the Dome. The roughly hemispherical AB-Dome is a gigantic inflated thin transparent film, located at altitude up to 1 - 5 kilometers, which converts the city into a closed-loop air system. The film may be armored with a basalt or steel grille or cloth pocket-retained stones that destroy (by collision or detonation) incoming rockets, shells and other projectiles. Such an AB-Dome would even protect the city in case of a third-party nuclear war involving temporary poisoning of the Earth atmosphere by radioactive dust. The building of the offered dome is easy; the film spreads on the ground, the fan engines turn on and the cover rises to the needed altitude and is supported there by a small internal overpressure.

The offered method is cheaper by thousands of times than protection of a city by current anti-rocket systems. The AB-Dome may be also used (height is up to 1-5 and more kilometers) for TV, communication, long distance location, tourism, suspended high speed and altitude windmills (energy), illumination and entertainment (projected messages and pictures).

The authors developed the theory of AB-Dome, made estimations, computations and computed a typical project. Discussion and results are at the end of the article.

Keywords: Protection from missile and projectile weapons, protection from chemical, biological weapon, inflatable structures, local weather control, mortar, rocket, Qassam defense, defense, isolation of GM crops, crop protection, pesticide free crops, biological isolation

---

* In this chapter the work of A. Bolonkin and J. Friedlander presented in http://Arxiv.org (2007) is used.

INTRODUCTION

One important problem in small countries with hostile borders (or larger countries with leaky borders open to inflitrators) is protection of fixed and mobile domestic targets from small rockets, missiles and mortar shells. For well over a hundred years there has been no satisfactory solution for this, which is why such weapons are favorites of guerilla groups. Israel, for example, has villages (Alumim, and dozens of others) towns (Sderot, Netivot and Qiryat Shemona), cities such as Ashqelon and numerous installations near unfriendly borders. For example, Sderot lies just one kilometer from the Gaza Strip and the town of Beit Hanoun. Since the beginning of the Second Intifada in October 2000, Sederot has been under constant rocket fire from Qassam rockets launched by various armed factions. Despite the imperfect aim of these homemade projectiles, they have caused deaths and injuries, as well as significant damage to homes and property, psychological distress and emigration from the city. Real estate values have fallen by about half. The Israeli government has installed a "Red Dawn" alarm system to warn citizens of impending rocket attacks, although its effectiveness has been questioned. Thousands of infamous Qassam rockets have been launched since Israel's disengagement from the Gaza Strip in September 2005, which essentially has killed popular support for any further withdrawals, particularly from West Bank areas near the heart of the country. Even pro 'land for peace' parties have no plausible answer to the question (in its various forms), "And what will you do when they start shooting at (the) Tel Aviv (stock exchange) and the (Ben Gurion International) airport?" Thus, even with perhaps 70 of 120 votes in the Israeli Parliament (Knesset) in favor of negotiations with the Palestinian Authority, no final 'end of conflict' agreement is seriously in prospect, nor is likely to be given the impossibility of guaranteeing a stoppage of the bombardments by *some* unsatisfied Palestinian faction—even one not yet on the scene.

In May 2007, a significant increase in shelling from Gaza prompted the temporary evacuation of thousands of residents. By November 23, 2007, 6,311 rockets had fallen on the city. The Israeli newspaper Yediot Aharonot reported that during the summer of 2007, 3,000 of the city's 22,000 residents (comprised mostly of the city's key upper and middle class residents, the heart of the economy, those most able to move,) had already left for other areas, out of Qassam rocket range. Russian-Israeli billionaire Arkady Gaydamak has in recent years supported relief programs for residents who cannot leave. [5]

On December 12, 2007, on a day during which more than *20* rockets landed in the Sderot area, including a direct hit to one of the main avenues, the mayor of Sderot, Eli Moyal (a well-known figure in Israeli media) unexpectedly announced his resignation from the job, citing the government's failure to stop the daily rocket attacks. "Maybe this will spark the government to launch an operation for the lives of its [Sderot's] residents. I can't keep making the decisions, they can't keep piling it all on me," Mr. Moyal reportedly said. Later under political pressure, he was asked to resume his job by key national leaders, as a duty rather than a choice.

Qassams were first fired at Israeli civilian targets in October 2001.The first Qassam to land in Israeli territory was launched on February 10, 2002. The first time an Israeli city was hit was on March 5, 2002, when two rockets struck Sderot. Some rockets have hit as far as the edge of Ashkelon. The total number of Qassam rockets launched exceeded 1000 by June 9, 2006. During the year 2006 alone, 1000+ rockets were launched. Tons of explosives have been intercepted at the Egyptian border; the uninterrupted shipments must be greater still, and the cumulative detonation yield has easily been in the tens of tons.

The introduction of the Qassam rocket took Israeli politicians and military experts by surprise. Reactions have been mixed. The Israeli Ministry of Defense views the Qassams as "more a psychological than physical threat." The rockets are fired largely at civilian populations. There is some evidence of psychological damage to children in the effected areas, particularly in Sderot. The IDF has reacted to the deployment of the Qassam rockets by deploying the Red Color early warning system in Sderot, Ashkelon and other at-risk targets. The system consists of advanced radar that detects rockets as they are being launched. Loudspeakers warn civilians to take cover approximately twenty seconds before impact in an attempt to minimize the threat posed by the rockets.

A rocket once fell into the electricity station in Ashkelon and caused electricity shortages in several areas, other time a rocket-similar to Qassam- fell inside an army base and injured more than 70 Israeli soldiers. The Ashkelon strike in particular was troubling as it added (by its radius) another 250,000 Israelis to the potential target list requiring defenses to be paid for, active or passive.

Some military bases of the USA in Afghanistan and Iraq (or in various parts of Asia and Latin America) are in the same situation. Any security consultant working to protect valuable installations in the more volatile corners of Africa, Latin America or Asia will recognize the dangers in the scenarios listed above. Rockets, and remotely triggered mortars, are man-portable and can be smuggled in, can be covertly emplaced and remotely fired with no appreciable warning, and endanger billions in investment with mere thousands in expenses. In Gaza, bonuses are allegedly given to impoverished children to retrieve the launchers, to reduce the expenses of replacing both rounds and launchers.

Israel Government Plan

(As reported at www.haaretz.com 24/12/07 http://www.haaretz.com/hasen/spages/937756.html), the plan of the Israeli government was to fund an anti-missile system, but not to reinforce the homes of all Sderot-area residents, which caused considerable local anger.

Many poorer residents don't have a reinforced secure room they can retire to when the alarm sounds, with metal in the walls, proof against shrapnel.

The Government's position was that the promise to armor all the homes was made before it was understood that the anti-rocket system would cost the same 1.5 billion NIS (New Israeli Shekels) that reinforcing all the homes would. Critics object that the anti-rocket system may fail in at least a fraction of the cases to work, at which point those on the ground without a passive defense (a safe room) would be in mortal danger.

The currently offered RAFAEL anti-rocket system is shown below:

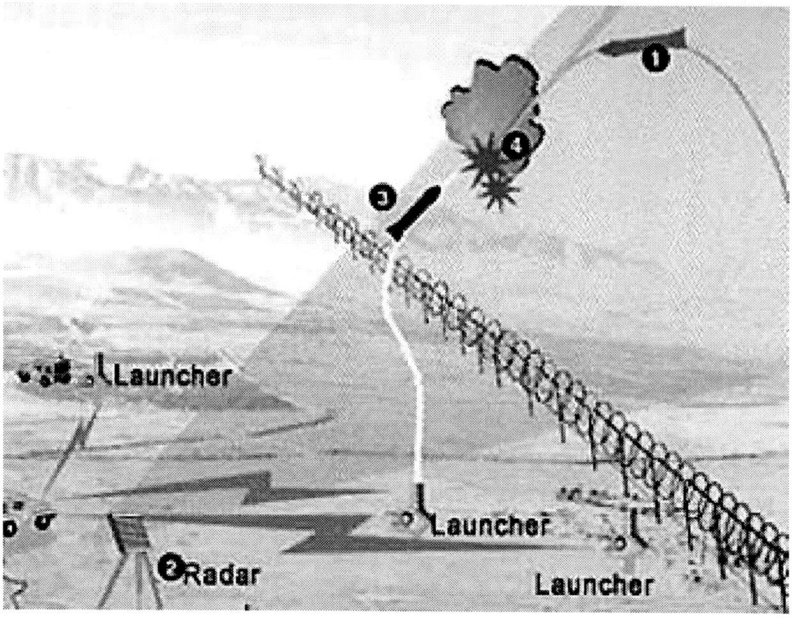

The RAFAEL anti-rocket system (Iron Dome) costs about 1 billion American dollars. It may be ready in 2011. Efficiency of operational system remains unknown.

The other system, such as "David's Sling" or C-RAM from Raytheon have perhaps 70 – 80% efficiency. That means every third to fifth missile would reach the target. From an investor's standpoint, this would be little better than unchecked bombardament; capital would still flee the targeted city. What is needed is a defense so thorough that the residents are *entirely unaware* of the bombardament other than possibly distant flashes and evening news summaries. Any difference in daily life that a bombardament causes ultimately limits the ability to conduct business as usual (with no risk premium). Anything less than this standard still makes the inhabitants feel under seige.

Quassam Rockets

Table 1. Specifications of several types of Qassam (home workshop class) rockets

	Qassam 1	Qassam 2	Qassam 3
Length (cm)	79	180	299+
Diameter (cm)	6	15	17
Weight (kg)	5.5	32	90
Explosives Payload kg)	0.5	5 - 7	10
Maximum range (km)	3	8-10	10

Qassam (Kassam) rocket at launch.

The Qassam rockets are tens (hundreds) times cheaper than the complex electronic anti-rockets and their delicate support system. The mortar shells are cheaper by hundreds of times. Attempted defense from them by conventional anti-rocket system may ruin any rich country, and in the end not work anyway, because the enemy always has the option of using large salvos, to probe the point where the system collapses trying to defeat X simultaneous launches.

Basalt Rock for Use as Blocking Grid

Basalt is volcanic magma (effusive rock) widely distributed cheap material in the Earth, One melts the temperature about 1500 K (1200 C), may be cast in grille and has good strength. Basalt has an extraordinarily low viscosity which is necessary for superior basalt castings. The mechanical properties of basalt is comparable to those of cast iron and many fine steels, and superior to aluminum, brass, bronze, and copper both in compression and shear strengths. Basalt is good matter for grills because it is very cheap (about $10-20 per ton) and may be cast in the shape of the needed grille. The basalt may be reinforced and secured in place by cheap basalt fiber having a very high tensile strength. A list of the properties of cast basalt is collected in Table 2. (This data is derived from a U.S. Government publication)

Description of AB-Dome and Innovations

The authors offer a new protection system against warheads from kilogram range all the way up to Hiroshima-yield nuclear weapons. That is the AB-Dome as described in works [1]-[4].

Table 2. Properties Of Cast Basalt

Physical properties	Average numerical value, MKS units
Density of solid	2900-2960 kg/m^3
Tensile strength	$3.5X10^7$ N/m^2
Compressive strength	$5.4X10^8$ N/m^2
Bending strength	$4.5X10^7$ N/m^2
Modulus of elasticity (Young's modulus)	$1.1X10^{11}$ N/m^2
Moh's hardness	8.5
Grinding hardness	$2.2X10^5$ m^2/m^3
Melting point	1400-1600 K
Hygroscopicity	0.1%

We can also use the plastic armour contains 50% clean granite, 43% limostone mineral and 7% bitumen. This cheap armour is successfully used in war W11 as 12 mm protection against aircraft guns in merchant ships.

The idea is a dome covering a city by a thin transparent film 2 (Figure 1). The film has thickness of 0.05 – 0.3 mm. This may be located at high altitude (0.2 – 0.5 kilometers—and if defense against Qassam rockets weapons is desired, the higher figure is definitely desired, and may even be increased). The film is supported at this altitude by a small additional air pressure produced by ground ventilators. That may be connected to Earth's ground by managing cables 5. The film may have a controlled transparency (option). The system may also have a second lower film 6, possibly in the form of a mesh, which protects the city from fragments of rockets (option). Cluster munitions, which may be, depending on the model, the size of soda cans, cell phones, baseballs, or D cell batteries, may have their outer carrier detonated on the outside of the dome and the hundreds of submunitions may roll down the exterior of the dome until their trailing bomblet tail, spinning, unscrews the firing pin and detonates each bomblet, kilometers distant from the city below. Options are listed below, including a mesh, for catching these small fragments.

Even if not caught, the fragments impact at mere terminal velocity, not their initial speed (perhaps 1000-1500 m/s a few meters from the detonation point.)

Terminal velocity on a raindrop of 1.5 mm radius is given at http://www.

gantless.com/paper.html as about 7 meters a second. Thus a small iron fragment should sting as Macklam, Janos et.al there conclude, or cut, or worse, but not do terrible damage in most cases to unshielded humans, certainly nothing in comparison to the horrific effect of a shredding cluster bomblet at ground level without an AB-Dome.

(The minimum 3-5% of submunitions that do not detonate would be found around the perimeter of the dome at ground level, concentrated for easy collection and disposal, instead of scattered cunningly, hanging from trees, inside barn windows, in ground crevasses, and other dangerous spots. In addition, there is the distinct possibility that the non-rigid nature of the AB-Dome would deform and rebound from the explosion, leaving much of the shrapnel on the outside. This possibility warrants further modeling.)

Means of repair are available to close the hole at the top armored film resulting from the exploded rocket without exposing the entire city to further attack.

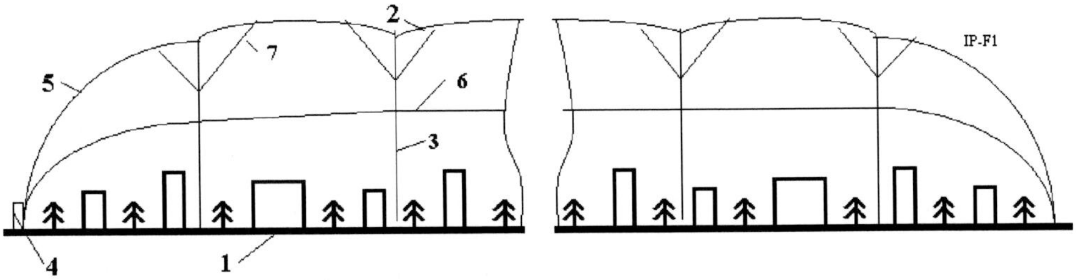

Figure 1. Film AB-Dome for given city. *Notations*: 1 - area, 2 - thin film cover with variable clarity (option), 3 – control support cable length (height) is 0.2 – 5 kilometers), 4 - exits and ventilators, 5 - border section (as opposed to interior section—part of the AB-Dome's periphery, requiring stouter construction), 6 – the second (lower) controlled reflectivity film cover (option), 7 – additional support cables.

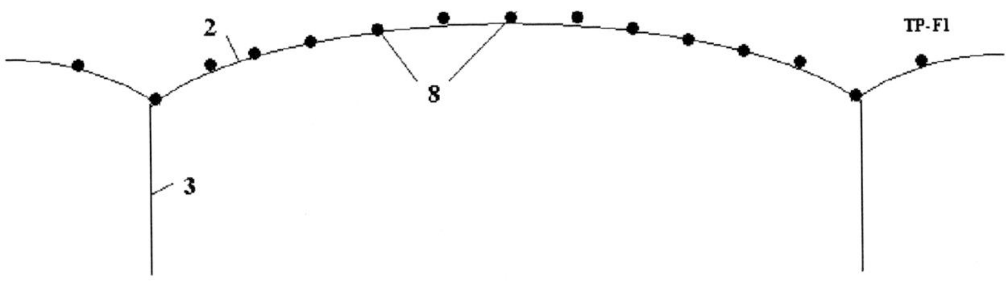

Figure 2. The top film is armored by basalt grille or bonded stones (possibly inside cloth retention pockets) or suspended basalt mesh 8.

Even a small additional overpressure creates a significant lift force. For example, the small overpressure of only only $p = 0.01$ atmosphere produces a lift force of about 100 kg/m^2. At altitude of $H = 5$ kilometers, owing to the lesser density of the air at that height, this force is more like 53 kg/m^2. The support cable has a weight of about 1-3 kg/m^2, the 1 m^2 of film weighs less than 0.05 - 0.5 kg (for example, kevlar film of thickness ~0.2 mm has the weight of 0.3 kg/m^2). *That means that every square meter of dome can keep at altitude a useful load from 50 - 95 kg*. At high altitude the useful load decreases, but if it is needed, we can increase the overpressure. As the fans are at ground altitude for easy maintenance, this is no technical problem (though an additional operating expense).

The top film may support, for example, a basalt grating 8 (Figure 2) on the upper surface. If the grille has a mesh of rod 1×2.5 cm of cross-section and a step (grid) distance of 10 cm, its weight will be only 16 kg/m^2. We can take more cross-section of rod (1.5×3 cm) but more step (up to 15 cm). The weight of grille will be 18 kg/m^2. The Dome cover can support 200 – 300 kg/m^2 and we can take more strong grilles and armed them by strong artificial fibers.

War time. The offered defense provides protection in the following way. The Qassam or other projectiles have a minimum diameter of 10-17 cm. They, by virtue of their minimum size, must strike the stones or basalt grille, are kinetically destroyed or their trigger is detonated from the shock. The city benefits from the saved lives of its' residents, and gets minimal damage in comparison with what might have occurred. (Even this is greatly reduced if a second optional mesh catches the fragments, which might rain down from an explosion at height. This mesh can be far less sturdy and far tighter than the detonation mesh of steel or basalt above; the object of this second layer, perhaps of Kevlar net, is to catch raining shrapnel as small as a few millimeters. (Even though mostly mesh, this layer can still provide lift by having thin film under parts of the mesh, as little as 10% in a criss-cross pattern; this provides a barrier to the upwelling air aimed at the higher dome and it balloons upwards. These strips should be at least tens of meters across to afford a sizeable stagnation front to the upwelling air.))

An optional third level of coverage would catch down-sifting dust such as anthrax spores from bomblets designed (in the inevitable measure-countermeasure wars) to pierce the outer layer and explode in the breach.

In *peace time* the offered dome produce a fine warm climate (weather) in a covered city. Certain layers of the AB-Dome's films can have a controlled transparency/reflectivity. That allows provision of different solar heating conditions in the city. These gigantic covers are composed of a cheap film, ideally having liquid crystal and conducting layers. The clarity of such may be controlled by electric voltage. They can pass or blockade the solar light (or parts of solar spectrum) and pass or blockade the Earth radiation. The outer and inside radiations have different wavelengths. That makes it possible to control of them separately and to control heating into (and re-radiation from) the Earth's surface. In conventional conditions about 50% of the solar energy reaches the Earth surface. The most part is reflected back to outer space by the white clouds.

In our closed system the clouds (and rain, or at least condensation based dripping) will occur at night when the temperature is low. That means that many cold regions (Alaska, Greenland, Siberia, Northern Antarctica) may absorb more net solar energy and become, within the bubbles, lands with a temperate or sub-tropic climate. That also means the Sahara desert (or locally in Israel, the Dead Sea, Jordan Valley and Negev regions) can be a prosperous area with a fine climate and with a closed-loop water cycle.

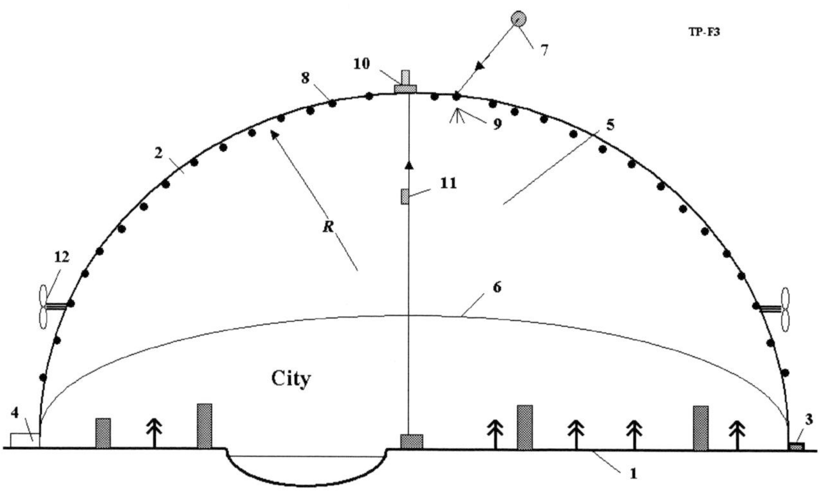

Figure 3. Spherical AB-Dome which may provide defense for a given city against rockets, missiles, aviation, chemical and biological weapons. *Notations*: 1 – protected area; 2 – thin film; 3 – ventilator (air pump); 4 – exit; 5 –spherical thin film AB-Dome; 6 – controlled reflectivity thin film (optional) fragment catcher inner layer (optional); 7 –incoming warhead, rocket, missile, bomb or weapon; 8 – strong grille supported by dome; 9 – fragments of destroyed warhead, rocket, missile or aircraft; 10 – TV, communication, telescope, locator emplacements, tourist observation deck; 11 – elevator; 12 – windmills.

The building of a film dome is rather easily accomplished. Do not think of the popular delusion of a science fiction dome city made of impervious thick crystal, which would be a huge construction macro-project. We simply spread out the film over Earth's surface, turn on the pumping fans and the film is raised by slight air over-pressures to needed altitudes, limited only by the support cables or air-inflated fabric ribs.

In case of obstacles in the terrain of the city itself, large areas of film can be readied on a relatively calm day under cable tension at one end of a city, on the edge in the fields, levitated by an overpressure under that film, then pulled across the city, high over head, by towing cables, then locked secure on the far side before final inflation.

Damage to the film is not particularly troublesome because the additional air over-pressure is very small and propeller pumps compensate for any air leakage. Unlike a space colony or planetary colony, the outside air is friendly and at worst we lose some heat (or cold) and water vapor.

The other advantages of the suggested method include the possibility to paint pictures on the sky (AB-Dome), to show films on the sky by projector inside the facility, to suspend illuminations, decorations, and air tramways and any other utilities and conveniences (and macro-engineering works) from this new lite-weight over-roof.

Long distance aircraft fly at altitudes between 8 – 11 kilometers and our AB-Dome (1- 5 kilometers or less) does not trouble them unless the AB-Dome is built on the edge of an airport's glide path! The support cables will have safety illumination lights (red, flashing, in a bead-like string) and internal helicopters will take normal flying precautions in avoiding physical contact with them.

More details on the offered AB-Dome are described in [1]-[4], [13]. Additional information is repeated below.

Our design for the AB-Dome is presented in Figures 1-3, which illustrates the thin inflated film dome. The *innovations* are listed here: (1) the construction is air-inflatable; (2) each AB-Dome is fabricated with very thin, transparent film (thickness is 0.05 to 0.3 mm, implying under 150-500 tons a square kilometer) having the control clarity quality without rigid supports; (3) the enclosing film can have (optionally) two conductive layers plus a liquid crystal layer between them which changes its clarity, color and reflectivity under an electric voltage; (4) the boundary section of the AB-Dome has a rounded form. The air pressure is more in these sections and they protect the central sections from the outer wind.

Figures 1-3 illustrate the thin transparent control dome cover we envision. The inflated textile shell—technical "textiles" can be woven or non-woven (films)—embodies the innovations listed: (1) the film is very thin, approximately 0.1 to 0.3 mm. A film this thin has never before been used in a major building; (2) the film has two strong nets, with a mesh of about 0.1 × 0.1 m and $a = 1 \times 1$ m, the threads are about 0.5 mm for a small mesh and about 1 mm for a big mesh. The net prevents the watertight and airtight film covering from being damaged by vibration; (3) the film incorporates a tiny electrically conductive wire net with a mesh about 0.1 x 0.1 m and a line width of about 100 μ (microns, thousandths of a millimeter) and a thickness near 10 μ (microns). The wire net is an electric (voltage) conductor. It can inform the dome supervisors concerning the place and size of film damage (tears, rips, etc.); (4) the film may be twin-layered with the gap — $c = 1$ m and $b = 2$ m— between covering's layers for heat saving. In polar regions this multi-layered low height covering is the main means for heat insulation and puncture of one of the layers won't cause a loss of shape because the film's second layer is unaffected by holing; (5) the airspace in the AB-Dome's covering can be partitioned, either hermetically or not; and (6) part of the covering can have a very thin shiny aluminum coating that is about 1μ (micron) thick for reflection of unnecessary solar radiation in equatorial or polar regions (without the liquid crystal layer, this aluminizing option is definitely recommended for Israeli use.—it can be preferentially deposited on sections that will receive insolation at higher summer angles, but not on those that will receive it at lower winter angles, just as old-fashioned awnings used to do to keep apartment windows in an acceptable heating range. In case of a near nuclear airburst, this would help reflect damaging thermal rays.) [1]-[4].

Theory and Computation (Estimation) of AB-Dome

a) General Information

Our AB-Dome cover (film) has 2 layers (figs. 1,3): top transparent layer 2, located at a maximum altitude (up to 4 –15 kilometers), and lower transparent layer 4 having controllable reflectivity, located at altitude of 1-3 kilometers (optional). The upper transparent cover has a thickness of about 0.05 – 0.3 mm and supports a network of pockets connected by containing hard stones (pebbles, shards of granite). 8. The stones have a mass 0.2 – 1 kg and located about every 0.5 m (step distance).

Brief Information about the Cover Film

If we want to control temperature in city, the top film must have some layers: transparent dielectric layer, conducting layer (about 1 - 3 μ), liquid crystal layer (about 10 - 100 μ), conducting layer (for example, SnO_2), and transparent dielectric (insulator) layer. Common thickness is 0.05 - 0.5 mm. Control voltage is 5 - 10 V. This film may be produced by industry relatively cheaply per unit area, given the quantities needed.

The conventional controlled clarity (transparency) film reflects superfluous energy back to space. If the film has solar cells integral to it, which will become cheaper as the years go on, then the extra solar energy may be partially converted into electricity.

1. Liquid crystals (LC) are substances that exhibit a phase of matter that has properties between those of a conventional liquid, and those of a solid crystal.

Liquid crystals find wide use in liquid crystal displays (LCD), which rely on the optical properties of certain liquid crystalline molecules in the presence or absence of an electric field. The electric field can be used to make a pixel switch between clear or dark on command. Color LCD systems use the same technique, with color filters used to generate red, green, and blue pixels. Similar principles can be used to make other liquid crystal based optical devices. Liquid crystal in fluid form is used to detect electrically generated hot spots for failure analysis in the semiconductor industry. Liquid crystal memory units with extensive capacity were used in Space Shuttle navigation equipment. It is also worth noting that many common fluids are in fact liquid crystals. Soap, for instance, is a liquid crystal, and forms a variety of LC phases depending on its concentration in water.

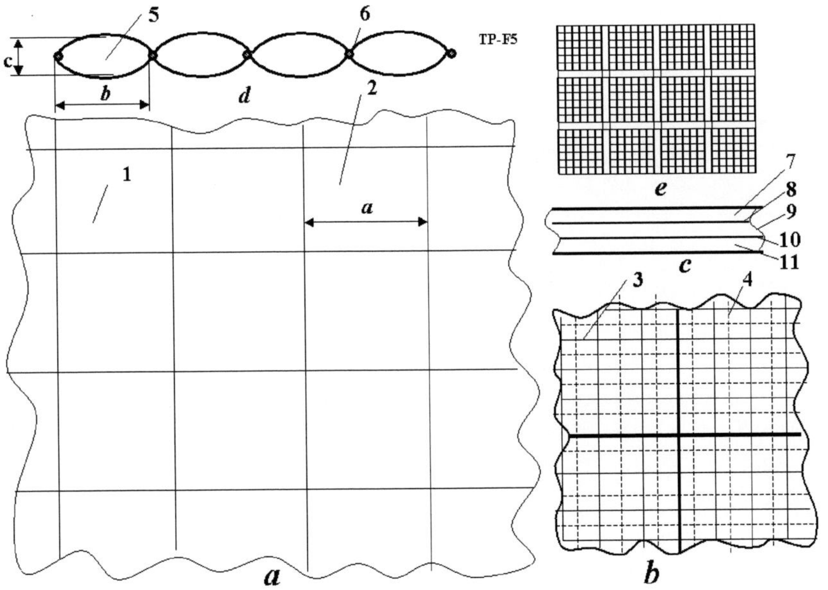

Figure 4-5. Design of covering membrane. *Notations*: (*a*) Large fragment of cover with controllable clarity (reflectivity, carrying capacity) and heat conductivity; (*b*) Small fragment of cover; (*c*) Cross-section of cover (film) having 5 layers; (*d*) Longitudinal cross-section of low height cover for cold and hot regions (optional); (*e*) Protection grilles. 1 - cover; 2 -mesh; 3 - small mesh; 4 - thin electric net; 5 - cell of cover; 6 - tubes; 7 - transparent dielectric layer, 8 - conducting layer (about 1 - 3 μ (micron)), 9 - liquid crystal layer (about 10 - 100 μ), 10 - conducting layer, and 11 - transparent dielectric layer. Common thickness is 0.1 - 0.5 mm. Control voltage is 5 - 10 V.

2. Transparency. In optics, transparency is the material property of allowing light to pass through. Though transparency usually refers to visible light in common usage, it may correctly be used to refer to any type of radiation. Examples of transparent materials are air and some other gases, liquids such as water, most glasses, and plastics such as Perspex and Pyrex. Where the degree of transparency varies according to the wavelength of the light. From electrodynamics it results that only a vacuum is really transparent in the strict meaning, any matter has a certain absorption for electromagnetic waves. There are transparent glass walls that can be made opaque by the application of an electric charge, a technology known as electrochromics.Certain crystals are transparent because there are straight lines through the crystal structure. Light passes unobstructed along these lines. There is a complicated theory "predicting" (calculating) absorption and its spectral dependence of different materials.

3. Electrochromism is the phenomenon displayed by some chemical species of reversibly changing color when a burst of charge is applied.

One good example of an electrochromic material is polyaniline which can be formed either by the electrochemical or chemical oxidation of aniline. If an electrode is immersed in hydrochloric acid which contains a small concentration of aniline, than a film of polyaniline can be grown on the electrode. Depending on the redox state, polyaniline can either be pale yellow or dark green/black. Other electrochromic materials that have found technological application include the viologens and polyoxotungstates. Other electrochromic materials include tungsten oxide (WO3), which is the main chemical used in the production of electrochromic windows or smart windows.

As the color change is persistent and energy need only be applied to effect a change, electrochromic materials are used to control the amount of light and heat allowed to pass through windows ("smart windows"), and has also been applied in the automobile industry to automatically tint rear-view mirrors in various lighting conditions. Viologen is used in conjunction with titanium dioxide (TiO2) in the creation of small digital displays. It is hoped that these will replace LCDs as the viologen (which is typically dark blue) has a high contrast to the bright color of the titanium white, therefore providing a high visibility of the display.

4. Film and cable properties [16]-[19] (see section "Material" in Attachment). Artificial fibers are currently being manufactured, which have tensile strengths of 3-5 times more than steel and densities 4-5 times less than steel. Although the described (1989) graphite fibers are strong ($\sigma/\gamma = 10\times10^6$), they are at least still ten times weaker than theory predicts. A steel fiber has a tensile strength of 5000 MPA (500 kg/sq.mm), the theoretical limit is 22,000 MPA (2200 kg/mm^2) (1987); the polyethylene fiber has a tensile strength 20,000 MPA with a theoretical limit of 35,000 MPA (1987). The very high tensile strength is due to its nanotube structure [19].

Traditional carbon fibers show high strength and stiffness, but fall far short of the theoretical, in-plane strength of graphite layers by an order of magnitude. Nanotubes come close to being the best fiber that can be made from graphite.

Specific strength (strength/density) is important in the design of the systems presented in this paper; nanotubes have values at least 2 orders of magnitude greater than steel. Traditional carbon fibers have a specific strength 40 times that of steel. Since nanotubes are made of graphitic carbon, they have good resistance to chemical attack and have high thermal stability. Oxidation studies have shown that the onset of oxidation shifts by about 100^0 C or higher in nanotubes compared to high modulus graphite fibers. In a vacuum, or reducing atmosphere, nanotube structures will be stable to any practical service temperature (in vacuum up 2800 ^{o}C. in air up 750°C).

About 60 tons/year of nanotubes are produced now (2007). Price is about \$100 - 50,000/kg. Experts predict production of nanotubes on the order of 6000 tons/year and with a price of \$1 – 100/kg to 2012.

Commercial artificial fibers are cheap and widely used in tires and countless other applications. The authors have found only older information about textile fiber for inflatable structures (Harris J.T., Advanced Material and Assembly Methods for Inflatable Structures, AIAA, Paper No. 73-448, 1973). This refers to DuPont textile Fiber B and Fiber PRD-49 for tire cord. They are 6 times strong as steel (psi is 400,000 or 312 kg/mm^2) with a specific gravity of only 1.5. Minimum available yarn size (denier) is 200, tensile module is 8.8×10^6 (B) and 20×10^6 (PRD-49), and ultimate elongation (percent) is 4 (B) and 1.9 (PRD-49). Some data are in Table 5 Attn.

Industrial fibers have up to σ = 500-600 kg/mm^2, γ = -1800 kg/m^3, and σ/γ = 2,78x10^6. But we are projecting use in the present projects the cheapest films and cables applicable (safety σ = 50 - 100 kg/mm^2).

5. Wind effect. As wind flows over and around a fully exposed, nearly completely sealed inflated AB-Dome, the weather affecting the external film on the windward side must endure positive air pressures as the wind stagnates. Simultaneously, low air pressure eddies will be present on the leeward side of the AB-Dome. In other words, air pressure gradients caused by air density differences on different parts of the AB-Dome's envelope is characterized as the "buoyancy effect". The buoyancy effect will be greatest during the coldest weather when the AB-Dome is heated and the temperature difference between its interior and exterior are greatest. In extremely cold climates such as the Arctic and Antarctic Regions the buoyancy effect tends to dominate AB-Dome pressurization.

6. *Solar radiation.* Solar radiation impinging the orbiting Earth is approximately 1400 W/m^2. The average Earth reflection by clouds and the sub-aerial surfaces (water, ice and land) is about 0.3. The Earth-atmosphere absorbs about 0.2 of the Sun's radiation. That means about q_0 = 700 W/m^{2}s of solar energy (heat) reaches our planet's surface in cloudy weather at the Equator. That means we can absorb about 30 - 80% of solar energy. It is enough for normal plant growth in wintertime (up to 40-50° latitude) and in circumpolar regions with a special variant of the AB-Dome design.

The solar spectrum is graphically portrayed in Figure 5 Ch.1A.

The visible part of the Sun's spectrum is only λ = 400 – 800 nm (0.4 to 0.8 μ.). Any warm body emits radiation. The emission wavelength depends on the body's temperature. The wavelength of the maximum intensity (see Figure 5) is governed by the black-body law originated by Max Planck (1858-1947):

$$\lambda_m = \frac{2.9}{T}, \quad [mm], \tag{1}$$

where T is body temperature, ^{o}K. For example, if a body has an ideal temperature ~20 ^{o}C (T = 293 ^{o}K), the wavelength is λ_m = 9.9 μ.

The radiation energy emitted by a body may be computed by employment of the Josef Stefan-Ludwig Boltzmann law:

$$E = \varepsilon\sigma_s T^4, \; [\mathrm{W/m^2}], \tag{2}$$

where ε is coefficient of body blackness (ε = 0.03 ÷ 0.99 for real bodies), σ_s = 5.67×10 $^{-8}$ [W/m^2.K] Stefan-Boltzmann constant. For *example*, the absolute black-body (ε = 1) emits (at T = 293 ^{0}K) the energy E = 418 W/m^2.

7. *Earth's atmosphere*. The property of Earth's atmosphere needed for computations are presented in Table 2 Ch.6 and in Useful Data in end of book.

4. *The thickness and weight of the AB-Dome envelope*, its sheltering shell of film, is computed by formulas (from equation for tensile strength):

$$\delta_1 = \frac{Rp}{2\sigma}, \quad \delta_2 = \frac{Rp}{\sigma}, \tag{3}$$

where δ_1 is the film thickness for a spherical dome, m; δ_2 is the film thickness for a cylindrical dome, m; R is radius of dome or radius of cover cell between cable (it may be half of distance between top cable, m; p is additional pressure into the dome, N/m^2, (p depends from altitude); σ is safety tensile stress of film, N/m^2.

For *example*, compute the film thickness for a dome having radius R = 100 m (distance between top cable 7 is 400 m), additional air pressure p = 0.01 atmosphere (p = 1000 N/m^2), safety tensile stress $\sigma = 50$ kg/mm^2 ($\sigma = 5\times10^8$ N/m^2), spherical dome. We receive $\delta_1 = 0.1$ mm. Distance between main cable 3 is $D = 0.8$ kilometers (Figure 6).

The computation for others case are presented in Figure 7 below.

The cover weight (mass) of 1 m^2 is computed by the equation:

$$m = \gamma\delta, \tag{4}$$

where m is 1 m^2 film mass, kg/m^2; γ is cover density, m. For *example*, if the cover thickness is $\delta = 0.2$ mm = 0.0002 m and $\gamma = 1500$ kg/m^3, the $m = 0.3$ kg/m^2.

Area S_c of semi-sphere diameter R, film cover mass M_f and cost C of Dome cover are

$$S_c = 2\pi R^2, \quad M_f = m_f S_c, \quad C = cS_c, \quad C = c_m M_f \quad , \tag{5}$$

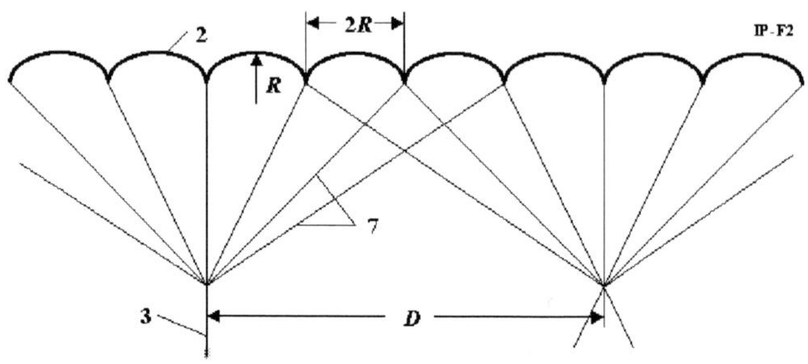

Figure 6. Cable support system. Radius R spherical cell of dome cover and distance D between main cable.

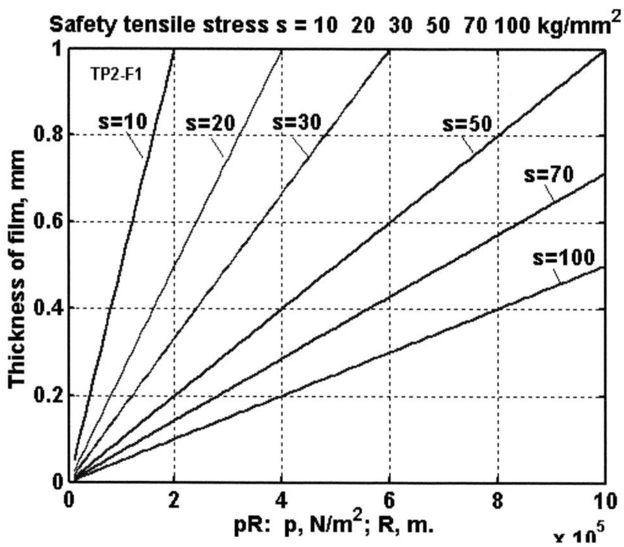

Figure 7. The thickness of top cover via the production of overpressure and radius spherical dome cell (distance between top cables for different safety film tensile stress.

where R is radius of semi-sphere, m; m_f is average cover area of 1 m^2 ; c is cost of 1 m^2, \$US/m^2; c_m is cover cost of 1 kg, \$US/m^2; C is cost of total cover, \$US.

Example. Let us take the hemi-sphere dome case. If m_f = 0.3 kg/m^2, film cost c = \$0.1 /m^2. The film mass covered of 1 km^2 of ground area is $M_1 = 2\times10^6\ m_c$ = 600 tons/km^2 and film cost is \$60,000/km^2. Fabrication costs may be somewhat more.

The area of city diameter 2 kilometers is 3.14 km^2. Area of a 2 kilometer hemi-spherical dome is 6.28 kilometers^2. The material cost of such an AB-Dome cover is theoretically 0.628 millions \$US.

The total cost of installation would be about 3-9 million \$US if the Government of Israel would make available free army labor to fabricate/assemble the shell and prepare ground installations.

That is less by nearly a hundred times, than the cost of an anti-rocket system (some hundreds of millions \$US), and by over a thousand times for an anti strategic nuclear missile system (billions of \$ US) The anti-rocket system is useless in peace time and it may be useless soon in war time because the offensive weapon is easily improved (multiple warheads, dispersing lethal substances, change of attack profile). The offered AB-Dome is very useful in peace time (controlling the weather and temperature inside, which could enable saving the AB-Dome cost in a few years of heating or air conditioning savings.), The AB-Dome defense may be upgraded to be proof also against any likely biological, chemical, radioactive dispersal weapons. The AB-Dome is a CLOSED-LOOP system (exclusive of leaks, but even 99% closure reduces outer-world threats by a hundred fold, and in practice even greater ratios should be obtainable).

The most extreme cases are deeply unpleasant, but must be discussed in the current Israeli threat environment. An incoming 20-kiloton missile warhead could be stopped by an AB-Dome and either impact-killed (relatively harmlessly) or detonated at a height too great to destroy the city below. In the latter case, assuming a 5 kilometers burst height and aluminized (summer sun inclination) upper dome sections, obviously the upper portions of the dome would be vaporized directly around the warhead, but the further sections would survive with progressively less damage, especially if the dome itself were sectioned internally (recommended). Because the AB-Dome contains positive overpressure internally, radioactive dust particles from the bomb or casing would tend to drift by even leaking sections of the AB-Dome outside the destroyed section. And the city below in the 'nuked' quarter would still be repairable and nearly uncontaminated by the standards of a low-altitude detonation (unstopped by the AB-Dome), which burst at optimal damage inflicting height (Mach effect enhanced).

Even more severe damage would have resulted from a surface detonation, which could easily have left a nearly 100 meter crater complex, which itself could cost more to decontaminate than the cost of the AB-Dome if there were *no* damage in the shielded city! (As well as unknown contamination plumes in groundwater and aquifers downstream)

(In addition, at 10 kilometers distance from a ground burst radiation from fallout would give an unshielded person a lethal dose *within 1 hour*. And crops from that land might be unsalable for generations. If a ground burst can be avoided, all these damages are greatly mitigated. (see for example http://www.epp.cmu.edu/domesticsecurity/KeithFlorig.pdf).

Consider the contrast in refugee problems: With undomed cities the entire small city is effectively rendered uninhabitable by an attack, (the Chernobyl-like panic about spreading radiation alone would see to that) and ironically, those fleeing the first blast may inadvertently head straight into a second target's fatal fallout plume!

By contrast, with a 20 kiloton explosion at 5 kilometers altitude impacting and detonating on the outer apex barrier of a 5 km high AB-Dome, winds at the surface underneath point zero (assuming they penetrated any optional lower barrier) would be no higher than 21 meters per second, or about 47 miles per hour. Because of resonance effects, it is quite possible that every window in the AB-Dome compartment under the detonation might be broken, but the much mocked 'duck and cover' training would literally be enough to avoid significant casualties (from heat pulse and flying glass). People could stay in their homes, sweep up, and await instructions. (In the compartment of the AB-Dome directly under the detonation, those instructions would presumably be to evacuate promptly to an adjoining compartment until the dome section could be repaired, because the city *in that compartment* is now open to a ground-level blast.

But the tens of thousands of dead, the unrecognizable charred bodies, the mass graves and all the other phenomena that Israelis are already quite familiar with from the Holocaust would not have taken place.

Obviously this is no magic solution or panacea for incoming nuclear missiles, (especially against ground-smuggled weapons—although the AB-Dome's restricted and monitored entrance points would aid in that other struggle).

People worry if any defense against missiles at all is developed, it will destabilize the concept of deterrence (or ignite an arms race) but we think a calm reading of the capabilities described would lead to the certain conclusion that only a crazy person would confuse 'a second chance at life' with 'immunity against all attacks'. To lend perspective to the matter, during the Cold War, a single US Titan II or a Soviet SS-9 could easily have destroyed the entire 5 km AB-Dome and underlying small city with a single shot. The lethal radius is the size of the entire AB-Dome! (This is one theoretically interesting case where the cubic kill capacity of a very large thermonuclear weapon would not be wasted, because the dome would be a target in three dimensions, not just two.)

To clarify: This is a partial, final layer defense only against low-yield weapons launched in anger or accident by novice nuclear powers with unstable command and control networks.

However, with India already with a hydrogen bomb, some Pakistanis calling to match it, China so armed now and the Japanese studying the situation, once can foresee at some point in the future a crazy world of the type Sir Martin Rees ("Our Final Hour") among others worries about; where apocalyptic war of some sort takes place between various super-weaponed regional contestants every other generation. In such a extreme case, like a early 1960's science fiction horror story, much of Earth could be poisoned by radioactive precipitations, tailored poison vapors and gases, genetically engineered killer microbes-- but a city (or even major portions of a country such as the best farmland) could continue to exist uncontaminated under AB-Domes and not only enable national survival and economic recovery, but hopefully be able to render some sort of aid to the survivors of the war—a lifeboat, one of many, on a national scale.

Disclaimer: Both the authors (and hopefully the reader) find the above scenario horrific and terrifying, and we suppose in today's academic climate it is actually necessary to state openly that we neither look forward to nor endorse crazy whole-population killing attacks or other such war-plans. Just for the record. But given the other record, the historical record of treatment of civilians of the last hundred years, and the scientific prospects for the next hundred, we do endorse having lifeboats on any vessel we sail on, even if those lifeboats never do leave Spaceship Earth.

5. The mass of support cable for 1 m^2 projection of dome cover. The mass of the support cable for every projection 1 m^2 of dome cover may be computed by the equation:

$$m_c \approx \gamma_c \frac{\overline{p}p}{\sigma} H, \quad M_c = \sum_S m_c, \quad M_c \approx \gamma_c \frac{p_a}{\sigma} S H_a, \tag{6}$$

where m_c is cable mass supported the projection one m^2 of the dome cover, kg/m^2; M_c is total mass of cable, kg; $S=\pi R^2$ is projection of cover on ground, m^2; γ_c is density of cable, kg/m^3; p is overpressure, N/m^2; σ is safety cable tensile stress, N/m^2; H is height of cable, m; $\overline{p}$ is relative air pressure at given altitude (Table 2); p_a is average overpressure, N/m^2; H_a is average height of the support cable, m.

We can design the dome cover without the support cable. In this case we compute the thickness of dome cover for radius R of a full Dome (see Eq.(3)). The cover thickness will be more. That is better for defense from rockets. If we want to compute more exactly and spend less cover mass, we must compute the variable overload (from altitude). In this case we must take into account the variable thickness of dome cover. If the dome doesn't have the support cable, one can have suspended cables, which allow reaching any part of Dome. The non-cable AB-Dome requires a (about 3-4 times) thicker cover that increases the impact force of the warhead and may make the stones unneeded.

For increased protection the dome cover may be an armored strong grille. The missile is destroyed from the shock of impact. It is not necessary to apply explosive of ours against it.

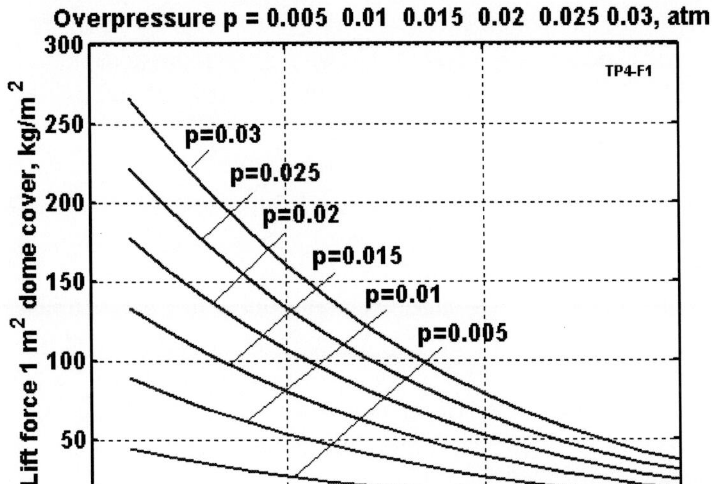

Figure 8. Lift force 1 m^2 of vertical projection the Dome cover versus altitude for different overpressures.

6. Lift force of 1 m² projection of dome cover. The lift force of 1 m² projection of dome cover is computed by equation:

$$L_1 = \bar{p}p\,, \tag{7}$$

where L_1 is lift force of 1 m² projection of dome cover, N/m²; p is overpressure, N/m²; $\bar{p}$ is relative air pressure at a given altitude (Table 2).

Example, for p = 0.01 atmosphere = 1000 N/m², H = 10kilometers ($\bar{p}$ = 0.261, Table 2) we get L_1 = 261 N = 26 kg.

Result of computation for different p is shown in Figure 11.

1. Leakage of air through hole. The leakage of air through hole, requested power of levitation fans (ventilator), and time of sinking of Dome cover (in case of large hole) may be estimated by the equation:

$$V = \sqrt{\frac{2p}{\rho}}, \quad M_a = \rho V S_h, \quad N = \frac{pVS_h}{\eta}, \tag{8}$$

where V is speed of air leakage, m/s; p is overpressure, N/m²; ρ is air density at given altitude, ρ = 1,225 kg/m³ at H = 0; S_h is area of hole, m; N is motor power, W; η is coefficient efficiency of motor.

Example. The area of hole equals S_h = 10² m² (10×10 m) at H = 0 m, p = 0.001 atmosphere = 100 N/m², η = 0.8. Computation gives the V = 12.8 m/s, M_a = 1568 kg/s, N =161 kW.

Let us estimate the time required for the AB-Dome cover to sink if no air is pumping it upward. Take the sphere radius R = 500 m. The volume of semi-sphere is $v = 262\times10^6$ m³, the air rate is $q = VS_h$ =12.8×100 = 1280 m³. The time of the full sinking is $t = v/q = 204\cdot10^3$ s = 57 hours.

Note, the overpressure-driven air will flow out from the dome. That means the radioactive (or chemical, biological, or other) bomb pollution cannot penetrate within the AB-Dome. The outer air pumped into the AB-Dome to levitate it can be filtered from radioactive or biological or chemical dust.

The repair of a dome hole is easy; no heroic measures are required at altitude. The support cable of the needed part of dome is reeled in and a film patch closes the hole. In case of a nuclear detonation at altitude this might effectively be most of a new dome section; presumably stockpiling repair materials en masse before a crisis will be a wise move.

9. Protection grille. For increased protection the dome cover may be armored with strong grilles (for example, steel rods having diameter 1 cm and step 10 cm), contain a mass 10 – 16 kg/m². All grilles are connected by flexible cables.

10. The wind dynamic pressure from wind is:

$$p_w = \frac{\rho V^2}{2}, \tag{9}$$

where $\rho = 1.225$ kg/m^3 is air density; V is wind speed, m/s.

For example, a storm wind with speed $V = 20$ m/s, standard air density is $\rho = 1.225$ kg/m^3. Than dynamic pressure is $p_w = 245$ N/m^2. That is four times less when internal pressure $p = 1000$ N/m^2 = 0,01 atm. When the need arises, sometimes the internal pressure can be voluntarily decreased, bled off.

MACRO-PROJECTS

Let us to consider the protection of a typical small city having diameter of 2 kilometers by semi-spherical AB-Dome, having radius (altitude) $R = 1$ kilometers. The covered region is $S = \pi R^2 = 3.14$ km^2, the semi-spherical area is $S_s = 2S = 6.28$ km^2.

The many computations for this AB-Dome are made as examples in the theoretical section above. We summarize these data in one project.

Let us assume that the standard over-pressure is 0.1 atmosphere at sea level. (Note that SCUBA divers routinely tolerate over-pressures of several atmospheres.) The Earth's atmosphere changes the pressure by some percentage points locally and we don't feel it. (Take a film having safety tensile stress $\sigma = 100$ kg/mm^2, local $R = 1000$ m. Dome is taken as a hemi-sphere without the support cables. Then the film has a thickness $\delta_1 = 0.5$ mm. The second lower film may be 0.1 – 0.2 mm thickness.

Area of semi-sphere is $S_s = 2\pi R^2 = 6.28$ km^2. If density of film equals $\gamma = 1500$ kg/m^3, the total mass of cover film is $M = \gamma\delta S_s = 4{,}700$ tons. If cost of film is \$1/kg, the total material cost of top cover is C = \$4.7 millions.

We armor the city from one side (area 3,14 km^2). The candidate materials are basalt grille (rebar mesh) or large pebble-sized hard stones (say 10 centimeters) suspended in cloth pockets tensioned by straps in a grid. The stones or grille may be armored by basalt fiber. Let us take the grid size of basalt grille as $s = 10$ cm, the rod cross-section 1.5×3 cm and the mass equal to $m_s = 27$ kg/m^2. The total mass of grilles is $M_s = m_s S_s = 85{,}000$ tons. If rebar price of the cast basalt (grille) is \$100 a ton (January 2008), then some 85,000 tons would cost approximately \$8.5M. If the stones' price is \$50/ton, then the stone ingredient costs <\$5 millions. The cloth pockets and straps may be \$1 million more. Adopting that macro-project building option the total cost of material becomes then about \$10.7 million from stones or \$14.2 with basalt grille.

The macro-project's total mass of construction is 54,700 – 85,000 tons and cost of construction material \$10.7 - \$14.2 million. The total cost of AB-Dome cover over a small city is about \$14 -17 million.

Our computation of this AB-Dome is far from being optimized at present. The average lift force (100 – 300 kg/m^2) is over in 4 - 12 times the weight of cover (0.75 kg/m^2) plus weight of stones (16 kg/m^2) or grille (27 kg/m^2). We can increase the mass (and number) of the pocketed stones or basalt impact grille by 3-5 times or decrease the thickness of the film cover.

Defense of Sederot

Notice:

Though this image is subject to copyright, the U.S. fair use laws cover its use, because:

- It illustrates an educational article with unique added content using the image as an unavoidable substrate, which is not replaceable with an uncopyrighted or freely copyrighted image of comparable educational value.
- It is a low-resolution image, deliberately decolorized, and thus not suitable for production of counterfeit goods.
- The logo is not used in such a way that a reader would be confused into believing that the article is written or authorized by the owner of the logo.

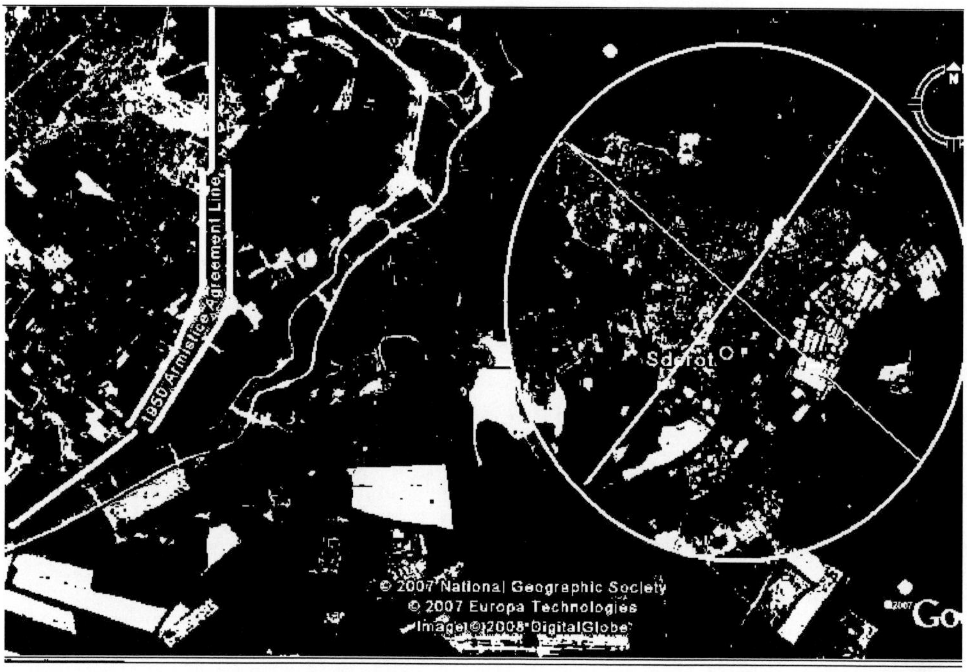

In the above edited Google Earth picture the 1950 Gaza border is seen at left. The dome on the right is 4.5 kilometers on a side; smaller domes might do better for subsections, but we compute very roughly the parameters needed for Sederot's defense.

A 4.5 kilometer wide dome, centered roughly at 31°31'19.75"N, 34°35'52.18"E as outlined above, would cover the widest portion of the built up area, with plenty of agricultural land under the dome as well.

The area relative to the previously computed project is ~5.07 the area and expense.

Thus, 1 layer of cover (dome material) is ~5.07 × 4.7 = 23.829 million $US. (2 layers are recommended for defense against 20-kiloton missile warheads, and for internal segregation)

Total material including stone cover is ~5.07 × 10.7 = 54.249 million $US.

Total cost of AB-Dome cover over Sederot is about ~5.07 * $14 -17 million or $~81.12 million $US. With the 20-kiloton protection option, the cost rises to above $105 million.

Even this is less than half the 1.5 billion NIS (New Israeli Shekels) allocated for the new missile defense system, and would enable an end to alerts, refugees and a city in Israel being held hostage at a cost of incalculable damage to Israel's deterrent power. With the missile defense system, the game simply goes on until the enemy figures out easily achieved new countermeasures. (Active offense nearly always beats out active defense; passive defense usually beats both in civil defense terms)

The authors are prepared to discuss the problems with organizations which are interested in research and development related projects.

Discussion

The cost of the offered passive AB-Dome Protection System macro-project may be less by hundred of times, then the cost of an *effective* active anti-rocket system (tens of billions $US). The anti-rocket system is useless in peacetime and it may be useless soon in wartime because of easily adopted enemy countermeasures. The offered AB-Dome is very useful in peacetime (especially for controlling the weather and temperature inside, which could enable saving the AB-Dome cost in a few years of heating or air conditioning savings.)

Additional benefits: Rainwater can run to collection troughs unmixed with sewer water or street pollution directly from the AB-Dome. This can be a few hundred thousand tons for a couple of square kilometers per year—of very high purity. This is no small matter in a rain-short country like Israel. Also, farms within will have evaporation of water from the fields during day—and condensation and collection at night—a closed cycle system, the water 'capital' subtracted from only by final export of the grown produce—a tiny percentage of the life cycle water intake of the crops. (For example a steer might drink 24 tons of water in its lifetime, be washed with 7, and be fed crops containing 5221 tons of water—but only the steer's final weight must leave the agricultural dome. Data is taken as to potential water savings (from http://www.waterfootprint.org/--900 liters of water to grow 1 kilogram of corn (maize); 16000 liters of water to grow 1 kg of beef. Instead of being a sign of waste, this would be the measure of conservation instead! Respectively,.9 tons and 16 tons of water would remain in the dome—1 kg of packaged product in each case would leave it.)

Besides this vast water savings (enabling desalination to actually be economical to grow crops in the present day (though recycling water, and only making up the net system leaks) tropical crops could be grown in temperate countries with such domes. Extended growing seasons through heat containment would be possible in most locations, and multiple growing seasons in many, doubling the productivity of the land. And the crops would be protected against external migrating insects such as locusts, crop and bird diseases, migrating pests or wildlife that eat crops, frosts, and being downwind of nuclear, biological or chemical contamination. (In the year of Chernobyl, 1986, many eastern Europe countries had great difficulty marketing even their *uncontaminated* crops to overseas buyers—people simply did not wish to take any chances, no matter how small, as we see in the present GM (genetically modified) crops controversy.

And incidentally, non-GM crops, or 100% GM crops, could be isolated within subsections of an AB-Dome, and be able to prove it to their target markets.) It is no exaggeration to say that in the futures farmers wanting bank financing will have to prove they are under a dome system *to make the odds of growing and marketing their crops successfully good enough to bank on.*

The AB-Dome City Defense System also provides protection from nearly all external terrorist threats imaginable, even a 9-11 type airliner incident, even a dirty bomb upwind (or fallout from a nuclear war around the world, would be kept literally kilometers away from the center of the city. Given the strong financial, water conservation and security incentives to dome over good croplands (and dome over cities, as well) –doing both would together provide a powerful degree of national continuity in an environmentally and security challenged place like Israel.

The control of regional and global weather of the Earth is an important problem of humanity. That ability dramatically increases the territory suitable for men to live in, the sown area, and crop capacity. In the long term, it allows us to convert all Earth land such as Alaska, North Canada, Siberia, deserts Sahara or Gobi to a prosperous garden. The suggested method is very cheap (cost of covering 1 m^2 is about 2 - 15 cents) and may be utilized at the present time. We can start from a small area, from small towns in bad climactic regions and extended to a large area.

Film domes can foster the fuller economic development of cold regions such as the Earth's Arctic and Antarctic and, thus, increase the effective area of territory dominated by humans. Normal human health can be maintained by ingestion of locally grown fresh vegetables and healthful "outdoor" exercise. The domes can also be used in the Tropics and Temperate Zone. Eventually, they may find application on the Moon or Mars since a vertical variant, inflatable towers to outer space, are soon to become available for launching spacecraft inexpensively into Earth-orbit or interplanetary flights. An AB-Dome can keep at high altitude a load up to 300 kg/sq.m. A launch site on such a lofted dome would have the same advantage as from a mountaintop, yet be buildable, for example, at the European Space Agency and the NASA tropical launch sites, easily accessible by cable-car from their existing surface bases. Because of the exponential nature of the rocket equation, by taking off from 5 km elevation, even a minor thrust augmentation because of the thinner atmosphere would probably enable enough extra fuel being loaded to nearly double a given launcher's payload to orbit. (When one additionally takes into account the reduced drag losses)

Lest it be objected that such domes as considered in this article would take impractical amounts of plastic, consider that the world's plastic production is today on the order of 100 million tons. If, with economic growth, this amount doubles over the next generation and the increase is used for doming over territory, at 300-500 tons a square kilometer 200,000 square kilometers could be roofed over annually. While small in comparison to the approximately 150 million square kilometers of land area, consider that 200,000 one kilometer sites scattered over the face of the Earth newly made productive and more habitable could revitalize vast swaths of land surrounding them—one square kilometer here could grow local vegetables for a city in the desert, one over there could grow biofuel, enabling a desolate South Atlantic island to become fuel independent; at first, easily a billion people a year could be taken out of sweltering heat, biting cold and slashing rains, saving the money buying and running heating and air conditioning equipment would require. Additionally, clean rain water could flow directly to cisterns, away from the pollution of the storm sewers. And if needed, bio-oil crops could be grown within to enable further dome production—*possibly using carbon dioxide exhaust from enclosed power stations* to hasten growth rates.

In effect, by doming over inhospitable land as specified, in exchange we get new territory for living with a wonderful climate, largely free from serious attacks, even if right near a hostile border.

The associated problems are researched in references [1]-[12].

RESULTS

Authors offer the cheap AB-Dome which protects the big cities from rockets, mortar shell, projectiles, chemical, biological weapon (bombs) delivered by rockets, missiles, guns and aviation. The offered AB-Dome is also very useful in peacetime because that protects the city from ambient weather and creates a very livable climate inside the Dome.

The principal advantages of the offered AB-Dome follow:

1. AB-Dome may be cheaper by hundreds to thousands of times than current comparable anti-rocket systems.
2. AB-Dome may not need high technology and can built by a poor country.
3. Additional yearly crops, high-yield growing conditions, conservation of scarce water, and freedom from pesticides, genetic drift and damaging factors for any crops grown within. (This also may enable its financing).
4. By making useless land valuable, the dome may be able to enable its own financing.
5. It is easy to build.
6. AB-Dome may be used in peacetime; it creates a fine climate (weather) inside the dome and thus can be sold to a national assembly on air conditioning and heating savings alone.
7. AB-Dome may protect from rocket, mortar, chemical, biological and worse weapons.
8. Dome protects the autonomous existence of the city population after third-party nuclear or biological war and spreading contamination of the regional atmosphere.
9. AB-Dome may be used for high antennae site for regional TV, for cell phone communication, for long distance location (differential GPS service), for astronomy (telescope above ~50% of air thickness). (This also may enable its financing).
10. Unlike any known active ballistic missile defense, this AB-Dome also can defend against submunitions and dispersed killing agents such as shrapnel.
11. AB-Dome may be used for high altitude tourism, or support for high altitude activities.
12. AB-Dome may be used for the high altitude windmills (getting of cheap renewable wind energy).
13. AB-Dome may be used for night illumination and entertainment or advertising (this also may enable its financing).

Additional applications of the offered AB-Dome the reader may find in [1]-[15].

REFERENCES

(The reader may find some of these articles at the author Bolonkin's web page: *http://Bolonkin.narod.ru/p65.htm*, *http://arxiv.org*, search "Bolonkin", in the book "*Non-Rocket Space Launch and Flight*", Elsevier, London, 2006, 488 pgs. and in book "*New Concepts, Ideas, Innovations in Aerospace, Technology and Human Science*", NOVA, 2008, 430 pgs.)

[1] Bolonkin A.A., Cathcart R.B., *Inflatable 'Evergreen' Polar Zone Dome (EPZD) Settlements,* 2006. http://arxiv.org search "Bolonkin".

[2] Bolonkin, A.A. and R.B. Cathcart, Inflatable 'Evergreen' Dome Settlements for Earth's Polar Regions. *Clean. Techn. Environ.* Policy. DOI 10.1007/s10098.006-0073.4.

[3] Bolonkin A.A., *Control of Regional and Global Weather.* 2006. *http://arxiv.org* search for "Bolonkin".

[4] Bolonkin A.A., AB *Method of Irrigation without Water (Closed-loop water cycle).* 2007. http://arxiv.org search for "Bolonkin".

[5] Bolonkin A.A., *Cheap Textile Dam Protection of Seaport Cities against Hurricane Storm Surge Waves, Tsunamis, and Other Weather-Related Floods,* 2006. *http://arxiv.org.*

[6] Bolonkin, A.A. and R.B. Cathcart, *Antarctica: A Southern Hemisphere Windpower Station?* http://arxiv.org, 2007.

[7] Cathcart R.B. and Bolonkin, A.A. *Ocean Terracing,* 2006. *http://arxiv.org.*

[8] Bolonkin, A.A. and R.B. Cathcart, *The Java-Sumatra Aerial Mega-Tramway,* 2006. *http://arxiv.org.*

[9] Bolonkin, A.A., "Optimal Inflatable Space Towers with 3-100 km Height", *Journal of the British Interplanetary Society* Vol. 56, pp. 87 - 97, 2003.

[10] Bolonkin A.A., Inflatable Dome for Moon, Mars, Asteroids and Satellites, *Presented as paper AIAA-2007-6262 by AIAA Conference "Space-2007",* 18-20 September 2007, Long Beach. CA, USA

[11] Bolonkin, A.A., *Electrostatic Space Towers,* 2007, *http://arxiv.org*, search "Bolonkin".

[12] Bolonkin A.A., *Non-Rocket Space Launch and Flight*, Elsevier, London, 2006, 488 ps.

[13] Bolonkin A.A., "*New Concepts, Ideas, Innovations in Aerospace, Technology and Human Science*", *NOVA,* 2008, 430 pgs.

[14] Bolonkin A.A., *Cheap Method of City Protection from Rockets and Nuclear Warheads, http://arxiv.org*, search "Bolonkin"

[15] Macro-Engineering: *A Challenge for the Future*. Springer, 2006. 318 ps. Collection articles.

[16] Galasso F.S., Advanced Fibers and Composite, Gordon and Branch Scientific Publisher, 1989.

[17] Carbon and High Performance Fibers Directory and Data Book, London-New. York: Chapmen& Hall, 1995, 6th ed., 385 p.

[18] Concise Encyclopedia of Polymer Science and Engineering, Ed. J.I.Kroschwitz, N. Y.,Wiley, 1990, 1341 p.

[19] Dresselhaus, M.S., Carbon Nanotubes, *b*y, Springer, 2000.

[20] *Wikipedia. http://wikipedia.org.* Israel.

References about Israel Protection System

Israeli Defense Company UTG-PRI Develops Futuristic Weapon Systems - India Defence, 07.01.2008 *http://lenta.ru/news/2007/02/12/arrow/*

Израиль закроется от ракет "Хизбаллы" "Щитом Давида" - Lenta.ru, 09.10.2006 (Russian).

Chapter 9

AB-NET METHOD OF PROTECTION FROM PROJECTILES* (CITY, MILITARY BASE, BATTLE-FRONT, ETC.)

ABSTRACT

The author suggests a low cost special AB-Net from artificial fiber, which may protect cities and important objects from rockets, artillery and mortar shells, projectiles, bullets, and strategic weapons. The idea is as follows: The offered AB-Net joins an incoming projectile to a small braking parachute and this incoming projectile looses speed by air-braking after a drag distance of 50 - 150 meters. A following interception net after the first may serve to collect the slowed projectiles and their fragments or bomblets so that they do not reach the aimpoint. The author offers the design of AB-Net, a developed theory of snagging with a small braking parachute by AB-Net; and sample computations. These nets may be used for defense of a town, city, military base, battle-front line, road (from terrorists), or any important objects or installations (for example nuclear electric station, government buildings, etc.). Computed projects are: Net to counter small rockets (for example, from the infamous Qassam), net to counter artillery projectile (caliber 76 mm), net to counter bullets (caliber 7.6 mm).

The offered method is cheaper by thousands of times than protection of a city by current anti-rocket systems. Discussion and results are at the end of the article.

Keywords: Protection from missile and projectile weapons, mortar, rocket, AB-Net, Qassam defense, incoming defense, armor

---

* Presented in http://arxiv.org 12 January 2008

INTRODUCTION

Review of Current Situation and Methods of Protection

Protection of cities, bases, and important objects. One important problem in small countries with hostile borders (or larger countries with leaky borders) is protection of abroad military bases, domestic targets from small rockets, missiles, mortar shells and terrorists. For well over a hundred years there has been no satisfactory solution for this, which is why such weapons are favorites of guerilla groups. Israel, for example, has villages (Alumim, and dozens of others) towns (Sderot, Netivot and Qiryat Shemona), cities such as Ashqelon and numerous installations near unfriendly borders. For example, Sderot lies one kilometer from the Gaza Strip and the town of Beit Hanoun. Since the beginning of the Second Intifada in October 2000, Sederot has been under constant rocket fire from Qassam rockets launched by various armed factions. Despite the imperfect aim of these homemade projectiles, they have caused deaths and injuries, as well as significant damage to homes and property, psychological distress and emigration from the city. Real estate values have fallen by about half. The Israeli government has installed a "Red Dawn" alarm system to warn citizens of impending rocket attacks, although its effectiveness has been questioned. Thousands of Qassam rockets have been launched since Israel's disengagement from the Gaza Strip in September 2005, which essentially has killed popular support for any further withdrawals, particularly from West Bank areas near the heart of the country.

In May 2007, a significant increase in shelling from Gaza prompted the temporary evacuation of thousands of residents. By November 23, 2007, 6,311 rockets had fallen on the city. The Israeli newspaper Yediot Aharonot reported that during the summer of 2007, 3,000 of the city's 22,000 residents (comprised mostly of the city's key upper and middle class residents, the heart of the economy, those most able to move) had already left for other areas, out of Qassam rocket range.

The total number of Qassam rockets launched exceeded 1000 by June 9, 2006. During the year 2006 alone, 1000+ rockets were launched. Tons of explosives have been intercepted at the Egyptian border; the uninterrupted shipments must be greater still, and the cumulative detonation yield has easily been in the tens of tons.

A rocket once fell into the electricity station in Ashkelon and caused electricity shortages in several areas, other time a rocket-similar to Qassam - fell inside an army base and injured more than 70 Israeli soldiers. The Ashkelon strike in particular was troubling as it added (by its radius) another 250,000 Israelis to the potential target list requiring defenses to be paid for, active or passive.

Some military bases of the USA in Afghanistan and Iraq (or in various parts of Asia and Latin America) are in the same situation. Any security consultant working to protect valuable installations in the more volatile corners of Africa, Latin America or Asia will recognize the dangers in the scenarios listed above. Rockets, and remotely triggered mortars, are man-portable and can be smuggled in, can be covertly emplaced and remotely fired with no appreciable warning, and endanger billions in investment with mere thousands in expenses. In Gaza, bonuses are allegedly given to impoverished children to retrieve the launchers, to reduce the expenses of replacing both rounds and launchers.

The Qassam rockets are tens (hundreds) times cheaper than complex electronic-guided anti-rockets and their delicate support system. The mortar shells are cheaper by hundreds of times. Attempted defense from them by conventional anti-rocket system may ruin any rich country, and in the end not work anyway, because the enemy always has the option of using large salvos, to probe the point where the system collapses trying to defeat X+1 simultaneous launches.

Protection by Armor

At present times for defense of small objects armor is employed. For defense of large objects (e.g., a city) from a nuclear warhead the very complex and very expensive anti-rocket systems under development may be used (missile interceptors). Obviously these would be economically infeasible to use against a conventional barrage.

Review of current armor and defense systems. Military vehicles are commonly armored to withstand the impact of shrapnel, bullets, missiles, or shells, protecting the personnel inside from enemy fire. Such vehicles include tanks, aircraft, and ships.

Civilian vehicles may also be armored. These vehicles include cars used by reporters, officials and others in conflict zones or where violent crime is common, and presidential limousines. Armored cars are also routinely used by security firms to carry money or valuables to reduce the risk of highway robbery or the hijacking of the cargo.

Armor may also be used in vehicles from threats other than deliberate attack. Some spacecraft are equipped with specialised armor to protect them against impacts from tiny meteors or fragments of space junk. Even normal civilian aircraft may carry armor in the form of debris containment walls built into the casing of their gas turbines to prevent injuries or airframe damage should the compressor/turbine wheel disintegrate.

The design and purpose of the vehicle determines the amount of armor plating carried, as the plating is often very heavy and excessive amounts of armor restrict mobility.

Vehicle armor is sometimes improvised in the midst of an armed conflict. In World War II, U.S. tank crews welded spare strips of tank track to the hulls of their Sherman, Grant, and Stuart tanks. In the Vietnam War, U.S. "gun trucks" were armored with sandbags and locally fabricated steel armor plate. More recently, U.S. troops in Iraq armored Humvees and various military transport vehicles with scrap materials: This came to be known as "haji" armor by Iraqis and "hillbilly" armor by the Americans.

For increasing vehicle protection the space armor, composed armor, active protection and other methods.

Spaced armor. Armor with two or more plates spaced a distance apart, called spaced armor, when sloped reduces the penetrating power of bullets and solid shot as after penetrating each plate they tend to tumble, deflect, deform, or disintegrate, when not sloped reduces the protection offered by the armor, and detonates explosive projectiles before they reach the inner plates. It has been in use since the First World War, where it was used on the Schneider CA1 and St. Chamond tanks. Many early-WWII German tanks had spaced armor in the form of armored skirts, to make their thinner side armor more effective against anti-tank fire.

Composite Armor. Composite armor is armor consisting of layers of two or more materials with significantly different chemical properties; steel and ceramics are the most common types of material in composite armor. Composite armor was initially developed in the 1940s, although it did not enter service until much later and the early examples are often ignored in the face of newer armor such as Clobham armor. Composite armor's effectiveness depends on its composition and may be effective against kinetic energy penetrators as well as shaped charge munitions; heavy metals are sometimes included specifically for protection from kinetic energy penetrators.

Plastic armor, called plastic protection in the United States, was a type of vehicle armor originally developed for merchant ships by the British Admiralty in 1940. The original composition was described as 50% clean granite of half-inch size, 43% of limestone mineral, and 7% of bitumen. It was typically applied in a layer two inches thick and backed by half an inch of steel.

Plastic armor was highly effective at stopping armor piercing bullets because the hard granite particles would turn the bullet which would then lodge between plastic armor and the steel backing plate. Plastic armor could be applied by pouring it into a cavity formed by the steel backing plate and a temporary wooden frame.

The armor was cheap and easy to install on ships, and the skills and equipment for installation came from the under-utilized road building industry.

Once installed on ships, plastic armor proved highly effective, when applied in sufficient thickness. Many anti aircraft guns such as the Oerlikon were fitted with only very thin plastic shields, which served mainly to improve the morale of the gunner. By some measures, it was as good as plate steel, and was widely adopted by allied ships. In the United States, some 3,000 merchant ships and 1,000 other ships were equipped with it, and in Britain and the Commonwealth some 7,000 ships were fitted.

An *active protection system*, or *APS*, protects a tank or other armored fighting vehicle from incoming fire *before* it hits the vehicle's armor. There are two general categories: *soft kill* systems, which use jamming or decoys to confuse a missile's guidance system, and *hard kill* systems, which attempt to detect and destroy incoming projectiles.

Soft-kill systems were (unsuccessfully) deployed by Iraq in the Gulf War. Iraqi tanks were fitted with strobe lights that masqueraded as the guidance beacon on the back of a TOW missile. The multinational force was aware of their use and adjusted the frequency of their guidance systems so they would not be confused. A soft-kill system currently in service is the Russian Shtora, deployed on Russian and Ukrainian tanks.

Hard-kill systems are activated when a millimetre-wavelength radar or other sensor detects an incoming projectile. In considerably less than a second, they launch a counter-projectile in an attempt to physically damage or destroy the incoming round. Examples include the TROPHY and Iron Fist from Israel and the Russian Drozd and Arena.

Attempts to use aircraft-mounted flak cannon as such an APS against anti-aircraft missiles proved ineffective, Anti-aircraft missiles are designed for effectiveness in a near-miss shot, making APS inefficient and unreliable. Among the effective countermeasures for aircraft are ECM, flares or anti-radar chaff.

DARPA is presently developing the High Energy Liquid Laser Area Defense System, which is planned to be capable of knocking out missiles, and may be used to actively defend future combat aircraft.

Warships have been equipped with similar systems (more frequently known as Close-In Weapon Systems, CIWS), which use small- to medium-caliber (12.7-76mm) guns and guided missiles to destroy inbound missiles and cannon shells. Examples include the US Phalanx

CIWS, Dutch Goalkeeper, Russian Kashtan, joint USA/German Rolling Airframe Missile, British Sea Wolf, Chinese Type 730 and Turkish Sea Zenith.

Shells can also be divided into three configurations: bursting, base ejection or nose ejection. The latter is sometimes called the shrapnel configuration. The most modern is base ejection, which was introduced in World War I. Both base and nose ejection are almost always used with airburst fuzes. Bursting shells use various types of fuze depending on the nature of the payload and the tactical need at the time.

Descriptions and Innovations

The main idea of the offered method is using the atmospheric friction gradient as armor against incoming hostile bodies. It is well-known the air produces a drag effect on moving objects, especially on a body which moves with high speed near sea level. For rockets and projectiles this added drag is mission-harmful because it significally decreases the projectile distance (range) and piercing force of projectiles, shells and bullets. At distances of ~ 500 – 1000m, as the projectile speed decreases aproximately by two times, the theoretical distance (in airless space) decreases by 2 – 4 times. Artillery designers try to decrease air drag by various clever techniques. Artillery designers model sharp forms for a projectile, specifically to decrease the cross-section area of said projectile.

Our purpose is precisely the opposite – maximally increasing the drag of ENEMY projectiles, decreasing their range and piercing force, changing their trajectory and not allowing the projectile to reach their (defended) aimpoint. The traditional way to affect this to armor the aimpoint. It is possible (in limited cases) when the target is a tank, car or ship. But it is impossible when the target is a city, military base, nuclear electric station or very large government building. For defending these targets we know one method – anti-rockets, ballistic missile defense (BMD). But their design needs very expensive R&D over many years, high technology industry, expensive production facilities and will probably be too expensive to test fully and therefore unlikely to work properly the first few times it is used! The cost of this BMD in many times more expensive than the cost of the intercepted projectile. It is not a practical solution for small or poor countries.

The suggested system, by contrast is cheap and can affordably protect a large territory.

There is a well-known method to increase the air drag – a parachute. The parachute is used for braking a parachutist, or braking an entire aircraft in a runway landing. The main problem in the offered system sounds impossible at first hearing – how to join and mount the parachute to the enemy projectile in just a moment of contact? The invented net for it is described below.

Two interactions may exist between a flying body and a net in the moment of impact (Figure 1):

If body mass and speed is small (<50 m/s) and an elastic net is strong, the body fully loses its velocity and is stopped by the net (Figure 1a). An example might be a badminton ball. The energy (speed) of the body is high. The body then tears the net and continues its' flight (Figure 1b). In this case (using special net and design) we can join to the projectile a small braking parachute. The mechanism of joining is simply that of being snagged in a resilent pocket whose sudden jerking deploys a retro-parachute. The projectile quickly dissipates the high speed within a distance of ~50 – 150m. The next elastic net located after the parachute-affixing net (Figure 1b, notation 9) can capture and collect the projectile and its bomblets, if any have dispersed. (option).

The design of parachute net is shown in Figure 2. That has a big strong net 10. The thin light sub-net 11 made from high-strength artificial fiber and located in a cell of the big net. The net 11 contains the packed small light parachute 12 and a copy of this is located into every cell of the big net. Internal subnets are weakly joined to the main big net. When an incoming projectile penetrates into a given subnet, it catches on the strong light elastic subnet together with its' rapidly deploying small light parachute and is snagged, ripping it off and triggering deployment. The parachute opens and brakes the projectile. The big net now has a small hole. The cell of subnet can have 1, 2, or 4 parachutes as it is shown in Figure 2b.

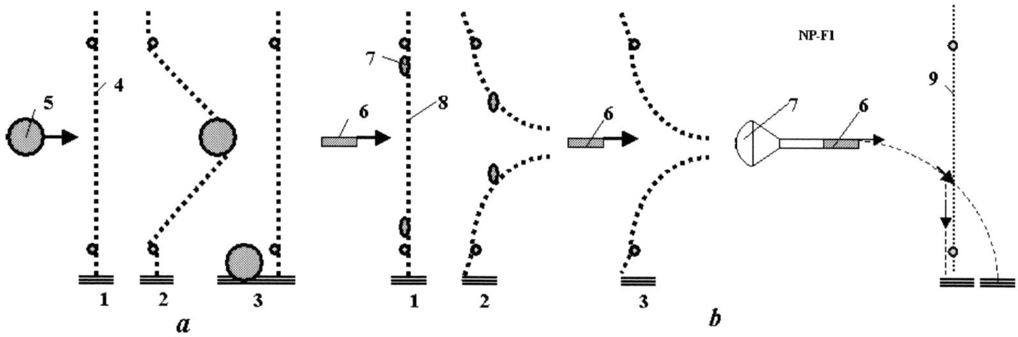

Figure 1. Two models of body impact in net: (***a***) Weak impact (the body speed is small, less than 20 – 50 m/s); (***b***) Strong impact (the body speed is high, more than 50 m/s). *Notations*: 1 – position of body and net before impact; 2 – position at impact; 3 - position after impact; 4 – brake net; 5 – ball; 6 – projectile (bullet); 7 – small braking parachute; 8 –parachute net; 9 – capture net.

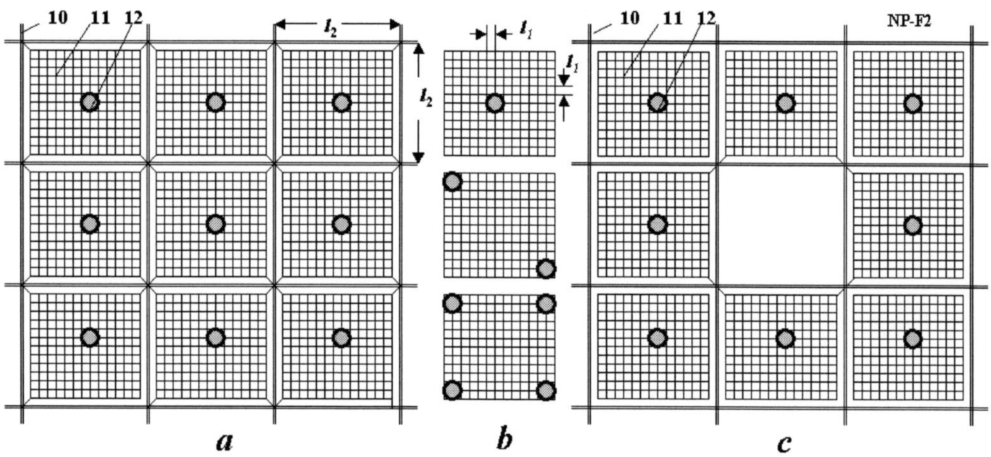

Figure 2. Design of parachute net. (***a***) Brake net before impact; (***b***) Location of small braking parachutes (1, 2, 4) in small section (cell) of the braking subnet; (***c***) braking net after projectile impact. *Notations*: 10 – strong cables of braking net; 11 – small section (cell) of braking subnet weak connected to strong cable; 12 – small braking parachute in compact form; l_1 is stepping distance of fibers (cell size) in subnet; l_2 is stepping distance of fibers (the cell size) in the main net.

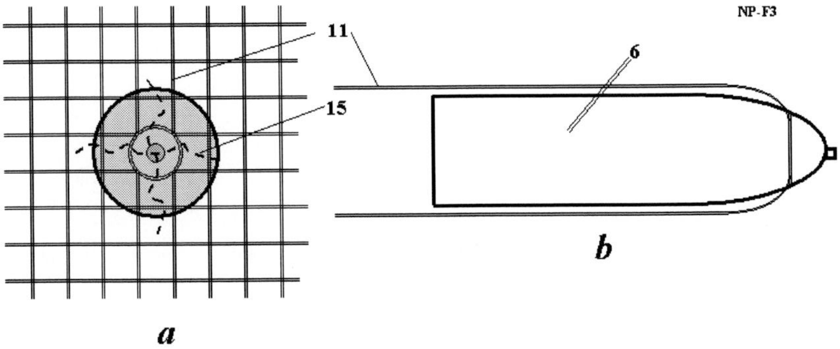

Figure 3. Impact of projectile in parachute net: (*a*) Forward view; (*b*) Side view. *Notations*: 11 – parachute subnet; 15 - possible places of subnet cable break-off-point of parachute subnet.

The impact of projectile into the subnet is shown in Figure 3. The cell size of the subnet must be less than the diameter of any likely projectile. (This obviously also states that there are different mesh sizes of AB-Nets for different incoming ammunition—one layer of AB-Net might pass bullets but stop small mortars, another might shield for yet larger incoming aerial bombs, etc.) The maximum figure for this cell size is one half of the projectile diameter. If strength of subnet cables is not enough, the possible places 15 of subnet cable break-off-points of parachute subnet is shown in Figure 3a.

The projectile can snap the break-off-point of between 2, 4 or 8 subcables. The exact figure depends upon the relation between subnet stepping distance (cell size) l_1 and the diameter and orientation of the incoming projectile.

The air drag of the parachute (and stress on it) may be very high in the first moment of opening. Although hard on the missile, this is also hard on the parachute! A better design is a slow opening parachute as shown in Figure 4. The first moment the parachute has demi-form (Figure 4a), when the speed and drag decreases, the parachute opens to full form (Figure 4b).

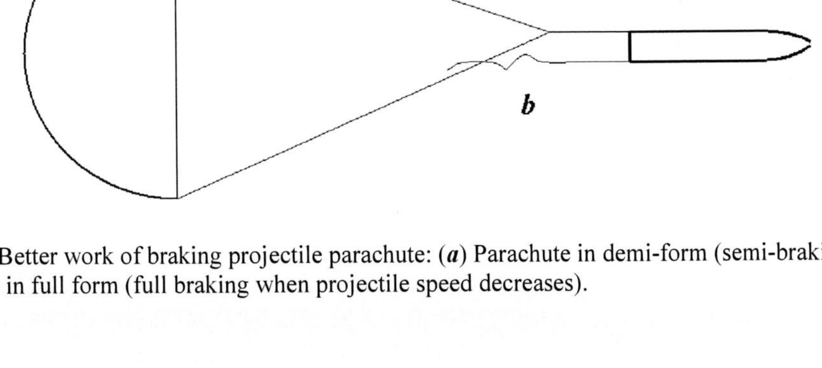

Figure 4. Better work of braking projectile parachute: (*a*) Parachute in demi-form (semi-braking); (*b*) Parachute in full form (full braking when projectile speed decreases).

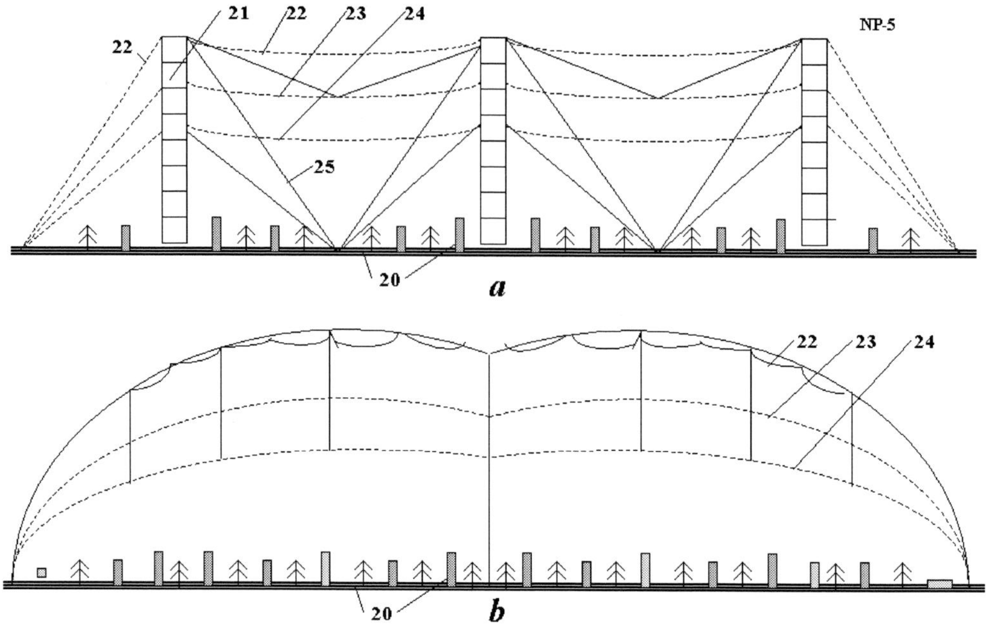

Figure 5. Protection by AB-Net of a city (base, big important object) from rockets and projectile. (***a***) Protection by nets suspended in the inflatable sectionated stacked balloon towers (masts) ; (***b***) Protection by nets suspended in inflatable dome. *Notations*: 20 – city (base or important object); 21 - inflatable sectionated stacked balloon towers (masts); 22 – parachute net; 23 – capture net; 24 – debris (fragments) net (if projectile will explode); 25 – bracing cable.

The protection (covering) of a big city by offered AB-Net is shown in Fig,5. Inflatable sectionated stacked balloon masts (imagine a stack of doughnut balloons) 21 (Figure 5a) (about 10 - 20m of diameter and height 300 -500m) are installed over a distance of about 0,5 – 1 km and supported by bracing cables. The masts support the parachute net 22. The capture net 23 is located below this parachute net by a margin of about ~150 – 200m. The debris net 24 is located below the capture net 23 by a margin of about 50m.

The AB-Net protection system works the following method. The upper parachute net 22 joins the braking parachutes to rockets and projectiles. In a braking distance of 150 – 200m, they decrease projectile speed to the range of 25 – 40 m/s. The lower capture net 23 captures the projectile. If it explodes,, the lower debris net 24 collects splinters shrapnel and fragments.

Note. The inflatable towers (masts) have sectionated stacked balloon design and small internal pressure. If some section is damaged, the leakage is easy compensated for by ground ventilators (increased fan flow)

The other protection method for an entire city by AB-Net is shown in Figure 5b. The city is covered by light inflatable AB-Dome film as described in [1] - [14]. Below this film is located the AB-Nets 22 - 24 suspended by mounting fastening hardpoints on the dome cover, as it is shown in Figure 5b. This design has advantages – it does not need support masts (instead hanging from the overpressure suspended overroof) and the climate inside the Dome is controlled [1]-[14].

The methods of protection in different cases are shown in Figure 6a-d. Figure 6a is protection of soldiers in the front line of battle. The inflatable (or steel) sectionated stacked balloon masts 3 are installed forward of our soldiers. The parachute net 2 is supported by masts. The AB-Net is designed thus (and this is part of the invention) that it joins the braking small parachutes to projectiles flying from the enemy side, but allows passage from the friendly side to the enemy side (the parachutes don't connect or not open). In sum, the enemy cannot kill the our soldiers, but our soldiers can repulse the attack. *Note*: Contrary to expectation, this protection is suitable for a fighting environment. The bullets and shells easily pass through the thin film of inflatable masts and produse only small holes equal in each case to the round's diameter. The air leakage throuth these holes is easily compensated for by small increases in the ventilator speed.

The figure 6b shows the protection of a vital road (highway) (or runway) from enemy sniping. The AB-Nets are installed along the dangerous part of road and protects the transport convoys (or expensive aircraft) from terrorist fire.

The AB-Net may be used in a counter-launch mode against small rockets and projectiles (Figure 6c). The locator system locates and computes the trajectory of incoming enemy projectiles. An anti-projectile cannon shoots a special shell in trajectory toward the enemy projectile. The anti-shell crosses the trajectory of the oncoming enemy shell. At the moment of contact, the shell snags the net and its' small brake parachute triggers. The parachute opens and brakes the enemy shell and does not allow the shell to reach its' aimpoint.

This system has considerable advantage in comparison to a more expensive rocket based APS. This system is cheaper because it doesn't need complex expensive self-guided missiles whose guidance system is destroyed after one use. The cannon doesn't need a precise pointing system. (The error margin is covered by the spreading net, not super-precise aiming, multiple rounds or a mid-course maneuver). The same system, though with a different net grid size, may be used against strategic rockets and their warheads having speed up 1500 m/s in end of trajectory (Figure 6d).

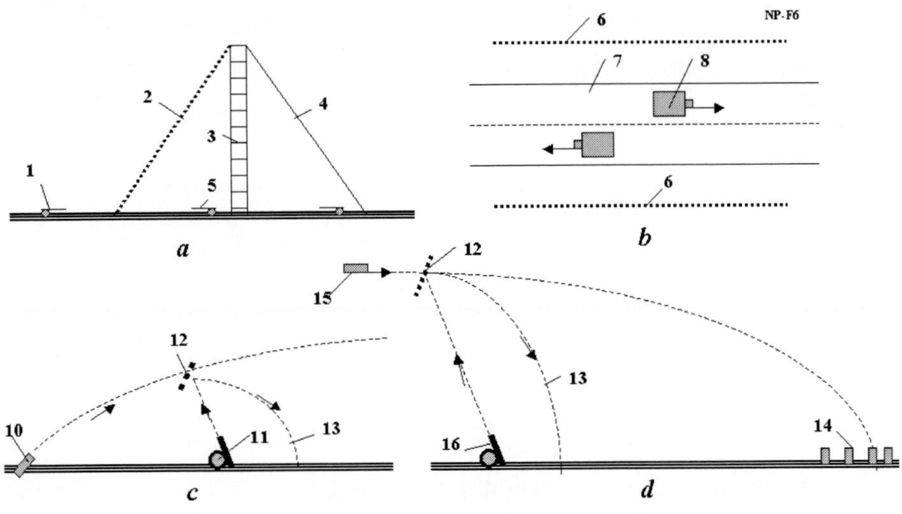

Figure 6. Protection from projectiles by AB-Nets: (*a*) Proiection of battle-front line; (*b*) Protection of the car (track) road; (c) Protection of object (for example, small town) from small (tactic) rockets; (*d*) Protection of a city from strategic missiles. *Notations*: 1 –enemy; 2 – parachute AB-Net; 3 – inflatable sectionated stacked balloon towers (or steel masts); 4 – bracing cable; 5 – our soldiers; 6 – ground parachute nets; 7 car (track) road; 8 – car, track; 10 – launch of enemy rocket; 11 – our cannon for shooting the parachute net; 12 – parachute air net; 13 – brake rocket trajectory; 14 - city; 15 - strategic missile or nuclear warhead; 16 – our cannon for shooting the parachute AB-Net.

THEORY AND COMPUTATION (ESTIMATION) OF AB-NET

a) General information

The efficiency of the offered net significantly depends from used artificial fiber.

1. Artificial fiber and cable properties [15]-[20] (see section "Nanotubes" in book ettachment). Cheap artificial fibers are currently being manufactured, which have tensile strengths of 3-5 times more than steel and densities 4-5 times less than steel.

For example, whiskers of Carbon nanotube (CNT) material have a tensile strength of 200 Giga-Pascals and a Young's modulus over 1 Tera Pascals (1999). The theory predicts 1 Tera Pascals and a Young's modules of 1-5 Tera Pascals. The hollow structure of nanotubes makes them very light (the specific density varies from 0.8 g/cc for SWNT's (Single Wall Nano Tubes) up to 1.8 g/cc for MWNT's, compared to 2.26 g/cc for graphite or 7.8 g/cc for steel). Tensile strength of MWNT's nanotubes reaches 150 GPa.

Some data are in Table 5 Attn.

Industrial fibers have up to σ = 500-600 kg/mm^2, γ = 1500 -1800 kg/m^3, and $\sigma\gamma$ = 2,78x10^6. But we are projecting use in the present projects the cheapest films and cables applicable (safety σ= 100 - 200 kg/mm^2).

2. Leakage of air through inflatable tower (mast) hole. The leakage of air through bullet holes, requested power of fans (ventilator), and time of sinking of tower (mast) cover (in case of large hole) may be estimated by the equation:

$$V=\sqrt{\frac{2p}{\rho}},\quad M_a=\rho VS_h,\quad N=\frac{pVS_h}{\eta}, \tag{1}$$

where V is speed of air leakage, m/s; p is overpressure, N/m^2; ρ is air density at given altitude, ρ = 1,225 kg/m^3 at H = 0; S_h is area of hole, m; N is motor power, W; η is coefficient efficiency of ventilator.

Example. The total area of holes equals S_h = 1 m^2 (10×10 m) at H = 0 m, p = 0.001 atmosphere = 100 N/m^2, η = 0.8. Computation gives the V = 12.8 m/s, M_a = 15.7 kg/s, N =1.6 kW.

3. The maximum specific wind dynamic pressure to inflatable mast (tower) from wind is:

$$p_w=\frac{\rho V^2}{2}, \tag{2}$$

where ρ = 1.225 kg/m^3 is air density; V is wind speed, m/s; p_w is specific dynamic wind pressure to 1 m^2, N/m^2.

For *example*, a storm wind with speed V = 20 m/s, standard air density is ρ = 1.225 kg/m^3. Than dynamic pressure is p_w = 245 N/m^2. That is four times less when internal pressure p = 1000 N/m^2 = 0,01 atm. When the need arises, sometimes the internal pressure can be voluntarily decreased, bled off.

b) Method of Computation and Estimation

1. Brake projectile distance having brake parachute. The motion of a projectile with parachute is described by differential equations:

$$\frac{dV}{dt} = -C_d \frac{\rho S}{2m} V^2 = -aV^2, \quad \frac{dL}{dt} = V, \quad \text{where} \quad a = C_d \frac{\rho S}{2m} \approx 0.6 \frac{S}{m}, \tag{3}$$

where V is projectile speed, m/s, $V \geq V_{\min} = \sqrt{g/a}$; t is time, sec; $C_d \approx 1$ is parachute drag coefficient; $\rho = 1.225$ kg/m^3 is standard air density; S is projection of parachute area, m^2; m is mass of system (projectile + parasite + connection cable (part of AB-subnet)), kg; L is distance of fly projectile, m; a is drag constant, 1/m.

After integration of equations (3) we get

$$L = -\frac{1}{a} \ln \frac{V}{V_0} \quad \text{or} \quad V = V_0 e^{-aL}, \tag{4}$$

where V_0 is initial speed of projectile, m/s.

The computation of equation (4) is presented in Figure 7.

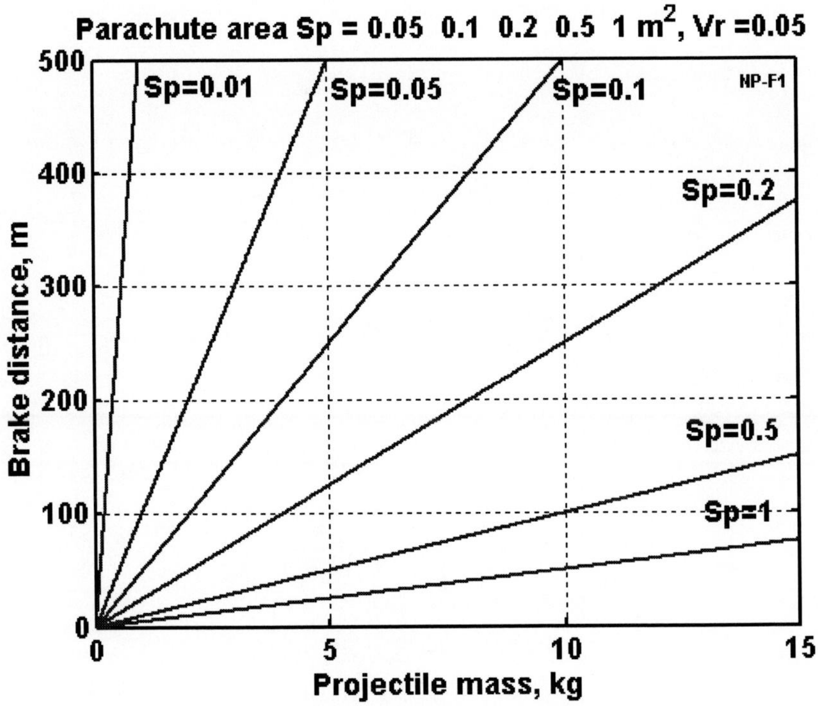

Figure 7. The brake distance of projectile versus the projectile mass (0 ÷ 15 kg) and area of brake parachute$S$ = Sp = (0.05 ÷ 1 m^2) for relative finish projectile speed.

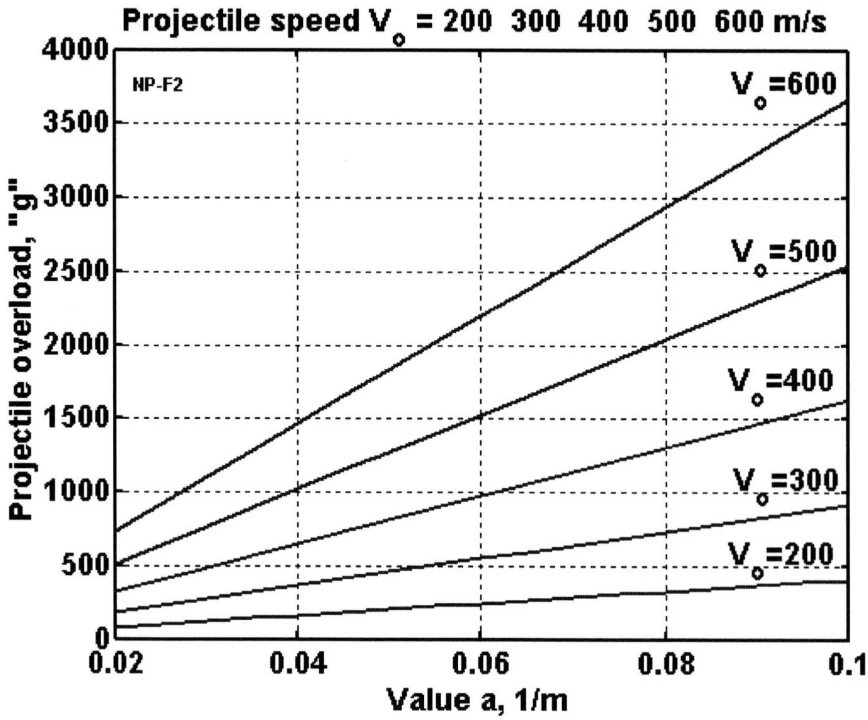

Figure 8. Maximal projectile overload (in "g") via projectile speed V_0 and coefficient a.

$V_r = (V/V_0) = 0.05$.

2. The maximal braking force and overload of projectile compute by equations:

$$P_m = amV_0^2, \quad n_m = \frac{P_m}{mg} = \frac{a}{g}V_0^2, \tag{5}$$

where P_m is maximal braking force of projectile (in moment of connection to parachute), N; n_m is maximal overload of projectile, in "g"; $g = 9.81$ m/s^2 is Earth's gravitation. The computation is shown in Figure 8.

3. The cross-section and diameter of subnet cable compute by equation:

$$s = \frac{P_m}{2\sigma} \approx 5\frac{mn_m}{\sigma}, \quad d = \sqrt{\frac{4s}{\pi}}, \tag{6}$$

where s is cross-section of subnet cables, m^2; σ is safety tensile stress of subnet cable, N/m^2; d is diameter of subnet cable, m. We assume that projectile tears (loads) ONLY TWO subnet cables. (A conservative estimate).

The computation is presented in Figure 9.

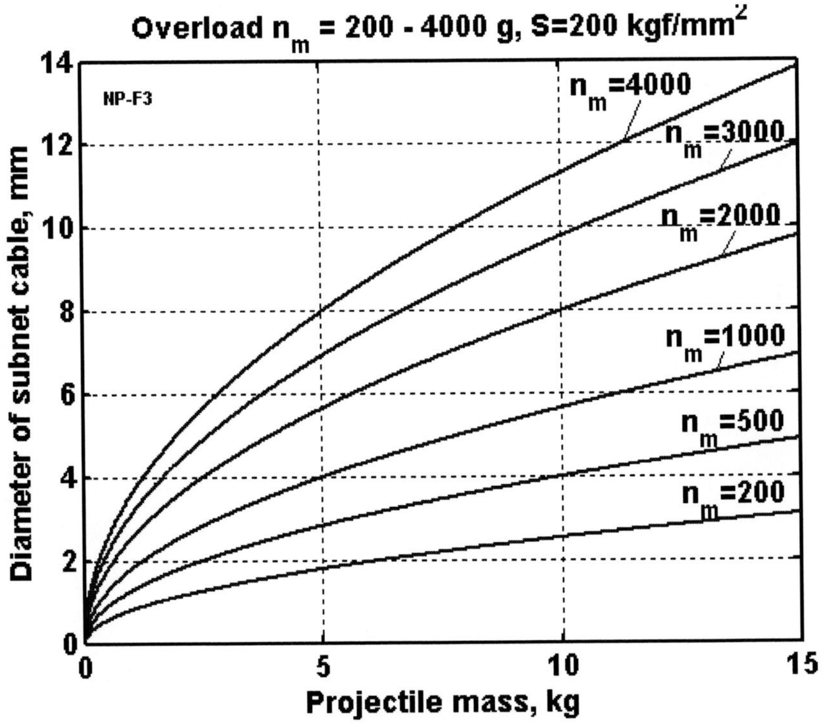

Figure 9. Diameter of subnet cable (mm) versus mass of projectile and its maximal overload.

4. 1 m2 mass of parachute net and mass of subnet may be estimated by the equations:

a) Mass of parachute

$$p = \frac{\rho V_0^2}{2} \approx 0.61 V_0^2, \quad r = \sqrt{\frac{S}{\pi}}, \quad \delta = \frac{pr}{2\sigma_p}, \quad m_1 = 2\gamma\delta S, \tag{8}$$

where m_1 is mass of parachute, kg; $\gamma \approx 1500$ kg/m^3 is specific density of parachute material, S is projection of parachute area, m^2; δ is thickness of parachute material, m; p is air dynamic pressure, N/m^2; r is radius of parachute, m; σ_p is safety stress of parachute, N/m^2. $\sigma_p \approx 100$ kgf/mm^2 = 10^9 N/m^2.

b) Mass of subnet cables:

$$m_2 = 2\gamma s_1/l_1. \tag{7}$$

where m_2 is mass of subnet cable, kg/m^2; $\gamma \approx 1500$ kg/m^3 is specific density of cable material, kg/m^3; s_1 cross-section of subnet cable, m^2; l_1 is step of subnet cables, m. 1 m^2 of parachute net for step of main net l_2 and cell area $S_b = l_2^2$ requests $N = 1/S_b$ parachutes/m^2.

c) Mass of main net cables is computed as:

$$m_3 = 2\gamma s_2/l_2 \tag{9}$$

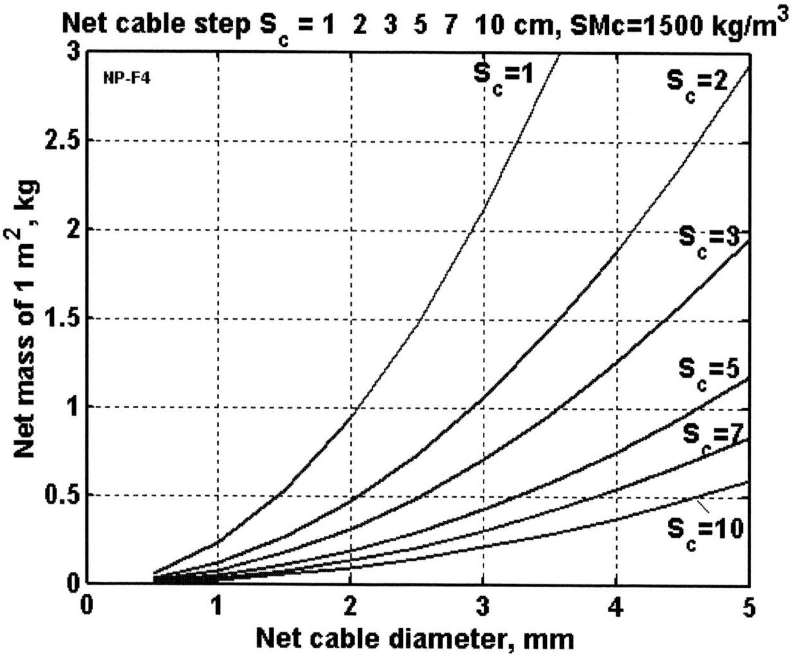

Figure 10. One m^2 net mass via diameter of the net cable (0.5 - 5 mm) and net mesh size (cable stepping distance) (1 – 10 cm).

where m_3 is mass of net main cable, kg/m^2; $\gamma \approx 1500$ kg/m^3 is specific density of cable material, kg/m^3; $s_2 \approx (2 \div 3)s_1$ cross-section of net main cable, m^2; l_2 is mesh size (step size) of big net (main cables), m.

The total 1 m^2 mass of parachute net

$$m_n = Nm_1 + m_2 + m_3, \tag{10}$$

Mass of one cell subnet cable and one parachute:

$$m_s = m_1 + m_2 S_b. \tag{11}$$

The result of computations yielding the mass of net per 1 m^2 is presented in Figure 10.

4. Tear out the subnet and parachute by projectile from AB-Net. The overload and force in impact moment may be estimated by formulas

$$n_s = \frac{V_0^2}{2gL_2}, \quad P_s = gn_sm_s, \quad \text{from} \quad P_s = P_m \quad \text{we get} \quad L_2 = \frac{m_s}{am} \approx \frac{m_s}{0.6S}, \tag{12}$$

where L_2 is tear out distance, m ($L_2 \approx (1 \div 2)\, l_2$); n_s is tear out overload, “g”; m_s is mass of one cell subnet cable + parachute, kg (Eq.(11)). We assume that projectile tears ONLY TWO subnet cables.

5. Cost of 1 m2 net. The cost of 1 m2 net approximately is

$C = cm_n$,

where C is cost of 1 m^2 net, \$/m^2; c is cost 1 kg artificial fiber, \$/kg; m_n is mass of 1 m^2 of net, kg/m^2 (Eq.(10)).

Computation of Lower (Capture) Net

The purpose of this capture net is to catch (collect) the projectiles of sufficiently low speed that capture can be accomplished without damage to the net. This net is needed in cases when the main (parachute dispensing slowdown net) is located such (for example, horizontal (as on a roof) and the already braked projectile must not reach the city after extinction of momentum because the round is still live and can have salvage fuzing with, for example, an ignition (detonation) delay.

The flexure of the capture net, maximal force, and projectile overload may be estimated by formulas

$$\text{From} \quad \frac{mV^2}{2} = k\frac{l_c^2}{2} \quad \text{we get} \quad l_c = \sqrt{\frac{m}{k}}V, \quad P = \sqrt{km}V, \quad n_c = \frac{P}{mg}, \tag{13}$$

where l_c is flexure of the capture net, m; k is net elasticity, N/m; m is projectile mass, kg; V is braked projectile speed, m/s; P is maximal net force, N; n_s is projectile overload, "g".

The cross-section area of net cable s_c, diameter of net cable d_c, tensile net stress F, and 1 m^2 cable mass m_c may be estimate the following equations (in metric system):

$$s_c = \frac{P}{2\sigma}, \quad d_c = \sqrt{\frac{4s_c}{\pi}}, \quad F \approx \frac{PL_c}{16l_c}, \quad m_c = 2\gamma\frac{s_c}{l_c}, \tag{14}$$

where L_c is approximately distance between the inflatable masts (towers), m. l_c is cable step in capture net.

We assume that projectile tears ONLY TWO subnet cables everywhere. This is the conservative case; more will be more effective in stopping the projectile.

MACRO-PROJECTS

Project #1. Protection City from Qassam (Kassam-3) Rockets (Figure 5a)

Let us to consider the protection of a typical small city from Kassam-3 rockets having diameter D = 17 cm, flight mass (after launch) m = 30 kg, and landing speed V_0 = 300 m/s (in reality $V_0 \approx$ 250 m/s). Assume the city has AB-Net roof, located on inflatable masts having radius R = 10 m and altitude 450 m. We calculate the data and cost for area 1 km^2 and 10 km^2. If you know the city area, you easily can recalculate the data for the new area.

1. Computation of parachute net. Let us take S = 0.5 m^2 and V/V_0 = 0.1, V = 30 m/s $\approx V_{min}$ = 31.3 m/s. From equations (4)-(14) we find:

1) Coefficient $a = 0.6S/m = 0.6\times0.5/30 = 0.01$ [1/m].

2) Braking distance $L = (-1/a) \times \ln(V/V_0) = (-1/0{,}02)\times\ln(0.1) = 230$ m.

3) The maximal braking force and overload of projectile is computed by the equation:

$$P_m = amV_0^2 = 0{,}01\cdot 30\cdot 300^2 = 2.7\cdot 10^4\ N,\quad n_m = \frac{P_m}{mg} = \frac{2.7\cdot 10^4}{30\cdot 9.81} = 90\ g\,.$$

4) The cross-section and diameter of subnet cable are computed by the equation (for σ = 200 kg/mm^2 = 2×10^9 N/m^2):

$$s_1 = \frac{P_m}{2\sigma} = \frac{2.7\cdot 10^4}{2\cdot 2\cdot 10^9} = 6{,}75\cdot 10^{-6}\ m^2 = 6.75\ mm^2,\quad d = \sqrt{\frac{4s_1}{\pi}} = 2.93\ mm\,.$$

5) 1 m^2 mass of subnet cables, mass of parachute, and cell mass of subnet, 1 m^2 mass of total parachute net (for γ = 1500 kg/m^2, l_1 = 0.1 m) are:

a) Mass of one parachute for safety parachute stress $\sigma_p = 10^9$ N/m^2:

$$p = \frac{\rho V_0^2}{2} = 0.61\cdot 300^2 = 5.51\cdot 10^4\ \frac{\text{N}}{\text{m}^2},\quad r = \sqrt{\frac{S}{\pi}} = 0.4\ \ \text{m},$$

$$\delta = \frac{pr}{2\sigma_p} = 1.1\cdot 10^{-5}\ \text{m},\ m_1 = 2\gamma\delta S = 2\cdot 1500\cdot 1.1\cdot 10^{-5}\cdot 0.5 = 0.0165\ \text{kg}$$

1 m^2 of parachute net for step of main net l_2 = 0.5 m and cell area $S_b = l_2^2 = 0.25\ \text{m}^2$ requests $N = 1/S_b = 4$ parachutes/m^2.

b) 1 m^2 mass of subnet cables having small net step l_1 = 0.1 m:

$$m_2 = \frac{2\gamma s_1}{l_1} = \frac{2\cdot 1500\cdot 6.75\cdot 10^{-6}}{0.1} = 0.2\ \text{kg/m}^2$$

c) Mass of 1 m^2 main net cables (for $s_2 = 2s_1$, S_b = 0.25 m^2) is:

$$m_3 = \frac{2\gamma s_2}{l_2} = \frac{2\cdot 1500\cdot 2\cdot 6.75\cdot 10^{-6}}{0.5} = 0.081\ \text{kg/m}^2$$

The total 1 m^2 mass of parachute net

$m_n = Nm_1 + m_2 + m_3 = 4\times0.0165 + 0.2 + 0.081 = 0.347$ kg/m^2, Mass of one cell subnet cable and one parachute (S_b = 0.25 m^2):

$m_s = m_1 + m_2 S_b = 0.0165 + 0.2\times0.25 = 0.0665$ kg.

6) Parameters of tearing out the subnet and parachute by projectile from AB-Net: The overload and force in impact moment is ($L_2 = (1 \div 2)\times l_2$):

$$n_s = \frac{V_0^2}{2gL_2} = \frac{300^2}{2\cdot 9.81\cdot 1} = 4.6\cdot 10^3\ g,$$

$$P_s = gn_s m_s = 9.81\cdot 4.6\cdot 10^3\cdot 0.0665 = 0{,}3\cdot 10^4\ N < P_m = 2.7\cdot 10^4\ N\ .$$

The difference between flight braking force P_m and tear out force P_s is very high. That means the parachute area is large; our design is far from optimum. But we cannot decrease the parachute area because the braking distance L = 230 m (the distance between parachute and capture nets) requires that much braking power

In any case the AB-Net requires testing for tuning the cross-section cable area and spacing of the net mesh for any given projectile. These tests are cheap.

2. Computation of capture net. Let us take the net elasticity k = 500 N/m and final braking speed of Qassam rocket V = 30 m/s (see above), m = 30 kg.

The flexure of capture net l_c, maximal force P, and projectile overload n_c may be estimated by formulas (Eq. (13)):

$$l_c = \sqrt{\frac{m}{k}}V = \sqrt{\frac{30}{500}}30 = 7.35\ m, \quad P = \sqrt{km}V = \sqrt{500\cdot 30}\cdot 30 = 3674\ N,$$

$$n_c = \frac{P}{mg} = 12.5\ g\ ,$$

,

where l_c is maximal flexure of the capture net, m; k is net elasticity, we take k = 500 N/m; m is projectile mass, kg; V is *braked* projectile speed, m/s; P is maximal net force, N; n_s is projectile overload, "g".

Let us take the distance between the support cable L_c = 100 m, the safety tensile stress $\sigma = 2\times 10^9$ N/m^2. The cross-section area of net cable s_c, diameter of net cable d_c, tensile net stress F, and 1 m^2 cable mass m_c may be estimate the following equations (in metric system) (Eq. (14)):

$$s_c = \frac{P}{2\sigma} = \frac{3674}{2\cdot 2\cdot 10^9} = 0.92\cdot 10^{-6}\ m^2, \quad d_c = \sqrt{\frac{4s_c}{\pi}} = 1.08\ mm,$$

$$F \approx \frac{PL_c}{16l_c} = \frac{3674\cdot 100}{16\cdot 7.35} = 3124\ N, \quad m_c = 2\gamma\frac{s_c}{l_c} = 2\cdot 1500\cdot\frac{0.92\cdot 10^{-6}}{0.1} =$$

$$= 0.0276\ kg/m^2\ .$$

The total mass of nets is

$$m_t = m_n + m_c = 0.347 + 0.028 = 0.375\ \text{kg/m}^2.$$

We assume, as before, that the projectile tears ONLY TWO subnet cables. This is the conservative case; more will be more effective in stopping the projectile.

3. Cost of net and protection. The price of the glass fiber is 0.7$/kg. If price of artificial fiber is c = 2$/kg, the cost of 1 m^2 net is about C = 0.75$/m^2. The net for 1 km^2 costs about $0.75 M/km^2, for small city having area 10 km^2 the net cost $7.5 million. We need 1- 2 masts in 1 km^2. If the inflatable support mast has diameter 20 m, height 500 m and cost about $10,000 each, the total mast cost is about $200,000. The protection from Qassam rockets of the small city will costs about $8 -10 millions.

Project #2. Protection of Battle-front from Shells

having mass m = 6.4 kg, diameter D = 76 mm and speed V_0 = 600 m/s, Figure 4a.

Let us to consider the protection of a battle-front from a modern typical 76 mm caliber cannon having 6,4 kg shell. The nozzle speed of the shell is about 600 ÷750 m/s, but at distance ~200-400m m the air drag decreases this speed to 600 m/s. That way we take in our computation the contact speed V_0 = 600 m/s.

We calculate the data and cost for front line 1 km. If you know the front length, you easily can recalculate the data for new length.

1. Computation of parachute net. Let us take the parachute area S = 0.15 m^2 and V/V_0 = 0.05, V = 30 m/s. The V_{min} = 26.7 m/s. From equations (4)-(14) we find:

1) Coefficient $a = 0.6S/m = 0.6\times0.15/6.4 = 0.014$ [1/m].

2) Braking distance $L = (-1/a) \times \ln(V/V_0) = (-1/0{,}014)\times\ln(0.05) = 214$ m.

3) The maximal braking force and overload of projectile compute by equations:

$$P_m = amV_0^2 = 0{,}014\cdot6.4\cdot600^2 = 3.22\cdot10^4\ N,\quad n_m = \frac{P_m}{mg} = \frac{3.22\cdot10^4}{6.4\cdot9.81} = 513\ g\,.$$

4) The cross-section and diameter of subnet cable are computed by equation (for σ = 200 kg/mm^2 = 2×10^9 N/m^2):

$$s_1 = \frac{P_m}{2\sigma} = \frac{3.22\cdot10^4}{2\cdot2\cdot10^9} = 8\cdot10^{-6}\ m^2 = 8\ mm^2,\quad d = \sqrt{\frac{4s}{\pi}} = 3.2\ mm\,.$$

5) 1 m^2 mass of subnet cables, mass of parachute, and cell mass of subnet,
1 m^2 mass of total parachute net (for γ = 1500 kg/m^2, l_1 = 0.04 m, l_2 = 0.2 m) are:
a) Mass of one parachute for safety parachute stress σ_p = 10^9 N/m^2:

$$p = \frac{\rho V_0^2}{2} = 0.61\cdot600^2 = 22\cdot10^4\ \frac{\text{N}}{\text{m}^2},\quad r = \sqrt{\frac{S}{\pi}} = 0.22\ \text{ m},\quad \delta = \frac{pr}{2\sigma_p} =$$
$$= 2.42\cdot10^{-5}\ \text{m},\quad m_1 = 2\gamma\delta S = 2\cdot1500\cdot2.42\cdot10^{-5}\cdot0.15 = 0.011\ \text{kg}$$

1 m^2 of parachute net for step of main net l_2 = 0.2 m and cell area $S_b = l_2^2 = 0.04$ m^2 requests $N = 1/S_b$ = 25 parachutes/m^2.

b) 1 m^2 mass of subnet cables having small net step $l_1 = 0.04$ m:

$$m_2 = \frac{2\gamma s_1}{l_1} = \frac{2 \cdot 1500 \cdot 8 \cdot 10^{-6}}{0.04} = 0.6 \text{ kg/m}^2 .$$

c) Mass of 1 m^2 main net cables (for $s_2 = 2s_1$, $S_b = 0.04$ m^2) is:

$$m_3 = \frac{2\gamma s_2}{l_2} = \frac{2 \cdot 1500 \cdot 2 \cdot 8 \cdot 10^{-6}}{0.2} = 0.24 \text{ kg/m}^2 .$$

The total 1 m^2 mass of parachute net

$m_n = Nm_1 + m_2 + m_3 = 25 \times 0.011 + 0.6 + 0.24 = 0.851$ kg,

Mass of one cell subnet cable and one parachute ($S_b = 0.04$ m^2):

$m_s = m_1 + m_2 S_b = 0.011 + 0.6 \times 0.04 = 0.035$ kg.

6) Tear out the subnet and parachute by projectile from AB-Net. The overload and force in impact moment is ($L_2 = (1 \div 2) \times l_2$):

$$n_s = \frac{V_0^2}{2gL_2} = \frac{600^2}{2 \cdot 9.81 \cdot 0.4} = 4.5 \cdot 10^4 \ g,$$

$$P_s = gn_s m_s = 9.81 \cdot 4.5 \cdot 10^4 \cdot 0.035 = 1.54 \cdot 10^4 \ N < P_m = 3.22 \cdot 10^4 \ N$$

Our net mass is high. We can decrease it up $m_n = 0.5$ kg/m^2 if we decrease the parachute area up $S = 0{,}1$ m^2. But braking distance increases up $L = 300$ m.

We assume that projectile tears ONLY TWO subnet cables.

3. Cost of net and protection. If price of artificial fiber is $c = 2$\$/kg, the cost of 1 m^2 net is about $C = 1.8$\$/m^2. The net for 1000×20 m^2 costs about \$36,000/km. If the support mast has height 20 m and cost about \$100 each, and distance between them is 50 m, the total mast cost is about \$2,000/km. The protection from bullets and small shells of the 1 km of battle-front will costs about \$60,000 - 80,000/km.

Project #3. Protection of Place, Road, Base, Meeting, Building, etc. from Kalashnikov Sub-machine Gun

The Kalashnikov sub-machine gun is widely used by terrorists. Caliber is 7.6 mm, the nozzle speed is 715 m/s but at distance 150-250 m the bullet speed is about $V_0 = 600$ m/s,. Assume the mass of bullet is m = 10 g =0.01 kg.

Let us to consider the protection of a battle-front from the Kalashnikov sub-machine gun. We calculate the data and cost for front line 1 km. If you know the battlefront's length, you easily may recalculate the data for the given installation.

1. Computation of parachute net. Let us take $S = 10$ cm^2 = 10^{-3} m^2 and $V/V_0 = 0.05$. From equations (4)-(14) we find:

1) Coefficient $a = 0.6S/m = 0.6\times10^{-3}/10^{-2} = 0.06$ [1/m], $V_{min} = 13$ m/s.

2) Braking distance $L = (-1/a) \times \ln(V/V_0) = (-1/0,06)\times\ln(0.05) = 50$ m.

3) The maximal braking force and overload of projectile compute by equations:

$$P_m = amV_0^2 = 0,06\cdot0.01\cdot600^2 = 2.16\cdot10^2\ N,$$

$$n_m = \frac{P_m}{mg} = \frac{2.16\cdot10^2}{10^{-2}\cdot9.81} = 2.16\cdot10^3\ g$$

4) The cross-section and diameter of subnet cable are computed by equation (for $\sigma = 100$ kg/mm^2 = 2×10^9 N/m^2):

$$s_1 = \frac{P_m}{2\sigma} = \frac{2.16\cdot10^2}{2\cdot10^9} = 1.08\cdot10^{-7}\ \mathrm{m}^2 = 0.108\ \mathrm{mm}^2,\quad d = \sqrt{\frac{4s}{\pi}} = 0.14\ \mathrm{mm}.$$

5) 1 m^2 mass of subnet cables, mass of parachute, and cell mass of subnet, 1 m^2 mass of total parachute net (for $\gamma = 1500$ kg/m^2, $l_1 = 0.004$ m, $l_2 = 0.03$ m) are:

a) Mass of one parachute for safety parachute stress $\sigma_p = 10^9$ N/m^2:

$$p = \frac{\rho V_0^2}{2} = 0.61\cdot600^2 = 22\cdot10^4\ \frac{\mathrm{N}}{\mathrm{m}^2},\quad r = \sqrt{\frac{S}{\pi}} = 0.018\ \ \mathrm{m},\quad \delta = \frac{pr}{2\sigma_p} =$$

$$= 2\cdot10^{-6}\ \mathrm{m},\quad m_1 = 2\gamma\delta S = 2\cdot1500\cdot2\cdot10^{-6}\cdot0.001 = 6\cdot10^{-6}\ \mathrm{kg}$$

1 m^2 of parachute net for step of main net $l_2 = 0.03$ m and cell area $S_b = l_2^2 = 0.0009$ m^2 requests $N = 1/S_b = 1111$ parachutes/m^2.

b) 1 m^2 mass of subnet cables having small net step $l_1 = 0.004$ m:

$$m_2 = \frac{2\gamma s_1}{l_1} = \frac{2\cdot1500\cdot1.08\cdot10^{-7}}{0.004} = 0.081\ \mathrm{kg/m}^2.$$

c) Mass of 1 m^2 main net cables (for $s_2 = 2s_1$, $l_2 = 0.03$ m^2) is:

$$m_3 = \frac{2\gamma s_2}{l_2} = \frac{2\cdot1500\cdot2\cdot.1.08\cdot10^{-7}}{0.03} = 0.022\ \mathrm{kg/m}^2.$$

The total 1 m^2 mass of parachute net

$m_n = Nm_1 + m_2 + m_3 = 1111 \times 6 \times 10^{-6} + 0.081 + 0.022 = 0.11$ kg, Mass of one cell subnet cable and one parachute ($S_b = 0.0009$ m^2):

$$m_s = m_1 + m_2 S_b = 6 \times 10^{-6} + 0.081 \times 0.0009 = 78 \times 10^{-6} \text{ kg}.$$

6) Tear out the subnet and parachute by projectile from AB-Net. The overload and force in *impact moment* is ($L_2 = (1 \div 2) \times l_2$):

$$n_s = \frac{V_0^2}{2gL_2} = \frac{600^2}{2 \cdot 9.81 \cdot 0.06} = 30 \cdot 10^4 \ g,$$

$$P_s = gn_s m_s = 9.81 \cdot 30 \cdot 10^4 \cdot 78 \cdot 10^{-6} = 230 \ N > P_m = 216 \ N$$

We receive $P_s > P_m$. That means the cell size 3×3 cm is small. One must decreases to 4×4 cm. We assume that projectile tears ONLY TWO subnet cables.

3. Cost of net and protection. If price of artificial fiber is c = 3\$/kg, the cost of 1 m^2 net is about C = 0.4\$/m2. The net for 1000×10 m^2 costs about \$4000/km and weighs <1350 kg. If the support mast has height of 10 m and cost about \$50 each, and the distance between them is 20 m, the total mast cost is about \$2500/km and each mast supports ~27 kg. The total protection from submachine gun and rifle of the 1 km of battle-front will cost about \$6500/km (without labor cost).

The author is prepared to discuss the problems with organizations which are interested in research and development related projects.

Discussion

For over a century has raged the competition between armor and shell. The tank armor has reached 200 mm thickness, the military ship armor 400 mm and more. The tank itself as a consequence became very expensive, heavy and awkward. The military ships became heavy, expensive and required deep water.

Entire human generations have become accustomed to the idea of steel or super-hard ceramic armor and when they hear the possibility bruited publicly that a raised thin net can offer protection to humans from shells or bullets they may feel a certain amount of intellectual and/or emotional disorientation.

But the author does not offer the thin net for mobile vehicles such as a tank or ship. Instead, he offers protection for large geographical territories, cities, military and industrial bases, roads, stable battle-fronts—the entire range of what are now known as "soft targets" from enemy (terrorists) attacks – with the requirement only that they be places which can be separated by sufficient (projectile braking) distance from planned and spontaneous enemy assaults. He offers the interesting new possibility of using the overlying atmosphere (and the property of high air drag) as armor.

There are problems in adopting the idea to Earthly reality: how to join the rocket, shell or bullet to the small parachute or piece of net. But even supposing the case where, for example, the projectile did not successfully snag onto the parachute, the projectiles' very impact against the net creates a short high overload, changing the projectile trajectory; and the projectile misses its aimpoint.

The net is very cheap and the idea may be easily field tested. The need, optimal parameters may be found by experiments.

The cost of the offered passive AB-Net Protection System may be less by hundred of times, then the cost of an *effective* active anti-rocket system (tens of billions $US). The anti-rocket system is useless in peacetime and it may be useless soon in wartime because of easily adopted enemy countermeasures.

The offered AB-Nets are different for different projectiles. The net against Kalashnikov automatic gunfire probably cannot join enough parachutes to a 76 mm gun shell. But in any case the impact to the net changes the projectile trajectory of projectile and one shot would miss its aim. A barrage, of course, as always, presents the target with greater dangers. For protection from different projectiles we can use the suitable nets located in ascending-size series.

The one lack of conventional AB-Nets would be the significant physical drawback that they also shield the assulating enemy from reciprocal fire from victims within the AB-Nets. But the author has also designed a specialized AB-Net which protects one side and still allows firing from the other side. He has numerous inventions concerning details of design: How the parachutes open, net form details, connection cables and the proper way to align them, etc.

The associated problems are researched in author references [1]-[14].

RESULTS

Authors offer the described inexpensive AB-Nets macro-project which protects cities, military bases and vital industial sites, and stabilized battle-fronts from rockets, mortar shell, rifle projectiles and bombs delivered by rockets, missiles, guns and aviation.

The principal listed advantages of the offered AB-Nets follow:

1. AB-Nets may be cheaper by hundreds to thousands of times than current comparable anti-rocket systems or defense systems.
2. The city employing the film AB-Dome for supporting the AB-Nets can has fine climate inside (the domed city)[1]. (This also may enable its financing).
3. AB-Nets may be installed in some hours (the current defense zone against incoming supersonic weapons is built over some years and cost some billions of dollars).
4. AB-Nets do not need high-technology and can be built by a poor country.
5. The AB-Net is easy field tested and evaluated at extremely low cost.
6. Author has a subdesign of AB-Nets which allow protecting army soldiers from enemy bullets and shells and simultaneously allows shooting towards the enemy through AB-Nets. (This prospect offers deterrence to all potential attackers if they are made aware by past unsuccessful attack experience.)
7. Height (up to 400-500 m) city AB-Nets may be used for high antennae site for regional TV, for cell phone communication, for long distance location (differential GPS service). (This also may enable its financing).

8. Unlike any known active ballistic missile defense, offered AB-Nets also can defend against submunitions and dispersed killing agents such as shrapnel.
9. The height over the city of suspended AB-Nets may be used for the high altitude windmills (and the getting of cheap renewable wind energy).
10. An AB-Net covered by thin film may be used for night illumination and entertainment or advertising (this also may enable its financing).
11. Additional applications of the closed author ideas the reader may find in [1]-[14].

REFERENCES

(The reader may find some of these articles at the author's web page: *http:// Bolonkin.narod.ru/p65.htm*, *http://arxiv.org*, search "Bolonkin", in the book "*Non-Rocket Space Launch and Flight*", Elsevier, London, 2006, 488 pgs. and in book "*New Concepts, Ideas, Innovations in Aerospace, Technology and Human Science*", NOVA, 2008, 430 pgs.)

[1] Bolonkin A.A., *Cheap Method for Shielding a City from Rocket and Nuclear Warhead Impacts. http://arxiv.org* search for "Bolonkin".

[2] Bolonkin A.A., Cathcart R.B., *Inflatable 'Evergreen' Polar Zone Dome (EPZD) Settlements,* 2006. *http://arxiv.org* search "Bolonkin".

[3] Bolonkin A.A., *Control of Regional and Global Weather.* 2006. *http://arxiv.org* search for "Bolonkin".

[4] Bolonkin A.A., *AB Method of Irrigation without Water* (Closed-loop water cycle). 2007. *http://arxiv.org* search for "Bolonkin".

[5] Bolonkin A.A., *Cheap Textile Dam Protection of Seaport Cities against Hurricane Storm Surge Waves*, Tsunamis, and Other Weather-Related Floods, 2006. *http://arxiv.org*.

[6] Bolonkin, A.A. and R.B. Cathcart, *Antarctica: A Southern Hemisphere Windpower Station? http://arxiv.org*, 2007.

[7] Cathcart R.B. and Bolonkin, A.A. *Ocean Terracing,* 2006. *http://arxiv.org*.

[8] Bolonkin, A.A. and R.B. Cathcart, T*he Java-Sumatra Aerial Mega-Tramway,* 2006. *http://arxiv.org*.

[9] Bolonkin, A.A., "Optimal Inflatable Space Towers with 3-100 km Height", *Journal of the British Interplanetary Society* Vol. 56, pp. 87 - 97, 2003.

[10] Bolonkin A.A., *Inflatable Dome for Moon, Mars, Asteroids and Satellites, Presented as paper AIAA-2007-6262 by AIAA Conference "Space-2007",* 18-20 September 2007, Long Beach. CA, USA

[11] Bolonkin, A.A., *Electrostatic Space Towers,* 2007, *http://arxiv.org*, search "Bolonkin".

[12] Bolonkin A.A., *Non-Rocket Space Launch and Flight,* Elsevier, London, 2006, 488 ps.

[13] Bolonkin A.A., *"*New Concepts, Ideas, Innovations in Aerospace, Technology and Human Science*"*, NOVA, 2008, 430 pgs.

[14] Bolonkin, A.A. and R.B. Cathcart, Inflatable 'Evergreen' Dome Settlements for Earth's Polar Regions. *Clean. Techn. Environ. Policy.* DOI 10.1007/s10098.006-0073.4.

[15] Macro-Engineering: *A Challenge for the Future*. Springer, 2006. 318 ps. Collection articles.

[16] Galasso F.S., Advanced Fibers and Composite, Gordon and Branch Sci. Publisher, 1989.

[17] *Carbon and High Performance Fibers Directory and Data Book,* London-New. York: Chapmen& Hall, 1995, 6th ed., 385 p.

[18] Concise Encyclopedia of Polymer Science and Engineering, Ed. J.I.Kroschwitz, N.Y.,Wiley, 1990, 1341 p.

[19] Dresselhaus, M.S., Carbon Nanotubes, by, Springer, 2000.

[20] *Wikipedia. http://wikipedia.org*. Israel.

Chapter 10

AB-SPACE PROPULSION*

ABSTRACT

On 4 January 2007 the article "Wireless Transfer of Electricity in Outer Space" in http://arxiv.org was published wherein was offered and researched a new revolutionary method of transferring electric energy in space. In next article (see http://arxiv.org) was developed the theory of new engine.

That Chapter describes a new engine which produces a large thrust without throwing away large amounts of reaction mass (unlike the conventional rocket engine). A sample computed project shows the big possibilities opened by this new "AB-Space Engine". The AB-Space Engine gets the energy from ground-mounted power; a planet's electric station can transfer electricity up to 1000 millions (and more) of kilometers by plasma wires. Author shows that AB-Space Engine can produce thrust of 10 tons (and more). That can accelerate a spaceship to some thousands of kilometers/second. AB-Space Engine has a staggering specific impulse owing to the very small mass expended. The AB-Space Engine reacts not by expulsion of its own mass (unlike rocket engine) but against the mass of its planet of origin (located perhaps a thousand of millions of kilometers away) through the magnetic field of it's plasma cable. For creating this plasma cable the AB-Space Engine spends only some kg of hydrogen.

Keywords: AB-Space Engine, AB propulsion, transferring of electricity in space

INTRODUCTION

General Information

A *rocket* is a vehicle, missile or aircraft which obtains thrust by the reaction to the ejection of fast moving fluid from within a rocket engine. Chemical rockets operate due to hot exhaust gas made from "propellant" acting against the inside of an expansion nozzle. This

---

* Presented in http://arxiv.org (1 March, 2008)

generates forces that both accelerate the gas to extremely high speed, as well as, since every action has an equal and opposite reaction, generating a large thrust on the rocket.

The history of rockets goes back to at least the 13th century, possibly earlier. By the 20th century it included human spaceflight to the Moon, and in the 21st century rockets have enabled commercial space tourism.

Rockets are used for fireworks and weaponry, as launch vehicles for artificial satellites, human spaceflight and exploration of other planets. While they are inefficient for low speed use, they are, compared to other propulsion systems, very lightweight, enormously powerful and can achieve extremely high speeds.

Chemical rockets contain a large amount of energy in an easily liberated form, and can be very dangerous, although careful design, testing, construction and use can minimise the risks.

A *rocket engine* is a jet engine that takes all its reaction mass ("*propellant*") from within tankage and forms it into a high speed jet, thereby obtaining thrust in accordance with Newton's third law. Rocket engines can be used for spacecraft propulsion as well as terrestrial uses, such as missiles. Most rocket engines are internal combustion engines, although non combusting forms also exist.

Transfer of electricity into space. The production, storage, and transference of large amounts of electric energy are an enormous problem for humanity--especially of energy transfers in the vacuum of outer space. Entire spheres of industry should be searching for, since they badly need, revolutionary ideas. If in the production of energy, space launch and flight we have new ideas (see [1]-[17]), but we have not seen revolutionary ideas in transferring and storage of energy except for reference [5].

However, if we solve the problem of transferring energy in outer space, then we solve some of the problems of manned and unmanned space flight. For example, spaceships can move long distances by using efficient electric engines, orbiting satellites can operate for unlimited time periods without falling prey to orbital decay and premature re-entry to Earth's atmosphere, communication satellites can transfer a strong signal directly to customers, the International Space Station's users can conduct many practical experiments and the global space industry can produce new materials. In the future, Moon and Mars outposts can better explore the celestial bodies on which they are placed at considerable expense [1].

Another important Earth mega-problem is efficient transfer of electric energy for long distances (intra-national, international, inter-continental). Nowadays, a lot of loss occurs from such energy transformation. The consumption of electric energy strongly depends on time (day or night), weather (hot or cold), and season (summer or winter). But an electric station can operate most efficiently in a permanent base-load generation regime. We need to transfer the energy long distance to any region that requires a supply in any given moment or to special pumped storage stations. One solution for this macro-problem is to transfer energy from Europe to the USA during nighttime in Europe and from the USA to Europe when it is night in the USA. Another solution is efficient energy storage, which allows people the option to save electric energy [1].

The storage of a big electric energy can help to solve the problem of cheap space launch. The problem of an acceleration of a spaceship can be solved by use of a new linear electrostatic engine suggested in [6]. However, the cheap cable space launch offered by author [5] requires the use of gigantic amounts of energy in short time period. (It is inevitable for any launch method because we must accelerate big masses to the very high speeds of 8 - 11 km/s). But it is impossible to turn off a whole state and connect an entire electric station to one customer. The offered electric energy storage can help solving this mega-problem for humanity [1]-[17].

Railgun

The scientists used a railgun for high acceleration of small conducting body. A railgun is a form of gun that converts electrical energy (rather than the more conventional chemical energy from an explosive propellant) into projectile kinetic energy. It is not to be confused with a coilgun (Gauss gun). Rail guns use magnetic force to drive a projectile. Unlike gas pressure guns, rail guns are not limited by the speed of sound in a compressed gas, so they are capable of accelerating projectiles to extremely high speeds (many kilometers per second).

A wire carrying an electrical current, when in a magnetic field, experiences a force perpendicular to the direction of the current and the direction of the magnetic field.

In an electric motor, fixed magnets create a magnetic field, and a coil of wire is carried upon a shaft that is free to rotate. An electrical current flows through the coil causing it to experience a force due to the magnetic field. The wires of the coil are arranged such that all the forces on the wires make the shaft rotate, and so the motor runs.

A railgun consists of two parallel metal rails (hence the name) connected to an electrical power supply. When a conductive projectile is inserted between the rails (from the end connected to the power supply), it completes the circuit. Electrical current runs from the positive terminal of the power supply up the positive rail, across the projectile, and down the negative rail, back to the power supply (Figure 1).

This flow of current makes the railgun act like an electromagnet, creating a powerful magnetic field in the region of the rails up to the position of the projectile. In accordance with the right-hand rule, the created magnetic field circulates around each conductor. Since the current flows in opposite direction along each rail, the net magnetic field between the rails (B) is directed vertically. In combination with the current (I) flowing across the projectile, this produces a Lorentz force which accelerates the projectile along the rails. The projectile slides up the rails away from the end with the power supply.

If a very large power supply providing a million amperes of current is used, then the force on the projectile will be tremendous, and by the time it leaves the ends of the rails it can be travelling at many kilometres per second. So far, about 20 kilometers per second has been achieved with experimental small projectiles explosively injected into the railgun. Although these speeds are theoretically possible, the heat generated from the propulsion of the object is enough to rapidly erode the rails. Such a railgun would require frequent and inordinately expensive replacement of the guide rails, or use a perhaps costly heat-resistant construction material that would be conductive enough to produce the same desired effect.

The need for strong conductive materials with which to build the rails and projectiles; the rails need to survive the violence of an accelerating projectile, and heating due to the large currents and friction involved. The force exerted on the rails consists of a recoil force - equal and opposite to the force propelling the projectile, but along the length of the rails (which is their strongest axis) - and a sideways force caused by the rails being pushed by the magnetic field, just as the projectile is itself. The affixed rails need to endure this without bending, and thus must be very securely mounted.

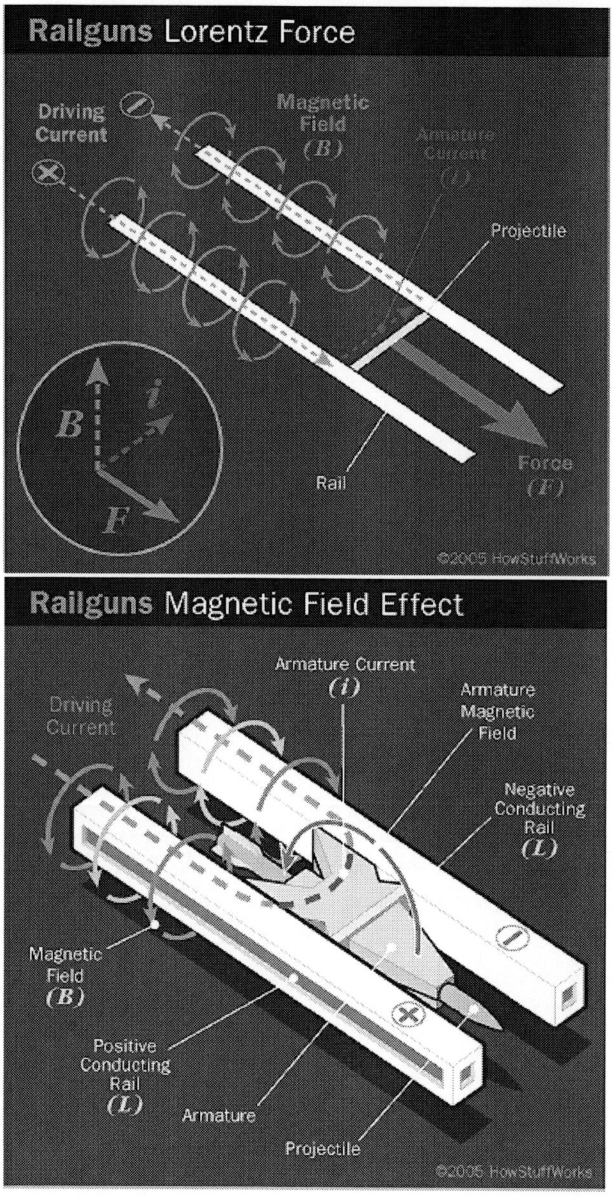

Figure 1. Schematic diagrams of a railgun.

The power supply must be able to deliver large currents, with both capacitors and compulsators being common.

The rails need to withstand enormous repulsive forces during use, and these firing forces will tend to push them apart and away from the projectile. As rail/projectile clearances increase, electrical arcing develops, which causes rapid vaporization and extensive damage to the rail surfaces and the insulator surfaces. This limits most research railguns to one shot per service interval.

Some persons have mentally speculated that there are fundamental physical limits to the projectile exit velocity due to the inductance of the system, and particularly of the rails; but the USA's federal government funded R&D has made significant progress in railgun design and has recently revealed designs of a railgun that would be used on a naval vessel. The designs for naval vessels, however, are limited by their required power usages for the magnets in the rail guns. This level of power is currently unattainable on a ship and currently reduces the usefulness of the warfare concept for military purposes somewhat.

Massive amounts of heat are created by the electricity flowing through the rails, as well as the friction of the projectile leaving the device. This leads to three main problems: melting of equipment, safety of personnel, and detection by enemy forces. As briefly discussed above, the stresses involved in firing this sort of device require an extremely heat-resistant material. Otherwise the rails, barrel, and all equipment attached would melt or be irreparably damaged. Current railguns are not sufficiently powerful to create enough heat to damage anything; however the military is pushing for more and more powerful prototypes. The immense heat released in firing a railgun could potentially injure or even kill bystanders. The heat released would not only be dangerous, but easily detectable. While not visible to the naked eye, the heat signature would be unmistakable to infrared detectors. All of these problems can be solved by the invention of an effective cooling method.

Railguns are being pursued as weapons with projectiles that do not contain explosives, but are given extremely high velocities: 3500 m/s (11,500 ft/s) or more (for comparison, the M16 rifle has a muzzle speed of 930 m/s, or 3,000 ft/s), which would make their kinetic energy equal or superior to the energy yield of an explosive-filled shell of greater mass. This would allow more ammunition to be carried and eliminate the hazards of carrying explosives in a tank or naval weapons platform. Also, by firing at higher velocities railguns have greater range, less bullet drop and less wind drift, bypassing the inherent cost and physical limitations of conventional firearms - "*the limits of gas expansion prohibit launching an unassisted projectile to velocities greater than about 1.5 km/s and ranges of more than 50 miles [80 km] from a practical conventional gun system.*"

If it is even possible to apply the technology as a rapid-fire automatic weapon, a railgun would have further advantages in increased rate of fire. The feed mechanisms of a conventional firearm must move to accommodate the propellant charge as well as the ammunition round, while a railgun would only need to accommodate the projectile. Furthermore, a railgun would not have to extract a spent cartridge case from the breech, meaning that a fresh round could be cycled almost immediately after the previous round has been shot.

Tests. Full-scale models have been built and fired, including a very successful 90 mm bore, 9 MJ (6.6 million foot-pounds) kinetic energy gun developed by DARPA, but they all suffer from extreme rail damage and need to be serviced after every shot. Rail and insulator ablation issues still need to be addressed before railguns can start to replace conventional weapons. Probably the most successful system was built by the UK's Defence Research Agency at Dundrennan Range in Kirkcudbright, Scotland. This system has now been operational for over 10 years at an associated flight range for internal, intermediate, external and terminal ballistics, and is the holder of several mass and velocity records.

The United States of America's military services are funding railgun experiments. At the University of Texas at Austin in the Institute for Advanced Technology there, military railguns capable of delivering tungsten armor-piercing bullets with kinetic energies of nine million joules have been developed and tested. Nine mega-joules is enough energy to deliver 2 kg of projectile at 3 km/s - at that velocity a tungsten or other dense metal rod could penetrate a well armored tank.

U.S. Navy Invents Railgun.

The United States Naval Surface Warfare Center Dahlgren Division demonstrated an 8 mega-joule rail gun firing 3.2 kilogram (slightly more than 7 pounds) projectiles in October of 2006 as a prototype of a 64 mega-joule weapon to be deployed aboard Navy warships. Such weapons are expected to be powerful enough to do a little more damage than a BGM-109 Tomahawk missile at a fraction of the projectile cost.

Due to the very high muzzle velocity that can be attained with railguns, there is interest in using them to shoot down high-speed missiles.

Offered Innovations and Brief Descriptions

1. Transfer of electricity by plasma cable. The author offers a series of innovations that may solve the many macro-problems of transportation, energy and thrust in space. Below are some of them.

1. Transfer of electrical energy in outer space using a conductive cord from plasma. Author has solved the main problem - how to keep the plasma cord from dissipation, and in compressed form. He has developed the theory of space electric transference, made computations that show the possibility of realization for these ideas with existing technology. The electric energy may be transferred for hundreds millions of kilometers in space (including Moon and Mars) [1].

2. Method of construction of space electric lines and electric devices.
3. Method of utilization and tapping of the plasma cable electric energy.
4. Two methods of converting the electric energy to impulse (thrust) motion of a spacecraft (these two means are utilization of the magnetic field and of the kinetic energy of ions and electrons of the electric current).
5. Design of a triple electrostatic mirror (plasma reflector), which can reflect the plasma flow [1].

Below are some succinct descriptions of some constructions made possible by these revolutionary macro-engineering ideas.

1. Transferring electric energy in Space. The electric source (generator, station) is connected to the distant location in space by two artificially generated rarefied plasma cables (Figure 1a). These cables can be created by a plasma beam [1, 8] sent from the Moon, Earth mounted super high tower, or from a space station in low Earth orbit, or a local base at the target location. If the plasma beam is sent remotely from the Earth, a local reflector station is required at the target site or at a third location to turn the circuit back toward its' starting point and closure.

The plasma cable, in radial direction, may also be constructed of ulta-cold plasma.

The plasma cable is self-supported in cable form by the magnetic field created by the electric current going through the plasma cable. The axial electric current produces an contracting magnetic pressure opposed to an expansive gas dynamic plasma pressure (the well-known theta-pinch effect)(Figure 3). The plasma has a good conductivity (equal to that of silver and more) and the plasma cable can have a very big cross-section area (up to thousands of square meters cross-section). The plasma conductivity does not depend on its' density. That way the plasma cable has no large resistance although the length of plasma cable is hundreds of millions of kilometers. The needed minimum electric current is derived from parameters of a plasma cable researched in the theoretical section of this article.

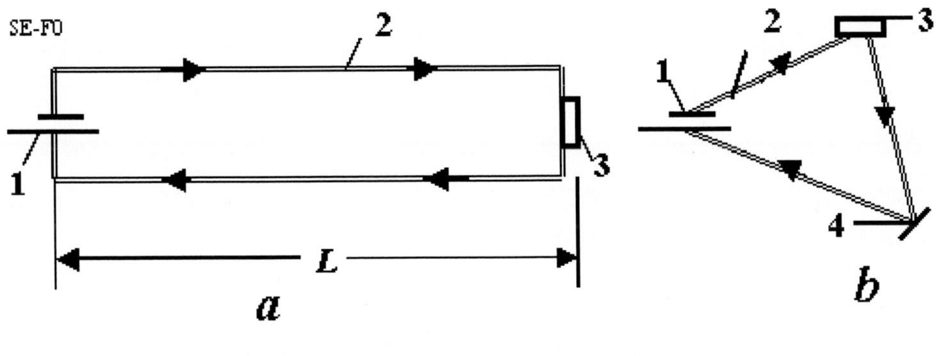

Figure 2. Long distance plasma transfer electric energy and thrust in outer space. ***a*** - plasma transfer with parallel plasma cable, ***b*** - plasma transfer with triangular (three-wire) plasma cable. *Notations*: 1 - current source (generator), 2 - plasma wire (cable), 3 - spaceship, orbital station or other energy destinations, 4 - plasma reflector located at planet, asteroid or space station.

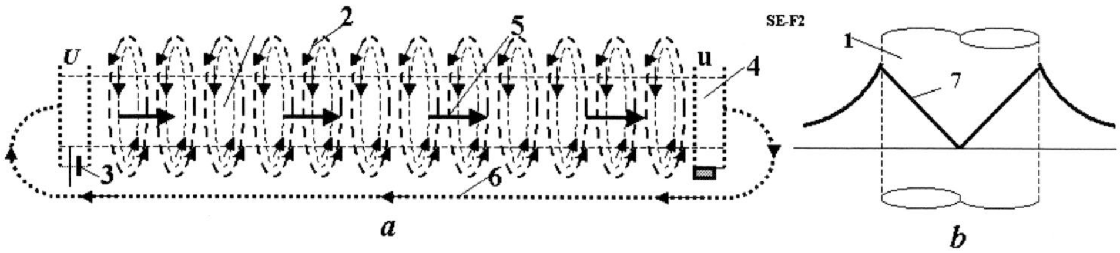

Figure 3. ***a***. A plasma cable supported by its' own magnetic field, ***b***. Magnetic intensity into and out of plasma cable. *Notations*: 1 -plasma cable, 2 - compressing magnetic field, 3 - electric source, 4 - electric receiver, 5 - electric current, 6 - back plasma line; 7 – magnetic intensity into and out of plasma cable.

The parallel cables having opposed currents repel each other (Figure 2a)(by magnetic force). This force may be balanced by attractive electric force if we charge the cables by electric charges (see theoretical section). They also can be separated by a special plasma reflector as it shown in figures 2b. The electric line can be created and exist independently. The spaceship connects to this line at a suitable point. By altering the diameter and direction of the plasma cable we can supply energy to a spacecraft. Though we must supply energy to accelerate the spacecraft we can also regenerate energy by its braking activity. At any time, the spaceship can disconnect from the line and can exist without line support (propulsion, electricity, etc). The apparatus can hook up to or disconnect from the plasma cable at will. But breaking (loss of continuity) of the plasma cable itself destroys the plasma cable line to the remote location! We must have additional (parallel) plasma lines and apparatus must disconnect from a damaged or occulted (for example on the far side of a remote planet) plasma line and connect to another line to keep the connection in existence. The same situation is true in a conventional electric net. The apparatus can also restore the damaged part of plasma line by own injected plasma, but the time for repairing is limited (by tens of minutes or a few hours). The original station can also to send the plasma beam which connects the ends of damaged part of the line.

The electric tension (voltage) in a plasma cable is between two ends (for example, as cathode- anode) of the conductor in the issuing electric station (electric generator) [1,8,9]. The plasma cable current has two flows: Electron (negative) flow and opposed ions (positive) flow in the same cable.

These flows create an electric current. (In metal we have only electron flow, in liquid electrolytes we have ions flowing).

The author offers methods (for extraction and inserting) of energy from the plasma electric cable (Figure 4) by customer (spacecraft, other energy destination or end user).

The double net can accelerate the charged particles and insert energy into plasma cable (Figure, 4a) or brakes charged particles and extract energy from electric current (Figure 4b). In the first case the two nets create the straight electrostatic field, in the second case the two nets create the opposed electrostatic field in plasma cable (resistance in the electric cable [1, 8, 9]) (Figures 2, 4c). This apparatus resistance utilizes the electric energy for the spaceship or space station. In the second case the charged particles may be collected into a set of thin films and emit (after utilization in apparatus) back into continued plasma cable (see [1,8,9]).

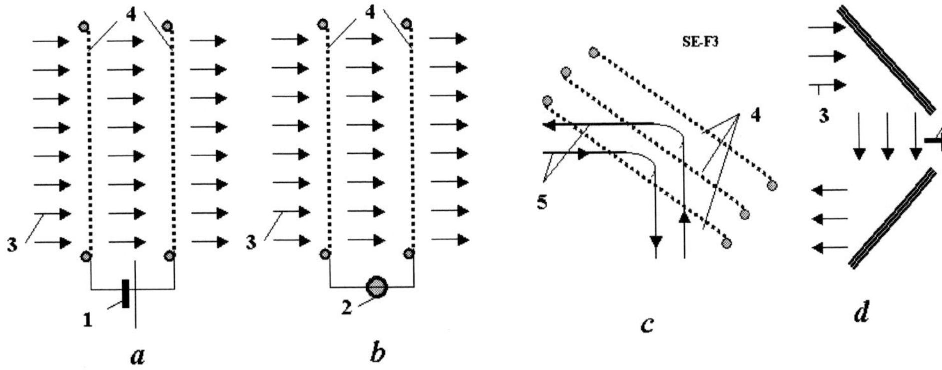

Figure 4. Getting and inserting in (off) plasma cable the energy and turning of plasma cable. ***a*** – inserting electric energy into plasma cable by means of two thin conducting nets or films; ***b*** - getting the energy from plasma cable by means of two thin conducting nets or films; ***c*** – offered triple net plasma reflector; ***d*** – double triple net plasma reflectors - the simplest AB thruster. *Notations*: 1 - spaceship or space station, 2 – receiver of energy, 3 - plasma cable, 4 - electrostatic nets, 5 – two opposed flows of charged plasma particles (negative and positive: electron and ions), 7 – thrust of AB-Space Engine.

Figure 4c presents the plasma beam reflector [1,8,9]. That has three charged nets. The first and second nets reflect (for example) positive particles, the second and third nets reflected the particles having an opposed charge.

Figure 5 shows the different design the plasma cable in space.

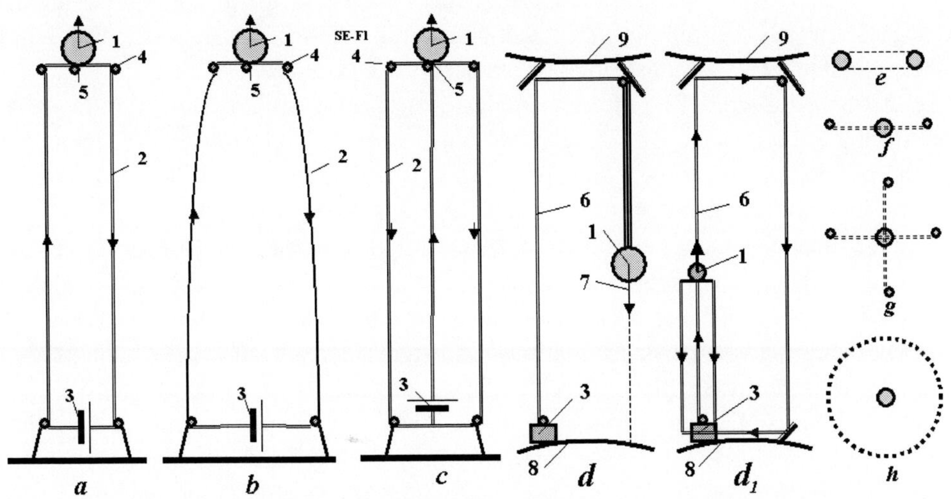

Figure 5. Transfer electricity and thrust by AB-Space Engine: ***a***. Two plasma parallel cables; ***b***. Curved cable; ***c***. Plasma multi-cables; ***d***. Transfer of back thrust through planet or asteroid; d_1. Using of ready plasma line; ***e – h***. Forms of straight and back plasma cables (cross-sections of cables). *Notations*: 1 – Space ship; 2 – plasma cable; 3 – source electricity; 4 - plasma injector; 5 – user of energy; 6 – double plasma line; 7 - thrust; 8 – Earth; 9 – planet or asteroid.

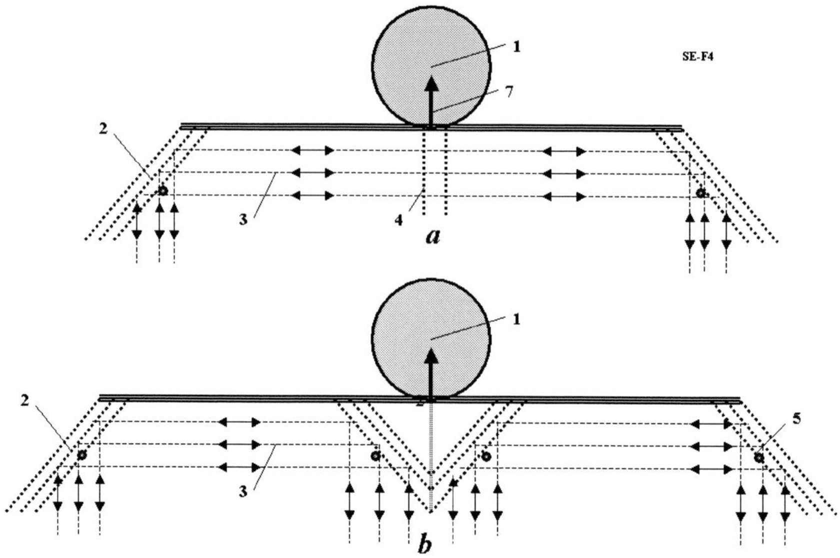

Figure 6. Some versions of AB-Space Engine (thruster). a. two cable AB-Space Engine; b. Three cable AB-Space Engine. Notations; 1 – space ship; 2 – offered special (three nets) electrostatic reflector; 3 – plasma cable; 4 – receiver or source of energy; 5 – injector of plasma, 7 – thrust.

Figure 5a shows two plasma parallel cables. Figure 5b two shows plasma parallel cables of a curved form of line. Figure 5c presents three plasma parallel cables, one to space ship and two for back (return) current. Figure 5d shows the transfer of the reverse impulse (or braking) thrust to space ship through planet or asteroid. Figures5e-h shows the different forms of the straight and back plasma cables (cross-section of cables).

2. AB-Space Engine. The offered simplest AB-Space Engine is shown in Figure 4d and more details in Figures 6a, 6-2. That includes two new triple electrostatic reflectors 2 which turn the plasma cables' (flow), (electric current 3) in back direction. The engine may also contain (optional) the plasma injectors 5 and electric generator (user) 4.

As feed material for the plasma may be used hydrogen gas, as plasma reflector may be used three conductivity nets connected to voltage sources, as generator - the double conductivity nets located into plasma flow and connected to voltage sources or users.

The other design of AB-Space Engine is shown in figure 6b. Here the central plasma flow divides in two side flows which go back to the electric station.

The AB-Space Engine works as follows. The electric current (voltage) produced by electric station (that may be located far from AB-Space Engine, for example, in orbit around the Earth or mounted on the Moon, Phobos or another space body) transfers by plasma cable to the AB-Space Engine. The power of the electric current in the plasma produces the power plasma flow of electrons and ions. The engine turns back the plasma flow (electric current) and returns it to the source electric station by the other plasma cable. The magnetic and centrifugal forces appear at the point of turning from outgoing to ingoing plasma paths place and create the thrust which can be used for movement (acceleration, braking) the space apparatus (or conventional vehicle or projectile).

Long-time readers of proposed space drive papers may suspect something fishy here. Don't worry: The AB-Space Engine doesn't violate Newton's third law of action and reaction.

The AB-Space Engine reacts against the (planet or station mounted) electric station which may be located hundreds of millions of kilometers away! No other engine has the same capability.

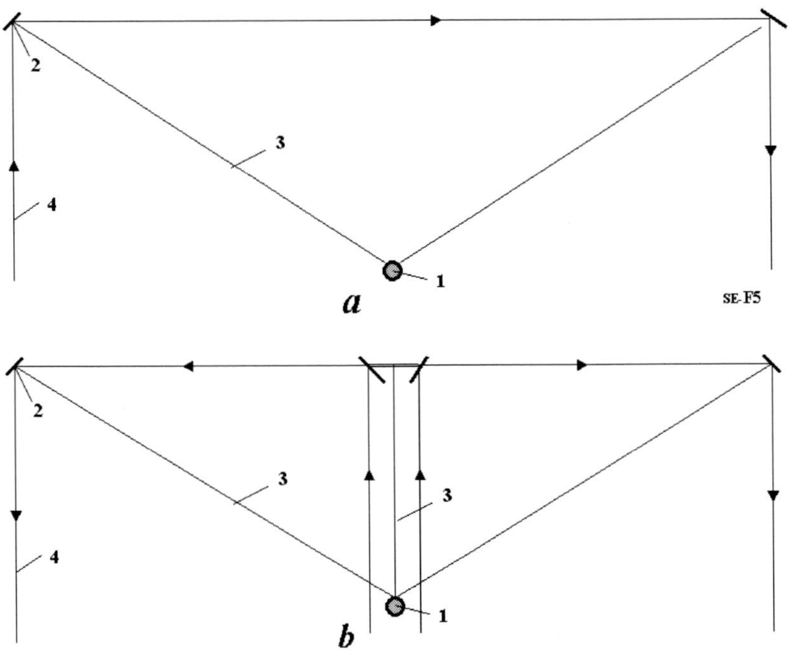

Figure 6-2. Some versions of widely (many km) AB-Space Engine (thruster). a. two cable AB-Space Engine; b. Three cable AB-Space Engine. Notations; 1 – space ship; 2 – offered special (three nets) electrostatic reflector; 3 – thin space cable connected the ship and reflector; 4 – plasma cable.

Your attention is also directed to the following differences between a railgun and an AB-Space Engine:

1) The railgun uses SOLID physical rails for delivery the electric current to conductivity projectile. These are easily damaged by huge electric current. The AB-Space Engine uses flexible plasma cables which can self-repair.
2) The railgun uses the rails which are of fixed construction (unalterable) and a spacecraft so launched can move solely in the rail direction. The AB-Space Engine creates the plasma cable in the course of apparatus movement and can select and change the apparatus' future direction.
3) Even a theoretical railgun girdling the globe of the Moon in vacuum (for star probe launch) would have a possible length of only some kilometers (as any solid construction). The plasma electric line (used byAB-Space Engine) can have a length (an acceleration path) of millions of kilometers (and thus may someday power manned craft on missions to near interplanetary space).

Theory of AB-Engine, Estimations and Computations

1. General Theory of AB-Engine and Transfer Electricity in Space

The magnetic intensity and magnetic pressure of an electric current reaches a maximum upon the surface of a plasma cable. Let us attempt to equate plasma gas pressure to a magnetic pressure and find the requested equilibrium electric current for a given (same) temperature of electrons and ions

$$P_g = 2nkT_k, \quad P_m = \frac{\mu_0 H^2}{2}, \quad H = \frac{I}{2\pi r},$$
$$P_m = P_g, \quad I = 4\pi r\left(\frac{knT_r}{\mu_0}\right)^{0.5}, \quad T_k = \frac{m_e u_r^2}{2k}, \tag{1}$$

where P_g is plasma gas pressure, N/m^2; P_m is magnetic pressure, N/m^2; n is plasma density (number of electron equals number of ions: $n = n_e = n_i$), 1/m^3; $k = 1.38\times10^{-23}$ is Boltzmann coefficient, J/K; $\mu_0 = 4\pi10^{-7}$ is magnetic constant, H/m; H is magnetic intensity, A/m; I is electric current, A; r is radius of plasma cable, m; T_r is plasma temperature in radial direction of plasma cable, K; $m_e = 9.11\times10^{-31}$ is electron mass, kg; u_r is average electron speed in radial direction of plasma cable, m/s.

Minimal Electric current. From (1) we receive relation between a minimal electric current I_{min}, gas density n and the radial temperature of electrons

$$I_m = 4\pi r\left(\frac{k\,nT_r}{\mu_0}\right)^{0.5} \approx 4.16\times10^{-8} r\sqrt{nT_r},$$
$$j_m = \frac{I}{\pi r^2} = 4\left(\frac{k}{\mu_0}\right)^{0.5}\frac{\sqrt{nT_r}}{r} \approx 1.33\cdot10^{-8}\frac{\sqrt{nT_r}}{r}, \tag{2}$$

where I_m is minimal electric current, A; j_m is density of electric current, A/m^2; $\pi r^2 = S$ is the cross-section area of plasma cable, m^2.

Assume the temperature (energy) of electrons equals temperature (energy) of ions. Let us to write well-known relations

$$j = en(u_i + u_e), \quad \frac{m_i u_i^2}{2} = \frac{m_e u_e^2}{2}, \tag{3}$$

where $e = 1.6\times10^{-19}$ C is charge of electron, C; $m_e = 9.11\times10^{-31}$ kg is mass of electron, kg; m_i is mass of ion, kg (for H_2 $m_i = 2\times1.67\times10^{-27}$ kg); u_i, u_e is speeds of ions and electrons respectively *along* cable axis produced by electric intensity (electric generator), m/s.

The computation j by Eq. (2) is presented in Figure 7.

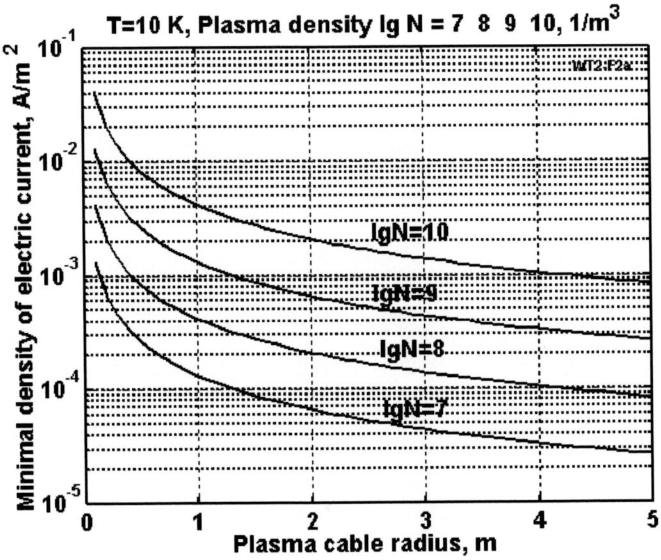

Figure 7. Minimal density of electric current in plasma cable for radial plasma temperature 10K.

From (3) we receive axial speeds of ions and electrons produced by electric intensity (electric generator)

$$u_e = \frac{j}{en(1+\sqrt{m_e/m_i})}, \quad u_i = \frac{j}{en(1+\sqrt{m_i/m_e})}. \tag{4}$$

or

$$u_e \approx 6{,}15\cdot 10^{18} j/n \quad for\ H_2 \quad u_i \approx 10^{17} j/n, \quad u_e >> u_i, \quad u = u_e + u_i \approx u_e, \quad or \quad u \approx j/en. \tag{4'}$$

Under electric intensity the electrons and ions have opposed speeds along cable axis. The computation of electron speed produced by minimal electric current is presented in Figure 8.

Temperature along plasma cable axis induces by minimal electric voltage is

$$T_k = \frac{m_e u^2}{2k} \approx 3.3\cdot 10^{-8} u^2 \ [K], \quad T = \frac{k}{e} T_k \approx 2.71\cdot 10^{-12} u^2 \quad [eV], \tag{5}$$

where T_k is induced temperature in K; T is this temperature along cable axis in eV. Computation is shown in Figure 9.

Specific Spitzer plasma resistance (the so-called Spitzer Conductivity) and typical resistance of a plasma cable can be computed by equations:

$$\rho = \eta_\perp = 1.03\times 10^{-4} Z \ln\Lambda T^{-3/2} \quad \Omega\cdot\text{m}, \quad R = \rho L/S, \tag{6}$$

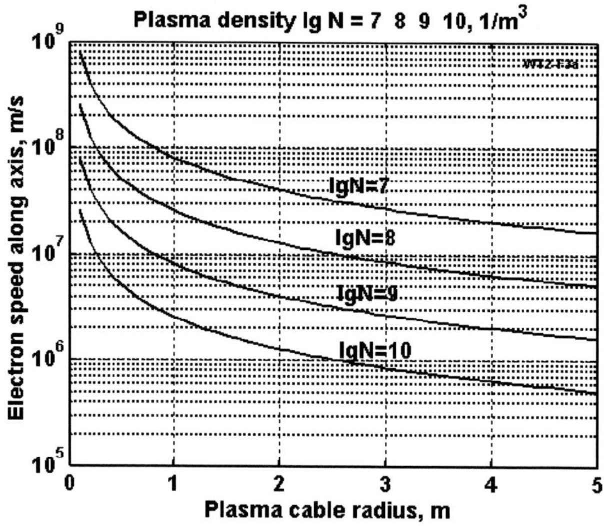

Figure 8. The electron speed produced by the electric current of the minimal current density versus plasma cable radius. The ions (H_2) speed is less 61.5 times and opposed the electron speed.

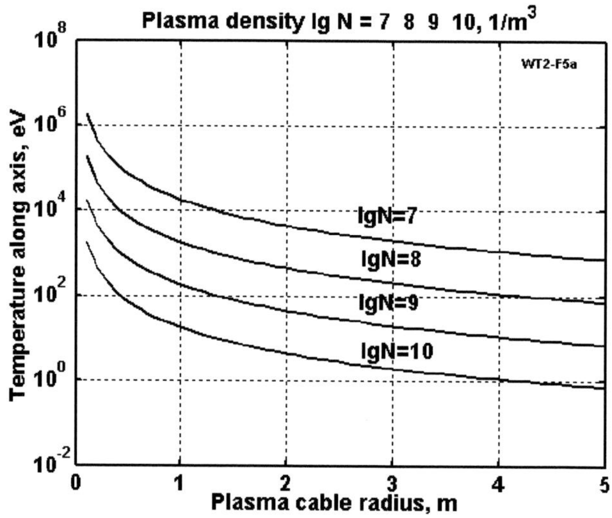

Figure 9. The temperature of electron and ions (H_2) (in eV) produced by the electric current in the minimal current density versus the plasma cable radius and different plasma density. It is assumed the ions (H_2) temperature equals the electron temperature.

where ρ is specific plasma resistance, Ωm; Z is ion charge state, ln $\Lambda \approx 5 \div 15 \approx 10$ is the Coulomb logarithm; $T = T_k k/e = 0.87\times10^{-4} T_k$ is plasma temperature along cable axis in eV; $e = 1.6\times10^{-19}$ is electron charge, C; R is electric resistance of plasma cable, Ω; L is plasma cable length, m; S is the cross-section area of the plasma cable, m^2.

The computation of the specific resistance of a plasma cable for minimal electric current is presented in Figure 10.

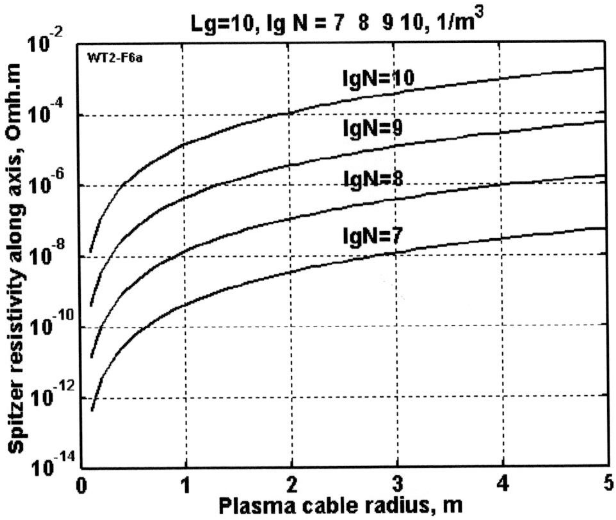

Figure 10. Specific (Spitzer) plasma resistance Ω·m of equilibrium plasma cable for the minimal electric current versus cable radius and different plasma density. Coulomb logarithm equals 10.

The requested minimum voltage, power, transmitter power and coefficient of electric efficiency are:

$$U_m = IR, \quad W_m = IU_m, \quad U = U_m + \Delta U, \qquad (7)$$
$$W = IU, \quad \eta = 1 - W_m / W = 1 - U_m / U$$

where U_m, W_m are requested minimal voltage, [V], and power, [W], respectively; U is used voltage, V; ΔU is electric voltage over minimum voltage, V; W is used electric power, W; η is coefficient efficiency of the electric line. If $\Delta U >> U_m$ the coefficient efficiency closed to 1.

Computation of loss voltage and power into plasma cable having length 100 million km is in Figures 11-12.

The mass M [kg] *of one plasma* cable is

$$M = \pi r^2 n m_i L, \qquad (8)$$

where m_i is ion mass of plasma, kg; L is length of plasma cable, m.

The mass of plasma cable is very small, about some grams for 100 millions km for $n < 10^{10}/m^3$. The mass of a plasma cable is close to zero for any practical case when R < 5 m.

The force acting in a particle (proton) moved in electric and magnetic fields may be computed by the equations:

$$\overline{F}_1 = \frac{m_i v^2}{r}. \quad \overline{F}_2 = e\overline{v}\overline{B}, \quad \overline{F}_3 = \frac{eQ_0}{4\pi\varepsilon_0 R^2}, \quad \overline{F}_4 = \gamma\frac{m_1 m_2}{R^2}, \quad \overline{F}_4 = g m_i \qquad (9)$$

where F_1, F_2, F_3, F_4 are centrifugal, Lorenz, electrostatic, and gravitational forces respectively (all vectors), N; $m_p = 1.67\times10^{-27}$ kg mass of proton (or ion m_i); v - speed of particle, m/s; e - electron (proton) charge; B - total magnetic induction (magnetic field strength), T; Q_0 - charge of central body, C; $\varepsilon_o = 8.85\times10^{-12}$ F/m - electric constant; m_1, m_2 are mass of bodies (central and particle), kg; γ – gravitational constant (for Earth $\gamma = 6.67\times10^{-11}$ m^3/kg s^2, $g_o = 9.81$ m/s^2; for Sun $g_o = 274$ m/s^2); r – radius curve, m; R – distance between charges (gravitational bodies), m.

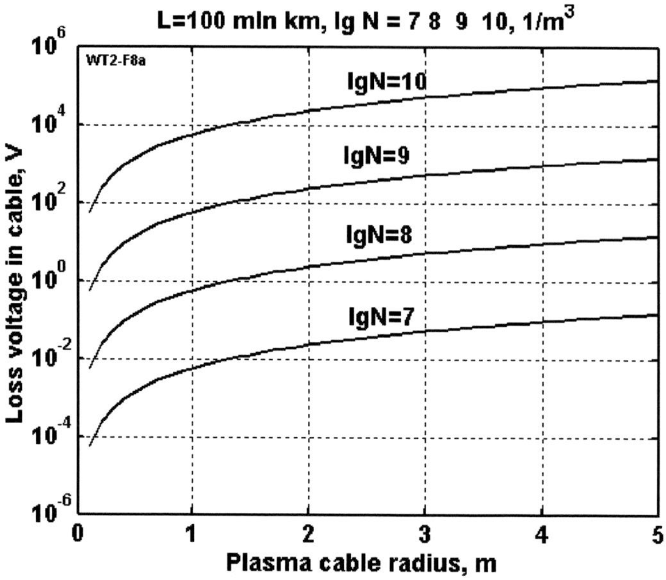

Figure 11. Loss voltage in plasma cable of 100 millions km length via cable radius for the minimal electric current and different plasma density.

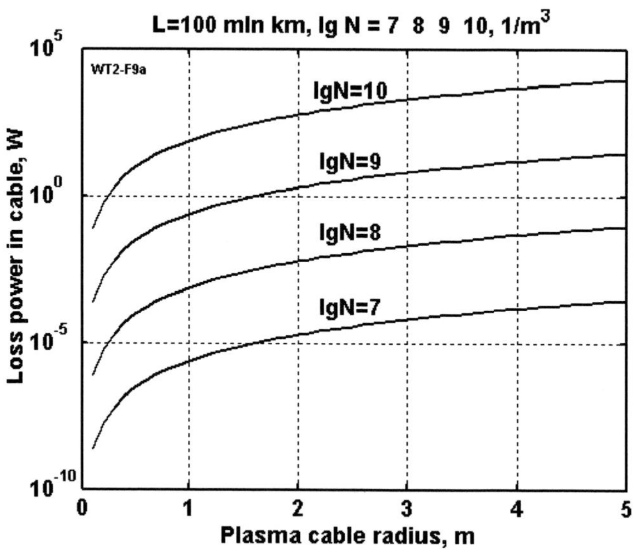

Figure 12. Power loss in plasma cable of 100 million km length via cable radius for the minimal electric current and different plasma density.

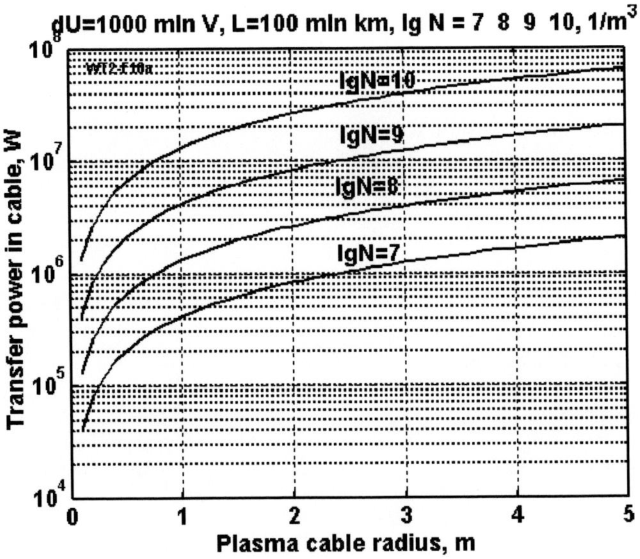

Figure 13. Electric power transfers by plasma cable of 100 millions km length via cable radius for the minimal electric current, over voltage 10^9 V and the different plasma density. Coefficient efficiency is about 0.9999.

The equilibrium condition is:

$$\sum_i F_i = 0 .$$

Magnetic Pressure (magnetic thrust) from the Plasma Cable. The plasma exerts a pressure within the plasma cable. This pressure is small, but the cable can has a large diameter (up 200 m or more) and this pressure acting over a long time can accelerate or brake a space apparatus with no reaction mass. This magnetic pressure P [N/m^2] from only one cable can be computed by equations:

$$P_m = \frac{\mu_0 H^2}{2}, \quad H = \frac{I}{2\pi r}, \quad P = \frac{1}{2} 2 P_m S = \frac{\mu_0}{4\pi} I^2 . \tag{10}$$

Estimation. For $I = 10^4$ A, the magnetic pressure equals 10 N; for $I = 10^5$ A, it equals 1000 N; for $I = 10^6$ A, the thrust of one cable is $P = 10^5$ N = 10 tons.

That is magnetic thrust from one cable. The AB-Space Engine has two cables (incoming and out coming), that means the magnetic thrust from two cable will be (at minimum) two times more. If we compute the horizontal part of plasma cable which is pressed by outer plasma magnetic field the full thrust is:

$$P = \frac{\mu_0}{2\pi} I^2 \ln \left| \left(\frac{d - r}{r} \right)^2 \right| , \tag{11}$$

where d is distance between centers of the incoming and out coming plasma cables.

Electric (kinetic) Pressure from the Plasma Cable. The high speed electrons and ions of electric current within plasma cable have kinetic energy. This energy produces kinetic (electric) pressure when space ship or final station uses the electric energy. Let us estimate the electric pressure.

Specific (kinetic) energy of electric current into plasma cable is

$$E = 0.5n(m_e u_e^2 + m_i u_i^2), \text{ [W/m}^3\text{]} \tag{12}$$

Substitute Eqs.(4) in (12) we have

$$E = \frac{j^2}{2ne^2}\left[\frac{m_e}{(1+\sqrt{m_e/m_i})^2} + \frac{m_i}{(1+\sqrt{m_i/m_e})^2}\right] \approx \frac{j^2}{2ne^2}\left[m_e + m_e\right] = \frac{m_e}{e^2}\frac{j^2}{n}. \tag{13}$$

But specific energy equals the specific pressure $P_e = E_s$ [N/m^2].

$$P_e = \frac{m_e}{e^2}\frac{j^2}{n} \approx 3.36\times 10^7 \frac{j^2}{n}. \tag{14}$$

Estimation: For j =100 A/m^2, n = 10^{10} 1/m^3 we get P = 35.6 N/m^2 (for comparison, 1 Pa = 1 N/m^2 = 10^{-5} bar).

Full kinetic energy of charged particles the plasma cable is $E = P_e sL$ [J], where s is cross-section area of plasma cable [m^2], L is length of plasma cable [m].

Additional Power from a Space Apparatus' Motion.

This power is:

$$W = PV, \tag{15}$$

where V is apparatus speed, m/s.

Estimation. For V = 11 km/s, P = 10^{-3} N, this power equals 11 W, for P = 1 N the power equals 11000 Watts. We spend this power when space apparatus moves away from the energy source ('launch point') and receive it when apparatus approaches to the energy station. ('landing site')

Track Length of Plasma Electrons and Ions.

The track length L and the track time τ of particles is

$$L = \upsilon_T/\nu, \quad \tau = 1/\nu, \tag{16}$$

where υ_T is particle velocity, cm/s; ν is particle collision rate, 1/s.

The electron, ion, and electron-ion *thermal* collision rate are respectively:

$$\nu_e = 2.91\times10^{-6} n_e \ln\Lambda T_e^{-3/2} \quad s^{-1}$$
$$\nu_i = 4.80\times10^{-8} Z^4 \mu^{-1/2} n_i \ln\Lambda T_i^{-3/2} \quad s^{-1}, \tag{17}$$
$$\nu_{ei} = 4.4\times10^{-6} n_i \lg\Lambda T^{-3/2}.$$

where Z is ion charge state, ln $\Lambda \approx 5 \div 15 \approx 10$ is Coulomb logarithm, $\mu = m_i/m_p$ is relative mass of ion; $m_p = 1.67\times10^{-27}$ is mass of proton, kg; n is density of electrons and ions respectively, 1/cm^3; T is temperature of electron and ion respectively, eV.

Electron and ion *thermal* velocity are respectively:

$$\upsilon_{Te} = (kT_e / m_e)^{1/2} = 4.19\times10^7 T_e^{1/2} \quad \text{cm/s}$$
$$\upsilon_{Ti} = (kT_i / m_i)^{1/2} = 9.79\times10^7 \mu^{-1/2} T_i^{1/2} \quad \text{cm/s}, \tag{18}$$

Substitute equations (12)-(13) in (11) we receive the length of electron and ion tracks:

$$L_e = 1.44\times10^{13} T_e^2 / n_e \ln\Lambda \quad \text{cm},$$
$$L_i = 2.04\times10^{13} T_e^2 / Z^4 n_e \ln\Lambda \quad \text{cm}, \tag{19}$$
$$L_{ei} = 0.95\times10^{13} T_e^2 / n_e \ln\Lambda \quad cm.$$

Estimation. For electron having $n = 10^5$ 1/cm^3, $T = 100$ eV, ln $\Lambda \approx 10$ we get $L = 2\times10^6$ km, $\tau \approx 300$ s.

That means the plasma electrons have very few collisions, small dispersion, (in our case) and it can have different average ELECTRON (relative to ion) temperature along the cable axis and perpendicular cable axis. It is not a surprise because the plasma can have different average temperatures of electron and ions. That also means that our assumption about the terminal and current electron velocities being the same is very limited and the parameters of a plasma electric system will often be better, than in our computation. The *plasma in our system may be very cool in a radial direction and simultaneously very hot in the axial direction.* That decreases the electric current needed for plasma compression and allows a transfer of the plasma beam, energy, and thrust to a great distance.

Magnetic force between two parallel cables. This force is

$$F_m = -\mu_0 \frac{i_1 i_2}{2\pi d} L, \tag{20}$$

where F_m is magnetic force, N (the force is repeal when currents are opposed, and attractive when currents have same directions); $\mu_0 = 4\pi10^{-7}$ is permeability constant, H/m; i is electric current in the 1-st and 2-nd cable respectively, A; d is distance between center of cables, m; L is length of cables, m.

This force for two cable line (Figure 5e) having current $I = 10^5$ A, distance $d = 1000$ km equals $F_m = 2$ N/km. But force decreases if we use multi-cable system: for three cables (Figure 5f)(3/8 F_m); for 5 cables (Figure 5g)(5/32 F_m); for multi-cables (Figure 5h)($F_m = 0$).

Electrostatic force between two parallel cables. This force is

$$F_e = k\frac{2\tau_1\tau_2}{d}L, \tag{21}$$

where F_e is electrostatic force, N (the force is attractive when charges is different and repeal when charges are same); $k = 1/4\pi\varepsilon_0 = 9\times10^9$ electrostatic constant, $\mathrm{Nm^2/C^2}$; τ is linear charge of the 1-st and 2-nd cable, C/m; d is distance between cables, m; L is length of cables, m.

Electrostatic force is attractive force for opposed charges. This force may be used for balance the electromagnetic force. From $F_m = F_e$, (20) = (21) we get for two line cable system

$$\tau = \frac{1}{2}\sqrt{\frac{\mu_0}{\pi k}}I, \quad \Delta U = 2k\tau\int^{d}\frac{dR}{R} = \sqrt{\frac{\mu_0 k}{\pi}}I\ln\left(\frac{d}{r}\right) = 60I\ln\left(\frac{d}{r}\right), \quad \Delta P = \Delta U\cdot I, \tag{22}$$

where r is plasma cable radius, m. Example: for $I = 10^4$ A, $d/r = 10$ we have $\Delta U = 1.38\times10^6$ V,

The linear charge appears on cable when is voltage between cables. The other way of balance is cable design in Figure 5h.

Electric capacity two parallel cables is

$$C = \frac{\pi\varepsilon_0}{\ln(d/r)}L, \tag{23}$$

where C is electric capacity, F; $\varepsilon_0 = 8.85\times10^{-12}$ is electrostatic constant, F/m; r is cable radius, m;

Energy of two parallel cables as electric condenser is

$$E = \frac{1}{2}CU^2 = \frac{1}{2}qU = \frac{1}{2}\frac{q^2}{C},$$

where E is energy in condenser, J; U is electric voltage, V; q is electric charge, C.

Example: for d = 100 km, r = 10 m, the electric capacity is C = 0.05 F/one million km, the energy is $E = 2.5\times10^6$ J/one million km.

Inductance of two parallel cables is

$$L_i = \frac{\mu_0}{\pi}\left(\frac{1}{2} + \ln\frac{d}{r}\right)L, \tag{24}$$

where L_i is inductance, H.

Inductance energy of two parallel cables is

$$E = L_i \frac{I^2}{2}, \tag{25}$$

where E is energy in a closed-loop contour, J. Example: for d = 100 km, r = 10 m, the inductance is $L_i = 3.9\times10^3$ H/one million km, the energy is $E = 1.94\times10^{15}$ J/one million km. This energy is high and the starting station (where the plasma cables originate) spends a lot of energy for creating the magnetic field.

Change electric current in closed-loop contour is

$$I = I_0 \exp\left(-\frac{t}{T}\right), \quad \text{where} \quad T = \frac{L_i}{R}, \tag{26}$$

where R is electric resistance of closed-loop electric contour, Ohm. T is time decreasing current by factor e = 2.71 times. Example: for two lines cable the length L = 100 millions of km and the electric resistance $R = 10^{-3}$ Ohm (see project below), $T \approx 3.9\times10^3/10^{-3} = 3.9\times10^6$ sec = 45 days. That means our electric line is a large storage reservoir of energy.

'Virtual' Specific Impulse of AB-Space Engine. Specific impulse of rocket engine is ratio of an engine thrust to fuel consumption per second. It is difficult to speak about specific impulse of AB-engine because AB-engine doesn't spend fuel for the thrust, but it does expend matter (for example hydrogen) for creating the plasma cables. That way we can take ratio of the thrust to mass expenditure per second for produce of new cable.

This 'virtual' specific impulse of the AB-Space Engine is

$$I_y = \frac{P}{m_s}, \quad P = \frac{\mu_0}{2\pi} I^2, \quad m_s = 2snm_i V, \quad I_y = \frac{\mu_0}{3\pi m_i} \frac{I^2}{snV}. \tag{27}$$

$$\text{For } H_2 \quad I_y = 6\cdot10^{19} \frac{I^2}{snV}$$

where I_y is specific impulse of AB-Space Engine, m/s; P is magnetic thrust, N; $s = \pi r^2$ is cross section area of plasma cable, m^2; $m_i = 2\times1.67\times10^{-27}$ is mass of one molecule of hydrogen, kg; V is speed of apparatus, m/s.

Estimation. Let us take $I = 10^6$ A, $s = 10^2$ m^2. $n = 10^{14}$ /m^3, $V = 10^4$ m/s. We get $I_y = 6\times10^{11}$ m/s. That is one gigantic specific impulse. No present *rocket engine* in the World has such a specific impulse and a rival is unlikely in the future.

For comparison the specific impulse are: conventional liquid-propellant rocket engine had maximum I_y = 4200 m/s; hydrogen rocket engine - I_y = 5180 m/s; thermonuclear rocket engine (H_2+H_3)(for 100% efficiency) has $I_y = 26\times10^6$ m/s; ideal laser engine has $I_y = 3\times10^8$ m/s; and the most power – annihilation rocket engine (for 100% efficiency) has theoretical impulse $I_y = 4.24\times10^8$ m/s.

Our AB-Space Engine has very high specific impulse and it may be a good candidate for interstellar flights. This system spends less mass for producing the plasma cable than any rocket engine spends for producing thrust. Another advantage is that it gets the energy from a (planet-mounted) station, i.e., the power source needn't travel with it and weigh it down. The

AB-Engine is very light, simple, safety, and reliable with comparison to any likely (or perhaps nearly any dream) nuclear engine. In most cases at least part of the cable mass can be injected from the planet-mounted energy station.

Coefficient efficiency of AB-Space Engines. Author offers the following the estimation efficiency of AB-Space Engines: the ratio of energy (power) getting by apparatus to energy (power) spending by station:

$$\eta = \frac{PV}{N}, \tag{28}$$

where η is coefficient efficiency; P is full thrust getting by apparatus, N; V is apparatus speed, m/s; N is electric station power, W. The formulas above allow us to compute it, but one is a distinctly variable value.

The other coefficient efficiency is the ratio the apparatus thrust to spending power of electric station [N/W]:

$$\eta_N = \frac{P}{N}, \quad P = \frac{\mu_0}{2\pi} I^2, \quad N = I^2 R, \quad R = \rho \frac{L}{s},$$
$$\rho = 1.03 \cdot 10^{-4} \ln \Lambda \cdot T^{-3/2}, \quad T = 2.71 \cdot 10^{-12} \left(\frac{j}{en} \right)^2 . \tag{29}$$

Substitute all equations (29) in the first equation (29) we receive

$$\eta_N = 1.06 \cdot 10^{36} \frac{s}{L \cdot \ln \Lambda} \left(\frac{j}{n} \right)^3 , \tag{30}$$

Estimation. For $n = 10^{14}$ /m^3, $j = 10^3$ A/m^2, $s = 10^2$ m^2, $L = 10^5$ m, $\ln\Lambda = 10$ we get $\eta_N =$ 0.1 N/W. However for high L the coefficient is very small. That is because the electric station spends a lot of energy for producing magnetic field of closed-loop cables.

MACRO-PROJECT

As an example, we estimate the parameters of the AB-Space Engine having the thrust from one plasma cable about $P = 10^5$ N = 10 tons. Our design is not optimal. That is only a very simple calculation.

Most our computation is made for one cable. In reality, for a two cable engine, the employer must at least exactly double all relevant values.

Let us take the following initial data: electric current $I = 10^6$ A, thrust of one cable $P = 10^5$ N = 10 tons, radius of plasma cable $r = 10$ m, plasma density $n = 10^{14}$/m^3. $I > I_{min} = 13$ A.

We derive the following calculational results:

1) Density of electric current and electron speed are

$$j = \frac{I}{\pi r^2} = \frac{10^6}{3.14 \cdot 10^2} = 3180 \quad A/m^2, \quad u \approx \frac{j}{en} = \frac{3180}{1.6 \cdot 10^{-19} 10^{14}} = 2 \cdot 10^8 \quad m/s,$$

2) Temperature electrons and ions along cable axis and Spitzer electric resistance (for $\ln\Lambda = 10$) are

$$T = 2.71 \cdot 10^{-12} u^2 = 1.08 \cdot 10^5 \quad eV,$$
$$\rho = 1.03 \cdot 10^{-4} \ln\Lambda \cdot T^{-3/2} = 0.92 \cdot 10^{-10.5} \quad \Omega \cdot m$$

3) Electric resistance, required voltage and electric power for one plasma cable and its length $L = 10^{12}$ m = 1000 million kilometers. (A gentle reminder: the actual distance from our Sun to our Earth is 150 million kilometers, from the Sun to Mars is about 228 million kilometers, from the Sun to gaseous giant Jupiter is about 778 million kilometers, and from the Sun to ringed Saturn is about 1427 million kilometers.)

$$R = \rho \frac{L}{s} = 0.92 \cdot 10^{-10.5} \frac{10^{12}}{3.14 \cdot 10^2} = 9.5 \cdot 10^{-2} \quad \Omega,$$
$$U = IR = 10^6 \cdot 9.5 \cdot 10^{-2} = 9.5 \cdot 10^4 \quad V, \quad N = IU = 10^6 \cdot 9.5 \cdot 10^4 = 9.5 \cdot 10^{10} \quad W.$$

The USA produced about 1022 GW electric energy in winter 2007. The 950 GigaWatts is about 20% power of all kinds of energy produced in the USA in 2005 or of electric power use in the World. Our thrust is very high. If we take the thrust 1,000 – 10,000 N the required electric power decreases by hundreds to thousands of times. The apparatus can reach any needed speed by increasing the acceleration distance.

4) The mass of one cable is

$$M = sLnm_i = 3.14 \cdot 10^2 10^{12} 10^{14} 2 \cdot 1{,}67 \cdot 10^{-27} = 105 \quad kg$$

For two (forward and back) cables the cable mass is 210 kg. That is small mass for a cable having cross-section s = 314 m^2 (diameter 10 m) and length L = 1000 millions of km. That mass may be significantly less if we take the less plasma density. Part of this mass (about half) may be ejected from start electric station.

5) The magnetic thrust of cables (Figure 5g, h) located about apparatus at distance d = 110 m is

$$P = \frac{\mu_0 I^2}{2\pi} \ln \left| \left(\frac{d-r}{r} \right)^2 \right| = \frac{4\pi 10^{-7} 10^{12}}{2\pi} \ln \left| \left(\frac{110-10}{10} \right)^2 \right| = 9.2 \cdot 10^5 \; N = 92 \; tons$$

That thrust is different from initial thrust of two single plasma cables (20 tons) because we take the distance between plasma cable d = 110 m and receive ln 100 = 4.6 (not one).

6) The kinetic thrust of charged particles of one plasma cable is

$$P_e = 3.36\cdot 10^7 \frac{j^2}{n}s = 3,36\cdot 10^7 \frac{I^2}{sn} = 3.36\cdot 10^7 \frac{10^{12}}{314\cdot 10^{14}} = 1.07\cdot 10^3 \ N \approx 107 \ kg$$

As you see the kinetic (electric) thrust is small (in comparison with magnetic thrust). We can neglect it.

The total kinetic energy (charged particles) of one cable the length L = 1000 millions of km is

$$E = P_e L = 1.07\cdot 10^3 10^{12} = 1.07\cdot 10^{15} \quad J.$$

7) The additional voltage between the two cable line for balance magnetic (repeal) and electric (attracted) force and a power which must use for this the space ship or electric station are

$$\Delta U = 60\cdot I\cdot \ln\left(\frac{d}{r}\right) = 60\cdot 10^6\cdot \ln 10 = 1.38\cdot 10^8 \ V,$$

$$\Delta P = \Delta U\cdot I = 1.38\cdot 10^8\cdot 10^6 = 1.38\cdot 10^{14} \ W$$

That is large power. It may be requested the lower current and thrust or use the figure 5h design.

8) Estimation of flight possibilities the spaceship having mass M_s = 92 tons, flight time 10 days = t = 8.64×10^5 sec, and computed AB-Engine. The acceleration of the space ship, speed and range are:

$$a = \frac{P}{M_s} = \frac{9.2\cdot 10^5}{9.2\cdot 10^4} = 10 \ m/s. \quad V = at = 10\cdot 8.64\cdot 10^5 = 8.64\cdot 10^6 \ m/s =$$

$$= 8.64\cdot 10^3 \ km/s,$$

$$L_s = \frac{at^2}{2} = \frac{10\cdot (8.64\cdot 10^5)^2}{2} = 3.73\cdot 10^{12} \ m = 3730 \ \textit{millions of km}.$$

This is ~ 50 times more than minimal distance from Earth to Mars.

Our estimation is not optimal, that is example of computation.

Note that the V displayed is something on the order of 2.88% of light's natural speed in outer space!

DISCUSSION

Advantages of AB-engine

1) The offered AB-Space Engine is very light, simple, safe, and reliable with comparison to any likely nuclear engine.
2) The AB-Space Engine has a gigantic 'virtual specific impulse', being more capable of realistic operation in a projectable near-future environment, than virtually any proposed means of thermonuclear or light-propulsion scheme the author is aware of.
3) The AB-Space Engine can accelerate a near-term space apparatus to very high speed (approaching light speed). At present time this is the single real method to be able to approach this 'ultimate' velocity.
4) At least part of the needed injected plasma cable mass and nearly all of the energy needed (and the cooling facilities needed to maintain that energy supply) can be from the planet-bound energy-supplying station, further improving the on-board ship 'mass ratio'.
5) The AB-Space Engine can use far cheaper energy from a planet-bound electric station.

The offered ideas and innovations may create a jump in space and energy industries. Author has made initial base researches that conclusively show the big possibilities offered by the methods and installations proposed. Further research and testing are necessary. Those tests are not expensive. As that is in science, the obstacles can slow, even stop, applications of these revolutionary innovations. For example, the plasma cable may be unstable. The instability mega-problem of a plasma cable was found in tokomak R&D, but it is successfully solved at the present time. The same method (rotation of plasma cable) can be applied in our case.

The other problem is production of the plasma cable in Earth's atmosphere. This problem may be sidestepped by operations from a suitably high super-stratospheric tower such as outlined in others of the author's works, or is no problem at all if the electric station of the plasma cables' origin is located on the Moon [8].

The author has ideas on how to solve this problem with today's technologies and to use the readily available electric stations found on this planet Earth. Inquiries from serious parties are invited.

SUMMARY

This new revolutionary idea – The AB-Space Engine and wireless transferring of electric energy in the hard vacuum of outer space – is offered and researched. A rarefied plasma power cord in the function of electric cable (wire) is used for it. It is shown that a certain minimal electric current creates a compression force that supports and maintains the plasma cable in its compacted form. Large amounts of energy can be transferred hundreds of millions of kilometers by this method. The requisite mass of plasma cable is merely hundreds of grams (some kg). A sample macroproject is computed: An AB-Space Engine having thrust = 10 tons. It is also shown that electric current in plasma cord can accelerate or slow various kinds of outer space apparatus.

Future space apparatus (credit NASA)

REFERENCES

(Reader finds some of author's articles in *http://Bolonkin.narod.ru/p65.htm*, *http://arxiv.org*, search "Bolonkin" and in books: "*Non-Rocket Space Launch and Flight*", Elsevier, 2006, 488 ps; and "*New Concepts, Ideas, and Innovations in Aerospace, Technology and Human Life*", NOVA, 2008, 400 ps.)

[1] Bolonkin A.A., Wireless Transfer of Electricity in Outer Space, *Presented as paper AIAA-2007-0590 to 45th AIAA Aerospace Science Meeting,* 8 - 11 January 2007, Reno, Nevada, USA. Published in *http://arxiv.org* of January 4, 2007, search "Bolonkin" and in International Journal "Aircraft Enginering and Aerospace Technology", Vol.78, No.3, 2007, pp.273-282, and.

[2] Bolonkin, A.A., Getting of Electric Energy from Space and Installation for It, *Russian patent application #3638699/25 126303*, 19 August, 1983 (in Russian), Russian PTO.

[3] Bolonkin, A. A., Method of Transformation of Plasma Energy in Electric Current and Installation for It. *Russian patent application #3647344 136681* of 27 July 1983 (in Russian), Russian PTO.

[4] Bolonkin, A. A., Transformation of Energy of Rarefaction Plasma in Electric Current and Installation for it. *Russian patent application #3663911/25* 159775, 23 November 1983 (in Russian), Russian PTO.

[5] Bolonkin A.A., *Non-Rocket Space Launch and Flight,* Elsevier, London, 2006, 488 ps.

[6] Bolonkin A.A., Micro-Thermonuclear AB-Reactors, *AIAA-2006-8104, 14th Space Planes and Hypersonic System Conference,* 6-9 November, 2006, Australia. http://arxiv.org.

[7] Bolonkin A.A., Utilization of Wind Energy at High Altitude, A*IAA Conference Guidance, Navigation, and Control, Rhode Island,* 16-19 August, 2004, AIAA-2004-5705. *http://arxiv.org.*

[8] Bolonkin A.A., Beam Space Propulsion, AIAA-2006-7492, *Conference Space-2006,* 18-21 Sept;, 2006, San Jose, CA, USA. *http://arxiv.org.*

[9] Bolonkin A.A., Electrostatic AB-Ramjet Space Propulsion, *AIAA/AAS Astrodynamics Specialist Conference,* 21-24 August 2006, USA. AIAA-2006-6173. http://arxiv.org. Journal AEAT, Vol. 79, No. 1, 2007, pp.3-16.

[10] Bolonkin A.A., Electrostatic Linear Engine, AIAA-2006-5229, *42nd Joint Propulsion Conference,* 9-12 June 2006, Sacramento, USA. Journal AEAT, Vol. 78, No.6, 2006, pp.502-508.

[11] Bolonkin A.A., High-Speed Solar Sail, AIAA-2006-4806, *42nd Joint Propulsion Conference*, 9-12 June 2006, Sacramento, USA. *http://arxiv.org*

[12] Bolonkin A.A., A New Method of Atmospheric Reentry for Space Shuttle, AIAA-2006-6985, *MAO Conference,* 6-9 Sept. 2006, USA. *http://arxiv.org.*

[13] Bolonkin A.A., Suspended Air Surveillance System, AIAA-2006-6511, *AFM Conference, 21-29 Aug. 2006,* Keystone, USA. *http://arxiv.org.*

[14] Bolonkin A.A., Optimal Inflatable Space Tower with 3-100 km Height, *Journal of the British Interplanetary Society,* Vol.56, No. 3/4, 2003, pp.97-107.

[15] Bolonkin A.A., Optimal Solid Space Tower, AIAA-2006-7717. *ATIO Conference, 25-27 Sept. 2006,* Wichita, Kansas, USA. *http://arxiv.org.*

[16] *Wikipedia.* Some background material in this article is gathered from Wikipedia under the Creative Commons license.

Chapter 11

THERMONUCLEAR REFLECT AB-REACTOR*

ABSTRACT

The authors offer a new kind of thermonuclear reflect reactor. The remarkable feature of this new reactor is a three net AB reflector, which confines the high temperature plasma. The plasma loses part of its' energy when it contacts with the net but this loss can be compensated by an additional permanent plasma heating. When the plasma is rarefied (has a small density), the heat flow to the AB reflector is not large and the temperature in the triple reflector net is lower than 2000 - 3000 K. This offered AB-reactor has significantly less power then the currently contemplated power reactors with magnetic or inertial confinement (hundreds-thousands of kW, not millions of kW). But it is enough for many vehicles and ships and particularly valuable for tunnelers, subs and space apparatus, where air to burn chemical fuel is at a premium or simply not available.

The author has made a number of innovations in this reactor, researched its theory, developed methods of computation, made a sample computation of typical project. The main point of preference for the offered reactor is its' likely cheapness as a power source.

Keywords: Micro-thermonuclear reactor, Multi-reflex AB-thermonuclear reactor, Self-magnetic AB-thermonuclear reactor, aerospace thermonuclear engine

INTRODUCTION

Brief Information about Thermonuclear Reactors

Fusion power is useful energy generated by nuclear fusion reactions. In this kind of reaction two light atomic nuclei fuse together to form a heavier nucleus and release energy. The largest current experiment, JET, has resulted in fusion power production somewhat larger than the power put into the plasma, maintained for a few seconds. In June 2005, the construction of the experimental reactor ITER, designed to produce several times more fusion

* Presented to http://arxiv.org

power than the power into the plasma over many minutes, was announced. The production of net electrical power from fusion is planned for the next generation experiment after ITER.

Unfortunately, this task is not easy, as scientists thought earlier. Fusion reactions require a very large amount of energy to initiate in order to overcome the so-called *Coulomb barrier* or *fusion barrier energy*. The key to practical fusion power is to select a fuel that requires the minimum amount of energy to start, that is, the lowest barrier energy. The best fuel from this standpoint is a one-to-one mix of deuterium and tritium; both are heavy isotopes of hydrogen. The D-T (Deuterium & Tritium) mix has a low barrier energy. In order to create the required conditions, the fuel must be heated to tens of millions of degrees, and/or compressed to immense pressures.

At present, D-T is used by two main methods of fusion: inertial confinement fusion (ICF) and magnetic confinement fusion (MCF)(for example, tokamak).

In i*nertial confinement fusion (ICF)*, nuclear fusion reactions are initiated by heating and compressing a target. The target is a pellet that most often contains deuterium and tritium (often only micro or milligrams). Intense laser or ion beams are used for compression. The beams explosively detonate the outer layers of the target. That accelerats the underlying target layers inward, sending a shockwave into the center of pellet mass. If the shockwave is powerful enough and if high enough density at the center is achieved some of the fuel will be heated enough to cause fusion reactions. In a target which has been heated and compressed to the point of thermonuclear ignition, energy can then heat surrounding fuel to cause it to fuse as well, potentially releasing tremendous amounts of energy.

Fusion reactions require a very large amount of energy to initiate in order to overcome the so-called *Coulomb barrier* or *fusion barrier energy*.

Magnetic confinement fusion (MCF). Since plasmas are very good electrical conductors, magnetic fields can also confine fusion fuel. A variety of magnetic configurations can be used, the basic distinction being between magnetic mirror confinement and toroidal confinement, especially tokamaks and stellarators.

Lawson criterion. In nuclear fusion research, the Lawson criterion, first derived by John D. Lawson in 1957, is an important general measure of a system that defines the conditions needed for a fusion reactor to reach ignition, that is, that the heating of the plasma by the products of the fusion reactions is sufficient to maintain the temperature of the plasma against all losses without external power input. As originally formulated the Lawson criterion gives a minimum required value for the product of the plasma (electron) density n_e and the "energy confinement time" τ. Later analyses suggested that a more useful figure of merit is the "triple product" of density, confinement time, and plasma temperature T. The triple product also has a minimum required value, and the name "Lawson criterion" often refers to this inequality.

The key to practical fusion power is to select a fuel that requires the minimum amount of energy to start, that is, the lowest barrier energy. The best fuel from this standpoint is a one-to-one mix of deuterium and tritium; both are heavy isotopes of hydrogen. The D-T (Deuterium & Tritium) mix has a low barrier.

Short history of thermonuclear fusion. One of the earliest (in the late 1970's and early 1980's) serious attempts at an ICF design was *Shiva,* a 20-armed neodymium laser system built at the Lawrence Livermore National Laboratory (LLNL) that started operation in 1978. Shiva was a "proof of concept" design, followed by the NOVA design with 10 times the power. Funding for fusion research was severely constrained in the 80's, but NOVA nevertheless successfully gathered enough information for a next generation machine whose goal was ignition. Although net energy can be released even without ignition (the breakeven point), ignition is considered necessary for a *practical* power system.

The resulting design, now known as the National Ignition Facility, commenced being constructed at LLNL in 1997. Originally intended to start construction in the early 1990s, the NIF is now six years behind schedule and overbudget by over $1.4 billion. Nevertheless many of the problems appear to be due to the "big lab" mentality and shifting the focus from pure ICF research to the nuclear stewardship program, LLNLs traditional nuclear weapons-making role. NIF is now scheduled to "burn" in 2010, when the remaining lasers in the 192-beam array are finally installed.

Laser physicists in Europe have put forward plans to build a £500m facility, called HiPER, to study a new approach to laser fusion. A panel of scientists from seven European Union countries believes that a "fast ignition" laser facility could make a significant contribution to fusion research, as well as supporting experiments in other areas of physics. The facility would be designed to achieve high energy gains, providing the critical intermediate step between ignition and a demonstration reactor. It would consist of a long-pulse laser with an energy of 200 kJ to compress the fuel and a short-pulse laser with an energy of 70 kJ to heat it.

Confinement refers to all the conditions necessary to keep a plasma dense and hot long enough to undergo fusion:

- *Equilibrium:* There must be no net forces on any part of the plasma, otherwise it will rapidly disassemble. The exception, of course, is inertial confinement, where the relevant physics must occur faster than the disassembly time.
- *Stability:* The plasma must be so constructed that small deviations are restored to the initial state, otherwise some unavoidable disturbance will occur and grow exponentially until the plasma is destroyed.
- *Transport:* The loss of particles and heat in all channels must be sufficiently slow. The word "confinement" is often used in the restricted sense of "energy confinement".

To produce self-sustaining fusion, the energy released by the reaction (or at least a fraction of it) must be used to heat new reactant nuclei and keep them hot long enough that they also undergo fusion reactions. Retaining the heat generated is called energy *confinement* and may be accomplished in a number of ways.

The hydrogen bomb weapon has no confinement at all and, not to be too darkly humorous, that is why it is easily noticeable by close observers! The fuel is simply allowed to fly apart, but it takes a certain length of time to do this, and during this time fusion can occur. This approach is called *inertial confinement* (figure 1). If more than about a milligram of fuel is used, the explosion would destroy the machine, so controlled thermonuclear fusion using inertial confinement causes tiny pellets of fuel to explode several times a second. To induce the explosion, the pellet must be compressed to about 30 times solid density with energetic beams. If the beams are focused directly on the pellet, it is called *direct drive*, which can in principle be very efficient, but in practice it is difficult to obtain the needed uniformity. An alternative approach is *indirect drive*, in which the beams heat a shell, and the shell radiates x-rays, which then implode the pellet. The beams are commonly laser beams, but heavy and light ion beams and electron beams have all been investigated and tried to one degree or another.

They rely on fuel pellets with a "perfect" shape in order to generate a symmetrical inward shock wave to produce the high-density plasma, and in practice these have proven difficult to produce. A recent development in the field of laser-induced ICF is the use of ultra-short pulse multi-petawatt lasers to heat the plasma of an imploding pellet at exactly the moment of greatest density after it is imploded conventionally using terawatt scale lasers. This research will be carried out on the (currently being built) OMEGA EP petawatt and OMEGA lasers at the University of Rochester and at the GEKKO XII laser at the Institute for Laser Engineering in Osaka Japan which, if fruitful, may have the effect of greatly reducing the cost of a laser fusion-based power source.

At the temperatures required for fusion, the fuel is in the form of a plasma with very good electrical conductivity. This opens the possibility to confine the fuel and the energy with magnetic fields, an idea known as *magnetic confinement* (figure 2).

Much of this progress has been achieved with a particular emphasis on tokamaks (figure 2).

In fusion research, achieving a fusion energy gain factor $Q = 1$ is called *breakeven* and is considered a significant although somewhat artificial milestone. *Ignition* refers to an infinite Q, that is, a self-sustaining plasma where the losses are made up for by fusion power without any external input. In a practical fusion reactor, some external power will always be required for things like current drive, refueling, profile control, and burn control. A value on the order of $Q = 20$ will be required if the plant is to deliver much more energy than it uses internally.

Figure 1. Laser installation for NOVA inertial thermonuclear reactor. Note the size of the man in comparison to the gigantic size of the class of laser installation needed for a reactor. Cost is some billions of dollars.

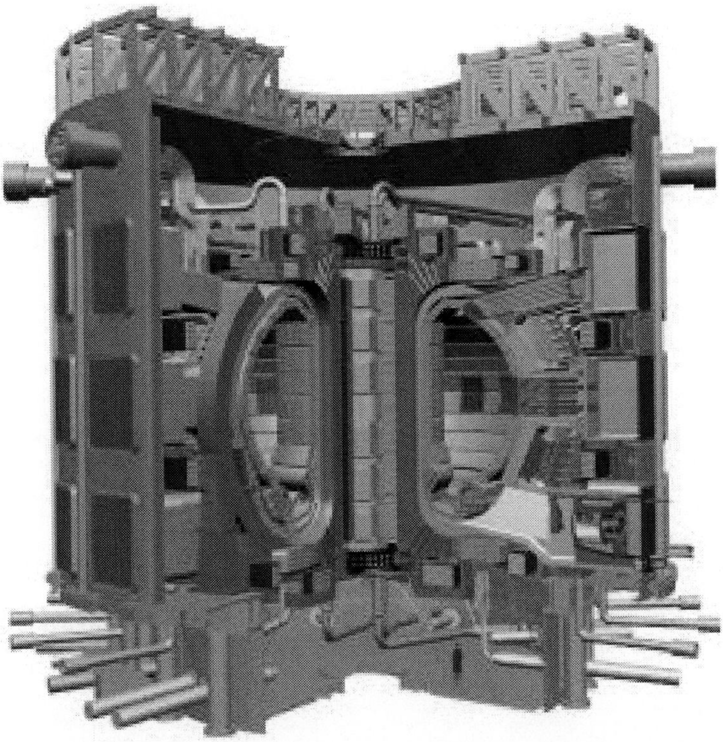

Figure 2. Magnetic thermonuclear reactor (ITER). The size of the installation is obvious if you compare it with the "Little Blue Man" inside the machine at the bottom. Cost is $12.8-billions USD.

In a fusion power plant, the nuclear island has a *plasma chamber* with an associated vacuum system, surrounded by a plasma-facing components (first wall and divertor) maintaining the vacuum boundary and absorbing the thermal radiation coming from the plasma, surrounded in turn by a blanket where the neutrons are absorbed to breed tritium and heat a working fluid that transfers the power to the balance of plant. If magnetic confinement is used, a *magnet* system, using primarily cryogenic superconducting magnets, is needed, and usually systems for heating and refueling the plasma and for driving current. In inertial confinement, a *driver* (laser or accelerator) and a focusing system are needed, as well as a means for forming and positioning the *pellets*.

The magnetic fusion energy (MFE) program seeks to establish the conditions to sustain a nuclear fusion reaction in a plasma that is contained by magnetic fields to allow the successful production of fusion power.

In thirty years, scientists have increased the Lawson criterion of the ICF and tokamak installations by tens of times. Unfortunately, all current and some new installations (ICF and totamak) have a Lawrence criterion that is tens of times lower than is necessary (figure 3).

Economics. It is far from clear whether nuclear fusion will be economically competitive with other forms of power. The many estimates that have been made of the cost of fusion power cover a wide range, and indirect costs of and subsidies for fusion power and its alternatives make any cost comparison difficult. The low estimates for fusion appear to be competitive with but not drastically lower than other alternatives. The high estimates are several times higher than alternatives.

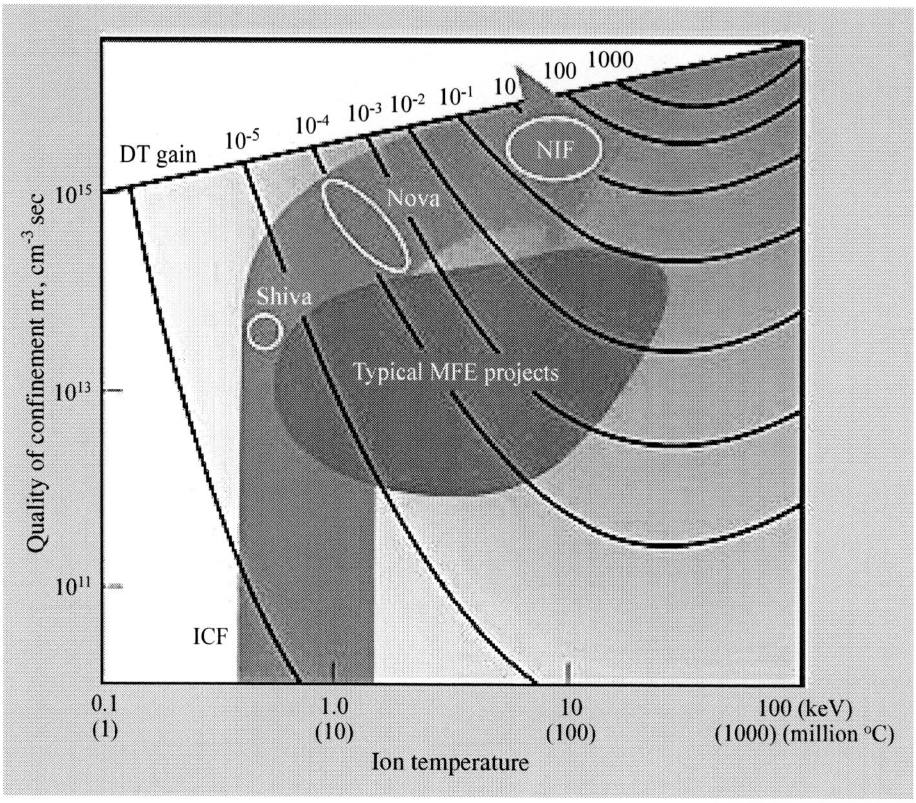

Figure 3. Parameter space occupied by inertial fusion energy and magnetic fusion energy devices. Theregime allowing thermonuclear ignition with high gain lies near the upper right corner of the plot.

While fusion power is still in early stages of development, vast sums have been and continue to be invested in research. In the EU almost € 10 billion was spent on fusion research up to the end of the 1990s, and the new ITER reactor alone is budgeted at € 10 billion. It is estimated that up to the point of possible implementation of electricity generation by nuclear fusion, R&D will need further promotion totalling around € 60-80 billion over a period of 50 years or so (of which € 20-30 billion within the EU). In the current EU research programme (FP6), nuclear fusion research receives € 750 million (excluding ITER funding), compared with € 810 million for all non-nuclear energy research combined, putting research into fusion power well ahead of that of any single rivaling technology.

Unfortunately, despite optimism dating back to the 1950s about the wide-scale harnessing of fusion power, there are still significant barriers standing between current scientific understanding and technological capabilities and the practical realization of fusion as an energy source. Research, while making steady progress, has also continually thrown up new difficulties. Therefore it remains unclear that an economically viable fusion plant is even possible." An editorial in New Scientist magazine explained that "if commercial fusion is viable, it may well be a century away."

An important aspect of fusion energy in contrast to many other energy sources is that the cost of production is inelastic. The cost of wind energy, for example, goes up as the optimal locations are developed first, while further generators must be sited in less ideal conditions. With fusion energy, the production cost will not increase much, even if large numbers of plants are built. It has been suggested that even 100 times the current energy consumption of the world is possible.

Some problems which are expected to be an issue in the next century such as fresh water shortages can actually be regarded merely as problems of energy supply. For example, in desalination plants, seawater can be converted into pure freshwater through a process of either distillation or reverse osmosis. However, these processes are energy intensive. Even if the first fusion plants are not competitive with alternative sources, fusion could still become competitive if large scale desalination requires more power than the alternatives are able to provide.

Despite being technically non-renewable, fusion power has many of the benefits of long-term renewable energy sources (such as being a sustainable energy supply compared to presently-utilized sources and emitting no greenhouse gases) as well as some of the benefits of such much more finite energy sources as hydrocarbons and nuclear fission (without reprocessing). Like these currently dominant energy sources, fusion could provide very high power-generation density and uninterrupted power delivery (due to the fact that they are not dependent on the weather, unlike wind and solar power).

Several fusion reactors have been built, but as yet none has produced more thermal energy than electrical energy consumed. Despite research having started in the 1950s, no commercial fusion reactor is expected before 2050. The ITER project is currently leading the effort to commercialize fusion power.

Summary. At present time the most efforts of scientists are directed toward very large, superpower thermonuclear stations (Shiva, NOVA, NIF, LLNL, HiPER, OMEGA EP, ITER, Z-machine, etc.). These stations request gigantic finances, years of development, complex technology. That is well for scientists seeking a stable career path over many years, (generations!) but not well for the technical progress of humanity. Governments spent billions of dollars for development of thermonuclear technology. However, we have not had achieved a stable long duration thermonuclear reaction after *50 years* of thermonuclear development. In the author's opinion, industrial thermonuclear electric stations may appear after 20 – 40 more years and their energy will be more expensive than a current technology conventional electric station.

The author offers to direct government attention toward development smaller cheaper thermonuclear installations, which don't require huge funds and decades for development. Smaller power units are actually more practical in an immediate way: The world economy depends on transport vehicles (cars, ships, aviation), which are the main users of oil fuel. They very much need small thermonuclear engines. (Large fixed conventional stations use coal, replacement of which does not necessarily relieve pressure on world oil prices.)

Humanity needs new sources of energy. The offered innovations and researches are in [1]-[43].

DESCRIPTION AND INNOVATION

1. AB Plasma Reflector

A three net plasma reflector was offered by the author in his early works (see, for example, [5-10]). The plasma reflector (figure 4) has three conductive nets (*1, 2, 3*) and electric voltage between them. *V*oltage between nets *1* and *2* nets is about 100 kV in our case, when plasma temperature is 50 keV. The electric intensity is taken so that *1, 2* reflect the positive charged particles (for example atoms of deuterium D and tritium T). The voltage between the nets *2* and *3* nets is about 200 kV in our case. The electric intensity is taken so the *2, 3* reflect the negative charged particles (electrons).

The reflector works in the following way. When plasma is inserted in the reflector confinement, the nets *1, 2* reflect the positive charged atoms back to the plasma if the electric voltage is more than plasma temperature in eV. The particle speed has a Maxwellian distribution. That way the voltage between nets must be a minimum of two times more than the average temperature of the plasma. In this case, most positive particles will be reflected back to the plasma. Simultaneously, the nets *1,2* accelerate the negative particles (electrons) and increase their energy (temperature) additional up to 100 keV. That way the voltage between nets 2, 3 must be more then *V* plus the electron temperature of plasma (about 200 keV > 100 + 50 keV in our case).

The plasma contacts with the nets and heats them. This heating significantly depends upon the density of the plasma. For example, the Earth's atmosphere over 160 km has temperature more than 1200 K. but Earth satellites fly some years at this atmosphere without any perceptible heating.

The inner Van Allen Belt extends from an altitude of 700–10,000 km (0.1 to 1.5 Earth radii) above the Earth's surface, and contains high concentrations of energetic protons with energies exceeding 100 MeV and electrons in the range of hundreds of kiloelectronvolts, trapped by the strong (relative to the outer belts) magnetic fields in the region.

The charged particles in Earth's radiation belts have a temperature of some billions degrees, but a space apparatus orbiting there has a low temperature. It is possible because the density of plasma at these altitudes is very low and a heat flow from the high temperature plasma is small. This effect is used in the offered thermonuclear AB-reactor.

We use plasma which has a density of a million times less than Earth's air and a hundred times less than a conventional tokomak machine. As a result, our net has an equilibrium temperature (after its own re-radiation of heat) less than the melting temperature of a refractory conductive material (for example, tungsten has a melting temperature of 3416 C = 3689 K).

Low density plasma produces low energy (a low power output). The current thermonuclear reactors under development are contemplated to cost milliards (billions of) USA dollars and would be reduced to operational practice as power electric stations having some millions of kW capacity. Therefore, this class of reactor, with a lower plasma density, is not acceptable for them. The authors' opinion is that such powerful thermonuclear stations would appear on the market not earlier than after 30 – 40 years. (Merely demonstrating breakeven does not mean we get practical thermonuclear energy. There are a lot of problems that we must solve before industrial-scale production of thermonuclear energy becomes practical: For example, protection of people from neutrons, converting the neutron energy into electricity, etc. I am sure the big thermonuclear energy for many years will be more costly then conventional heat energy. For example, the first conventional (uranium) nuclear energy was released about 50 years ago, but for years it was more expensive than unsubsidized coal power. But with this practical, buildable new approach we should be able to develop a usable thermonuclear reaction within 3 – 5 years.

The developing the small power thermonuclear installations (hundreds and thousands kW, as offered by author) is useful for vehicles, sea ships, and space apparatus and will help to solve many future problems of the big thermonuclear stations on a smaller, more realizable scale. Developing small reactors should cost thousands of times less than building ITER or NOVA.

The computations show the main contribution to net heating may not be contact of the plasma to the net, but a product of the thermonuclear reaction – neutrons (14 MeV) and alpha particles (3.5 MeV).

The neutrons have full reflection if the angle between trajectory and surface is 10 – 12 degrees. That way the author offers to use the cross-section (wedge) form of net wire (figure 2b, form 2). This form will reflect the part of high energy particles and has more wire surface for thermal re-radiation.

We can also use the form figure 2b, 3 which has an internal channel for cooling liquid.

The second heating of net (less, than contact or thermonuclear particles) is produced by the Bremsstrahlung, or braking radiation. The contact heating is about 5 – 25% of the full energy.

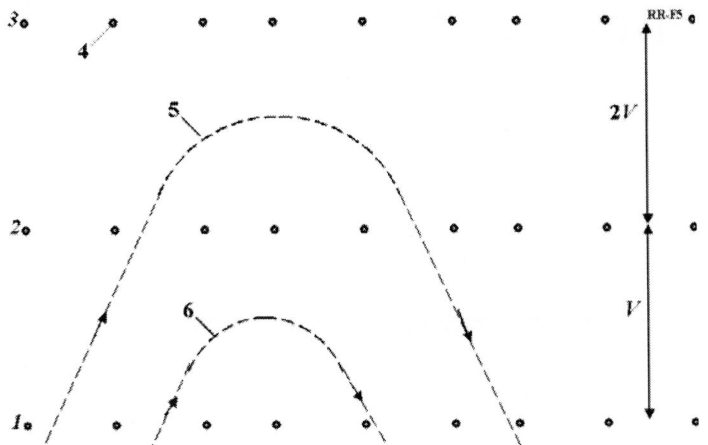

Figure 4. Three net plasma AB-reflector. *Notation*: *1, 2, 3* – the first, second, and third nets; 4 - wire of net; 5 – trajectory of negative charged particles; 6 – trajectory of positive charged particles. *V* is voltage between nets *1* and *2* nets (about 100 kV in our case). The electric intensity is taken so the *1, 2* reflect the positive charged particles (for example, atoms D and T). 2*V* is the voltage between the nets *2* and *3* nets (about 200 kV in our case). The electric intensity is taken so the *2, 3* reflect the negative charged particles (electrons).

Note that most heating occurs at the first net, which contacts with the nuclear particles. The second net only has contact with electrons. The mass of electrons and density of electron gas is less by about 2000 - 5000 times, than the mass of nuclear particles D, T. (The Deuterium and Tritium, or mix of heavy hydrogen isotopes that serve as nuclear fuel) That way we cannot be troubled by heating of the second net. The third net heats only by reaction product (α, n) and by Bremsstrahlung radiation as well.

We can also use hard material for nets, for example, tungsten, having a melting temperature of 3416 C. One tungsten atom has an atomic mass of about 184. The deuterium D has the atomic mass of 2 and tritium T has an atomic mass of 3. That means the light atoms hit on the heavy atom and passes to the heavy atom only a small part of its kinetic energy. The other important requirement of the net material must be a good tensile stress in high temperature. The tungsten has such a good property mix. Carbide of tantalum and zirconium has the melt temperature up to 3500 – 3900 C. The nanotube has good stability up to 2300 C in vacuum and excellent tensile stress.

Innovations in the plasma reflector in comparison with a conventional 'particle mirror' are the following:

1. The AB plasma reflector has three nets and it can reflect the plasma. Conventional electrostatic mirrors have two nets and it can reflect only charged particles of one sign.
2. Net wires can have an internal cooling (figure 5b-3).
3. Cross-section of net wires can have a special form (figure 5b-2) for reflecting the particles.
4. Net wires are made from heat-resistant electro-conductive material having good tensile stress in high temperature.
5. These net wires can be electronically cooled.

2. Space thermonuclear AB reactor. Scheme of this reactor is presented in figure 5. This incarnation uses only charged α-particles and lets out (sets free) the neutrons and Bremstrahlung (X-soft ray) radiation (it doesn't protect from them). That significantly simplifies the reactor and decreases its mass. But this design can be used in space apparatus when the reactor is located separately (and far) from space ship (figure 7). The manned space cabin can have additional light protection.

The reactor has the spherical form and contains: AB spherical reflector enclosed D+T plasma, spherical conductivity cover, two source of electric voltage (about 100 kV and 200 kV), outer electric user, plasma injector, and plasma electric heater.

Spherical conductivity cover is a thin (0.1 mm) aluminum film which collects the positive charged high energy (3.5 MeV) (α) particles, created by thermonuclear reaction. When they move between net 3 and reactor cover (3, figure 5), they are braked and accumulated by the cover. As the result the thermonuclear reactor produces electric current-- about 3 MV in voltage. Part of this energy is used for heating the plasma, fuel, and support of the voltage in plasma reflector.

The loss of plasma temperature from contact plasma with reflector net and Bremstrahlung (X-ray) radiation is compensated by a plasma heater (figure 5, notation 7) by passing the electric current through the plasma (inventions).

The reactor works continuously, producing tens to hundreds of kW of electric energy. That output harnesses only 20% of the liberated thermonuclear energy. (Efficiency is sacrificed for the sake of lightness, vital for space engines).

3. Earthbound thermonuclear reactor. The space reactor is not acceptable for Earth's biosphere because that version produces neutrons and X-ray radiation. The neutrons create radioisotopes which damage biological creatures. The Earthbound reactor needs a special protection for men and environment from neutrons and, as well, Bremsstrahlung radiation. This protection increases the reactor mass by some (4-8) times. But that allows utilizing the full energy of reactor and produces the tritium – the second important component of thermonuclear fuel. The full energy received from the thermonuclear reactor increases by 4 times. About 40% this additional energy can be converted to electric energy (total is 60%) and the rest may be utilized for home heating, production of freshwater, etc.

The Earth-bound macro-engineering version of the thermonuclear reactor is shown in figure 6. Reactor contains: installation of figure 5 inserted into protection cover. Protective cover has cooling tubes 9, which connects to vapor turbine 11. The turbine drives the electric generator 12, producing useful energy. The vapor after turbine flows to the heat exchanger 13 which produces hot water (vapor) 14 for home heating (freshwater production, etc.).

The reactor has lithium blankets 10 for reproduction of tritium.

4. Thermonuclear fuel. The reactor fuel is deuterium and tritium, D + T. This reaction needs a high temperature about 1 - 5 keV.

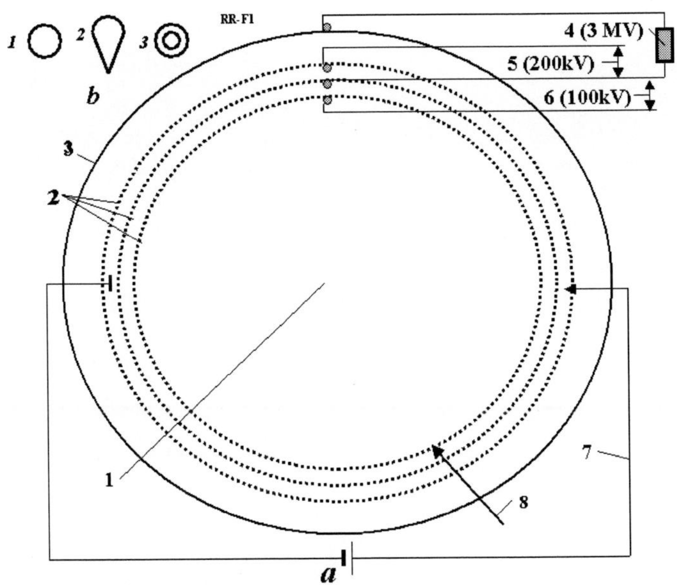

Figure 5. Scheme of Space Thermonuclear reflector AB-reactor, having spherical form. *Notations*: a) Reactor: 1 – plasma; 2 – three net AB-reflector; 3 – conductivity cover; 4 – voltage between net and cover; 5 – voltage between the second and third nets; 6 – voltage between the first and the second net; 7 – plasma electric heater; 8 – fuel injector. b) Cross section of the net wire. *Notation*: 1 – round; 2 – spherical (wedge) form; 3 – tube.

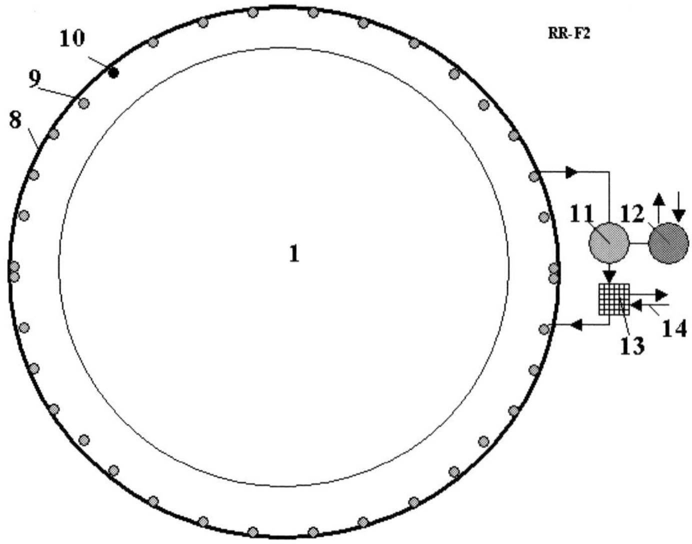

Figure 6. Scheme of Earth-based thermonuclear reflector AB-reactor. *Notations*: 1 – reactor of figure 5; 8 – protection against neutrons (neutron moderator); 9 – cooling (turbine) tubes; 10 – blanket for production of tritium; 11 - vapor turbine; 12 – electric generator; 13 – heat exchanger; 14 – hot water (vapor) for home heating (freshwater production, etc.).

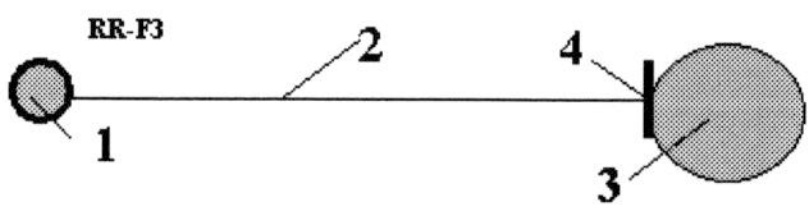

Figure 7. Reflector AB-reactor stationed in space. *Notation*: 1 – space reactor; 2 – connection to space ship; 3 – space ship; 4 – protection from neutrons and reactor radiation.

Deuterium (symbol D), also called *heavy hydrogen*, is a stable isotope of hydrogen with a natural abundance in the oceans of Earth of approximately one atom in 6500 of hydrogen. Deuterium thus accounts for approximately 0.015% (on a weight basis, 0.030%) of all naturally occurring hydrogen in the oceans on Earth. The nucleus of deuterium, called a *deuteron*, contains one proton and one neutron. Deuterium occurs in trace amounts naturally as deuterium gas, written $^2H_2$ or D_2, Density: 0.180 kg/m³ at STP (0 °C, 101.325 kPa). Data at approximately 18 K for D_2 (triple point): Density: liquid - 162.4 kg/m^3, gas - 0.452 kg/m^3. Viscosity: 12.6 μPa·s at 300 Kelvin (gas phase). Specific heat capacity at constant pressure c_p: solid 2950 J/(kg·K), gas 5200 J/(kg·K)

Tritium (symbol *T* or ^3H) is a radioactive isotope of hydrogen. It is a gas (T_2 or $^3H_2$) at standard temperature and pressure. Tritium combines with oxygen to form a liquid called tritiated water T_2O or partially tritiated THO. Tritium occurs naturally due to cosmic rays interacting with atmospheric gases. Because of tritium's relatively short half-life, however, tritium produced in this manner does not accumulate over geological timescales, and its natural abundance is negligible. Industrially, tritium is produced in nuclear reactors by neutron activation of lithium-6. Tritium is also produced in heavy water-moderated reactors when deuterium captures a neutron; however, this reaction has a much smaller cross section and is only a useful tritium source for a reactor with a very high neutron flux. It can also be produced from boron-10 through neutron capture. The tritium has half-life time only 12.32 years. That way the tritium is absent in the Earth. One is in Moon in small amount because the Moon doesn't have atmosphere and cosmic rays reach its surface and produce the tritium.

Current nuclear fusion research is focused on the D + T thermonuclear fusion reaction

$$D + T \rightarrow {}^4He\ (3.5\ MeV) + n\ (14.1\ MeV),$$

This reaction can occur in high-temperature deuterium-tritium plasma. Most energy released by the reaction is converted to the kinetic energy of the neutron. Since the neutron is not confined or reflected by a magnetic or electrostatic field it leaves, going outwards to surrounding space or hits the screen or vessel wall (or blanker) immediately after reaction. In last instance, the neutron kinetic energy is converted to heat. The heat is taken away from the screen by direct radiation or and indirect circulating coolant and can be used to run an electric generator. If we add ^6Li inside the blanket, then tritium can be produced by reaction

$$n + {}^6Li \rightarrow {}^4He\ (2.1\ MeV) + T\ (2.7\ MeV)$$

with it then used subsequently as the fuel. Another reaction product is the alpha (α) particles ^4He carrying 3.5 MeV which can be directed or confined by electro-magnetic field.

The reaction that produces only charged particles are best for the proposed propulsion system and generator. Unfortunately, these reactions are not great (see Table 1). All of them request the very high temperature for ignition and has the low cross-section of reaction (they produced low energy in volume unit).

The *D-T* reaction is favored since it has the largest fusion cross-section (~ 5 barns peak) and reaches this maximum cross-section at the lowest energy (~ 65 keV center-of-mass) of any potential fusion fuel.

According to IEER's 1996 report about the United States Department of Energy, only 225 kg of tritium has been produced in the US since 1955. Since it is continuously decaying into helium-3, the stockpile was approximately, 75 kg at the time of the report.

Table 1. Aneutronic reactions

$D+{}^3He \rightarrow {}^4He(3.6)+p(14.7)$
$D+{}^6Li \rightarrow 2{}^4He(22.4)$
${}^3He+{}^3He \rightarrow {}^4He(4.3)+2p(8.6)$
${}^3He+{}^6Li \rightarrow 2{}^4He(1.9)+p(16.9)$
$p+{}^{11}B \rightarrow 3{}^4He(8.7)$
$p+{}^6Li \rightarrow {}^4He(1.7)+{}^3He(2.3)$
$p+{}^7Li \rightarrow 2{}^4He(17.2)$

THEORY, COMPUTATIONS, AND ESTIMATIONS

The estimation and computation of the offered AB-reactor are made better in this way:

Assign the radius of the first net r = 0 .5 – 10 m, plasma pressure $n = 10^{18} - 10^{20}$ per m^3, and temperature of plasma T =1 – 50 keV. *Note*: the plasma density is about $n = 10^{21}$ per m^3 in a conventional big tokomak.

For our example, let us take: r = 3 m, $n = 10^{19}$ 1/m^3, T = 50 keV $=5\times10^4\times1.16\times10^4 = 5.8\times10^8$ K.

The surface and volume of a sphere are:

$$S = 4\pi r^2, \quad V = \frac{4}{3}\pi r^3, \quad S = 113\ \mathrm{m}^2, \quad V = 113\ \mathrm{m}^3, \tag{1}$$

Thermonuclear energy of reaction D+T for T = 50 keV released in form of charged particles is (in 1 m^3):

$$\begin{aligned} P_{DT} &= 5.6\cdot10^{-13} 0.25\cdot n^2 \left(\overline{\sigma v}\right)_{DT} = 1.4\cdot10^{-13}\cdot10^{26} 8.7\cdot10^{-16} = \\ &= 1.22\cdot10^{-2}\quad W/cm^3 = 1.22\cdot10^4 \quad \mathrm{W/m^3}, \end{aligned} \tag{2}$$

where n is 1/cm^3 and $\left(\overline{\sigma v}\right)_{DT}$ is taken from Table 2 below:

Table 2. Reaction rates $\left(\overline{\sigma v}\right)_{DT}$ (in cm^{3}s^{-1}), averaged over Maxwellian distribution

Temperature, keV	Reaction D - D	Reaction D - T	Reaction D – He^3
5	1.8×10^{-19}	1.3×10^{-17}	6.7×10^{-21}
10	1.2×10^{-18}	1.1×10^{-16}	2.3×10^{-19}
20	5.2×10^{-18}	4.2×10^{-16}	3.8×10^{-18}
50	2.1×10^{-17}	8.7×10^{-16}	5.4×10^{-17}
100	4.5×10^{-17}	8.5×10^{-16}	1.6×10^{-16}

Source: AIP, Physics Desk Reference, 3-rd Edition, p. 644.

The additional reactions noted in Table 2 are:

$$\begin{aligned} &\mathrm{D} + \mathrm{D} \rightarrow {}^3\mathrm{H} + {}^1\mathrm{H} + 4.033\ \mathrm{MeV}\ 50\%, \\ &\qquad \rightarrow {}^3\mathrm{He} + \mathrm{n} + 3.27\quad \mathrm{MeV}\ 50\%, \\ &\mathrm{D} + \mathrm{He}^3 \rightarrow {}^4\mathrm{He} + {}^1\mathrm{H} + 18.354\ \mathrm{MeV}, \\ &\qquad \rightarrow {}^6\mathrm{Li} + \gamma + 16.388\quad \mathrm{MeV}, \end{aligned} \tag{3}$$

The power density released in the form charged particle is:

$$P_{DD} = 3.3 \cdot 10^{-13} n_D^2 (\overline{\sigma v})_{DD}, \quad P_{DHe^3} = 2.9 \cdot 10^{-12} 0.25 \cdot n^2 (\overline{\sigma v})_{DHe^3}, \text{ W/cm}^3, \tag{4}$$

where n is 1/cm^3 and $(\overline{\sigma v})_{DT}$ is taken from Table 2.

Full thermonuclear energy is

$$P_{DT,F} = P_{DT} V = 1.22 \cdot 10^4 \cdot 113 = 1.38 \cdot 10^6 \quad \text{W}. \tag{5}$$

3. Energy from surface 1 m^2:

$$P_{DT,S} = r P_{DT} / 3 = 1.22 \cdot 10^4 \quad \text{W/m}^2. \tag{6}$$

4. Number reaction in 1 m^3

$$N = \frac{P_{DT}}{17.5 \ MeV} = \frac{1.22 \cdot 10^4}{17.5 \cdot 10^6 \cdot 1.6 \cdot 10^{-19}} = 4.36 \cdot 10^{15} \quad 1/\text{m}^3, \tag{7}$$

5. Fuel consumption (D+T) by volume 1 m^3 in day

$$\begin{aligned} C_{f,1} &= N \cdot m_i = 4.35 \cdot 10^{15} \cdot 5 \cdot 1.67 \cdot 10^{-27} = \\ &= 1.82 \cdot 10^{-7} \ \text{kg/s} \cdot \text{m}^3 = 0.157 \cdot 10^{-2} \ \text{g/day} \cdot \text{m}^3, \end{aligned} \tag{8}$$

Full fuel consumption is $C_f = C_{f,1} \times V$.

6. Expense power for fuel heating

$$p_f = NVkT = 4.36 \cdot 10^{15} \cdot 113 \cdot 1.38 \cdot 10^{-23} \cdot 5.8 \cdot 10^8 = 3.81 \cdot 10^3 \ \text{W}, \tag{9}$$

where $k = 1.38 \times 10^{-23}$ is Bolzmann's constant, J/K.

7. Computation of reflector net.

a) Plasma pressure

$$p = nkT = 10^{19} \cdot 1.38 \cdot 10^{=23} 5.8 \cdot 10^8 = 8 \cdot 10^4 \quad \text{N/m}^2. \tag{10}$$

b) Pressure in one net is

$$p_1 = \frac{1}{3} p = 2.67 \cdot 10^4 \quad \frac{\text{N}}{\text{m}^2}. \tag{11}$$

c) Thickness of net as continues cover

$$\delta = \frac{r p_1}{2\sigma} = \frac{3 \cdot 2.67 \cdot 10^4}{2 \cdot 4 \cdot 10^8} = 10^{-4} \text{ m} = 0.1 \text{ mm}, \tag{12}$$

where $\sigma = 4\times10^8$ N/m^2 = 40 kg/mm^2 is safety tensile stress of net wire.

d) Diameter net wire for net cell L = 100×100 mm (Distance between nets is 0.5 m) is

$$D = 2\sqrt{\frac{\delta L}{\pi}} = \sqrt{\frac{0.1 \cdot 100}{3.14}} = 3.57 \text{ mm}. \tag{13}$$

e) Transparency of net is

$$\xi = \frac{2D}{L} = \frac{2 \cdot 3.57}{100} = 0.0714. \tag{14}$$

Coefficient of the net transparency is a very important value for computation of the reactor efficiency. If net transparency is less, the loss of plasma energy is smaller. If net transparency is high, the required compensation energy may be more than the received thermonuclear energy. We can decrease it by using a conductive material having a high safety tensile stress at high temperature. We can also increase the size L of the net cell. But this method requires some increase in the size of reactor.

8. Volume Bremsstrahlung radiation of 1 m^3 is

$$P_B = 5.34 \cdot 10^{-37} n^2 T^{0.5} = 5.34 \cdot 10^{-37} \cdot 10^{38} \cdot 50^{0.5} = 3.78 \cdot 10^2 \text{ W/m}^3. \tag{15}$$

Here n is in 1/m^3, T is in keV.

Total Bremsstrahlung radiation is

$$P_{B,T} = P_B V = 3.78 \cdot 10^3 \cdot 113 = 4.27 \cdot 10^4 \text{ W}. \tag{16}$$

Surface Bremsstrahling radiation of 1 m^2 is

$$P_{B,S} = r \cdot P_B / 3 = 3.78 \cdot 10^2 \text{ W/m}^2. \tag{17}$$

9. Contact heat transfer is

$$P_c = \frac{\rho}{\rho_a} k_1 T = 2\times10^3 \ \frac{\text{W}}{\text{m}^2}, \quad \text{where} \quad \rho = n\mu m_p, \quad \rho_a = n_a \mu_a m_p. \tag{18}$$

Here ρ is density of plasma, kg/m^3; ρ_a = 1.225 kg/m^3 is standard density of atmosphere air; k_1 = 100 W/m^{2}K – heat transfer coefficient from gas to a solid well (one is right for T equals some thousands K, but comparing with other estimations shown that may be applied for T millions K); $\mu = m_i/m_p \approx 30$ for air and $\mu \approx (2 + 5)/2 = 2.5$ for D + T plasma. For our example, $P_c = 2\times10^3$ W/m^2.

10. Full energy fall into 1 m^2 of net surface is (sum of Bremsstrahlung radiation + plasma contact transfer)

$$q = \frac{P_{B,S}}{\pi} + 2\cdot P_c = \frac{3.78\cdot10^2}{3.14} + 2\cdot2\cdot10^3 = 4\cdot10^3 \quad \frac{\text{W}}{\text{m}^2}. \tag{19}$$

Temperature of the first net is

$$T = 100\sqrt[4]{\frac{q}{C_S}} = 517 \text{ K}, \quad \text{where} \quad C_S = 5.67 \frac{\text{J}}{\text{m}^2\text{K}^4} - \text{heat coefficient.} \tag{20}$$

Energy is getting by Space reactor from electric generator for the efficiency coefficient η = 0,9:

$$E_e = \eta\cdot V\cdot P_{DT} = 0.9\cdot113\cdot1.22\cdot10^4 = 1{,}24\cdot10^6 \quad \text{W}. \tag{21}$$

Less of energy for heating 3 nets is:

$$E_N = 3\xi\cdot VP_{DT} = 3\cdot0.0714\cdot113\cdot1.22\cdot10^4 = 2.95\cdot10^5 \quad \text{W}. \tag{22}$$

Total less for supporting the continuous thermonuclear reaction (compensation of plasma loss for heating nets, fuel and radiation loss):

$$L_s = P_{B,T} + E_N + p_f = 4.27\cdot10^4 + 2.95\cdot10^5 + 0.381\cdot10^4 = 3.38\cdot10^5 \text{ W}. \tag{23}$$

Useful electric power is:

$$P = E_e - L_s = 1.24\cdot10^6 - 3.38\cdot10^5 = 0.902\cdot10^6 \text{ W} = 902 \text{ kW}. \tag{24}$$

Mass of Space reactor is:

Mass of three nets

$$M_1 = 3\gamma\delta S = 3\cdot19.34\cdot10^{3\cdot}\cdot10^{-4}\cdot113 = 656 \text{ kg}. \tag{25}$$

Here $\gamma = 19340$ kg/m^3 is density of tungsten.

Mass of aluminum cover having $\delta_2 = 0.1$ mm, $\gamma_2 = 2700$ kg/m^3 is

$$M_2 = \gamma_2\delta_2 S = 2.7\cdot10^{3\cdot}\cdot10^{-4}113\cdot1{,}2 = 36.6 \text{ kg}. \tag{26}$$

The radius of cover is about 4.2 m. Here coefficient 1.2 increases the mass.

Total mass of Space reactor is about 700 – 800 kg.

17. Powerful power of Earth's version of AB Reactor is significantly more.

$$P = \frac{17.5}{3.5} P_{DT,F} - p_f = 5(1.38 \cdot 10^6 - 3.81 \cdot 10^3) \approx 6.9 \cdot 10^6 \text{ W} = 6.9 \text{ MW}. \quad (27)$$

In this version there is only the less for fuel heating. The neutron, Bremsstrahlung and nets radiation not leave the reactor. About 55 – 60% of this energy may be utilized as electric energy. The rest may be used as heat energy.

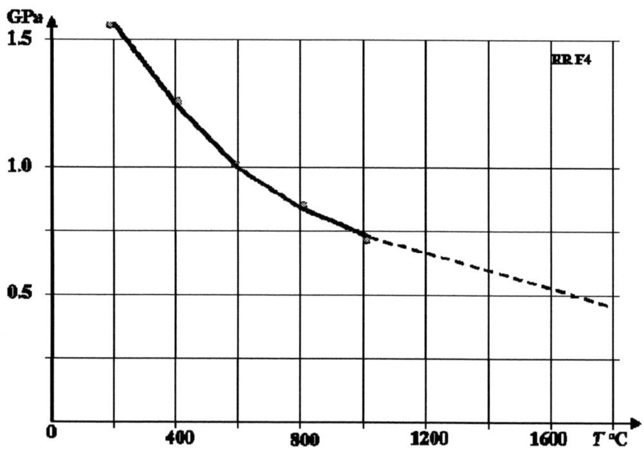

Figure 8. Tensile stress of tungsten via temperature in C. Continuous curve is experiment, broken curve is extrapolation to the melting point 3416° C.

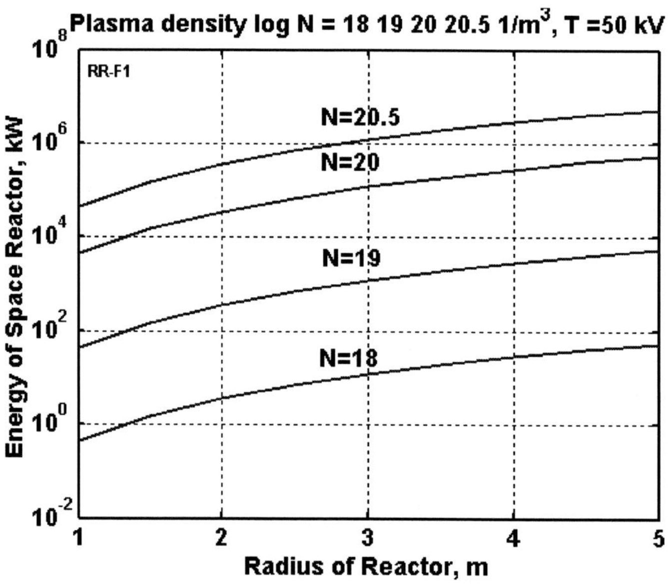

Figure 9. Electric power of Thermonuclear reflector AB-reactor (18% from total power) via radius of reactor and plasma density. Initial data for computation: coefficient of electric efficiency is $\eta = 0.9$.

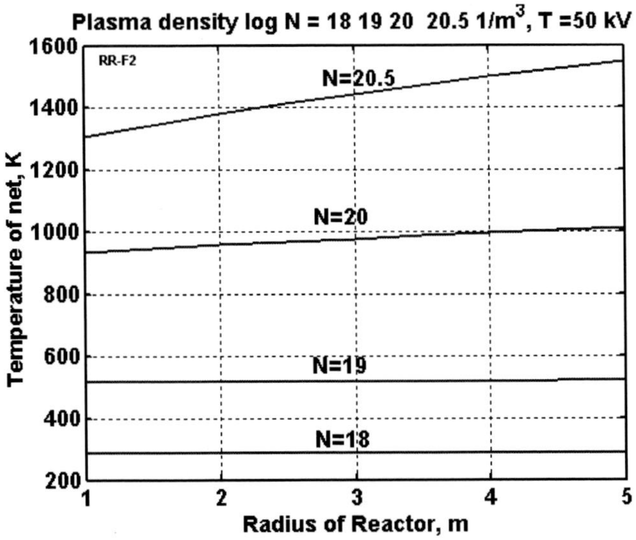

Figure 10. Temperature (K) of the first net via reactor radius and plasma density. Initial data for computation: plasma temperature 50 keV; net cell size 100 mm, form of net wire is round; specific density of tungsten is 19340 kg/m^3, safety tensile stress of net for T = 1500 K is $\sigma = 2\times10^8$ N/m^2.

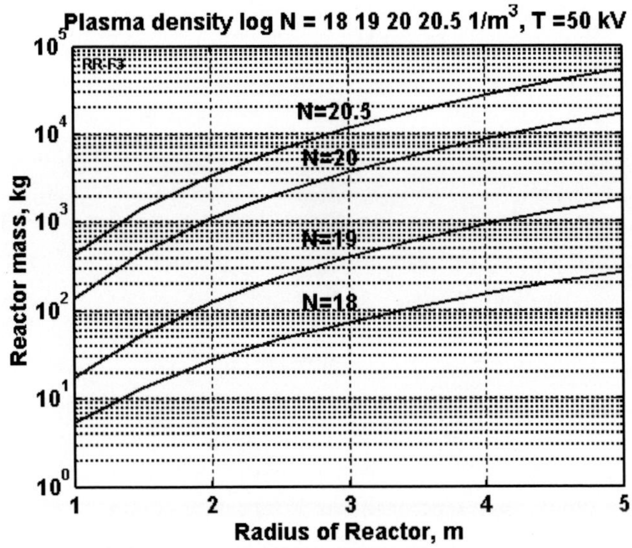

Figure 11. Reactor mass (kg) versus reactor radius and plasma density. Initial data for computation: plasma temperature 50 keV; net cell size L = 100×100 mm, form wire is round; specific density of tungsten is γ_1 = 19340 kg/m^3, safety tensile stress of net for T = 1500 K is $\sigma = 4\times10^8$ N/m^2; specific density of aluminum is γ_2 = 2700 kg/m^3, thickness of cover is 0.1 mm.

The mass of the Earth-stationed version is about 6 – 8 tons.

The results of computation via reactor radius for different plasma density are presented in Figures 9 – 11. Figure 8 shows the maximal tensile stress of the tungsten via temperature.

Macro-project

As Space project of Thermonuclear AB Reflect Reactor may be used the computation above for

initial data: $r = 3$ m, $n = 10^{19}$ 1/m^3, $T = 50$ keV $= 5.8\times10^8$ K. We get the useful electric power 902 kW and reactor mass about 400 kg. Full diameter of reactor is about 8.5 m. Distance between nets is about 0.5 m.

Selected reactor parameters are not the best possible nor fully optimized. That is merely an example computation. The using of materials having more safety stress at high temperatures significantly decreases the net loss, (literally; the loss from the nets) and increases the useful electric power. The increasing of plasma density significantly increases the reactor power and decreases the reactor size (diameter) and mass.

The same notes apply to the Earthbound AB-reactor. That is more complex, has more mass, but it produces significantly more energy.

Discussion

The low density plasma radically decreases the heat flow to solid surface. The high temperature solid surface intensive radiates the heat and balance may be lower than the melting temperature of solid conductive material. The entire magnetic bottle difficulty is avoided!

The confinement nets contact to a rare thermonuclear plasma and absorb a part of the plasma energy. This energy compensates (as X-radiation and heating of fuel) an electric energy produced by reactor. This total loss may be 5 – 30% of electric energy created by the charged α particle. If we compensate the loss of plasma energy by the electric (or high frequency) heating, we can receive a stable plasma which will permanently produce thermonuclear energy.

The disadvantage of the offered method is the high size of the reactor (from 1 through 8 m of diameter). That is consequence of a used low density plasma (low produced energy of rarefied plasma per volume) and required for a given engine power. But, there are a lot of potential reactor users who will be fully satisfied by the useful energy output of 100–2000 kW coming from a functioning reactor that is 1–8 m in diameter.

At present time the most efforts of scientists are directed toward very large, superpower thermonuclear stations (Shiva, NOVA, NIF, LLNL, HiPER, OMEGA EP, ITER, Z-machine, etc.). These stations request gigantic finances, years of development, complex technology. That is well for scientists seeking a stable career path over many years, (generations!) but not well for the technical progress of humanity. Governments spent billions of dollars for development of thermonuclear technology. However, we have not had achieved a stable long duration thermonuclear reaction after *50 years* of thermonuclear development. In the author's opinion, industrial thermonuclear electric stations may appear after 20 – 40 more years and their energy will be more expensive than a current technology conventional electric station.

The authors offer to direct government attention toward development smaller cheaper thermonuclear installations, which don't require huge funds and decades for development. Smaller power units are actually more practical in an immediate way: The world economy depends on transport vehicles (cars, ships, aircraft), which are the main users of oil fuel. They very much need small thermonuclear engines. (Large fixed conventional stations use coal, replacement of which does not necessarily relieve pressure on world oil prices.)

In the offered macro-project, as in any innovation, the obstacles may appear daunting as, for example, in the heating of the plasma. But there are some not yet fully developed research ideas for solution of even these macro-problems!

RESULTS

Authors offer a new (reflect reactor) confinement for thermonuclear low density plasma (not magnetic or inertial). He offered the series of innovations which allow applying this concept and develops the method for estimation and computation of a new class of small thermonuclear reactor. This reactor may be used for space, Earth transportation and small energy stations.

Authors made series of inventions which solve the problems of developing this reactor.

ATTACHMENT

This is how much energy the United States of America used from 2002 to 2006. Notice that solar is 0.1%. Nuclear increase from 2002 to 2006 was equal to the total amount of all solar power. (even though that was just operating efficiency and some small nuclear uprates).

Oil and fossil fuel usage was increasing. Petroleum (oil) was the primary source. 21 million barrels per day or about 7.4 billion barrels per year.

Twenty times as much solar power as there was in 2006 would be 1.2 quads. It would be nice but 5% of the coal usage. Increasing wind by ten times 2006 would be 2.6 quads. Combined it would be equal to about what one would expect to be the business as usual increase in energy consumption. All of the old coal and oil would stay in place.

Table 3. Oil usage in the USA

T = travel; F = freight; H = heat; M = misc.

Code	Where and How Used	Type of Oil	Barrels/day	Share of total, %
T	Cars (includes SUV's, minivans, pickups)	Gasoline (99%)	7,855,000	40.7%
F	Trucks (> 8500 lbs)	Diesel (80%)	2,460,000	12.7%
M	Raw material for plastics, chemicals, etc.	Feedstocks, Liquefied Petroleum Gas (LPG)	1,993.000	10.3%
T	Air travel (passenger; freight shown separately)	Jet fuel	1,287,000	6.7%
H	Process heat for factories	Various grades	956,000	5.0%
H	Heat + hot water for homes, offices, stores	Distillate, Resid (residual oil), LPG	942,000	4.9%
H	Energy to run oil refineries	Still gas	639,000	3.3%
T	Road pavement	Asphalt	537,000	2.8%
F	Waterborne freight (domestic + international)	Resid, distillate	478,000	2.5%
M	Agriculture (drying crops, farm machinery, etc.)	LPG, diesel	432,000	2.2%
H	Electricity generation	Resid (>90%)	313,000	1.6%
M	Construction machinery	Diesel	310,000	1.6%
M	Military (mostly jets)	Jet fuel, mostly	298,000	1.5%
F	Rail freight	Diesel	239,000	1.2%
F	Air freight	Jet fuel	212,000	1.1%
T	Recreational vehicles (boats, ATV's, etc.)	Gasoline	203,000	1.0%

See "Notes following main text for source of figures and key assertions in report.

REFERENCES

(The reader may find some of the author's works at *http://Bolonkin.narod.ru/p65.htm*, *http://Arxiv.org* Search term: "Bolonkin" and books: Bolonkin A.A., "*Non-Rocket Space Launch and flight*", Elsevier, 2006, 488 pgs., Bolonkin A.A., "*New Concepts, ideas, and Innovations in Technology and Human Science*", NOVA, 2008, 400 pg.)

[1] Bolonkin, A.A., (1982a), Installation for Open Electrostatic Field, *Russian patent application #3467270/21* 116676, 9 July, 1982 (in Russian), Russian PTO.

[2] Bolonkin, A.A., (1982b), Radioisotope Propulsion. *Russian patent application #3467762/25 116952,* 9 July 1982 (in Russian), Russian PTO.

[3] Bolonkin, A.A., (1982c), Radioisotope Electric Generator. *Russian patent application #3469511/25 116927.* 9 July 1982 (in Russian), Russian PTO.

[4] Bolonkin, A.A., (1983a), *Space Propulsion Using Solar Wing and Installation for It, Russian patent application #3635955/23* 126453, 19 August, 1983 (in Russian), Russian PTO.

[5] Bolonkin, A.A., (1983b), Getting of Electric Energy from Space and Installation for It, *Russian patent application #3638699/25 126303,* 19 August, 1983 (in Russian), Russian PTO.

[6] Bolonkin, A.A., (1983c), Protection from Charged Particles in Space and Installation for It, *Russian patent application #3644168 136270,* 23 September 1983, (in Russian), Russian PTO.

[7] Bolonkin, A. A., (1983d), Method of Transformation of Plasma Energy in Electric Current and Installation for It. *Russian patent application #3647344 136681* of 27 July 1983 (in Russian), Russian PTO.

[8] Bolonkin, A. A., (1983e), Method of Propulsion using Radioisotope Energy and Installation for It. of Plasma Energy in Electric Current and Installation for it. *Russian patent application #3601164/25 086973* of 6 June, 1983 (in Russian), Russian PTO.

[9] Bolonkin, A. A.,(1983f), Transformation of Energy of Rarefaction Plasma in Electric Current and Installation for it. *Russian patent application #3663911/25 159775, 23 November 1983* (in Russian), Russian PTO.

[10] Bolonkin, A. A., (1983g), Method of a Keeping of a Neutral Plasma and Installation for it. *Russian patent application #3600272/25 086993,* 6 June 1983 (in Russian), Russian PTO.

[11] Bolonkin, A.A.,(1983h), Radioisotope Electric Generator. *Russian patent application #3620051/25 108943*, 13 July 1983 (in Russian), Russian PTO.

[12] Bolonkin, A.A., (1983i), Method of Energy Transformation of Radioisotope Matter in Electricity and Installation for it. *Russian patent application #3647343/25 136692,* 27 July 1983 (in Russian), Russian PTO.

[13] Bolonkin, A.A., (1983j), Method of stretching of thin film. *Russian patent application #3646689/10 138085*, 28 September 1983 (in Russian), Russian PTO.

[14] Bolonkin A.A., (1983k), Light Pressure Engine, *Patent (Author Certificate) No. 11833421*, 1985 USSR (priority on 5 January 1983).

[15] Bolonkin, A.A., (1987), "New Way of Thrust and Generation of Electrical Energy in Space". *Report ESTI,* 1987, (Soviet Classified Projects).

[16] Bolonkin, A.A., (1990), "Aviation, Motor and Space Designs", Collection *Emerging Technology in the Soviet Union*, 1990, Delphic Ass., Inc., pp.32–80 (English).

[17] Bolonkin, A.A., (1991), *The Development of Soviet Rocket Engines,* 1991, Delphic Ass.Inc.,122 p. Washington, (in English).

[18] Bolonkin, A.A., (1992a), "A Space Motor Using Solar Wind Energy (Magnetic Particle Sail)". *The World Space Congress*, Washington, DC, USA, 28 Aug. – 5 Sept., 1992, IAF-0615.

[19] Bolonkin, A.A., (1992b), "Space Electric Generator, run by Solar Wing". *The World Space Congress,* Washington, DC, USA, 28 Aug. – 5 Sept. 1992, IAF-92-0604.

[20] Bolonkin, A.A., (1992c), "Simple Space Nuclear Reactor Motors and Electric Generators Running on Radioactive Substances", *The World Space Congress,* Washington, DC, USA, 28 Aug. – 5 Sept., 1992, IAF-92-0573.

[21] Bolonkin, A.A. (1994), "The Simplest Space Electric Generator and Motor with Control Energy and Thrust", *45th International Astronautical Congress,* Jerusalem, Israel, 9–14 Oct.,1994, IAF-94-R.1.368.

[22] Bolonkin A.A. (2004a), Light Multi-reflex Engine, *JBIS,* vol. 57, No. 9/10, 2004, pp. 353-359.

[23] Bolonkin A.A., (2004b), Utilization of Wind Energy at High Altitude. *Presented in International Energy Conversion Engineering Conference at Providence.,* RI, Aug.16-19. 2004. AIAA-2004-5705, AIAA-2004-5756, USA. *http://arxiv.org.*

[24] Bolonkin A.A., (2004c), *Design of Optimal Regulator. http://arxiv.org.*

[25] Bolonkin A.A.(2004d), Multi-reflex Space Propulsion, *JBIS,* Vol. 57, No. 11/12, 2004, pp. 379-390.

[26] Bolonkin A.A. (2006a), *Non-Rocket Space Launch and Flight*, Elsevier, London, 2006, 488 pgs.

[27] Bolonkin A.A. (2006b), New Thermonuclear Reactor, *AIAA-2006-7225,* Conference "Space-2006", USA.

[28] Bolonkin A.A., (2006c). Micro -Thermonuclear AB-Reactors for Aerospace, *Presented as paper AIAA-2006-8104 in 14th Space Plane and Hypersonic Systems Conference,* 6-8 November, 2006, USA. *http://arxiv.org.*

[29] Bolonkin A.A., (2006-1). *New AB-Thermonuclear Reactor for Aerospace,* http:arxiv.org.

[30] Bolonkin, A.A. (2006-2). *Electrostatic AB-Ramjet Space Propulsion,* AIAA-2006-6173. *http://arxiv.org.*

[31] Bolonkin A.A., (2006d). *Beam Space Propulsion, AIAA-2006-7492. http://arxiv.org.*

[32] Bolonkin A.A., (2006f). *Electrostatic Linear Engine, AIAA-2006-4806. http://arxiv.org.* AEAT, Vol.78, No. 6, 2006, pp. 502-508.

[33] Bolonkin A.A., (2006g). *Suspended Air Surveillance System, AIAA-2006-6511. http://arxiv.org.*

[34] Bolonkin A.A., (2006h). *Optimal Solid Space Tower* (Mast), *http://arxiv.org*

[35] Bolonkin A.A., (2006i). T*ransfer of Electricity into Outer Space. http://arxiv.org.*

[36] Bolonkin A.A., (2006j). *Theory of Space MagSail Some Common Mistakes and Electrostatic MagSail. http://arxiv.org.*

[37] Bolonkin A.A., (2006k). *High Speed AB-Solar Sail. http://arxiv.org.*

[38] Bolonkin A,A,, (2006 l). *New Method of Atmospheric Reentry of Space Ships. http://arxiv.org.*

[39] Bolonkin A.A., (2006m). *Simplest AB-Thermonuclear Space Propulsion and Electricity Generator. http://arxiv.org*

[40] Bolonkin A.A., (2006n). *Control of regional and Global Weather. http://arxiv.org.*

[41] Bolonkin A.A., (2006 o). *Cheap Textile Dam Protection of Seaport Cities against Hurricane Storm Serve Waves, Tsunamis, and Other Weather-Related Flood*s. *http://arxiv.org.*

[42] Bolonkin A.A., Cathcart R.B., (2006p) *A Low-Cost Natural Gas Aerial Pipeline. http://arxiv.org.*

[43] Bolonkin A.A., Cathcart R.B., (2006q) *The Java-Sumatra Aerial Mega-Terracing, http://arxiv.org.*

[44] *Wikipedia.* Some background material in this article is gathered from Wikipedia under the Creative Commons license.

Chapter 12

AB Electronic Tubes and Quasi-Superconductivity at Room Temperature*

Abstract

Authors offer and research a new macro-engineering idea - filling tubes by electronic gases. Shown: If the insulating envelope (cover) of the tube is charged positively, the electrons within the tube are not attracted to covering. Tube (as a whole) remains a neutral (uncharged) body. The electron gas in the tube has very low density and very high conductivity, close to superconductivity. If we take the density (pressure) of electron gas as equal to atmospheric pressure, the thickness of insulator film may be very small and the resulting tube is very light.

Author shows the offered tubes can be applied to many technical fields. For example:

(1) Transfer of energy over very long distance with very small electric losses. (2) Design of cheap high altitude electric lines without masts. (3) Transfer of energy from one continent to another continent through the ionosphere. (4) Transfer of a plasma beam (which can convey thrust and energy) from Earth surface to a space ship. (5) Observation of the nighttime sky by telescope without twinkling atmospheric hindrances. (6) Dirigibles (air balloons) of the highest lift force. (7) Increasing of gun range severalfold. (8) Transfer of matter.

Keywords: AB tubes, electronic tubes, superconductivity, transmission energy

Introduction

The first innovations in electrostatic applications were developed in 1982-1983 [1]-[2].

Later the series articles of this topic were published in [3]-[17]. In particular, in the work [3] was developed theory of electronic gas and its application to building (without space flight!) inflatable electrostatic space tower up to the stationary orbit of Earth's satellite (GEO).

In given work this theory applied to special inflatable electronic tubes made from thin insulator film. It is shown the charged tube filled by electron gas is electrically neutral, that can has a high internal pressure of the electron gas.

The main property of AB electronic tube is a very low electric resistance because electrons have small friction on tube wall. (In conventional solid (metal) conductors, the electrons strike against the immobile ions located in the full volume of the conductor.). The abnormally low electric resistance was found along the lateral axis only in nanotubes (they have a tube structure!). In theory, metallic nanotubes can have an electric current density (along the axis) more than 1,000 times greater than metals such as silver and copper. Nanotubes have excellent heat conductivity along axis up 6000 W/m·K. Copper, by contrast, has only 385 W/m·K. The electronic tubes explain why there is this effect. Nanotubes have the tube structure and electrons can free move along axis (they have only a friction on a tube wall).

More over, the moving electrons produce the magnetic field. The author shows - this magnetic field presses against the electron gas. When this magnetic pressure equals the electrostatic pressure, the electron gas may not remain in contact with the tube walls and their friction losses. The electron tube effectively becomes a superconductor for any surrounding temperature, even higher than room temperature! Author derives conditions for it and shows how we can significantly decrease the electric resistance.

DESCRIPTION, INNOVATIONS, AND APPLICATIONS

Description. An electronic *AB-Tube* is a tube filled by electron gas (figure 1). Electron gas is the lightest gas known in nature, far lighter than hydrogen. Therefore, tubes filled with this gas have the maximum possible lift force in atmosphere (equal essentially to the lift force of vacuum). The applications of electron gas are based on one little-known fact – the electrons located within a cylindrical tube having a positively charged cover (envelope) are in neutral-charge conditions – the total attractive force of the positive envelope plus negative contents equals zero. That means the electrons do not adhere to positive charged tube cover. They will freely fly into an *AB-Tube.* It is known, if the Earth (or other planet) would have, despite the massive pressures there, an empty space in Earth's very core, any matter in this (hypothetical!) cavity would be in a state of weightlessness (free fall). All around, attractions balance, leaving no vector 'down'.

Analogously, that means the *AB-Tube* is a conductor of electricity. Under electric tension (voltage) the electrons will collectively move without internal friction, with no vector 'down' to the walls, where friction might lie. In contrast to movement of electrons in metal (where moving electrons impact against a motionless ion grate). In the *AB-Tube* we have only electron friction about the tube wall. This friction is significantly less than the friction electrons would experience against ionic structures—and therefore so is the electrical resistance.

---

* Presented to http://arxiv.org on 8 April, 2008.

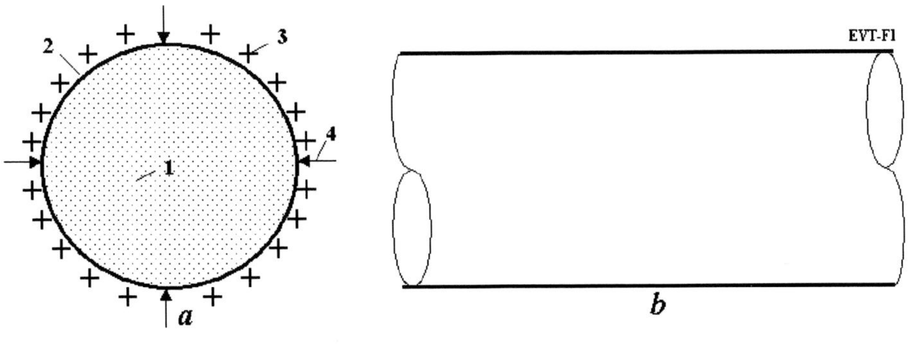

Figure 1. Electronic vacuum *AB-Tube.* ***a***) Cross-section of tube. ***b***) Side view. *Notation*: 1 – Internal part of tube filled by free electrons; 2 – insulator envelope of tube; 3 – positive charges on the outer surface of envelope (over this may be an additional film-insulator); 4 – atmospheric pressure.

When the density of electron gas equals $n = 1.65\times10^{16}/r$ per m^3 (where r is radius of tube, m), the electron gas has pressure equals atmospheric pressure 1 atm (see research below). In this case the tube cover may be a very thin—though well-sealed-- insulator film. The outer surface of this film is charged positively by static charges equal the electron charges and *AB-Tube* is thus an electrically neutral body.

Moreover, when electrons move into the *AB-Tube*, the electric current produces a magnetic field (Figure 2). This magnetic field compresses the electron cord and decreases the contact (and friction, electric resistance) electrons to tube walls. In the theoretical section is received a simple relation between the electric *current* and linear tube charge when the magnetic pressure equals to electron gas pressure $i = c\tau$ (where i is electric *current*, A; $c = 3\times10^8$ m/s – is the light speed; τ is tube linear electric charge, C/m). In this case the electron friction equals zero and *AB-Tube* becomes *superconductive at any outer temperature*. Unfortunately, this condition requests the electron speed equals the light speed. It is, however, no problem to set the electron speed very close to light speed. That means we can make the electric conductivity of *AB-Tube*s very close to superconductivity almost regardless of the outer temperature.

Example of electric line using the electron tubes is presented in figure 3. That consists the conventional constant-voltage electric generator 1, two nets 2 accelerating the electrons, two nets 4 braking the electrons and the electron tubes 3 connecting the electric source and customer.

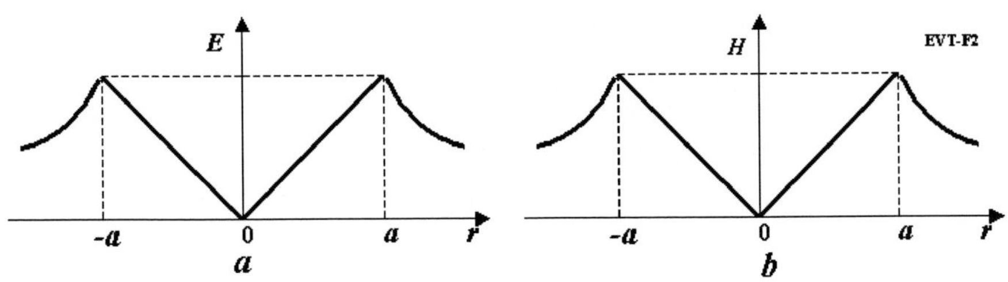

Figure 2. Electrostatic and magnetic intensity into *AB-Tube.* ***a***) Electrostatic intensity (pressure) via tube radius. ***b***) Magnetic intensity (pressure) from electric *current* versus rube radius.

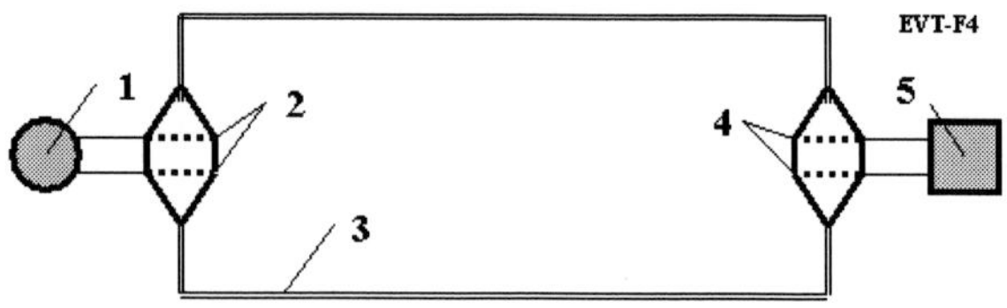

Figure 3. Example of electric line using the *AB-Tube* s. *Notations*: 1 – Electric constant-potential generator; 2 – acceleration nets; 3 – electric line from electronic tubes; 4 –brake nets; 5 – customer (user) of electric energy.

A matter worthy of your special attention: The insertion of electric energy into electronic *AB-Tubes* is accomplished not by the conventional method (direct contact to contactor), but the acceleration of electron flow by electric field between two nets. Accordingly, the extraction of electric energy from electron flow is also by means of the two nets by braking the electron flow.

Notes: 1) We can change the charges in offered *AB-Tube:* Put into tube an ION gas and block the internal charge by an opposed outer charge. In this case we can transfer matter by the offered *AB-Tube.*

2) The electron gas has a very small density and produces significant pressure (the electrostatic force (pressure) is more by thousands times than gas pressure!). This electric charge may be active and effect force over a long distance.

That way the *AB-Tube* has a maximal lift force in the Earth atmosphere in comparison of the helium or hydrogen air balloons.

Innovations. The offered AB electron tube has principal differences from the used cathode-anode lamp (rectifier):

1) *AB-Tube* has permanent electronic gas sealed in the tube (high negative charge inside tube);
2) *AB-Tube* has positive electric charge in the outer surface of tube, which neutralizes the internal negative charge into tube.
3) *AB-Tube* does not have a hot cathode and anode, which are wasteful of power.
4) *AB-Tube* can have any length (up some thousands km). The cathode lamp has distance between the cathode and anode equal to some millimeters. They cannot work if this distance is great.
5) Cathode tubes continually inject the electrons from cathode and adsorb them by anode. The *AB-Tube* has permanent electron gas.
6) Electric *current* in cathode tube can flow only from cathode to anode and flow only when cathode is hot. Electric *current* into *AB-Tube* can flow in any direction and with any temperature of tube parts.
7) If we change the electronic gas inside the *AB-Tube* with an ion gas, we can transfer the matter by offered tube. This is not a possibility in a classical cathode tube.
8) The offered *AB-Tube* has very high conductivity in comparison with a cathode tube.

Applications

There are numerous applications of offered *AB-Tube*s. Some of them are below:

1. Transfer electric energy in a long distance (up 10,000 km) with a small electric loss.
2. Superconductivity or close to superconductivity, 'Quasi-superconductivity'. The offered *AB-Tube* may have a very low electric resistance for any temperature because the electrons in the *AB-Tube* do not have large numbers of encounters with ions and do not lose energy from impacts with ions. The impact of electron to electron does not change the total impulse (momentum) of electron pairs and electron flow. If this idea is proved in experiments, that will be a big breakthrough in many fields of technology.
3. Cheap electric lines suspended at a high altitude (because the *AB-Tube*s have lift force in atmosphere and do not need electric masts and ground support paraphernalia]) (Figure 4a).
4. The big diameter *AB-Tube*s (including the electric lines) can be used as tramway for transportation (Figure 4a).
5. *AB-Tube*s can be used as vacuum tubes for an exit from the Earth's surface to outer space (a direct conduit outside of Earth's atmosphere from the surface!) (figure 4c). That may be used by Earth telescopes for observation of sky without atmospheric hindrances, or sending a plasma beam to space ships without atmospheric hindrances [8-9].
6. Transfer the electric energy from continent to continent through the Earth's ionosphere (Figure 4d) [11, 15].
7. Air balloons and dirigibles without expensive lifting gas.
8. Inserting an anti-gravitator cable into a vacuum-enclosing *AB-Tube* for near-complete elimination of air friction [4-5]. Same application for transmission of mechanical energy for long distance with minimum friction and losses. [4].
9. Increasing by some times the range of a conventional gun. They can shoot through the vacuum tube (to top of thinner layer of the atmosphere, up to 4-6 km) and projectile will fly in rare atmosphere where the air drag is small.

Attachment: In [3] author offered and applied this idea toward research the super high space tower (up to GEO, altitude 37,000 km). Below, is a basis for it and some reliable significant results.

1. Electron gas and space tower. The electron gas consists of conventional electrons. In contrast to molecular gas, the electron gas has many surprising properties. For example, electron gas (having same mass density) can have different pressure in the given volume. Its pressure depends from electric intensity, but electric intensity is different in different part of given volume (Figure 2a). For example, in our tube the electron intensity is zero in center of cylindrical tube and maximum near and at the tube surface.

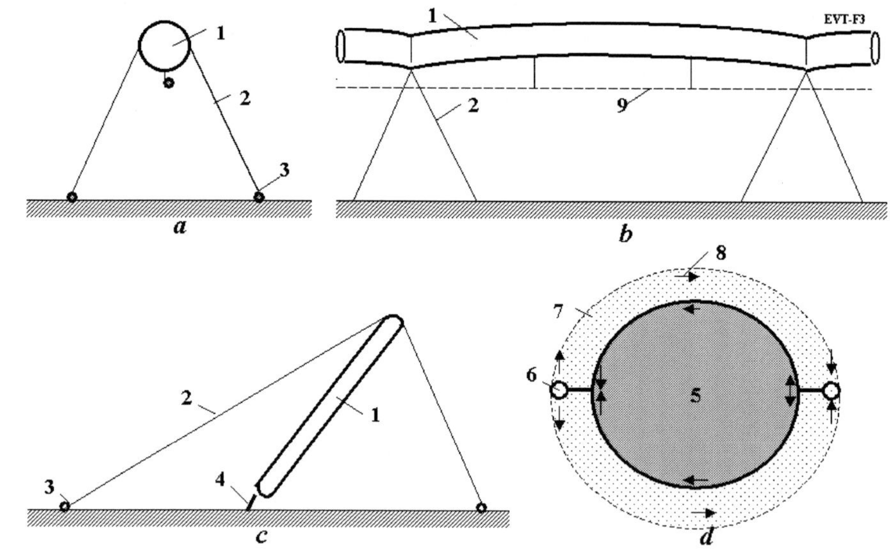

Figure 4. Some application of AB-Tubes. *a-b*) High efficiency electric long distance lines or (also) tramway (*a* - front view, *b* – side view). *c*) Earth's telescope without atmosphere hindrances, or plasma beam without atmosphere hindrances, *d*) Transfer of long distance of electric energy through Earth ionosphere.
Notations: 1 – electronic tube; 2 – guy lines; 3 – winch; 4 – telescope, or plasma beam injector, or gun; 5 – Earth; 6 – high altitude (100 km) electrostatic tower; 7 – ionosphere; 8 – electric current; 9 – tramway.

The offered *AB-Tube* is the main innovation in the suggested tower. It has positive control charges, isolated thin film cover, and electron gas inside. The positive cylinder creates a net zero electric field inside the tube and electrons conduct themselves as conventional molecules that are equal in mass density at any point. When kinetic energy of electrons is less than the energy of negative ionization of the dielectric cover; or the material of the electric cover does not accept the negative ionization; the electrons are reflected from the cover. In the other case the internal cover layer is saturated by negative ions and begins also to reflect electrons. Important also is that the offered AB electrostatic tube has a neutral summary charge in outer space.

Advantages of electrostatic tower. The offered electrostatic tower has very important advantages in comparison with conventional space elevator architecture commonly pictured today:

1. Electrostatic AB tower (mast) may be built from Earth's surface without rockets. That fact alone decreases the cost of a electrostatic mast by thousands of times.
2. One can have any height and payload capacity.
3. In particle, electrostatic tower can have the height of a geosynchronous orbit (37,000 km) WITHOUT the additional mandatory multi-hundred ton counterweight of a 'conventional' space elevator (and the long tether it must be suspended by to 120,000 - 160,000 km) [4], Ch.1.
4. The offered mast has less total mass by tens of times than conventional space elevator.

5. The offered mast can be built from less strong material than space elevator cable (see the computation here and in [4] Ch.1).
6. The offered tower can have high-speed electrostatic climbers impelled by high voltage electricity from Earth's surface [10, 6].
7. The offered tower is safer against the expected onslaught of meteorites (reentering space junk particles) than the cable of a 'conventional' space elevator, because any small meteorite damaging the cable may lead to a runaway collapse catastrophe for a space elevator, but it only creates small holes in a electrostatic tower. The electron escape may be compensated by automatic electron injection.
8. The electrostatic mast can bend in a needed direction when we give the appropriate electric voltage in needed parts of the mast.

The electrostatic tower of heights up to 100 - 500 km may be built from current artificial fiber material available at the present time. The geosynchronous electrostatic tower needs stronger material having a strong coefficient $K \geq 2$ (whiskers or nanotubes, see below).

2. Other applications of the AB-Tube.

The offered *AB-Tube* with the positive charged cover and the electron gas inside may find the many applications in other technical fields. For example:

1. *Air dirigible.* (1) The airship from the thin film filled by an electron gas has 30% more lift force than conventional dirigible filled by costly helium. (2) An electron dirigible is significantly cheaper than the equivalent helium-filled dirigible because the helium is a very expensive gas (~$7/kg) and is becoming more so over time. (3) One does not have a problem with changing the lift force for fine control because it is no problem to add or to delete electrons.
2. *Long arm.* The offered electron control tube can be used as a long control work arm for performing work on a planet or other celestial body, rescue operations, repairing of other spaceships and so on. For details see [4] Ch.9.
3. *Super-reflectivity.* If free electrons are located between two thin transparent plates, that may be usable as a super-reflectivity mirror for a widely spectrum of radiation. That is necessary in many important technical fields as a light engine, multi-reflection propulsion [4] Ch.12 and thermonuclear power [16].

Other applications of electrostatic technology are Electrostatic solar wind propulsion [4] Ch.13, Electrostatic utilization of asteroids for space flight [4] Ch.14, Electrostatic levitation on the Earth and artificial gravity for spaceships and asteroids [4, 5 Ch.15], Electrostatic solar sail [3] Ch.18, Electrostatic space radiator [4] Ch.19, Electrostatic AB-Ramjet space propulsion [4], etc.

THEORY AND COMPUTATION

Below the interested reader may find the evidence of main equations, estimations, and computations.

1. Relation between the linear electric charge of tube and electron gas pressure on tube surface:

$$p = \frac{\varepsilon_0 E^2}{2}, \quad E = k\frac{2\tau}{r}, \quad \varepsilon_0 = \frac{1}{4\pi k}, \quad \tau = \sqrt{\frac{2\pi r p}{k}}, \tag{1}$$

where p is electron pressure, N/m^2; $\varepsilon_0 = 8.85\times10^{-12}$ F/m –electrostatic constant; $k = 9\times10^9$ Nm^2/C^2 is electrostatic constant; E is electric intensity, V/m; τ is linear charges of tube, C/m; r is radius of tube, m.

Example, for atmospheric pressure $p = 10^5$ N/m^2 we receive $E = 1.5\times10^8$ V/m, N/C, the linear charge $\tau = 0.00833r$ C/m.

2. Density of electron (ion) in 1 m^3 in tube.

$$n = \frac{\tau}{\pi r^2 e} = \frac{1}{2\pi e k}\frac{E}{r} = 1.1\cdot10^8\frac{E}{r},$$

$$M_e = m_e n, \quad M_i = \mu n_p n, \quad \mu = \frac{m_i}{m_p}, \tag{2}$$

where n is charge (electron or ion) density, 1/m^3; $e = 1.6\times10^{-19}$ C is charge of electron; $m_e = 9.11\times10^{-31}$ is mass of electron, kg; $m_p = 1.67\times10^{-27}$ is mass of proton, kg; M_e is mass density of electron, kg/m^3; M_i is mass density of ion, kg/m^3.

For electron pressure 1 atm the electron density (number particles in m^3) is $n = 1.65\times10^{16}/r$.

3. Electric resistance of AD-tube. We estimate the friction of electron about the tube wall by gas-kinetic theory

$$F_B = \eta_B SV, \quad \eta_B = \frac{1}{6}\rho V, \quad \rho = m_e n,$$

$$\overline{F} = \frac{F}{S} = \frac{1}{6}m_e n V^2, \quad V = \frac{j}{en} = \frac{i}{en\pi r^2}, \tag{3}$$

where F_B is electron friction, N; η_B is coefficient of friction; S is friction area, m^2; V is electron speed, m/s; ρ is density of electron gas, kg/m^3; $\overline{F}$ is relative electron friction, N/m^2; j is *current* density, A/m^2.

4. Electric loss. The electric loss (power) into tube is

$$P_T = \overline{F}_B SV, \quad S = 2\pi r L, \quad P_T = \frac{1}{3}\pi m_e n r L V^3,$$

$$P_T = \frac{m_e}{3e^3\pi^2}\frac{i^3 L}{n^2 r^5} = 7.5\cdot10^{24}\frac{i^3 L}{n^2 r^5} \quad [W], \tag{4}$$

where P_T is electric loss, W; L is tube length, m; i is electric current, A.

5. *Relative electric loss* is

$$\overline{P}_T = \frac{P_T}{P}, \quad P = iU, \quad \overline{P}_T = \frac{m_e}{3\pi^2 e^3}\frac{i^2}{n^2 r^5}\frac{L}{U} = \\ = 7.5\cdot 10^{24}\frac{i^2}{n^2 r^5}\frac{L}{U} = 7.4\cdot 10^{25}\frac{j^2}{n^2 r}\frac{L}{U}, \qquad (5)$$

Compare the relative loss the offered electric (tube) line and conventional electric long distance line. Assume the electric line have length L = 2000 km, electric voltage $U = 10^6$ V, electric current i = 300 A, atmospheric pressure into tube. For offered line having tube r = 1 m the relative loss equals $\overline{P}_T$ = 0.005. For conventional electric line having cross section copper wire 1 cm^2 the relative loss is $\overline{P}_T$ = 0.105. That is in 21 times more than the offered electric line. The computation of Equation (5) for atmospheric pressure and for ratio L/U = 1 are presented in figure 5. As you see for electric line L = 1000 km, voltage U = 1 million V, tube radius r = 2.2 m, the electric current i = 50 A, the relative loss of electric power is one/millionth (10^{-6}), (only 50 W for transmitted power 50 millions watt!).

Moreover, the offered electric line is cheaper by many times, may be levitated into the atmosphere at high altitude, does not need a mast and ground, doesn't require expensive copper, does not allow easy surface access to line tapping thieves who wish to steal the electric energy. And this levitating electric line may be suspended with equal ease over sea as over land.

6. *Lift force of tube* ($L_{F,1}$, kg/m) and mass of 1 m length of tube (W_1. kg/m) is

$$L_{F,1} = \rho v = \rho\pi r^2, \quad W_1 = 2\pi r^2\gamma\delta, \qquad (6)$$

where ρ is air density, at sea level ρ =1.225 kg/m^3; v is volume of 1 m of tube length, m^3; γ is density of tube envelope, for most plastic γ = 1500 ÷ 1800 kg/m^3; δ is film thickness, m.

Example. For r = 10 m and δ = 0.1 mm, the lift force is 384 kg/m and cover mass is 11.3 kg/m.

7. *Artificial fiber and tube (cable) properties* [18]-[21]. Cheap artificial fibers are currently being manufactured, which have tensile strengths of 3-5 times more than steel and densities 4-5 times less than steel. There are also experimental fibers (whiskers) that have tensile strengths 30-100 times more than steel and densities 2 to 5 times less than steel.

Although the described (1989) graphite fibers are strong ($\sigma/\gamma = 10\times10^6$), they are at least still ten times weaker than theory predicts. A steel fiber has a tensile strength of 5000 MPA (500 kg/sq.mm), the theoretical limit is 22,000 MPA (2200 kg/mm^2) (1987); polyethylene fiber has a tensile strength 20,000 MPA with a theoretical limit of 35,000 MPA (1987). The very high tensile strength is due to its nanotube structure [21].

Apart from unique electronic properties, the mechanical behavior of nanotubes also has provided interest because nanotubes are seen as the ultimate carbon fiber, which can be used as reinforcements in advanced composite technology. Early theoretical work and recent experiments on individual nanotubes (mostly MWNT's, Multi Wall Nano Tubes) have confirmed that nanotubes are one of the stiffest materials ever made. Traditional carbon fibers show high strength and stiffness, but fall far short of the theoretical, in-plane strength of graphite layers by an order of magnitude. Nanotubes come close to being the best fiber that can be made from graphite.

For example, whiskers of Carbon nanotube (CNT) material have a tensile strength of 200 Giga-Pascals and a Young's modulus over 1 Tera Pascals (1999). The theory predicts 1 Tera Pascals and a Young's modules of 1-5 Tera Pascals. The hollow structure of nanotubes makes them very light (the specific density varies from 0.8 g/cc for SWNT's (Single Wall Nano Tubes) up to 1.8 g/cc for MWNT's, compared to 2.26 g/cc for graphite or 7.8 g/cc for steel). Tensile strength of MWNT's nanotubes reaches 150 GPa.

In 2000, a multi-walled carbon nanotube was tested to have a tensile strength of 63 GPa. Since carbon nanotubes have a low density for a solid of 1.3-1.4 g/cm³, its specific strength of up to 48,000 kN·m/kg is the best of known materials, compared to high-carbon steel's 154 kN·m/kg.

The theory predicts the tensile stress of different types of nanotubes as: Armchair SWNT - 120 GPa, Zigzag SWNT – 94 GPa.

Specific strength (strength/density) is important in the design of the systems presented in this paper; nanotubes have values at least 2 orders of magnitude greater than steel. Traditional carbon fibers have a specific strength 40 times that of steel. Since nanotubes are made of graphitic carbon, they have good resistance to chemical attack and have high thermal stability. Oxidation studies have shown that the onset of oxidation shifts by about 100^0 C or higher in nanotubes compared to high modulus graphite fibers. In a vacuum, or reducing atmosphere, nanotube structures will be stable to any practical service temperature (in vacuum up 2800 °C. in air up 750°C).

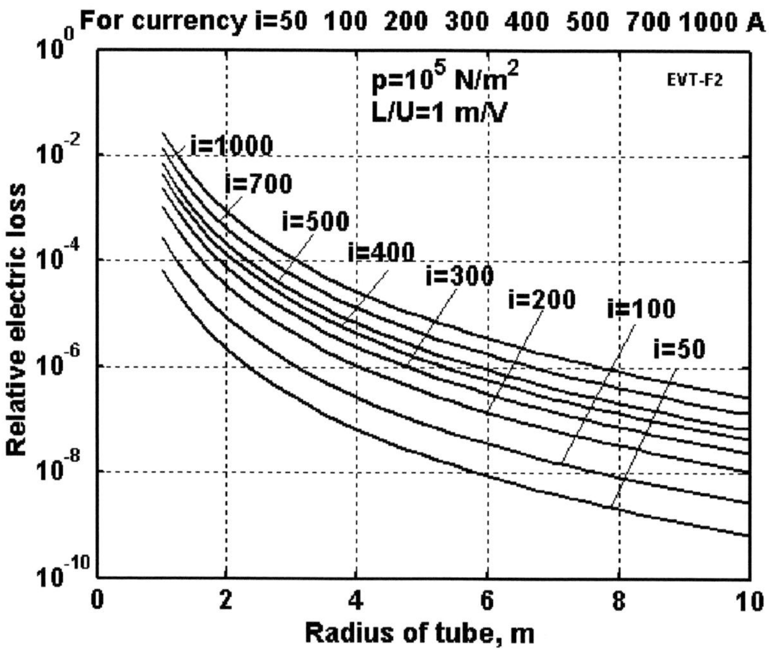

Figure 5. Relative electric loss via radius of tube for electric *current* $i = 50 \div 1000$ A, the atmospheric pressure into tube and ratio $L/U = 1$.

In theory, metallic nanotubes can have an electric current density (along axis) more than 1,000 times greater than metals such as silver and copper. Nanotubes have excellent heat conductivity along axis up 6000 W/m·K. Copper, by contrast, has only 385 W/m·K.

About 60 tons/year of nanotubes are produced now (2007). Price is about \$100 - 50,000/kg. Experts predict production of nanotubes on the order of 6000 tons/year and with a price of \$1 – 100/kg to 2012.

Commercial artificial fibers are cheap and widely used in tires and countless other applications. Some data are in Table 5 Attn. and section "Nanotibes" in Attachment to book.

Industrial fibers have up to σ = 500 - 600 kg/mm^2, γ = 1500 -1800 kg/m^3, and σ/γ = 2,78x10^6. But we are projecting use in the present projects the cheapest films and cables applicable (safety σ= 100 - 200 kg/mm^2).

8. Dielectric strength of insulator. As you see above, the tube needs film that separates the positive charges located in conductive layer from the electron gas located in the tube. This film must have a high dielectric strength. The current material can keep a high *E* (see Table 4 in Attn. taken from [10]).

Note: Dielectric constant ε can reach 4.5 - 7.5 for mica (*E* is up 200 MV/m), 6 -10 for glasses (*E* = 40 MV/m), and 900 - 3000 for special ceramics (marks are CM-1, T-900) [17], p. 321, (*E* =13 -28 MV/m). Ferroelectrics have ε up to 10^4 - 10^5. Dielectric strength appreciably depends from surface roughness, thickness, purity, temperature and other conditions of materials. Very clean material without admixture (for example, quartz) can have electric strength up 1000 MV/m. As you see, we have the needed dielectric material, but it is necessary to find good (and strong) isolative materials and to research conditions which increase the dielectric strength.

9. Tube cover thickness. The thickness of the tube's cover may be found from Equation

$$\delta = \frac{rp}{\sigma}, \tag{7}$$

where *p* is electron pressure minus atmospheric pressure, N/m^2. If electron pressure is little more then the atmospheric pressure the tube cover thickness may be very thin.

10. Mass of tube cover. The mass of tube cover is

$$M_1 = \delta\gamma, \quad M = 2\pi r L \gamma \delta,, \tag{8}$$

where M_1 is 1 m^2 cover mass, kg/m^2; *M* is cover mass, kg.

11. The volume V and surface of tube s are

$$V = \pi r^2 L, \quad s = 2\pi r L, \tag{9}$$

where *V* is tube volume, m^3; *s* is tube surface, m^2.

12. Relation between tube volume charge and tube liner charge for neutral tube is

$$E_V = \frac{\rho r}{2\varepsilon_0}, \quad E_s = \frac{\tau}{2\pi\varepsilon_0 r}, \quad E_V = E_s, \quad \tau = \pi\rho r^2, \quad \rho = \frac{\tau}{\pi r^2}, \tag{10}$$

where ρ is tube volume charge, C/m^3; τ is tube linear charge, C/m.

13. General charge of tube. We got equation from

$$\tau = 2\pi\varepsilon\varepsilon_0 Er, \quad Q = \tau L, \quad Q = 2\pi\varepsilon\varepsilon_0 ErL, \tag{11}$$

where Q is total tube charge, C; ε is dielectric constant (see Table 2).

14. Charging energy. The charged energy is computed by equation

$$W = 0.5QU, \quad U = \delta E, \quad W = 0.5Q\delta E, \tag{12}$$

where W is charge energy, J; U is voltage, V.

15. Mass of electron gas. The mass of electron gas is

$$M_e = m_e N = m_e \frac{Q}{e}, \tag{13}$$

where M_e is mass of electron gas, kg; $m_e = 9.11\times10^{-31}$ kg is mass of electron; N is number of electrons, $e = 1.6\times10^{-19}$ is the electron charge, C.

16. Transfer of matter (Matter flow of ion gas). If we change the electron gas by the ion gas, our tube transfer charged matter with very high speed

$$M = M_i \pi r^2 V, \quad M_i = \mu m_p n,$$
$$V = \frac{i}{en\pi r^2}, \quad M = \frac{m_p}{e}\mu = 1.04\cdot10^{-8}\mu i, \tag{14}$$

where M is the mass flow, kg/s; M_i is the gas ion density, kg/m^3; $\mu = m_i/m_p$; V is ions speed, m/s.

Example: We want to transfer to a remote location the nuclear breeder fuel – Uranium-238. ($\mu = 238$) by line having $i = 1000$ A, $r = 1$ m, ion gas pressure 1 atm. One day contains 86400 seconds.

The equation (12) gives $M = 214$ kg/day, speed $V = 120$ km/s. The AB-tubes are suitable for transferring small amounts of a given matter. For transferring a large mass the diameter of tube and electric current must be larger.

We must also have efficient devices for ionization and utilization of the de-ionization (recombination) energy.

The offered method allows direct conversion of the ionization energy of the electron gas or ion gas to light (for example, by connection between the electron and ion gases).

17. Electron gas pressure. The electron gas pressure may be computed by equation (1). This computation is presented in figure 6.

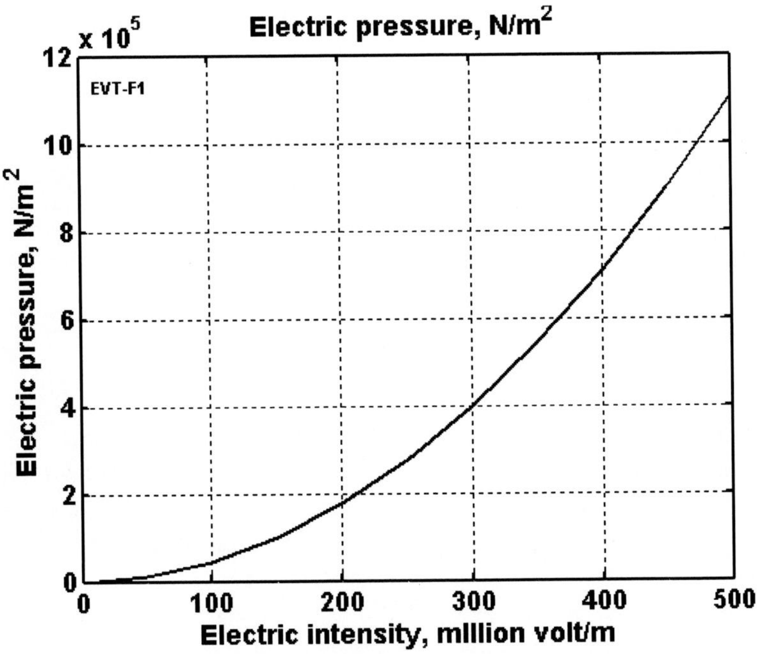

Figure 6. Electron pressure versus electric intensity.

As you see the electron pressure reaches 1 atm for an electric intensity 150 MV/m and for negligibly small mass of the electron gas.

18. Power for support of charge. Leakage current (power) through the cover may be estimated by equation

$$I = \frac{U}{R}, \quad U = \delta E = \frac{r\varepsilon_0 E}{\sigma}, \quad R = \rho\frac{\delta}{s}, \quad I = \frac{sE}{\rho}, \quad W_l = IU = \frac{\delta s E^2}{\rho}, \tag{15}$$

where I is electric current, A; U is voltage, V; R is electric resistance, Ohm; ρ is specific resistance, Ohm·m; s is tube surface area, m^2.

The estimation gives the support power has small value.

QUASI-SUPERCONDUCTIVITY OF AB-TUBE

The proposed AB-Tube may become what we may term 'quasi-superconductive' when magnetic pressure equals electrostatic pressure. In this case electrons cannot contact with the tube wall, do not experience resistance friction and the AB-Tube thus experiences this 'quasi-superconductivity'. Let us to get this condition:

$$P_e = \frac{\varepsilon_0 E^2}{2}, \quad P_m = \frac{B^2}{2\mu_0}, \quad P_e = P_m, \quad c = \frac{1}{\sqrt{\mu_0 \varepsilon_0}}, \quad E = cB, \tag{16}$$

where P_e is electronic pressure, N/m^2; P_m is magnetic pressure, N/m^2; B is magnetic intensity, T; E is electric intensity, V/m; c is light speed, $c = 3\times10^8$ m/s; ε_0, $\mu_0 = 4\pi\times10^{-7}$ are electrostatic and magnetic constants. The relation $E = cB$ is important result and condition of tube superconductivity. For electron pressure into tube 1 atm, the $E = 1.5\times10^8$ V/m (see above) and $B = 0.5$ T.

From Eq. (16) we receive the relation between the electric current and the tube charge for AB-Tube 'quasi-superconductivity' as

$$E = cB, \quad E = \frac{1}{2\pi\varepsilon_0}\frac{\tau}{r}, \quad B = \frac{\mu_0 i}{2\pi r}, \quad i = c\tau \quad , \tag{17}$$

where i is electric current, A; τ is liner charge of tube, C/m.

For electron pressure equals 1 atm and r = 1m the linear tube charge is $\tau = 0.00833$ C/m (see above) and the request electric current is $i = 2.5\times10^6$ A (j = 0.8 A/m^2). For r = 0.1 m the current equals $i = 2.5\times10^5$ A. And for r = 0.01 m the current equals $i = 2.5\times10^4$ A.

Unfortunately, the requested electron speed (for true and full normal temperature 'superconductivity') equals light speed c. That means we cannot exactly reach it, but we can came very close and we can have very low electric resistance of AB-Tube.

Information about high speed of electron and ion beam. Here $\gamma = (1-\beta^2)^{-1/2}$ is the relativistic scaling factor; quantities in analytic formulas are expressed in SI or cgs units, as indicated; in numerical formulas I is in amperes (A), B is in gauss (G, 1 T = 10^4 G), electron linear density N is in cm^{-1}, temperature, voltage, and energy are in MeV, $\beta_z = v_z/c$, and k is Boltzmann's constant.

For computation electrostatic and magnetic fields about light speed are useful the equations of relativistic theory (Lorenz's Equations):

$$\begin{aligned} &E_x = E_{1x}, && H_x = H_{1x}, \\ &\sqrt{1-\beta^2}\,E_y = E_{1y} - vB_{1z}, && \sqrt{1-\beta^2}\,H_y = H_{1y} + vD_{1z}, \\ &\sqrt{1-\beta^2}\,E_z = E_{1z} + vB_{1y}, && \sqrt{1-\beta^2}\,H_z = H_{1z} - vD_{1y}, \end{aligned} \tag{18}$$

where lower index "$_1$" means the immobile system coordinate, E is electric intensity, V/m; H is magnetic intensity, A/m; v is speed of mobile system coordinate along axis x, m/s; D is electric displacement. C/m^2; $\beta = v/c$ is relative speed one system about the other.

Relativistic electron gyroradius [22]:

$$r_e = \frac{mc^2}{eB}(\gamma^2 - 1)^{1/2} \ (\text{cgs}) = 1.70\cdot10^3(\gamma^2-1)^{1/2}B^{-1} \quad \text{cm}. \tag{19}$$

Relativistic electron energy:

$$W = mc^2\gamma = 0.511\gamma \quad \text{MeV}. \tag{20}$$

Bennett pinch condition:

$$I^2 = 2Nk(T_e + T_i)c^2 \quad (cgs) == 3.20 \cdot 10^{-4} N(T_e + T_i) \quad A^2. \tag{21}$$

Alfven-Lawson limit:

$$I_A = (mc^3 / e)\beta_z \gamma \quad (cgs) = (4\pi mc / \mu_0 e)\beta_z \gamma \quad (SI) = 1.70 \cdot 10^4 \beta_z \gamma \quad A. \tag{22}$$

The ratio of net current to I_A is

$$\frac{I}{I_A} = \frac{\nu}{\gamma}. \tag{23}$$

Here $\nu = Nr_e$ is the Budker number, where $r_e = e^2 / mc^2 = 2.82 \cdot 10^{-13}$ cm is the classical electron radius. Beam electron number density is

$$n_b = 2.08 \cdot 10^8 J\beta^{-1} \quad cm^{-3}, \tag{24}$$

where J is the current density in A cm-2. For a uniform beam of radius a (in cm):

$$n_b = 6.63 \cdot 10^7 Ia^{-2}\beta^{-1} \quad cm^{-3} \tag{25}$$

and

$$\frac{2r_e}{a} = \frac{\nu}{\gamma}, \tag{26}$$

Child's law: nonrelativistic space-charge-limited current density between parallel plates with voltage drop V (in MV) and separation d (in cm) is

$$J = 2.34 \cdot 10^3 V^{3/2} d^{-2} \quad A\ cm^{-2} \tag{27}$$

The condition for a longitudinal magnetic field B_z to suppress filamentation in a beam of current density J (in A cm^{-2}) is

$$B_z > 47\beta_z(\gamma J)^{1/2} \ G. \tag{28}$$

Kinetic energy necessary to accelerate a particle is

$$K = (\gamma - 1)mc^2. \tag{29}$$

The de Broglie wavelength of particle is $\lambda = h/p$, where $h = 6.6262\times10^{-34}$ J·s is Planck constant, p is particle momentum. Classical radius of electron is 2.8179×10^{-15} m.

CONCLUSION

The offered inflatable electrostatic AB tube has indisputably remarkable operational advantages in comparison with the conventional electric lines.

The main innovations and applications of *AB-Tube*s are:

1. Transferring electric energy in a long distance (up 10,000 km) with a small electric loss.
2. 'Quasi-superconductivity'. The offered AB-Tube may have a very low electric resistance for any temperature because the electrons in the tube do not have ions and do not lose energy by impacts with ions. The impact the electron to electron does not change the total impulse (momentum) of couple electrons and electron flow. If this idea is proved in experiment, that will be big breakthrough in many fields of technology.
3. Cheap electric lines suspended in high altitude (because the AB-Tube can have lift force in atmosphere and do not need ground mounted electric masts and other support structures) (Figure 4a).
4. The big diameter AB-Tubes (including the electric lines for internal power can be used as tramway for transportation (figure 4a).
5. AB-Tube s can be used as vacuum tubes for an exit from the Earth's surface to outer space (out from Earth's atmosphere) (figure 4c). That may be used by an Earth telescope for observation of sky without atmosphere hindrances, or sending of a plasma beam to space ships without atmosphere hindrances [8-9].
6. Transfer of electric energy from continent to continent through the Earth's ionosphere (figure 4d) [11, 15].
7. Inserting an anti-gravitator cable into a vacuum-enclosing AB-Tube for near-complete elimination of air friction [4-5]. Same application for transmission of mechanical energy for long distances with minimum friction and losses. [4].
8. Increasing in some times the range of a conventional gun. They can shoot through the vacuum tube (up 4-6 km) and projectile will fly in the rare atmosphere where air drag is small.
9. Transfer of matter a long distance with high speed (including in outer space, see other of author's works).
10. Interesting uses in nuclear and high energy physics engineering (inventions).

The offered electronic gas may be used as filling gas for air balloons, dirigibles, energy storage, submarines, electricity-charge devices (see also [1]-[17, 23]).

Future spaceships

ESA

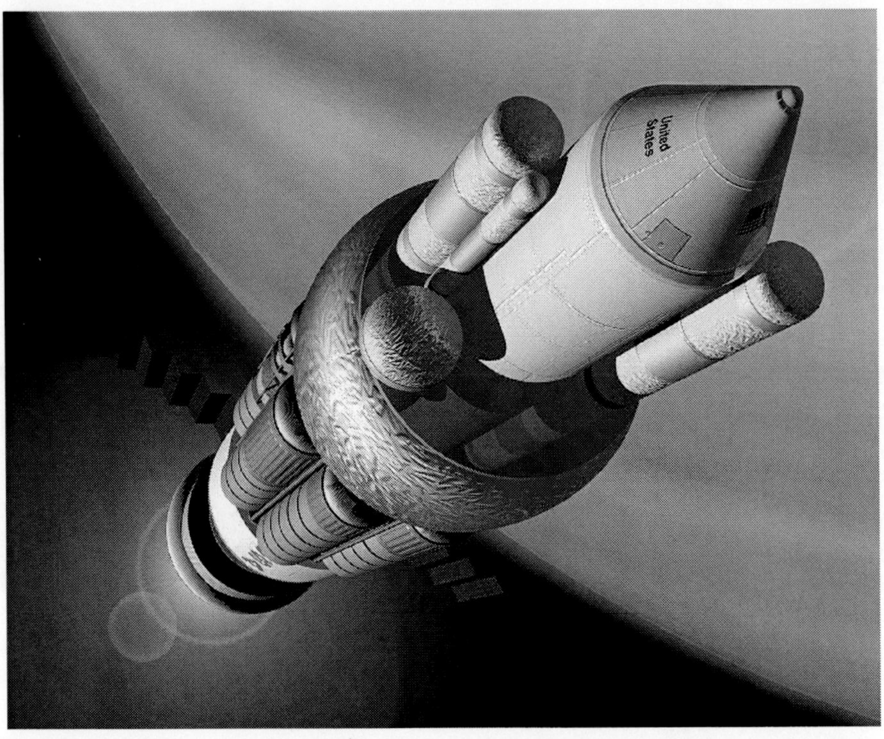

NASA

NASA. Ships Orion and Altair.

Possible spaceship.

REFERENCES

(Part of these articles the reader can find in author WEB page: ht*tp://Bolonkin.narod.ru/ p65.htm*, *http://arxiv.org*, search "Bolonkin", and in the books "*Non-Rocket Space Launch and Flight*", Elsevier, London, 2006, 488 pgs., "*New Concepts, Ideas, and Innovations in Aerospace, Technology and Human Sciences*", 2008, 400 pgs.)

[1] Bolonkin, A.A., (1982), Installation for Open Electrostatic Field, *Russian patent application #3467270/21 116676,* 9 July, 1982 (in Russian), Russian PTO.

[2] Bolonkin, A.A., (1983), Method of stretching of thin film. *Russian patent application #3646689/10 138085,* 28 September 1983 (in Russian), Russian PTO.

[3] Bolonkin A.A., Optimal Electrostatic Space Tower, *Presented as paper AIAA-2007-6201 to Space-2007 Conference,* 18-20 September 2007, Long Beach, CA, USA. http://arxiv.org, search "Bolonkin"

[4] Bolonkin A.A., Non-Rocket Space Launch and Flight, Elsevier, London, 2006, 488 ps.

[5] Book (2006),: *Macro-Engineering - A challenge for the future*. Collection of articles. Eds. V. Badescu, R. Cathcart and R. Schuiling, Springer, (2006). (Collection contains Bolonkin's articles: Space Towers; Cable Anti-Gravitator, Electrostatic Levitation and Artificial Gravity).

[6] Bolonkin A.A., Linear Electrostatic Engine, *The work is presented as AIAA-2006-5229 for 42 Joint Propulsion Conference,* Sacramento, USA, 9-12 July, 2006. Reader finds it in *http://arxiv.org*, search "Bolonkin".

[7] Bolonkin A.A., (2005), *Problems of Electrostatic Levitation and Artificial Gravity, AIAA-2005-4465. 41 Propulsion Conference,* 10-12 July, 2005, Tucson, Arizona, USA.

[8] Bolonkin A.A., (2006), Electrostatic AB-Ramjet Space Propulsion, AIAA/AAS Astrodynamics Specialist Conference, 21-24 August 2006, USA. AIAA-2006-6173. *Journal "Aircraft Engineering and Aerospace Technology",* Vol.79, #1, 2007. *http://arxiv.org*, search "Bolonkin".

[9] Bolonkin A.A., AB-Beam Space Propulsion, *Presented as paper AIAA-206-7492 to Conference "Space-2006",* 19-21 September, 2006, San-Jose, CA, USA.

[10] Bolonkin A.A., Electrostatic Climber for Space Elevator and Space Launcher, *presented as paper AIAA-2007-5838 for 43 Joint Propulsion Conference,* Cincinnati, Ohio, USA, 9-11 July, 2007. See also *http://arxiv.org* search "Bolonkin".

[11] Bolonkin A.A., Transfer of Electricity in Space, *Journal "Aircraft Engineering and Aerospace Technology",* Vol. 79, No. 3, 2007, pp. 273-282.

[12] Bolonkin A.A., Electrostatic Solar Wind Propulsion, *Presented as paper AIAA-2005-3857 at the 41-st Propulsion Conference,* 10-13 July 2005, Tucson, AZ, USA.

[13] Bolonkin A.A., Electrostatic Utilization of Asteroid for Space Flight, *Presented as paper AIAA-2005-4032 at 41-st Propulsion Conference,* 10-13 July 2005, Tucson, AZ, USA.

[14] Bolonkin A.A., *Electrostatic Solar Sail, Book* "Non-Rocket Space Launch and Flight", Ch.18. pp. 317-320.

[15] Bolonkin A.A., *New Concepts, Ideas, Innovations in Aerospace, Technology and Human Science, NOVA, N.Y.,* 2008, 400 pgs.

[16] Bolonkin A.A., *AB Reflector Thermonuclear Reactor, http://arxiv.org*, search "Bolonkin".

[17] Bolonkin A.A., Human Immortality and Electronic Civilization, Lulu, 2007, 66 pgs. (*http://www.lulu.com*, search "Bolonkin").

[18] Galasso F.S., Advanced Fibers and Composite, Gordon and Branch Science Publisher, 1989.

[19] Kikoin, I.K., (ed.), Tables of physical values. Atomuzdat, Moscow, 1976 (in Russian).

[20] *Carbon and High Perform Fibers, Directory,* 1995.

[21] Dresselhous M.S., Carbon Nanotubes, Springer, 2000.

[22] AIP. Physics desk reference, 3-rd Edition, Springer, 2003.

[23] Bolonkin A.A., Human Immortality and Electronic Civilization, Lulu, 3-d Ed., 2006. *http://www.lulu.com*, search “Bolonkin”.

Chapter 13

MAGNETIC PROPELLER FOR UNIFORM MAGNETIC FIELD LEVITATION*

ABSTRACT

Three new approaches to generating thrust in uniform magnetic fields are proposed. The first direction is based on employing Lorentz force acting on partial magnetically shielded 8-shaped loop with current in external magnetic field, whereby a net force rather than a torque origins.

Another approach, called a Virtual Wire System, is based on creating a magnetic field having an energetic symmetry (a virtual wire), with further superposition of external field. The external field breaks the symmetry causing origination of a net force. Unlike a wire with current, having radial energetic symmetry, the symmetry of the Virtual Wire System is closer to an axial wire.

The third approach refers to the first two. It's based on creation of developed surface system, comprising the elements of the first two types. The developed surface approach is a way to drastically increase a thrust-to-weight ratio.

The conducted experiments have confirmed feasibility of the proposed approaches.

INTRODUCTION

Overcoming force of gravity was, and remains even today, an ancient human dream. The great portion of energy a human being needs daily is spent for relocation within our Earth's gravity field. Ancient people intuitively understood that overcoming mystical gravity would drastically relieve their survival in any hostile Earthly environment: predators, unfriendly tribes, need to kill animals for nourishment and so on. People envied birds and, understanding that the birds repeal gravity temporarily, somehow, while in the air, people long dreamed about having wings themselves.

This is how concept of levitation settled into human consciousness.

---

* This Chapter was written by Dr. Krinker together with A. Bolonkin. Article presented as paper AIAA-2008-4610 to 44th Joint Propulsion Conference, 20-24 July, 2008, Hartford, CT, USA.

Levitation: rise or cause to rise despite gravity (WEBSTER DICTIONARY). Technical progress has partially made this dream a truth. However, since the time of considerable progress in understanding electricity and magnetism in 19th century, humans revealed another media where something (magnetic field) acts on another something (conductor with electric current), causing this the other something to move. Then new dream settled in human consciousness: flying in magnetic field. Unlike air, Earth's magnetic field extends to 1500 km and even more outwards from the planet's surface. Harnessing this force could open the gate for unusual flights. Despite the fact that the force acting on a powered conductor in Earth's magnetic field is insufficient for realizing this dream, works on magneto-levitation already enriched mankind with beneficial achievements.

Known today, magnetic levitation is based on electromagnetic induction in non-uniform magnetic field having a flux density gradient $\nabla\vec{B}$. A time-varying magnetic flux induces eddy current in a conductive object, what results in originating magnetic moment $\vec{m}$ of the object. As a result, a net force $\vec{F} = \vec{m} \cdot \nabla\vec{B}$ origins. The profound review of this technology is made in [1].

However, technical applications of such the method are limited by short-range of the operating non-uniform field as well as the non-uniformity itself.

It is the levitation in weak uniform permanent magnetic fields that holds the great interest today. In particular, levitation in Earth's and cosmic magnetic fields could open the way for a new generation of thrusters and propellers.

However, the induction of Earth's field is pretty insignificant – 30-60μT, that does not allow to develop needed thrust at reasonable consumption of energy. Moreover, loops with electric current, widely used today for rotational motion in heavy fields, produce a torque rather than a needed net force.

The proposed work shows ways of overcoming such obstacles.

Michael Faraday (1791-1867) was one of the first persons who, almost intuitively, understood that phenomenon of a force, acting on a conductor with electric current in magnetic field, can be explained by distortion of originally symmetric energetic patterns around the wire. Today, we say that the broken symmetry derives free energy and stimulates motion of the active body for minimization of the free energy, - a manifestation of the Le-Shatelier-Brown principle. Origination of well-known forces, acting either on a moving charge or a portion of a conductor with a current is explained by breaking the energetic symmetry. However, due to closed nature of electric current, in a uniform magnetic field a loop with current produces a torque rather than a net force needed for the propelling. Moreover, obtaining needed thrust-to-weight ratio for such the systems are accompanied with generating extremely heavy currents, which is energetically inefficiently. On the other hand, total heavy elemental Ampere's currents in permanent magnets also produce a torque rather than a net force in uniform fields, which block their direct application for obtaining thrust.

In early 70's, the attempt was done to overcome this principal obstacle:

The concept of an apparatus to fly in Earth's magnetic field, comprising rotated dielectric discs with embedded electrets was proposed [11]. As the discs, driven by a turbine, spin in Earth's magnetic field, the Lorentz force, acting on the electrets, origins. The Lorentz force has an opposite direction at opposite ends of the discs. To eliminate the unwanted opposite force, the discs are partially shielded in D-shaped ferromagnetic caves. To compensate an angular momentum, derived by high-speed spinning discs, the discs are rotated in opposite direction. Said system has the following disadvantages:

a) necessity to employ considerable electric charges in electrets, to gain needed Lorentz's force in a weak (30-60uT) Earth's magnetic field. The total charge needed for that counts for coulombs, that produce forces of electrostatic repulsion which can break the discs;
b) mechanical instability of the system, rotating heavy discs at high angular speed;
c) gradual discharge of the electrets by ionizing radiation;
d) limited altitude achieved by the apparatus over Earth, because of inability a gas turbine to operate in a thin air.

The proposed methods and apparatuses eliminate shown drawbacks and make magnetic propelling feasible.

1. Current Contour Having Magnetically Shielded Portion in External Magnetic Field

The method shown in Figure 1 is based on the following: close electric contour 10 with current, driven by the source 11, has its portion inside a hollow ferromagnetic body 12, [2]. The external magnetic field 13 acts only on the external portion of the contour 10 – there is actually no field inside the hollow ferromagnetic body 12. As the results, the Lorentz force 14 origins. The force acts only on the exposed portion of the contour 10. The value of the Lorentz force is $\vec{F} = I\left[\vec{l} \times \vec{B}\right]$, where I – current, l –length of exposed portion of the conductor 10. In particularly, if the external magnetic field is formed by Earth, having induction B = 30-60μT, then each 1m of the exposed conductor with 1A of the flowing current experiences action of $(3 \div 6) \cdot 10^{-5} N$, depending on the longitude.

1.1. Developed Surface 8-Shaped System

Taking into consideration that thrust is developed by 2 field-exposed conductors 10, each having length l and weight Pc, while the total weight depends on 4 involved conductors, the efficiency of the system shown above is $\dfrac{2IlB}{4P_c + P_t} = \dfrac{2IlB}{P}$, where Pt is a weight of the shielding tube 12.

Value of the total weight $P = 4\pi r^2 l \rho_1 g + 2\pi r_1 l d \rho_2 g = 2\pi g l\left(2r^2 \rho_1 + r_1 d \rho_2\right)$. The efficiency equals

$$\frac{IB}{\pi g\left(2r^2 \rho_1 + r_1 d \rho_2\right)}, \qquad (3)$$

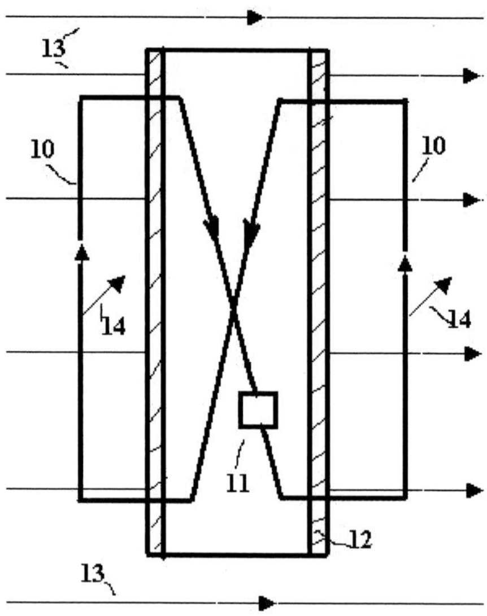

Figure 1. General concept of generating net force acting on a closed contour with current in external magnetic field. *Notations:* 10 - twisted, 8-shaped conductor with current; 11- source of current; 12- hollow tube-like ferromagnetic body;13 - external magnetic field; 14 - net Lorentz force. To increase the thrust, the whole contour 10, or at least its portion, which is exposed to magnetic field, can be made of a mu-metal, having considerable magnetic permeability. The mu-metal wire produces additional magnetic field around it and, therefore, enforces the thrust.

where $g, r, r_1, \rho_1, \rho_2, d$ are free-fall acceleration, radius of wire 10, radius of tube 12, density of wire 10, density of tube 12, and thickness of the tube 12, respectively.

Considering the minimal radius of the tube $r_1 = 2r$ and its thickness $d = r$, the efficiency can be shown as

$$\frac{IB}{\pi g\left(2r^2\rho_1 + r_1 d\rho_2\right)} = \frac{IB}{2\pi r^2 g(\rho_1 + \rho_2)} \qquad (4)$$

From here, for this system to be efficient, the condition $I_{cr} \geq 2\pi r^2 g(\rho_1 + \rho_2)B^{-1}$ has to be observed. Having $\rho_1 = 8.93 \cdot 10^3 kg/m^3$ (copper), $\rho_2 = 7{,}88 \cdot 10^3 kg/m^3$ (iron), $B = 5 \cdot 10^{-5} T$ (Earth), and the radius of the wire 10, $r = 10^{-5} m$, the critical is 2.1A. The less diameter of the wire 10, the less is critical current.

The phenomenon is explained due to the fact that the acting force is proportional to the surface on which it acts, while the weight is proportional to the volume. The smaller the radius, the more surface-to-volume ratio resulting. This is a developed surface system. The apparatus of Figure 2 can be made of arrays of the developed surface modules, like the one calculated above, fed by a common power bus. Beside, the contour 10 and the tube 12 can be made of a mu-metal to increase the magnetic field of the contour 10 as well as efficiently shielding the external field 13.

1.2. General Concept of the Apparatus Having Partially Shielded Current Contour in External Magnetic Field

Figure 2 shows a general concept of the aircraft employing the contour, having the magnetically shielded portion. The apparatus to fly in external magnetic field comprises a vertical thrust platform 20 with the exposed to the field portions of the 8-shaped contour with currents 21, the horizontal thrust platform 22 with the exposed portions of the 8-shaped contour with currents 23, and magnetic stabilizers 24 to align the aircraft for positioning the contours 21 and 22 normally to magnetic lines of force of the field 25.

As the result, two Lorentz forces origin: force 26 acts on the contour 21 vertically, while the force 27 acts on the contour 22 horizontally. Therefore, both vertical and horizontal thrusts origin. The sources of the currents 21 and 23 are located inside the platforms 20 and 22 (the sources are not shown here). The currents 21 and 23 form 8-shaped contour of the Figure 1. Therefore, the total shape of the currents 21 and 23 does not correspond to shape of the platforms 20 and 22.

The stabilizers 24 are heavy field magnet, having lines of force aligned along North-South direction. Actually, it conducts itself like a needle of a giant compass, always keeping position of the apparatus stable in Earth's field.

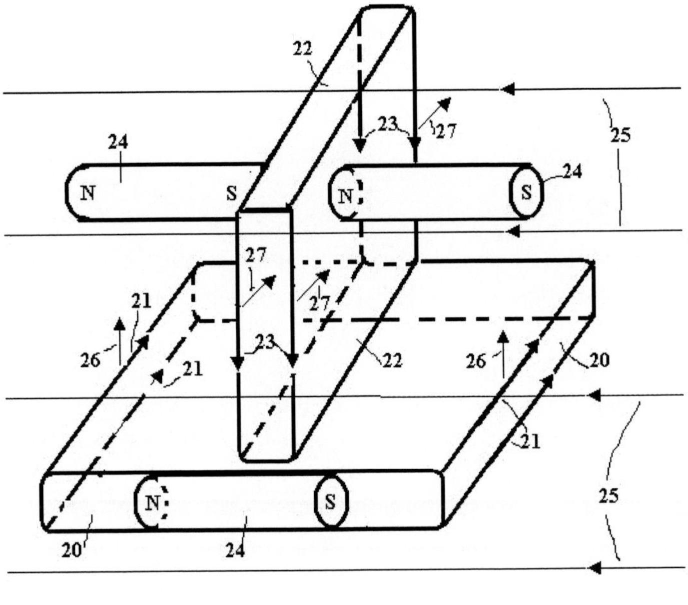

Figure 2. General concept of a thruster employing current contours having magnetically shielded portions in external magnetic field. *Notations:*
20-horizontal platform comprising 8-shaped powered conductors having magnetically-shielded internal portions; 21- exposed to external magnetic field portions of the 8-shaped horizontal conductors; 22-vertical platform comprising 8-shaped powered conductors having magnetically-shielded internal portions; 23- exposed to external magnetic field portions of the 8-shaped vertical conductors;
24- magnetic stabilizers, having magnetic moment, which is anti-collinear to the external magnetic field; 25- external magnetic field; 26- vertical net force acting on the field-exposed conductors 21 of the horizontal platform 20; 27- horizontal net force acting on the field-exposed conductors 23 of the vertical platform 22.

1.3. Magnetic Shielding. Materials for Shielding Device

Operation of the 8-shaped system of the Figure 1 depends on shielding of the portion of the current contour 10. First of all, it has to be said that unlike electrostatic shielding there is no ultimate magnetic shielding. This fundamental difference is caused by a different physical nature of conservative electrostatic field and solenoid magnetic field. There are no "magnetic charges" at which magnetic lines of force start and end like it has place when shielding conservative electric fields by conductive materials having free electric charges.

The shielding action of magnetic screens is based on refraction of magnetic lines of force due to magnetization of the material of the shield in external magnetic field. The magnetic field developed by the shield opposites the original external field in vicinity of the walls of the shield. Due to closed nature of magnetic lines of force, their superposition in external portion of the shield results in that the strength of the resulting field is reduced in immediate vicinity of the shield while the field inside the walls of the shield is elevated. It looks like refraction of the magnetic lines of force.

Successful screening external magnetic field implies immediate reaction of the magnetic shield to external field strength: the magnetic inductance of the material of the shield has to follow changes of the external field.

Because the induction of the field inside the material is directly proportional to relative magnetic permeability μ_r, then the capability to magnetic shielding depends on this property of the material:

$$\vec{B} = \mu_0 \vec{H} + \vec{M} = \mu_0 \mu_r \vec{B},$$

where M is a magnetization of the material of the shield.

However, magnetization experiences saturation in heavy external fields.

Due to that, the permeability is a sub-linear function of the magnetic field strength.

From this, it follows, that saturated magnetic shield can't perform its function.

Selecting materials for magnetic shielding depends on the fact that a relative magnetic permeability depends on a flux density, the induction, and changes with it. For this reason, relative permeability also changes: the raise of the temperature usually decreases the relative permeability. For instance, if the temperature increases from 20 to 80 centigrade, then a typical ferrite can suffer a 25% permeability drop.

Figure 3 shows permeability and inductance versus field strength for iron. As seen from the figure, magnetic permeability is maximal at intensive increase of inductance vs. field strength. This is why the relative permeability can be evaluated as a derivative: $\mu_r = \dfrac{dB}{dH}$.

Magnetic permeability also depends on a previous mechanical working, which defines a microstructure. It's especially considerable for transformer steel. The mechanical working orients grains along one direction (anisotropy) giving increased permeability.

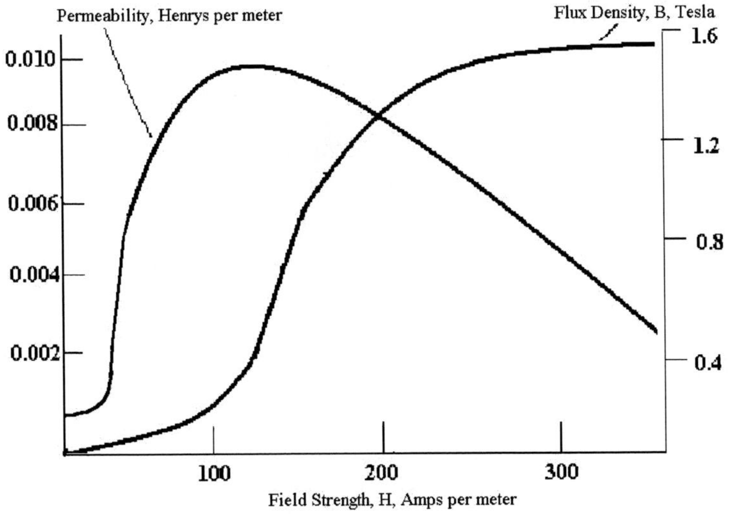

Figure 3. Permeability and induction vs. field strength for iron.

The table bellow shows both dimensional magnetic permeability $\mu(H \cdot m^{-1})$, (Henrys per meter) and relative magnetic permeability $\mu_r = \mu / \mu_0$.

Approximate maximum permeabilities

Material	μ/(H m^{-1})	μ_r	Application
Ferrite U 60	1.00E-05	8	UHF chokes
Ferrite M33	9.42E-04	750	Resonant circuit RM cores
Nickel (99% pure)	7.54E-04	600	-
Ferrite N41	3.77E-03	3000	Power circuits
Iron (99.8% pure)	6.28E-03	5000	-
Ferrite T38	1.26E-02	10000	Broadband transformers
Silicon GO steel	5.03E-02	40000	Dynamos, mains transformers
supermalloy	1.26	1000000	Recording heads

Selection materials for magnetic shielding have to take into consideration operating field strength developed by the current contour and visa versa. Therefore, the field strength developed by a shielded portion of the contour 10, Figure 1 has not to exceed some limit, depending on the material. Otherwise, the shield 12 becomes transparent for external magnetic field and the net force 14 experiences reduction. However, too weak field, as this follows from the Figure, is also unacceptable because reduced magnetic permeability.

Operating field, developed by conductors 10 of 8-shaped system counts tens-hundreds of A/m. Based on the table shown directly above, it follows that the best materials for the shielding are those having industrial power rates: ferrite N41, iron, ferrite T38, silicon steel. Supermalloy simply will lose its splendid shielding properties in such fields. However, this does not deny possibility of employing this material in the lower field of the Developed Surface 8-Shaped System.

2. Virtual Wire System: Employing Magnetic Field of an Unpowered Object

Another method to develop lift and moving forces in magnetic field is the developing of the difference of density of magnetic energy in vicinity of an object by means of employing magnetic field of the object rather than employing powered wire, Figure 4, [3,12]. The systems, realizing this approach can be subdivided for symmetric and asymmetric ones in terms of its physical structures.

2.1. Symmetric Virtual Wire Systems

Breaking energetic symmetry is a cause of originating forces.

Force F, acting on a border of two bodies, having different densities of energy w_1 and w_2, is $F_n = \iint_S \Delta w(x,y)dS$ where S is a total area of surface of the adjacent areas of energy.

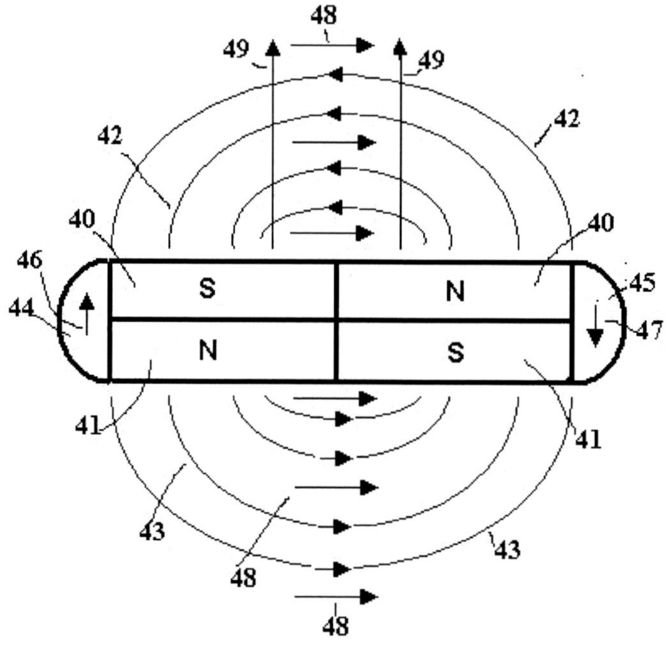

Figure 4a. Physical base of the virtual wire system: *Notations*: 40 - bar magnet; 41 - bar magnet, having magnetic moment, which is anti-collinear to the magnetic moment of the magnet 40; 42 - magnetic field of the magnet 40; 43 - magnetic field of the magnet 41; 44 - left magnetic shunt for "absorbing" end magnetic field of the magnets 40-41; 45 - right magnetic shunt for "absorbing" end magnetic field of the magnets 40-41; 46 = left-end magnetic field inside the shunt; 47 - right-end magnetic field inside the shunt; 48 - external magnetic field; 49 - net force. The magnetic assembly of the magnets 40-41 produces opposite-directed fields 42-43, which are symmetric in terms of energy while no external filed is applied. Superposition of the external field 48 breaks the energetic symmetry, developing the net force 49.

Origination of this force is based on the Le Chatelier- Braun principle: When a constraint is applied to a dynamic system in equilibrium, a change takes place within the system, opposing the constraint and tending to restore equilibrium. As far as the considered system is concerned, the equilibrium implies equality of density of magnetic energy around the object.

Figure 4a shows a system, comprising two magnets 40 and 41, arranged in anti-parallel way, producing magnetic fields 42 and 43, respectively. The fields 42 and 43 have equal induction BM. To eliminate penetrating unwanted opposite magnetic fields from the ends of the magnets 40 and 41 into external media, magnetic shunts 44 and 45 are employed. The fields 46 and 47 are canalized inside the shunts 44 and 45, which made of ferromagnetic materials. The external magnetic field 48, having induction BE, is subtracted from the filed 42 and is added to the field 43.

Figure 4b illustrates breaking initial axial energetic symmetry in vicinity of the Virtual Wire System as the external field *Be* is applied.

As seen from the Figures 4a and 4b, the intrinsic vector field *Bm* of the magnetic assembly 40-41and the external field *Be* may be added or subtracted.

Let's consider energetic conditions of the upper and lower planes of the Figure 4 taking into consideration that angle α between *Bm* and *Be* is a function of a coordinate on *x-y* plane.

The resulting field under the lower plane is

$$B_{r+} = \sqrt{B_m^2 + B_e^2 + 2B_m B_e \cos\alpha_{x,y}}$$

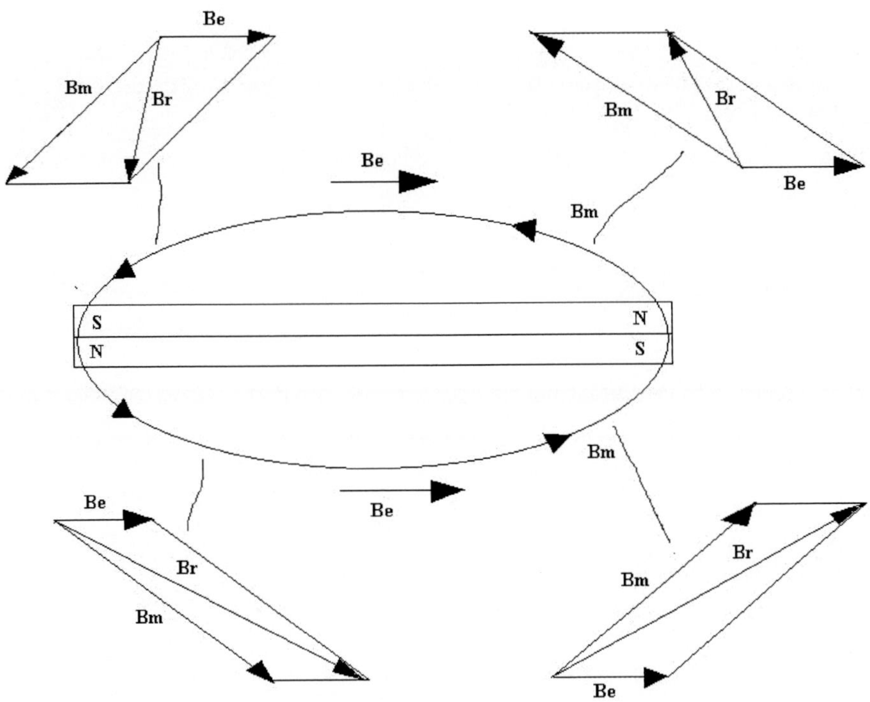

Figure 4b. Breaking initial axial energetic symmetry of the virtual wire system as the external field Be is applied. The resultant vector Br is a result of superposition of the vectors Be and Bm.

The resulting field over the upper plane is

$$B_{r-} = \sqrt{B_m^2 + B_e^2 - 2B_m B_e \left|\cos\alpha_{x,y}\right|}$$

The gained excessive density of magnetic energy under the lower plane is

$$w_+ = \frac{B_{r+}^2}{2\mu_0\mu_r} = \frac{B_m^2 + B_e^2 + 2B_m B_e \cos\alpha_{x,y}}{2\mu_0\mu_r}$$

The depleted density of magnetic energy over the upper plane is

$$w_- = \frac{B_{r-}^2}{2\mu_0\mu_r} = \frac{B_m^2 + B_e^2 - 2B_m B_e \left|\cos\alpha_{x,y}\right|}{2\mu_0\mu_r}$$

The difference of the densities, taken normally to *x-y* plane along *Z*-axis is

$$\Delta w = w_+ - w_- = \frac{2B_m B_e \left|\cos\alpha_{x,y}\right|}{\mu_0\mu_r}$$

The value of Δw is a density of free energy of the system. It derives a net force F directed into area of the depleted energy.

Generally, the distribution of the free energy depends on angle α between the external field 48 and intrinsic fields 42 and 43 of the magnet:

$$\Delta w(\alpha) = \frac{2}{\mu_0\mu_r} B_E B_M \left|\cos\alpha\right| \tag{5}$$

The difference of densities of energy Δw reaches maximum at the center of the system. Said the difference of densities of magnetic energy in vicinity of the magnets 40 and 41 results in origination of normally-acting force 49, which can be calculated as

$$F_n = \iint_S \Delta w(x,y)\, dS = F = \frac{kB_E B_M S}{\mu_0\mu_r}, \tag{6}$$

where $1 \le k \le 2$, depending on proportion of the object.

2.2. Equilibrium of the Virtual Wire System in Magnetic Field

Beside the produced net force, virtual wire systems have own magnetic moments, which develop torques in external magnetic fields. This affects the equilibrium of the magnetic propeller and has to be taken into consideration.

Any rotation in external magnetic field is accompanied by variation of projection of the vector of the external field onto direction of the internal field of the magnets of Figure 4. As this takes place, acting value of the external vector is $\vec{H} = \vec{H}_0 \cos\alpha$, where α is an angle between magnetization and external field during rotation. Energetically, the process of the rotation is asymmetric due to hysteresis, Figure 6. The area of hysteresis loop is an additional energy, which contributes into general energetic asymmetry of the virtual wire defining the equilibrium angle.

2.2.1. Equilibrium in a Vertical Plane in Gravity Field

According to Le Chatelier- Braun principle, the system of Figure 4 can restore the initial equilibrium, in which it rested before superposition of the field 48, by a linear motion or self-turning a system until the densities of magnetic energies get equal.

Figure 5 shows the system of Figure 4, having horizontal axis or rotation in a gravity field. The magnetic system 50, having its own fields 51, B_e, placed in external field 52, B_m. The free energy of the system originates in the volume of the magnets 50. The free energy of the system 50 forces it to turn around an axis 53. Then, its original center of mass 54 shifts to new position 55. The center of mass is shifted from the axis 53 for a distance 56, h. The value of its vertical shift is Δh. Then, the potential energy increases for $\Delta W = mg\Delta h$, which depends on the equilibrium angle β. The equilibrium takes place at the equal free energy and additional potential energy. From here

$$\frac{kB_E B_M S}{\mu_0 \mu_r} \cos\beta = F\cos\beta = 2mg\sin^2\left(\frac{\beta}{2}\right). \tag{7}$$

From here, the originated magnetic force can be calculated as

$$F = \frac{2mg\sin^2\left(\frac{\beta}{2}\right)}{\cos\beta}. \tag{8}$$

Asymmetry of energy caused by hysteresis processes taking place during rotation was ignored here. However, it's justified here taking into consideration that gain of gravitational energy exceed that of the hysteresis process. However, the hysteresis is a major factor as it comes to rotation in a horizontal plane, where no shift of the center of mass takes place.

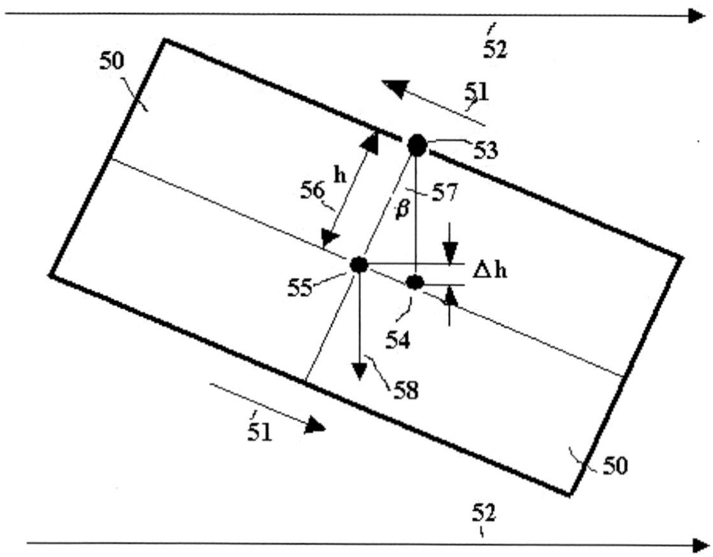

Figure 5. Equilibrium of the virtual wire system in gravity field (vertical plane). *Notations*: 50 - assembly of anti-collinear bar magnets; 51 - intrinsic magnetic fields of the assembly; 52 - external magnetic field; 53 - axis of rotation; 54 - position of center of mass before rotation; 55 - position of the center of mass after rotation- equilibrium position; 56 - distance between the center of mass and the axis of rotation; 57 - angle of rotation – the equilibrium angle; 57 - gravity force applied to the center of mass.

2.2.2. Rotation around Vertical Axis

Application of external magnetic field *H* causes rotating anti-parallel magnetic duplex (virtual wire system) in a horizontal plane until it reaches some equilibrium angle α between direction of magnetization *M* of the magnets 40-41 and the external field *H*. The process is initiated by a field-induced braking symmetry of magnetic moments. In reality, magnetic moments of the magnets 40-41 are equal only until the external field is applied. The following energetic situation can be illustrated by Figure 6.

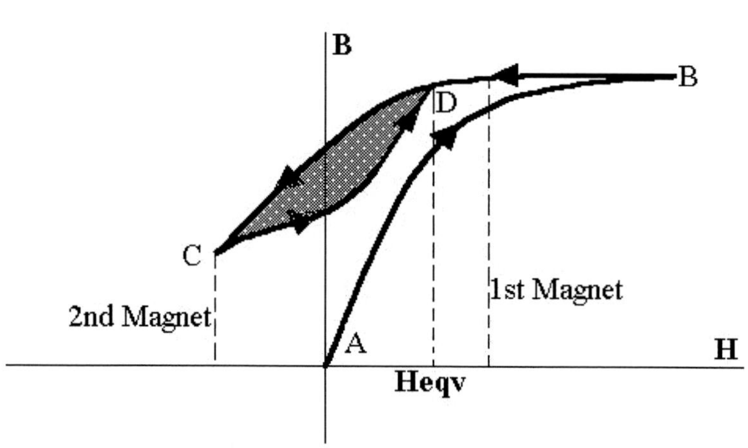

Figure 6. Influence of hysteresis of magnetic assembly on equilibrium of the virtual wire system.

The first magnet (41 of Figure 4) runs way BD during rotation, while the second magnet (40 of Figure 4) runs the lower branch of CD. Due to the hysteresis, their inductions meet each other at some point D, which corresponds to the equilibrium field strength $\vec{H}_{eqv} = \vec{H} \cdot \cos\alpha_{eqv}$. The diagram explains why the equilibrium angle between the field and the magnetic moments can't be 90 degrees ($H = 0$): the inductions are not equal at this angle.

The first magnet, 41, is in energetically unprofitable condition: it has a positive potential energy $U = -\vec{m} \cdot \vec{B}$ in the external field, while the second magnet, 40, has a negative potential energy, resting in energetically profitable condition before combining both the magnets into a whole system. As it follows from Figure 4 magnetic moment m of the first magnet exceeds that of the second magnet that causes the system to start turning for minimizing its free energy. As placing in the external field H, the second magnet runs a path BC on the Figure 6. As rotating in the external field, the second magnet 40 moves from the point C in B-H plane, while the first magnet, 41, experiences the path BD. Due to nonlinear properties of a hysteresis, the opposite paths of the both magnets do not co- inside until some point is reached D, where their inductions become equal. At this point *Heqv* their energies get equal and the system stops the rotation. While rotating, the magnetic duplex performs work $W = \int_C^D B_1 H dH - \int_D^C B_2 H dH$ within contour CD, where B_1 and B_2 are hysteresis branches for the magnets 1 and 2 respectively. The field $H = H_{\max} \cos\alpha$ and, therefore, the equilibrium angle α depends exquisitly on magnetic characteristics of the employed magnets. From the specific properties of the hysteresis, it follows that α never can be neither 90 nor 0 degrees.

2.2.3. Developed Surface Virtual Wire System

Force, produced in the virtual wire system depends on a total border area between the object and the media of its motion. The less physical size R of the object, the more surface area-to-volume ratio, $1/R$. On the other side, weight is a function of a volume. Therefore, to make a thrust-to-weight ratio more efficient, surface-to-volume ratio has to be as maximal as possible.

Development of this approach is shown in Figure 7. This system comprises an array of oblong film-like permanent magnets arranged in anti-parallel way. Their $l >> p$. Because of the total energy of end area drastically less than that of major portion of the film magnets, there is no need in magnetic shunts for elimination of end field in surrounding media. The total area of surface of a stratified system is $S = Nns$, where N-is a number of layers, n is a number of magnetic pairs in one layer, and s is an area of one magnetic pair. For one anti-parallel magnetic pair, thrust-to-weight ratio can be shown as

$$\frac{F}{W} = \frac{2B_M H_E S}{V\rho g} = \frac{2B_M H_E lw}{lwd\rho g} = \frac{2B_M H_E}{d\rho g}, \tag{9}$$

where l, w, d, ρ and g are a length, width, thickness, density and acceleration of free fall, respectively. For this system being feasible, F/W has to exceed 1. Therefore, at permanent B

and *H*, critical thickness $d_{cr} = \dfrac{2B_M H_E}{\rho g}$. From here, for iron–film ($\rho = 7.8\cdot10^3 kg/m^3$) magnets, having induction in immediate vicinity of the surface, *B* = 0.1 T and Earth's field strength 30 A/m, $d_{cr} = 7.7\cdot10^{-5} m = 77\mu m$, that is, if *d* is less than this value, the system can float in Earth's magnetic field. However, the calculation implies that length *l* >> *w,d*, otherwise magnetic shunts has to be employed to eliminate unwanted side field.

The system's payload, sometimes called "carrying capacity", can be found as

$$P = F - W = 2NnsB_M H_E - Nns\rho d g = Nns(2B_M H_E - \rho d g). \tag{10}$$

Here *N* and n are numbers of the layers and number of magnetic duplets in each the layer, while s is an area of each the duplet. For the system, having *N* = 1000 and *n* = 1000, with $s = 1m\cdot10^{-3} m$, $d = 3\cdot10^{-5} m$, and the rest of parameters like shown above, $P = 3.66\cdot10^3 N = 370\ kg$. Area of one layer is $1m^2$. The layers are arranges one under another. The following fact has to be taken into consideration:

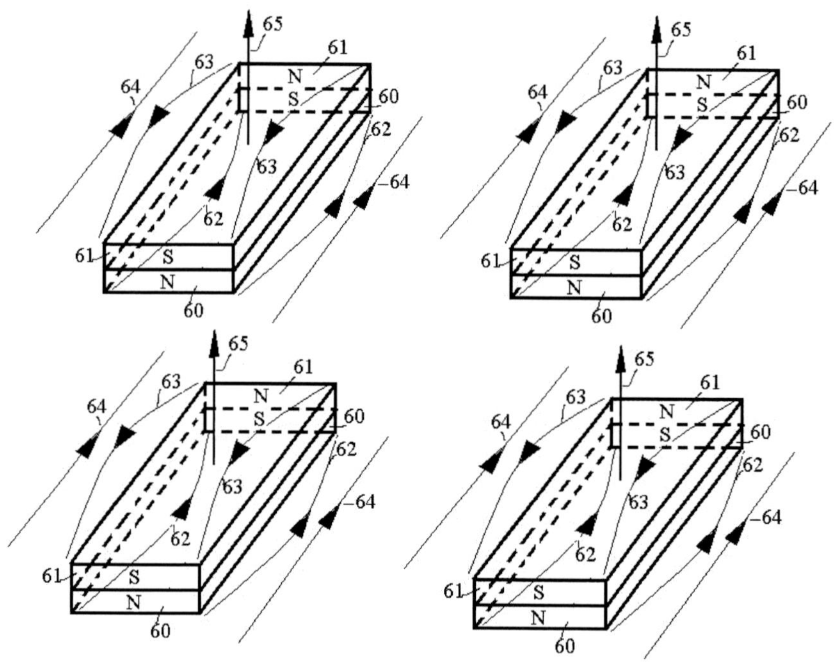

Figure 7. Developed Surface Virtual Wire System. The film-like oblong magnets are arranged layer by layers on a magnetically neutral film-like base (not shown here). The sketch shows two layers. *Notations*: 60 - film-like magnets; 61 - film-like magnets having magnetic moments, which are anti-collinear to the magnetic moments of the magnets 60; 62 - magnetic fields of the magnets 60; 63 - magnetic fields of the magnets 61; 64 - external magnetic fields; 65 - net forces acting on the assemblies of the magnets 60-61.

The $1m^2$ *base* comprises batch-arranged $N = 1000$ layers, while each of the layer counts $1m^2$. So, the total surface "hidden" over $1\,m^2$ - base is $10^3 m^2$. *This explains the seeming paradox of the abnormal thrust: it's actually calculated for* $10^3 m^2$ *effective area.*

So, one square meter of surface of the Developed Surface Magnetic System, having a vertical batch of 1000 layers, can lift 370 kg of payload in Earth's magnetic field if the induction developed by film-like magnets 60-61 is 0.1T.

The value of this induction is not abnormal for rare Earth magnet. However, the film-like magnets have to be mono-crystal rather than amorphous structures to develop this induction. The total thickness of 1000 of $10^{-4} m$ layers (including thickness of a plastic carrier) is 10 cm.

2.2.4. Asymmetric Virtual Wire System

Shown in Figure 8 is a ramification of a virtual wire system, while the shown in Figure 9, induction *B* vs. magnetizing field strength *H*, explains specific features of the system, explaining its operation. The basic ramification system comprises two separated magnetic coils 70 and 71, comprising two adjacent ferromagnetic bodies: an active body 72 and a magnetic shunt 73. The coils 70 and 71 are fed with direct current, producing magnetic field 74 between them. The ferromagnetic bodies 72 and 73 are different in that the body 72 is magnetically saturated in the field 74 of the coils 70-71, while the body 73 (the magnetic shunt) is unsaturated, Figure 7. Moreover, the active body 72 has a higher magnetic induction than the magnetic shunt 73. The system rests in the external magnetic field 75. Mechanical wholeness of the system is provided by supports 76 connected to a base 77. The base 77 also serves as a source of current for the magnetizing coils 70-71. The net force 78, produced due to energetic asymmetry of the system, acts normally to the system.

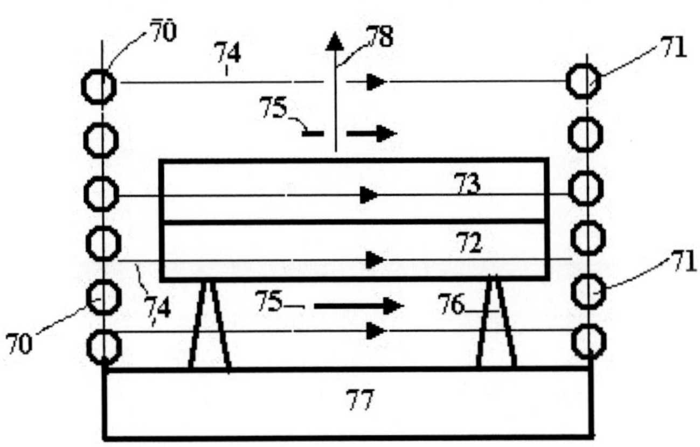

Figure 8. General conception of the asymmetric virtual wire system. Notations: 70 - left magnetizing coil; 71- right magnetizing coil; 72- active body, having saturated magnetization; 73- shunt body, having unsaturated magnetization; 74- magnetic field of the coils 70-71; 75- external magnetic field; 76- mechanical supports providing wholeness of the system; 77- base comprising sources of current for the coils 70-71; 78- net force acting on the system.

The net force originates due to difference of densities of magnetic energies in immediate vicinity of the active body and the shunt. Said difference origins because of different magnetic properties of the active body and the shunt in the control field 74 of the coils 70 - 71. If w_a, w_s, B_a, B_s and B_e are densities of magnetic energies and inductions of the active body, the shunt and the induction of the external filed, then the following takes place:

$$w_a = \frac{1}{2\mu\mu_0}\left(B_a^2 + 2B_aB_e + B_e^2\right) \; w_s = \frac{1}{2\mu\mu_0}\left(B_s^2 + 2B_sB_e + B_e^2\right),$$

$$\Delta w = w_a - w_s = \frac{1}{2\mu\mu_0}\left(B_a^2 - B_s^2 + 2B_e\left(B_a - B_s\right)\right).$$

Density of the free energy of the system is

$$w_f = \frac{B_e}{\mu\mu_0}(B_a - B_s) = \frac{B_e}{\mu_0\mu}(M_a(H) - M_s(H)),$$

where M_a and M_s are magnetizations, which depend on the operating field strength H of the driving coils 70 - 71.

The net force is $F = \iint_S w_f dS$.

As seen from the Figure 9, realization of this approach requires materials with characteristics like shown in this figure.

As this took place for the symmetric virtual wire system, the asymmetric one is maximally efficient as the developed surface system. Then, the elements 60s of the Figure 7 are the active bodies, while the elements 61s are magnetic shunts. Said elements made as ferromagnetic films.

3. The Experiment

The feasibility of the proposed systems was proved experimentally. Special attention was devoted to uniformity of magnetic field.

3.1. The 8-Shaped Wire System

The 8-Shaped Wire System was tested in Earth's field as well as in uniform artificial field. Both mechanical and electronic balances were employed to measure the net force.

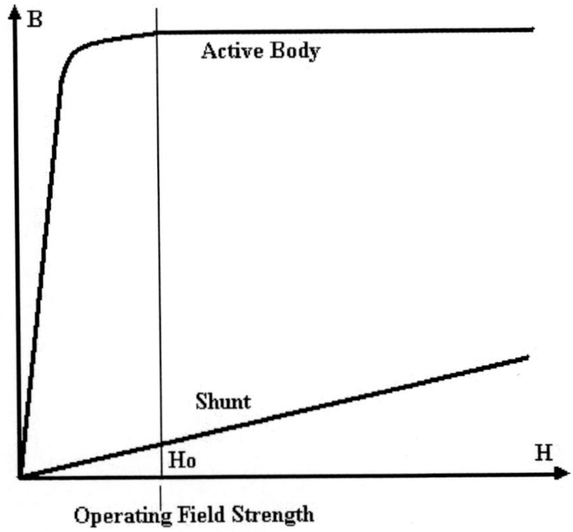

Figure 9. General magnetic properties of active body and magnetic shunt for asymmetric virtual wire system in external magnetic field.

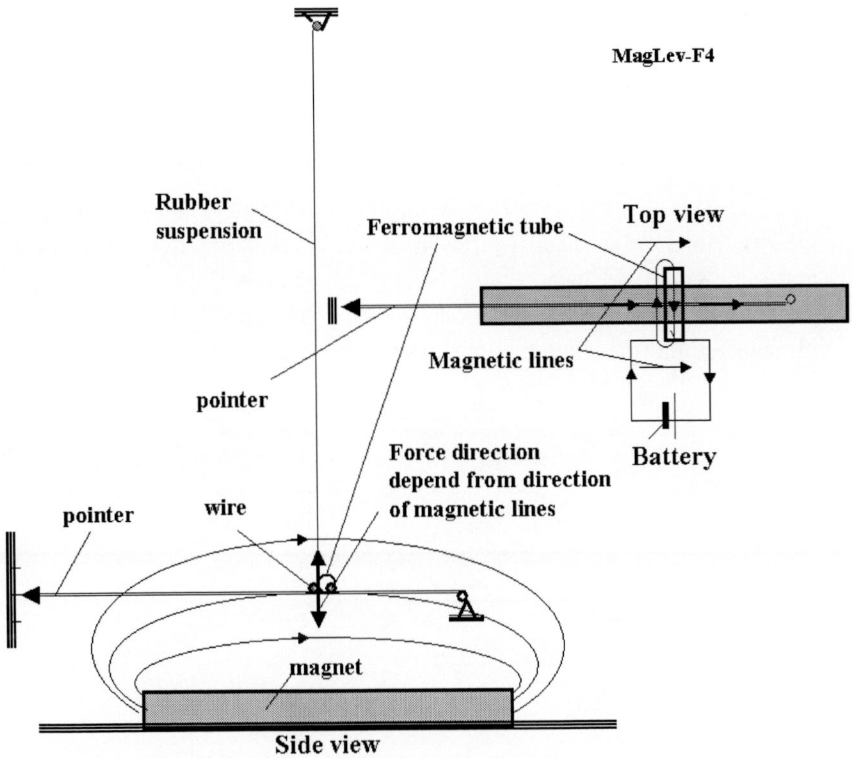

Figure 10. Sketch of Krinker's installation for demonstration the magnetic net force when a back current carrying wire is protected from the outer magnetic field. Data: ferromagnetic cylinder has: 28x18x10 mm; total wire of spool has length 2 m, 10 A current passed through the circuit during 7s; Magnetic field is about 0.01 T.

Figure 11. Installation for Krinker's experiment. In this version, the magnet (the round body on a back plane) was installed behind the wire system. Its lines of force lay in a horizontal plane. The uniformity of the field within 10 mm in a vertical plane was tested with a Hall-effect sensor. A hinge in a lower right portion secures motion in a horizontal plane. The right portion of the installation is suspended on a rubber string. After the circuit was activated (10A current, 7 seconds duration), the observed stretch of the rubber string corresponded to $\sim 10^{-1} N$ net force, what complies with preliminary calculations. For correctness of measurements, the circuit was triggered by a light signal after preliminary turning-on a switch in the right portion.

A simple experiment was made by physicist Dr. Mark Krinker showing origination of magnetic net force as it appears when a back current carrying wire is shielded in a ferromagnetic tube. The sketch of Krinker's installation is shown in Figure 10.

The picture of Krinker's installation is shown in Figure 11.

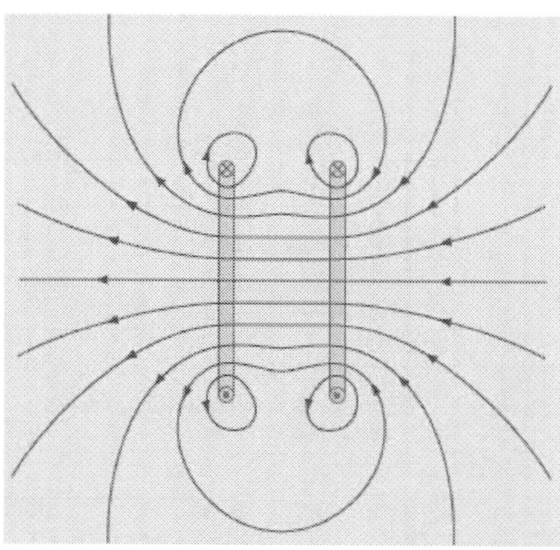

Figure 12. Magnetic Field Generated by Helmholtz Coils. The field is uniform in a central portion between the coils. The magnets and scales are located in the central portion.

3.2. Virtual Wire System

Special attention was devoted to uniformity of the field. For this reason, the Helmholtz coils were applied, Figure 12. However, the uniformity takes place only in central potion of the coils, as seen from the Figure 12. This is why magnet assembly of the virtual wire system was placed on a spacer resting on the scale in a center of the coils, Figure 13. For the Virtual Wire experiment the coils with diameters 400 and 120mm respectively were employed. The results of the experiments with both the coils were video-documented.

Both the coils have 100 turns of 0.2 mm of copper wire. The big coils were supplied with DC current up to 5A and generated up to 1500μT field in the center of the installation.

The small coils, supplied with 3.6A current, generated $3.78 \times 10^{-3} T$ field.

The field was measured with preliminary calibrated UGN3503U linear Hall-effect sensor, Allegro Company. Beside that, the field was calculated analytically. The matching between analytical and experimental estimation was pretty good. The magnetic assembly, having total dimension 51×25×7mm, Figure 10, was made of 4 magnets. The Neodymium rare Earth magnets, having 1T internal induction were employed. The magnets were encapsulated in a plastic envelope. Two opposite directed red arrows on both sides show direction of the intrinsic magnetic fields of each the side.

3.2.1. Active Area

Not all the surface of the magnetic assembly takes participation in origination of the net force: only the active area, producing horizontal lines of force in the center contributes in the effect. Those end lines of the system, which have no magnetic shunt, opposites to the effect and reduces it.

The magnetic assembly was composed of 2 pairs of thick-magnetized magnets having opposite magnetic moments (each of the pairs, in turn, was

composed of two magnets). Its magnetic lines of force are shown in Figure 4a, b. The system had no magnetic shunts.

To study distribution of the magnetic field, iron filings feelings in a glycerin were employed ("Doodle Pro", Fisher-Price Company). The magnetic patterns, developed by this system are shown in Figure 14a and 14b. As seen from the picture, the visible active horizontal portion (the feeling lay horizontally) counts just 5-6 mm. Because the shunts were not employed, the opposite-directed end fields existed. As shown in the Figure 14, their length equals 4 mm.

From here it follows that the length of the active area is ~2 mm. Therefore, the value of S in the expression $F = \dfrac{kB_E B_M S}{\mu_0 \mu_r}$ has to be taken as a difference of the central and end portions of the magnetic patterns.

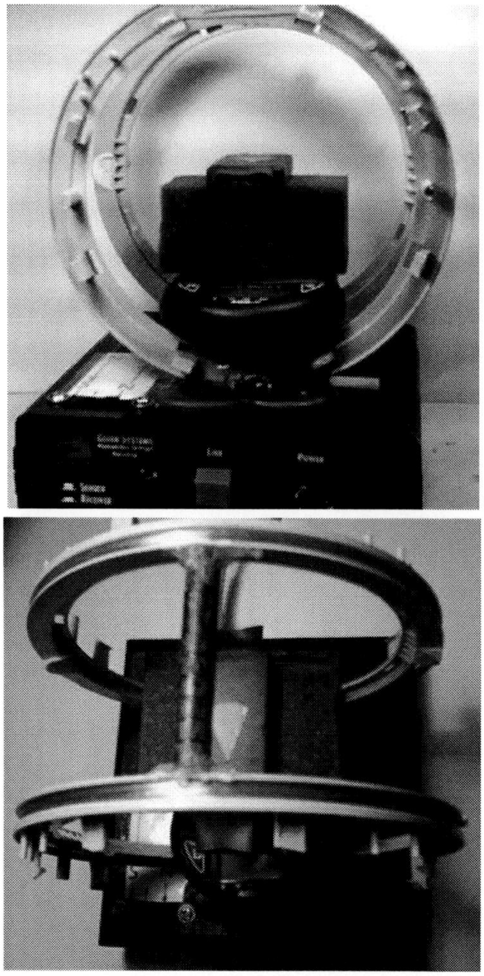

a.

b.

Figure 13. Experimental Installation made by Dr. Krinker for the Virtual Wire System.
The Virtual Wire Magnetic Assembly (central brick-like body on a spacer) rests on the Electronic Balance. The red arrow indicates direction of the intrinsic field of the upper side of the magnetic assembly. The direction of the field of the lower side is opposite to that of the upper side.

Figure 14a. Patterns of intrinsic magnetic field of the magnetic assembly of Figure 13, top view, obtained with iron fillings in a glycerin.

The horizontally- oriented fillings in the center show the active area, contributing into the net force, while edge horizontal lines indicate unwanted opposite-directed field. Vertically oriented fillings (light areas) show the "idling" portion of the field.

Figure 14b. Opposite-directed fields along the Magnetic Assembly. Two compasses show opposite directed fields over the assembly.

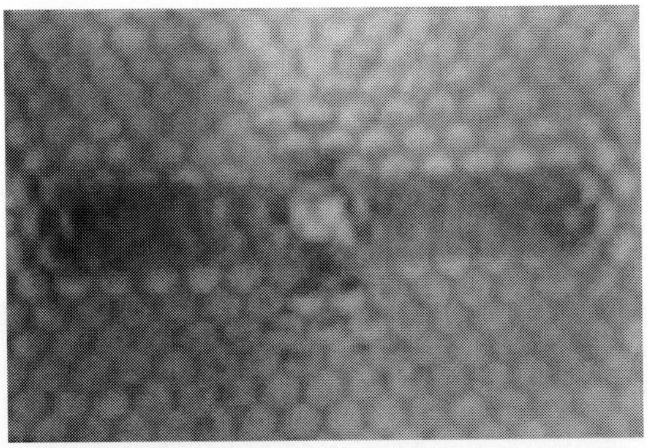

Figure 14c. Side magnetic imprint of the assembly of Figure 13. The iron feelings align along closed magnetic lines of force around the center. Analogy with a powered wire is obvious.

3.2.2 Analysis of the Experiment

The magnetic assembly of the Figure 13 weighs 41.7 G. Superposition field in the installation Figure 13 caused variation of the weight for 1.2 G. The sign of the variation depends on mutual orientation of the fields of the installation and the assembly.

The expected error of the experiment is $\frac{\Delta F}{F} = 2\frac{\Delta B}{B} + \frac{\Delta S}{S}$, as it follows from equation (6).

The parameters of the experiment were as follows:

The active surface: As it was shown above, after subtraction of the length of the end fields, the resulting length of the active area is 2 mm. So, the active area is $S = 25 \cdot 10^{-3} m \times 2 \cdot 10^{-3} m = 5 \cdot 10^{-5} m^2$.

The inductions: the installation - $3.78 \times 10^{-3} T$. The measured tangential component of the induction at the surface of the magnets – 0.11T;

The following error was estimated for the experiment:

$\frac{\Delta B}{B} = 0.2$ This is a general error for measuring induction in this experiment. This error includes inaccuracy of the performed calibration as well as the error of the measuring instruments (sensor, ammeter, mili-voltmeter, caliper and so on) $\frac{\Delta S}{S} = 0.15$. As seen from the Figure 14a, measuring the active area can bring a considerable error because of illegible contours of Figure 14a. This value was obtained due to analysis of the magnetic patterns picture.

Therefore, the error for calculating net force is $\frac{\Delta F}{F} = 0.55$.

The theoretical calculation of the net force according to (6) at $k = 1$ returns

$$F = \frac{3.78 \cdot 10^{-3} T \times 1.1 \cdot 10^{-1} T \times 5 \cdot 10^{-5} m^2}{1.26 \cdot 10^{-6} H \cdot m^{-1}} = 1.65 \cdot 10^{-2} N \approx 1.68 \pm 0.93 G$$

(Here $\mu_0 = 4\pi \cdot 10^{-7} H/m \approx 1.26 \cdot 10^{-6} H/m$, while the value of the relative permeability for air $\mu_r \approx 1$)

As it follows from the data, the experimental result matches the calculation within the error of the experiment.

The divergence of the experimental result and the theoretically predictable one can be explained by inaccurate experimental values of the inductions, the active area and the geometric coefficient *k*.

Figure 15. The Magnetic Shunt atop the thick-magnetized magnet. The non-shielded opposite side produces magnetic field under the magnet.

Note: In motion magnets intersect the outer magnetic lines and discharge. They may be restored by a wire winding having electric current from an internal electric source.

Figure 16a. Influence of the Magnetic Shunt. The Shunt "Absorbs" magnetic lines of force compared to Figure 15.

3.3. Magnetic Shunt

As this seen from the Figure 14, opposite-directed side fields occupy considerable area of the magnetic assembly. Elimination of the side fields promise to increase the efficiency of the generated net force. Figures 15-16 show effect of the magnetic shunting of the side fields. The simple magnetic shunt is made of ferromagnetic tubes at the end of the thick-magnetized magnet.

Figure 16b. Imprint, left by the magnetic shunt after the shunt was removed. The magnetic lines-free area is clearly seen at the end.

GREEN-AIRWAYS
PHOTO CREDIT: Courtesy of Delft University of Technology

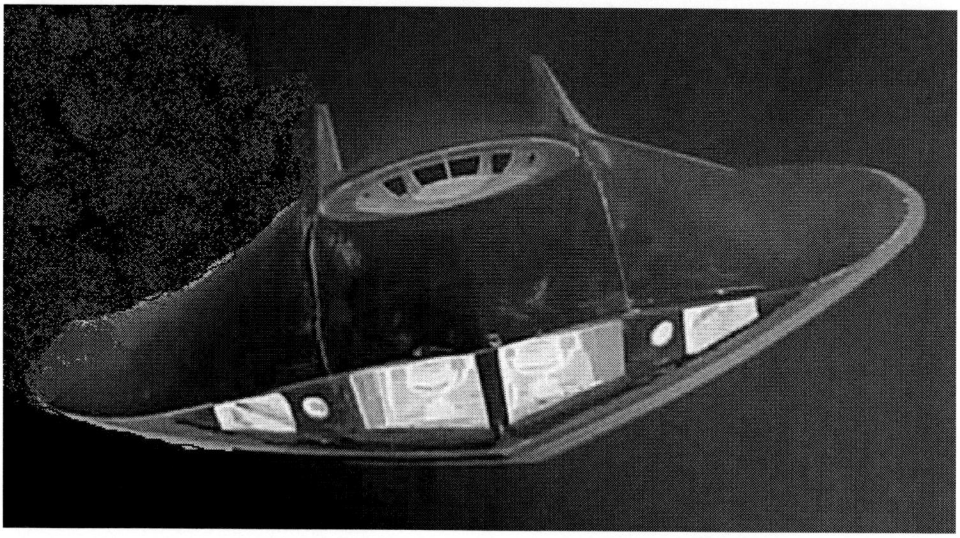

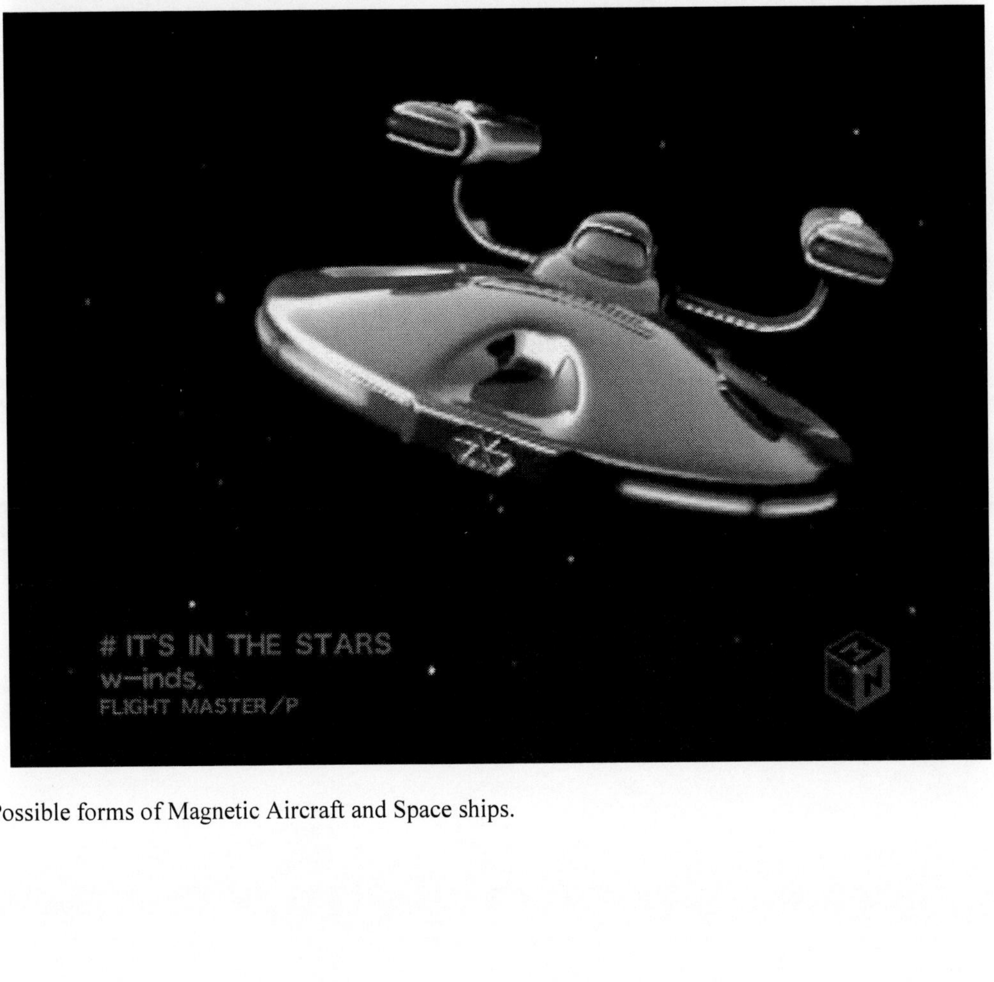

Possible forms of Magnetic Aircraft and Space ships.

REFERENCE

[1] G. Wouch and A.E. Lord, Jr. Levitation, Electromagnetic, Encyclopedia of Physics, p. 627, VCH Publishers, INC, Second Edition.

[2] Krinker, M., Magnetofly-1, Disclosure Document No. 269248, US Patent &Trademark Office, 1990.

[3] Krinker, M., Permanent Magnets Based Maglev, Disclosure Document No. 516184, US Patent & Trademark Office, 2002.

[4] Bolonkin, A.A., "Theory of Flight Vehicles with Control Radial Force". Collection *Researches of Flight Dynamics.* Mashinostroenie Publisher, Moscow, 1965, pp.79-118 (in Russian). International Aerospace Abstract A66-23338# (in English). The work contains theory of flight vehicles with control gravity or radial Magnetic force.

[5] Bolonkin, A.A., "Electrostatic Levitation on the Earth and Artificial Gravity for Space Ships and Asteroids". Paper AIAA-2005-4465, 41-st Propulsion Conference, 10-13 July 2005, Tucson, AZ, USA. The work contains theory of electroctatic levitation of flight vehicles and creating artificial gravity in space ships without rotation of apparatus (see also [6], Ch. 15).

[6] Bolonkin, A.A., *Non-Rocket Space Launch and Flight*, Elsevier, London, 2006, 488 pgs. The book contains theories of the more then 20 new revolutionary author ideas in space and technology.

[7] Bolonkin, A.A., AB Levitator and Electricity Storage, This work presented as paper AIAA-2007-4612 to 38th AIAA Plasmadynamics and Lasers Conference in conjunction with the16th International Conference on MHD Energy Conversion on 25-27 June 2007, Miami, USA. See also *http://arxiv.org* search "Bolonkin".

[8] Bolonkin, A.A., AB Levitrons and their Applications to Earth's Motionless Satellites. See: *http://arxiv.org* search "Bolonkin", July, 2007.

[9] Bolonkin, A.A., *New Concepts, Ideas, Innovation in Aerospace, Technology and Human Life*, NOVA, 2008, 320 pgs. Ch.1, Part 1. AB Levitation and Electricity Storage. Ch. 12, Part 2. AB Levitation and their Application.

[10] Bolonkin, A.A., AB Electron Tube and Semi-Superconductivity at Room Temperature. See *http://arxiv.org* search "Bolonkin".

[11] *Technology to Youth Magazine,* Reports of Inversor Laboratory, Moscow April, 1971.

[12] Krinker, M., To Possibility of Human Levitation. Published in Bio-Field Interactions and Medical Technologies. International Science Conference, pp.29-33, Moscow, April 2008.

PART II. ENVIRONMENT

Chapter 1

CHEAP ARTIFICIAL AB-MOUNTAINS, EXTRACTION OF WATER AND ENERGY FROM ATMOSPHERE AND CHANGE OF REGIONAL CLIMATE*

ABSTRACT

Authors suggest and researches a new revolutionary method for changing the climates of entire countries or portions thereof, obtaining huge amounts of cheap water and energy from the atmosphere. In this paper is presented the idea of cheap artificial inflatable mountains, which may cardinally change the climate of a large region or country. Additional benefits: The potential of tapping large amounts of fresh water and energy. The mountains are inflatable semi-cylindrical constructions from thin film (gas bags) having heights of up to 3 - 5 km. They are located perpendicular to the main wind direction. Encountering these artificial mountains, humid air (wind) rises to crest altitude, is cooled and produces rain (or rain clouds). Many natural mountains are sources of rivers, and other forms of water and power production - and artificial mountains may provide these services for entire nations in the future. The film of these gasbags is supported at altitude by small additional atmospheric overpressure and may be connected to the ground by thin cables. The authors have shown (in previous works dealing with the AB-Dome) that this closed AB-Dome allows full control of the weather inside the Dome (the day is always fine, the rain is only at night, no strong winds) and influence to given region. This is a realistic and cheap method of economical irrigation, getting energy and virtual weather control on Earth at the current time.

Keywords: Local weather control, gigantic film AB-Dome, converting a dry region to subtropics, converting desolate wilderness to a prosperous region, macroprojects. Induced rainfall, rainmaking, Green power supply

---

* Presented to http://arxiv.org (2007)

INTRODUCTION

1. Short History. A particularly ambitious proposal evaluated in the Farmhand *Fencepost* (Australia) was to build a new mountain range (from soil and stone!) in the West of Australia, so as to create rain in the dry interior of Australia (Engineering the Weather, Part 10, p.115). The macro-project idea, first suggested by Lawrence James Hogan in his 1979 Australia-published book *Man Made Mountain*, is to build a very large mountain range 4 km tall, 10km wide at the base, with a 2km-wide plateau at the summit and covering a distance of 2,000 km. Its purpose is to markedly change the dry climate in Australia, a continent-spanning ecosystem-nation. A trench approximately 100 meters deep, 200 km wide and 2,000 km long would be necessary to provide the amount of piled earthen material to build the mountain range (400,000 km^3). It has been estimated that the construction costs of such a mountain would be more than $100,000 billion (HWA, 2003). Given Australia's entire national budget is <$200 billion, the price tag would be about 500 times the national budget. If the new mountain were to create an extra 50 million ML of runoff each year, then the cost of the water would be more than $100,000 per ML ($100/kL) (HWA,2003). That was a visionary macro-engineering concept but an impractical macro-project proposal.

Joseph Friedlander (of Shave Shomron, Israel) suggested (2008) to the authors to apply the AB-Dome to this somewhat visionary macro-problem. In this section of the book, the authors develop a theory for estimation and computation of the effect and cost of artificial inflatable AB-Mountains. The result is this article in http://Arxiv.org. This theory and supporting mathematical computations may be applied to any particular macro-project in any country.

Common Information

2. Climate. Climate is the average and variations of weather in a region over long periods of time. Climate zones can be defined using parameters such as temperature and rainfall.

3. Precipitation. General information about precipitation. The extant amount of water in Earth's hydrosphere in the current era is constant. The average annual layer of Earth's precipitation is about 1000 mm or 511,000 km^3. 21% of this (108,000 km^3) falls on land and 79% (403,000 km^3) on oceans. Most of it falls between latitudes 20° North and 20° South. Both polar zones collect only 4% of Earth's precipitation. The evaporation from the World-Ocean equals 1250 mm (450,000 km^3). 1120 mm returns back as precipitation and 130 mm by river inflow. The evaporation from land equals 410 mm (61,000 km^3), the precipitation is 720 mm. The land loses 310 mm as river flow to the oceans (47,000 km^3). These are average data. In some regions the precipitation is very different (Figure 1).

In meteorology, precipitation (also known as one class of hydrometeors, which are atmospheric water phenomena) is any product of the condensation of atmospheric water vapor that is deposited on the earth's surface. It occurs when the atmosphere (being a large gaseous solution) becomes saturated with water vapour and the water condenses and falls out of solution (i.e., precipitates). Air becomes saturated via two processes, cooling and adding moisture.

Precipitation that aches the surface of the earth can occur in many different forms, including rain, freezing rain, drizzle, snow, sleet, and hail. Virga is precipitation that begins falling to the earth but evaporates before reaching the surface. Precipitation is a major component of the hydrologic cycle, and is responsible for depositing most of the fresh water on the planet. Given the Earth's surface area, that means the globally-averaged annual precipitation is about 1 m, and the average annual precipitation over oceans is about 1.1 m.

Air contains moisture, measured in grams of water per kilogram (or m^3) of dry air (g/kg) or g/m^3, but most commonly reported as a relative humidity percentage. How much moisture a parcel of air can hold before it becomes saturated (100% relative humidity) depends on its temperature. Warmer air has a higher capacity for holding moisture than cooler air. Because of this property of the air, one way to saturate a parcel of air is to cool it. The dew point is the temperature that a parcel needs to be cooled to for saturation to occur.

Some cooling mechanisms include:

- Lift (convective, mechanical, positive vorticity advection)
  - Conductive cooling (warm air moves over a cool surface)
  - Radiational cooling (heat radiates off into space at night)
  - Evaporative cooling (air temperature falls as liquid water uses the energy to change phase to vapour).

The other way to saturate an air parcel is to add moisture to it, by:

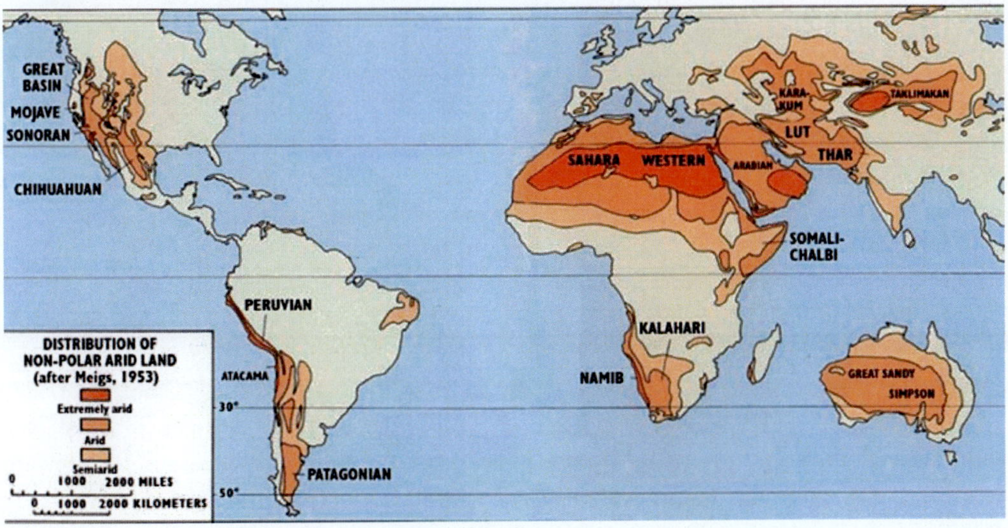

Figure 1. Distribution of non-polar arid land. 78 kb.

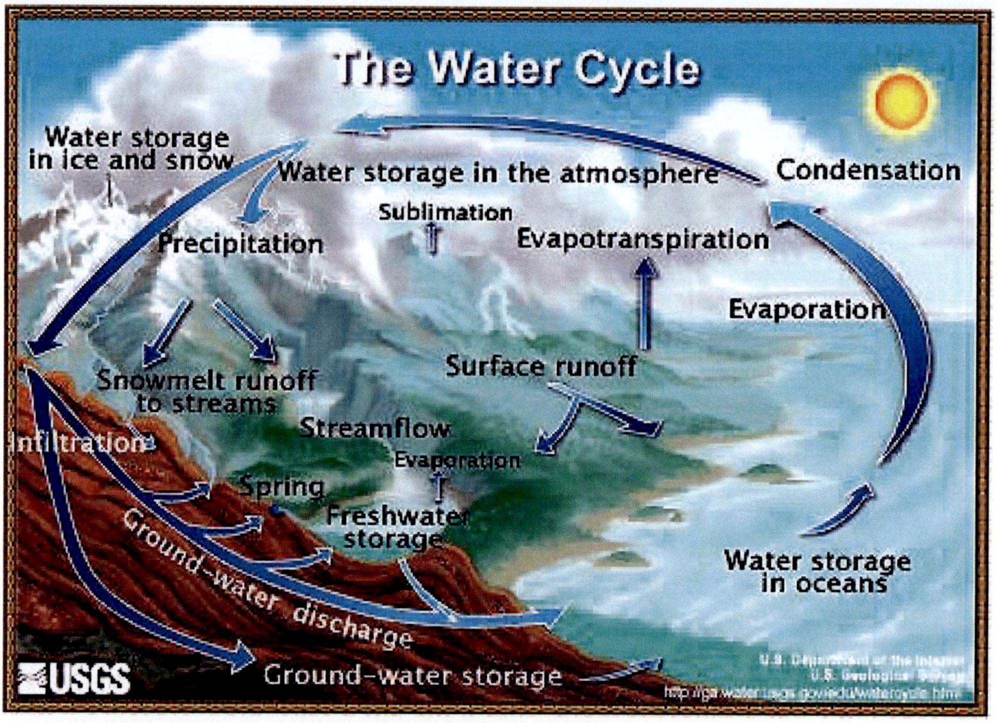

Figure 2. Nature water cycle. Credit: USGS. 186 kb.

- Precipitation falling from above (stratus forming in the rain under a higher cloud)
- Daytime heating evaporating water from the surface of oceans/lakes
- Drier air moving over open water (snow streamers off the Great Lakes in winter)

4. Orographic effects. Orographic precipitation occurs on the windward side of mountains and is caused by the rising air motion of a large-scale flow of moist air across the mountain ridge, resulting in adiabatic cooling and condensation.

In mountainous parts of our world subjected to relatively consistent winds (for example, the more or less steady trade winds), a moister climate usually prevails on the windward side of a mountain than on the leeward (downwind) side. Moisture is removed by orographic lift, leaving drier air, katabatic windsm, flowing over the descending (generally warming), leeward side where a rain shadow-effect is observable.

Orographic precipitation is well-known on oceanic islands, such as the Hawaiian Islands, where much of the rainfall received on an island is on the windward side, and the leeward side tends to be quite dry, almost desert-like, by comparison. This phenomenon results in substantial local gradients of average rainfall, with coastal areas receiving on the order of 500 to 750 mm per year (20 to 30 inches), and interior uplands receiving over 2.5 m per year (100 inches). Leeward coastal areas are especially dry 500 mm per year (20 inches) at Waikiki), and the tops of moderately high uplands are especially wet – ~12 m per year (~475 inches) at Wai'ale'ale on Kaua'i).

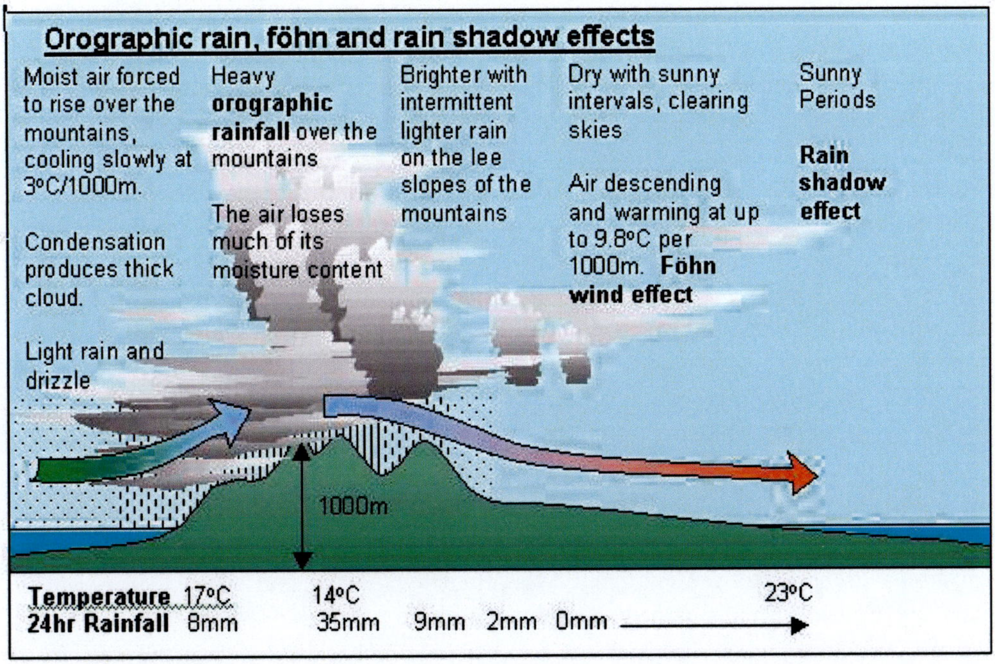

Figure 3. Orographic precipitation. After encounter with mountain the atmospheric temperature is 5 – 7 C more than the air temperature before encounter because the air temperature increases (when air pressure increases) at low altitude and vapor condences at high altitude. That air warming is an important benefit for countries which use the artificial mountains for protection from cold polar winds. 64 kb.

Twenty years ago, a record 683 inches of rain fell over a 12-month period on Mount Waialeale. Arguably, it may be the wettest place on Earth, with an average annual rainfall of 460 inches. Here's why raindrops fall on this 5,208-foot peak.

Mount Waialeale received an average annual rainfall of 460 inches over a 32-year period. However, measurements over 38 years at Mawsynram, India, give it an edge of 7.4 inches per year; in 1860, nearby Cherrapunji recorded an unofficial 1,000 inches. When warm air reaches a mountain range, it is lifted up the mountain slope, cooling as it rises. This process is known as orographic lifting, a process common to all mountain ranges.

Highest Average Annual Precipication

Contiment and Place	High everage precipitation, m/year (inch/year)	Elevation m (feet)
Asia, Mawsynram, India	11.9 (467.4)	1402 (4597)
Oceania, Mount Waialeale, Kaual	11.7 (460)	1570 (5138)
Australia, Relenden Ker, Qeensland	8.64 (340)	1556 (5105)
Europe, Crkvica, Rosnia - Herzegovina	4.65 (184)	1018 (3340)

It is the main rainfall-producing mechanism in Hawaii. Here, orographic lifting of moisture-laden northeast tradewinds over the windward slopes of each island creates wet windward weather and dry leeward weather. Just 15 miles west of Mount Waialeale, the Kekaha coastal region receives less than 20 inches of rain a year. This condition is called a "rainshadow."

In South America, the Andes mountain range blocks most of the Atlantic moisture that arrives in that continent, resulting in a desert-like climate on the Pacific coast of Peru and northern Chile, since the cold Humboldt Current ensures that the air off the Pacific is dry as well. On the leeward side of the Andes is the Atacama Desert of Chile. It is also blocked from moisture by mountains to its west as well. Not coincidentially, it is the driest place on earth. The Sierra Nevada range creates the same effect in North America forming the Great Basin desert, Mojave Desert and Sonoran Desert.

5. Wind. Wind is the flow of Earth's air, a fundamental resource for all life-forms. More speficically, it is the flow of the gases which compose an atmosphere; since wind is not unique to Earth. Simply it occurs as air is heated by the sun and thus rises. Cool air then rushes in to occupy the area the now hot air has moved from. It could be loosely classed as a convection current.Winds are commonly classified by their spatial scale, their speed, the types of forces that cause them, the geographic regions in which they occur, or their effect.

There are global winds, such as the wind belts which exist between the atmospheric circulation cells. There are upper-level winds which typically include narrow belts of concentrated flow called jet streams. There are synoptic-scale winds that result from pressure differences in surface air masses in the middle latitudes, and there are winds that come about as a consequence of geographic features, such as the sea breezes on coastlines or canyon breezes near mountains. Mesoscale winds are those which act on a local scale, such as gust fronts. At the smallest scale are the microscale winds, which blow on a scale of only tens to hundreds of meters and are essentially unpredictable, such as dust devils and microbursts.

Forces which drive wind or affect it are the pressure gradient force, the Coriolis force, buoyancy forces, and friction forces. When a difference in pressure exists between two adjacent air masses, the air tends to flow from the region of high pressure to the region of low pressure. On a rotating planet, flows will be acted upon by the Coriolis force, in regions sufficiently far from the equator and sufficiently high above the surface.

The three major driving factors of large scale global winds are the differential heating between the equator and the poles (difference in absorption of solar energy between these climate zones), and the rotation of the planet.Winds can shape landforms, via a variety of aeolian processes.

Some local winds blow only under certain circumstances, i.e., they require a certain temperature distribution. Differential heating is the motive force behind land breezes and sea breezes (or, in the case of larger lakes, lake breezes), also known as on- or off-shore winds. Land absorbs and radiates heat faster than water, but water releases heat over a longer period of time. The result is that, in locations where sea and land meet, heat absorbed over the day will be radiated more quickly by the land at night, cooling the air. Over the sea, heat is still being released into the air at night, which rises. This convective motion draws the cool land air in to replace the rising air, resulting in a land breeze in the late night and early morning. During the day, the roles are reversed. Warm air over the land rises, pulling cool air in from the sea to replace it, giving a sea breeze during the afternoon and evening.

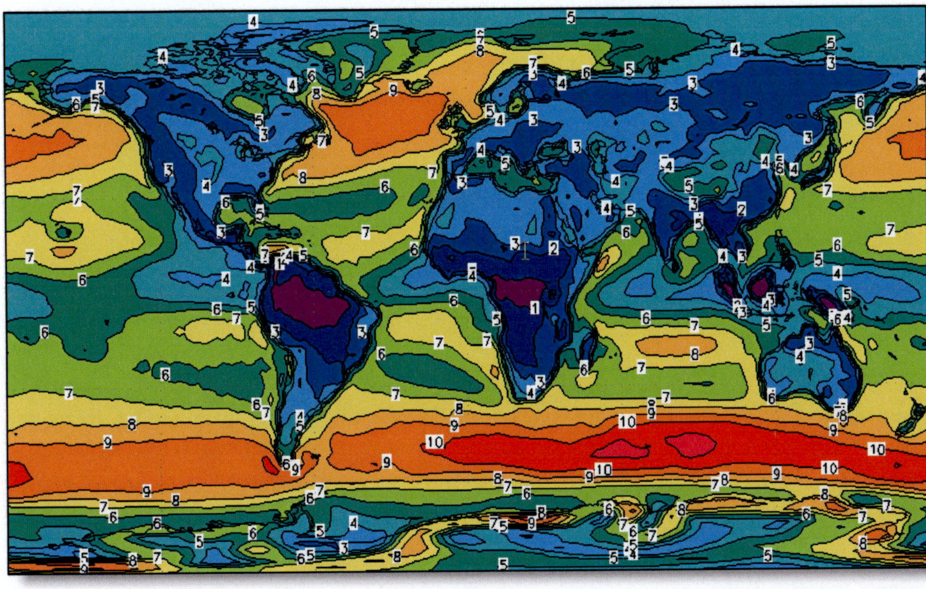

Figure 4. The map shows the mean wind speed in ms^{-1} @ 10 m a.g.l. for the period 1976-95, according to the NCEP/NCAR reanalysis data set. Even though the wind climate is known in broad outline for most of the world, more detailed and reliable information is usually required for wind resource assessments. This section lists a number of wind atlases and other wind investigations for different countries around the world; many of which contains measured and/or modeled data sets for application in wind resource assessment. 73 kb.

Figure 5. A lenticular cloud above New Mexico, USA.

Mountain breezes and *valley breezes* are due to a combination of differential heating and geometry. When the sun rises, it is the tops of the mountain peaks which receive first light, and as the day progresses, the mountain slopes take on a greater heat load than the valleys. This results in a temperature inequity between the two, and as warm air rises off the slopes, cool air moves up out of the valleys to replace it. This upslope wind is called a *valley breeze*. The opposite effect takes place in the afternoon, as the valley radiates heat. The peaks, long since cooled, transport air into the valley in a process that is partly gravitational and partly convective and is called a *mountain breeze*.

Mountain breezes are one example of what is known more generally as a katabatic wind. These are winds driven by cold air flowing down a slope, and occur on the largest scale in Greenland and Antartica...Most often, this term refers to winds which form when air which has cooled over a high, cold plateau is set in motion and descends under the influence of gravity. Winds of this type are common in regions of Mongolia and in glaciated locations.

Because *katabatic* refers specifically to the vertical motion of the wind, this group also includes winds which form on the lee side of mountains, and heat as a consequence of compression. Such winds may undergo a temperature increase of 20 °C (36 °F) or more, and many of the world's "named" winds (see list below) belong to this group. Among the most well-known of these winds are the chinook of Western Canada and the American Northwest, the Swiss föhn, California's infamous Santa Ana wind, and the French Mistral.The opposite of a katabatic wind is an anabatic wind, or an upward-moving wind. The above-described *valley breeze* is an anabatic wind. A widely-used term, though one not formally recognised by meteorologists, is *orographic wind*. This refers to air which undergoes orographic lifting. Most often, this is in the context of winds such as the chinook or the föhn, which undergo lifting by mountain ranges before descending and warming on the lee side.

A *sea-breeze* (or *onshore breeze*) is a wind from the sea that develops over land near coasts. It is formed by increasing temperature differences between the land and water which create a pressure minimum over the land due to its relative warmth and forces higher pressure, cooler air from the sea to move inland.

Land breezes. At night, the land cools off quicker than the ocean due to differences in their specific heat values, which forces the dying of the daytime sea breeze. If the land cools below that of the adjacent sea surface temperature, the pressure over the water will be lower than that of the land, setting up a land breeze as long as the environmental surface wind pattern is not strong enough to oppose it. If there is sufficient moisture and instability available, the land breeze can cause showers or even thunderstorms, over the water. Overnight thunderstorm development offshore can be a good predictor for the activity on land the following day, as long as there are no expected changes to the weather pattern over the following 12-24 hours. The land breeze will die once the land warms up again the next morning.

6. Mountains. A mountain is a landform that extends above the surrounding terrain in a limited area. A mountain is generally steeper than a *hill*, but there is no universally accepted standard definition for the height of a mountain or a hill although a mountain usually has an identifiable summit. Mountains cover 54% of Asia, 36% of North America, 25% of Europe, 22% of South America, 17% of Australia, and 3% of Africa. As a whole, 24% of the Earth's land mass is mountainous. 10% of people live in mountainous regions. Most of the world's rivers are fed from mountain sources, and more than half of humanity depends on mountains for water.

Sufficiently tall mountains have very different climatic conditions at the top than at the base, and will thus have different life zones at different altitudes. The flora and fauna found in these zones tend to become isolated since the conditions above and below a particular zone will be inhospitable to those organisms. These isolated ecological systems are known as sky islands and/or microclimates. Tree forests are forests on mountain sides which attract moisture from the trees, creating a unique ecosystem. Very tall mountains may be covered in ice or snow.

Air as high as a mountain is poorly warmed and, therefore, cold. Air temperature normally drops 1-2 degrees Celsius for each 300 meters of altitude.

7. Humidity. The term humidity is usually taken in daily language to refer to relative humidity. Relative humidity is defined as the amount of water vapor in a sample of air compared to the maximum amount of water vapor the air can hold at any specific temperature. Humidity may also be expressed as Absolute humidity and specific humidity. Relative humidity is an important metric used in forecasting weather. Humidity indicates the likelihood of precipitation, dew, or fog. High humidity makes people feel hotter outside in the summer because it reduces the effectiveness of sweating to cool the body by preventing the evaporation of perspiration from the skin. This effect is calculated in a heat index table. Warm water vapor has more thermal energy than cool water vapor and therefore more of it evaporates into warm air than into cold air.

Figure 6. At the famous "Blue Hole," Mount Waialeale drops off precipitously.

8. A desert is a landscape form or region that receives very little precipitation. Deserts are defined as areas that receive an average annual precipitation of less than 250 mm. In the Köppen climate classification system, deserts are classed as (BW).

Deserts take up one-third of the Earth's land surface. They usually have a large diurnal and seasonal temperature range, with high daytime temperatures (in summer), and low night-time temperatures (in winter down to 0 °C) due to extremely low humidity. Water acts to trap infrared radiation from both the sun and the ground, and dry desert air is incapable of blocking sunlight during the day or trapping heat during the night. Thus during daylight all of the sun's heat reaches the ground. As soon as the sun sets the desert cools quickly by radiating its heat into space. Urban areas in deserts lack large) daily temperature ranges, partially due to the urban heat island effect.

Many deserts are shielded in rain by rain shadows, mountains blocking the path of precipitation to the desert. Deserts are often composed of sand and rocky surfaces. Sand dunes called ergs and stony surfaces called hamada surfaces compose a minority of desert surfaces. Exposures of rocky terrain are typical, and reflect minimal soil development and sparseness of vegetation.

Bottomlands may be salt-covered flats. Eolian processes are major factors in shaping desert landscapes. Cold deserts (also known as polar deserts) have similar features but the main form of precipitation is snow rather than rain. Antarctica is the world's largest cold desert (composed of about 98 percent thick continental ice sheet and 2 percent barren rock). The largest hot desert is the Sahara. Deserts sometimes contain valuable mineral deposits that were formed in the arid environment or that were exposed by erosion.

Rain does fall occasionally in deserts, and desert storms are often violent. A record 44 millimeters of rain once fell within 3 hours in the Sahara. Large Saharan storms may deliver up to 1 millimeter per minute. Normally dry stream channels, called arroyos or wadis, can quickly fill after heavy rains, and flash floods make these channels dangerous.

Though little rain falls in deserts, deserts receive runoff from ephemeral, or short-lived, streams fed considerable quantities of sediment for a day or two. Although most deserts are in basins with closed or interior drainage, a few deserts are crossed by 'exotic' rivers that derive their water from outside the desert. Such rivers infiltrate soils and evaporate large amounts of water on their journeys through the deserts, but their volumes are such that they maintain their continuity. The Nile River, the Colorado River, and the Yellow River are exotic rivers that flow through deserts to deliver their sediments to the sea. Deserts may also have underground springs, rivers, or reservoirs that lay close to the surface, or deep underground. Plants that have not completely adapted to sporadic rainfalls in a desert environment may tap into underground water sources that do not exceed the reach of their root systems.

Lakes form where rainfall or meltwater in interior drainage basins is sufficient. Desert lakes are generally shallow, temporary, and salty. Because these lakes are shallow and have a low bottom gradient, wind stress may cause the lake waters to move over many square kilometers. When small lakes dry up, they leave a salt crust or hardpan. The flat area of clay, silt, or sand encrusted with salt that forms is known as a playa. There are more than a hundred playas in North American deserts. Most are relics of large lakes that existed during the last ice age about 12,000 years ago. Lake Bonneville was a 52,000 square kilometer lake almost 300 meters deep in Utah, Nevada, and Idaho during the Ice Age. Today the remnants of Lake Bonneville include Utah's Great Salt Lake, Utah Lake, and Sevier Lake. Because playas are arid landforms from a wetter past, they contain useful clues to climatic change.

When the occasional precipitation does occur, it erodes the desert rocks quickly and powerfully. Winds are the other factor that erodes deserts—they are slow yet constant.

A desert is a hostile, potentially deadly environment for unprepared humans. The high heat causes rapid loss of water due to sweating, which can result in dehydration and death within days. In addition, unprotected humans are also at risk from heatstroke and venomous animals. Despite this, some cultures have made deserts their home for thousands of years, including the Bedouin, Touareg and Puebloan people. Modern technology, including advanced irrigation systems, desalinization and air conditioning have made deserts much more hospitable. In the United States and Israel, desert farming has found extensive use.

Australia's Great Sandy Desert gets nearly all its rainfall from monsoonal thunderstorms or the occasional tropical cyclone rain depression. Thunderstorm days average 20-30 annually through most of the area (Burbidge 1983) although the desert has fairly high precipitation rates due to the high rates of evaporation this area remains an arid environment with vast areas of sands.

Other areas of our world, which are affected by these rare precipitation events in drylands, are Northwest Mexico, South West America, and South West Asia. In North America in the Sonoran and Chihuahuan desert have received some tropical rainfall in the last decade. Tropical activity is rare in all deserts but what rain does arrive here is important to the delicate ecosystem existing.

Figure 7. Typical sandy desert landscape.

9. Aridity. In general terms, the climate of a locale or region is said to be *arid* when it is characterized by a severe lack of available water, to the extent of hindering or even preventing the growth and development of plant and animal life. As a result, environments subject to arid climates tend to lack vegetation and are called xeric or desertic.

The expression 'available water' refers to water in the soil in excess to the wilting point. The air over a hot desert may actually contain substantial amounts of water vapor but that water may not be generally accessible to plants, except for very specialized organisms (such as some species of lichen). 'Lack of water' refers to use by plants. The water that is actually present in the environment may be sufficient for some species or usages (such as climax vegetation), and grossly insufficient for others. Aridity, the characteristic nature of arid climates, may thus depend on the use of the land. Regards to the presence of life, what is more important than the degree of rainfall is the fraction of precipitation that is not quickly lost through evaporation or runoff. Attempts to quantitatively describe the degree of aridity of a place has often led to the development of aridity indexes. There is no universal agreement on the precise boundaries between classes such as 'hyper-arid', 'arid', 'semi-arid', etc.

If different classification schemes and maps differ in their details, there is a general agreement about the fact that large areas of the Earth are considered arid. These include the hot deserts located broadly in sub-tropical regions, where the accumulation of water is largely prevented by either low precipitations, or high evaporation, or both, as well as cold deserts near the poles, where water may be permanently locked in solid forms (snow and ice). Other arid regions include areas located in the rain shadows of major mountain ranges or along coastal regions affected by significant upwelling (such as Chile's Atacama Desert).

The distribution of aridity observed at any one point in time is largely the result of the general circulation of the atmosphere. The latter does change significantly over time through climate change. In addition, changes in land use can result in greater demands on soil water and induce a higher degree of aridity.

10. Control of local weather. Governments spend billions of dollars merely studying the weather. The many big government research scientific organizations and perhaps a hundred thousand of scientists have been studying Earth's weather for more than a hundred years. There are gigantic numbers of scientific works about weather control. Most of them are impractical. We cannot exactly predict weather at long period, to avert a rain, strong wind, storm, tornado, or hurricane. We cannot control the clouds, temperature and humidity of the atmosphere, nor the power of rain. We cannot make more tolerable a winter or summer. We cannot convert a cold region to subtropics, a desolate wilderness to a prosperous region. We can only observe the storms and hurricanes and approximately predict their direction of movement. It is as if all the police department did was announce which neighborhoods were infested with killers and best avoided! Every year terrible storms, hurricanes, strong winds and rains and inundations destroy thousands of houses, kill thousands of men.

INNOVATIONS AND DESCRIPTION

1. Idea and Innovations. The idea here is for the creation of a cheap range of inflatable 'mountains' (really immense gasbags) (Figures 8, 9) from a thin film or densely-woven tensioned textile whose presence forces humid air (a wet wind) to rise to high altitude. It is widely-known that air expands and cools at altitude. The air humidity decreases, exceeds the maximal saturation level and superfluous water vapor condenses in various forms, including rain or rain clouds.

The top of the gasbags' film is located at an altitude of ~3-5 km. It is supported at this altitude by a small additional air pressure produced by ground ventilators. The film is connected to Earth's ground by controlled cables, which allow some change in the height and orientation of the artificial mountain(s). The gasbag's external surface may require double-layer film. We can control the heat conductivity of the dome cover by pumping an air between two layers of the dome cover and change the solar heating (solar radiation) by control of cover clarity or pumping a warm air between layers if icing-over or show is at the dome top. That allows selecting for different conditions (solar heating) in the covered area and by pumping air into the dome.

Special film may be used in a more complex dome design if we want to have finely control conditions inside the AB-Dome [1]. Envisioned is a cheap film having liquid crystal and conducting layers. The clarity is controlled by electric voltage. These layers, by selective control, can pass or blockade the solar light (or parts of solar spectrum) and pass or blockade the Earth's radiation. The incoming and outgoing radiations have different wavelengths. That makes control of them separately feasible and therefore possible to manage the heating or cooling of the Earth's surface under this film. In conventional conditions about 50% of the solar energy reaches the Earth surface. Much is reflected back to outer space entirely surrounding Earth by the white clouds. In our closed water system [1] the rain (or at least condensation) will occur at night when the temperature is low. In open atmosphere, the Sun heats the ground; the ground must heat the whole troposphere (6 – 10 km) before stable temperature rises are achieved. In our case the ground heats ONLY the air into the dome (as in a hotbed). We have a literal Greenhouse Effect, for the 'over-roof' prevents the hottest air from escaping to the Earth-atmosphere above. That means that many cold regions (Alaska, Siberia, North Canada) may absorb more solar energy and became a temperate climate or sub-tropic climate (under the dome, as far as plants are concerned). That also means, by increasing the albedo of the gas bags, that the Sahara and other deserts can be a prosperous area with a fine growing and living climate and with a closed-loop water cycle.

The building of a film dome is remarkably very easy. We simply spread out the collapsed film over Earth's worksite surface, turn on the air-pumping propellers and the film is raised gradually and safely by air over-pressure to the needed altitude, only limited by the restraining cables. Damage to the film is not a major trouble because the additional air pressure is very small (0.005 - 0.05 atmosphere) and air leakage is compensated for by speeding the turning air impellers. Unlike a space colony, or planetary surface colony, Earth's outside air is life-form friendly and, at worst, we lose some heat (or cold) and water vapor.

The first main innovation of the offered AB-Mountain (and main difference from a conventional hotbed, or greenhouse) is the inflatable HIGH span of the closed cover (up to 3 – 5 km). The great vertical scale-height of the enclosed volume aids organizing the rise of humid air at high altitude. The air is cooling and during that process it is also producing a lot of freshwater which may be collected by rain channels on the AB-Mountain at high altitude and make much energy in hydro-electric turbines located on our Earth's surface as the waters descend rapidly. This freshwater may be used for irrigation or sale [1]. The rest of the water will be precipitating as rain in near regions and can change a dry environment for a given region to a subtropical climate (the size of the region depends upon the immensity of the artificial mountain range).

The other profit from the offered AB-Domes (units of which make up the visible mountain range) is a closed loop water cycle, which prevents escape of water vaporized by plants; and returns this water in the nighttime when the air temperature decreases and internal condensation recurs. That allows us to perform irrigation in the gigantic portion of Earth's land area that does not have enough water for agriculture. We can convert the desert and desolate wildernesses into gardens without expensive delivery of remote freshwater. The initial amount of water for water cycle may be collected from atmospheric precipitation in some period or delivered. Only losses and exports of final products, typically one part in a thousand of all the water required during growing, need be replaced. The availability of good soil is not a necessity because hydroponics will permit farmers to achieve record crop harvests on any soil, which is employed mostly just to hold plant roots.

The second important innovation is the active control of dome height. That allows reaching a maximum efficiency of freshwater extraction and allows some effective 'control' of regional weather.

The third innovation is using the high altitude of collected freshwater for production of useful hydroelectric energy.

The fourth important innovation is the use of cheap thin film and/or dense tension-favorable textiles for building of mountains. This innovation decreases the construction cost by many millions of times in comparison to making artificial mountains from heaped soil and broken stones! Also, in case of improved climate modeling, when possibly we find that moving the 'mountain range' just a little bit geographically would be a vast placement improvement—the very subject is not impossible even to consider at the moment.

The fifth innovation is using a cheap controlled heat conductivity, double-layer cover (controlled clarity is optionally needed for some regions). This innovation allows conserving solar heat (in cold regions), to control the temperature in the dome (in hot climates). That allows users to harvest two to three rich crops annually in middle latitudes and to convert the Earth's coldest zones (Siberia, northern Canada, Alaska, the tip of South America, or elsewhere) to good single-crop regions.

The sixth innovation is using the cool water from the artificial mountain for cooling of buildings, crops, condensers, etc.

The seventh innovation is using the artificial mountain for high altitude windmills having high power (tens kW/m).

The eighth innovation is using the high water pressure (up 300 - 500 atm) for delivery of freshwater over long distances by tubes without pump stations.

The ninth innovation is control of cover reflectivity which allows users to influence to climate of region outside the cover.

Lest it be objected that such domes would take impractical amounts of plastic, consider that the world's plastic production is today on the order of 100 million tons. If, with economic growth, this amount doubles over the next generation and the increase is used for doming over territory, at 300 - 500 tons a square kilometer 200,000 square kilometers could be roofed over annually. While small in comparison to the approximately 150 million square kilometers of land area, consider that 200,000 1 kilometer sites scattered over the face of the Earth newly made habitable could revitalize vast swaths of landscape surrounding them—one square kilometer could grow local vegetables for a city in the desert, one over there could grow biofuel, enabling a desolate South Atlantic island to become fuel independent; at first, easily a billion people a year could be taken out of sweltering heat, biting cold and slashing rains, saving the money buying and running heating and air conditioning equipment would require.

Our design of the mountain-dome is presented in Figures 8, 8a, 9 that include the thin inflated film mountain-dome. The innovations are listed here: (1) the construction is air-inflatable; (2) each mountain-dome is fabricated with very thin, transparent film (thickness is 1 to 5 mm) having controlled clarity and controlled heat conductivity without rigid supports; (3) the enclosing film has two conductivity layers plus a liquid crystal layer between them which changes its clarity, color and reflectivity under an electric voltage (Figure 10); (4) the bound section of dome has a semi-cylindrical form (Figures 8, 8a, 9). The air pressure is more in these sections and they protect the central sections from the wind outside.

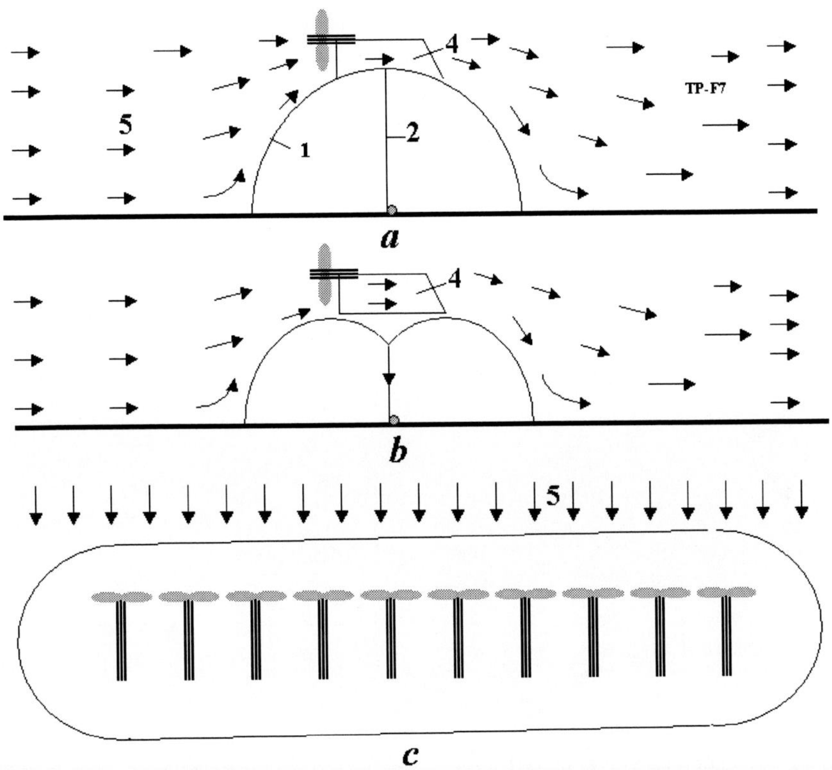

Figure 8. Artificial Inflatable Mountain. (*a*) Cross-section of mountain in position maximal altitude; (*b*) Cross-section of mountain in position decreased altitude; (*c*) Top view of cylindrical mountain ridge. *Notations:* 1 – Inflatable Mountain AB-Dome; 2 – cable of height control; 4 – water collector; 5 – wind. 32 kb.

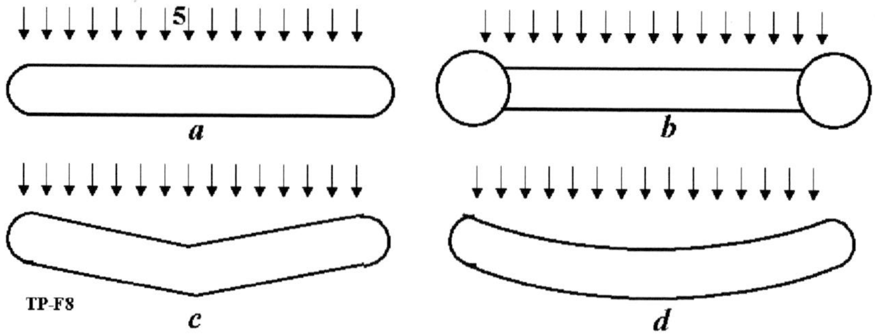

Figure 8a. Form of artificial mountains ridge: (a) straight; (b) Straight with spherical end; (c) Angle; (d) Concave. 5 – wind.

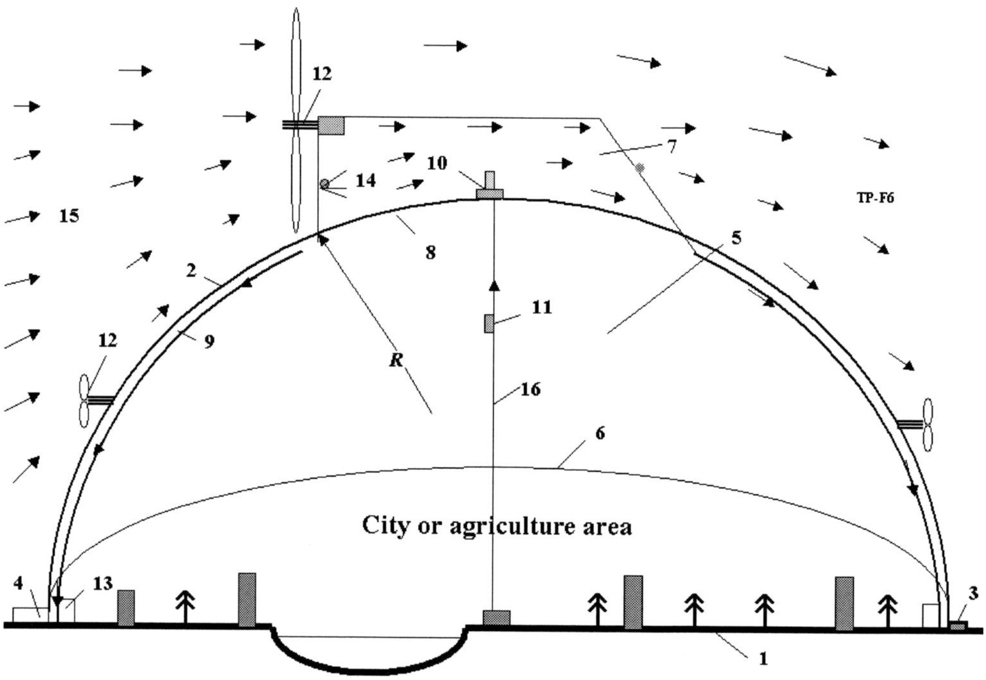

Figure 8b. Cross-section of Mountain AB-Dome used for big city or agriculture area; Notations: 1 – covered area; 2 – thin film; 3 – ventilator (air pump); 4 – exit; 5 –semi-cylindrical thin film AB-Dome; 6 – control reflectivity thin film (optional); 7 –collector of water; 8 – strong cables; 9 – water tube; 10 – rooftop utilities and businesses: TV, communication, telescope, locator service (differential GPS), tourists; 11 – elevator; 12 – windmills; 13 water turbines and electric generator; 14 – injector of water collector; 15 – wind; 16 – height control cable.

Figure 4 Ch.1A illustrates the thin transparent control dome cover we envision. The inflated textile shell—technical "textiles" can be woven or non-woven (films)—embodies the innovations listed: (1) the film is very thin, approximately 0.1 to 5 mm. A film this thin has never before been used in a major building; (2) the film has two strong nets, with a mesh of about 0.1×0.1 m and $a = 1 \times 1$ m, the threads are about additional 0.5 mm for a small mesh and about 1 mm for a big mesh. The net prevents the watertight and airtight film covering from being damaged by vibration; (3) the film incorporates (optional) a tiny electrically conductive wire net with a mesh about 0.1 x 0.1 m and a line width of about 100 μ (microns) and a thickness near 10 μ. The wire net is electric (voltage) control conductor. It can inform the dome maintenance engineers concerning the place and size of film damage (tears, rips, etc.); (4) the film may be twin-layered with the gap — $c = 1$ m and $b = 2$ m—between film layers for heat insulation. In polar (and hot) regions this multi-layered covering is the main means for heat isolation and puncture of one of the layers won't cause a loss of shape because the second film layer is unaffected by holing; (5) the airspace in the dome's covering can be partitioned, either hermetically or not; and (6) part of the covering can have a very thin shiny aluminum coating that is about 1μ (micron) for reflection of unnecessary solar radiation in equatorial or retention of additional solar radiation in the polar regions [1-16].

The author offers a method for moving off the snow and ice from the film at high altitude or in polar regions. After snowfall we decrease the heat cover protection, heating the show (ice) by warm air flowing into channels 5 (Figure 10) (between cover layers), and water runs down in tubes 3 (Figure 9).

The town cover may be used as a screen for projecting of pictures, films and advertising on the cover at night times.

2. Brief information about cover film. Our complex dome cover (film) may have 5 layers (Figure 10c): transparant dielectric layer, conducting layer (about 1 - 3 μ), liquid crystal layer (about 10 - 100 μ), conducting layer (for example, SnO_2), and transparant dielectric layer. Altogether thickness is ~ 0.1 - 0.5 mm. Control voltage is 5 - 10 V. This film may be produced by industry relatively cheaply.

Eikos Inc of Franklin, Massachusetts and Unidym Inc. of Silicon Valley, California are developing transparent, electrically conductive films of carbon nanotubes to replace indium tin oxide (ITO). Carbon nanotube films are substantially more mechanically robust than ITO films.

3. Liquid crystals (LC) are substances that exhibit a phase of matter that has properties between those of a conventional liquid, and those of a solid crystal.

Liquid crystals find wide use in liquid crystal displays (LCD), which rely on the optical properties of certain liquid crystalline molecules in the presence or absence of an electric field. The electric field can be used to make a pixel switch between clear or dark on command. Color LCD systems use the same technique, with color filters used to generate red, green, and blue pixels. Similar principles can be used to make other liquid crystal based optical devices. Liquid crystal in fluid form is used to detect electrically generated hot spots for failure analysis in the semiconductor industry. Liquid crystal memory units with extensive capacity were used in Space Shuttle navigation equipment. It is also worth noting that many common fluids are in fact liquid crystals. Soap, for instance, is a liquid crystal, and forms a variety of LC phases depending on its concentration in water.

The conventional control clarity (transparancy) film reflected excess solar energy back to space. If practical in the future, film of a future generation may incorporate thin-film solar cells that convert the solar energy into electricity.

4. Transparency. In optics, transparency is the material property of allowing light to pass through. Though transparency usually refers to visible light in common usage, it may correctly be used to refer to any type of radiation. Examples of transparent materials are air and some other gases, liquids such as water, most glasses, and plastics such as Perspex and Pyrex. Where the degree of transparency varies according to the wavelength of the light.

From electrodynamics it results that only a vacuum is really transparent in the strict meaning, any matter has a certain absorption for electromagnetic waves. There are transparent glass walls that can be made opaque by the application of an electric charge, a technology known as electrochromics.Certain crystals are transparent because there are straight lines through the crystal structure. Light passes unobstructed along these lines. There is a complicated theory "predicting" (calculating) absorption and its spectral dependence of different materials.

THEORY AND COMPUTATION OF THE AB-DOME AND MOUNTAIN SYSTEM

As wind flows over and around a fully exposed, nearly completely sealed inflated dome, the weather affecting the external film on the windward side must endure positive air pressures as the wind stagnates. Simultaneously, low air pressure eddies will be present on the leeward side of the dome. In other words, air pressure gradients caused by air density differences on different parts of the dome's envelope is characterized as the "buoyancy effect". The buoyancy effect will be greatest during the coldest weather when the dome is heated and the temperature difference between its interior and exterior are greatest. In extremely cold climates such as the Arctic and Antarctic Regions the buoyancy effect tends to dominate dome pressurization.

A reader can derive the equations below from well-known physical laws [17]. Therefore, the author does not give detailed explanations of these.

1. Amount of water in atmosphere. Amount of water in atmosphere depends upon temperature and humidity. For relative humidity 100%, the maximum partial pressure of water vapor is shown in Table 1.

Table 1. Maximum partial pressure of water vapor in atmosphere for given air temperature

t, C	-10	0	10	20	30	40	50	60	70	80	90	100
p,kPa	0.287	0.611	1.22	2.33	4.27	7.33	12.3	19.9	30.9	49.7	70.1	101

The amount of water in 1 m^3 of air may be computed by equation

$$m_W = 0.00625[p(t_2)h - p(t_1)], \qquad (1)$$

where m_W is mass of water, kg in 1 m^3 of air; *p(t)* is vapor (steam) pressure from Table 1, kPa; $h = 0 \div 1$ is relative humidity; *t* is temperature, C.

The computation of equation (1) is presented in Figure 9. Typical relative humidity of atmosphere air is 0.5 - 1.

$H = 0$ km, $t_1 = 0$ °C. 36 kb.

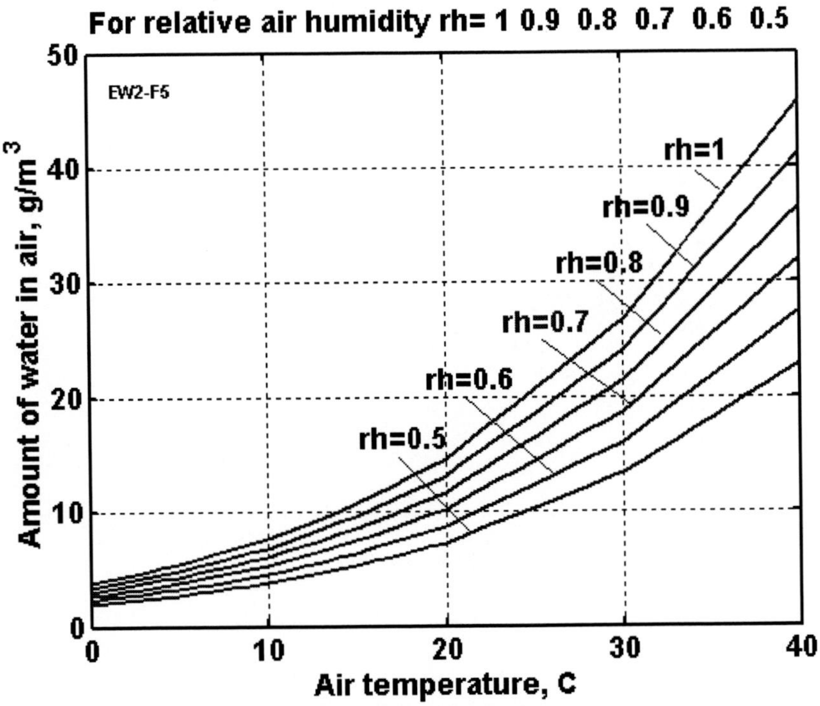

Figure 9. Amount of water in 1 m^3 of air versus air temperature and relative humidity (rh).

2. Air temperature, density and pressure versus altitude in troposphere. Standard atmosphere for our planet is to be found in our Table 2.

Temperature, relative air density and pressure of troposphere (up 10 km) versus the altitude computed by equations:

$$T = T_0 - 0.0065H, \quad \bar{\rho} = \frac{\rho}{\rho_0} = \left(1 - \frac{H}{44300}\right)^{4.265}, \quad \bar{p} = \frac{p}{p_0} = \left(1 - \frac{H}{44300}\right)^{5.265}, \tag{2}$$

where $T_0 = 15$ C, $\rho_o = 1.225$ kg/m^3, $p_o = 10^5$ N/m^2 are air temperature, density and pressure at sea level, $H = 0$; H is altitude, m; T, ρ, p are air temperature, density and pressure at altitude H, m. The computation of temperature related to altitude are presented in Figures 10, 11.

Table 2. Standard Earth atmosphere. $\rho_o = 1.225$ kg/m^3

H, km	0	0.4	1	2	3	4	5	6
t, °K	288.2	285.6	281.9	275.1	268.6	262.1	265.6	247.8
t, °C	15	12.4	8.5	2	-4.5	-11	- 17.5	-24
ρ/ρ_o	0	0.907	0.887	0.822	0.742	0.669	0.601	0.538

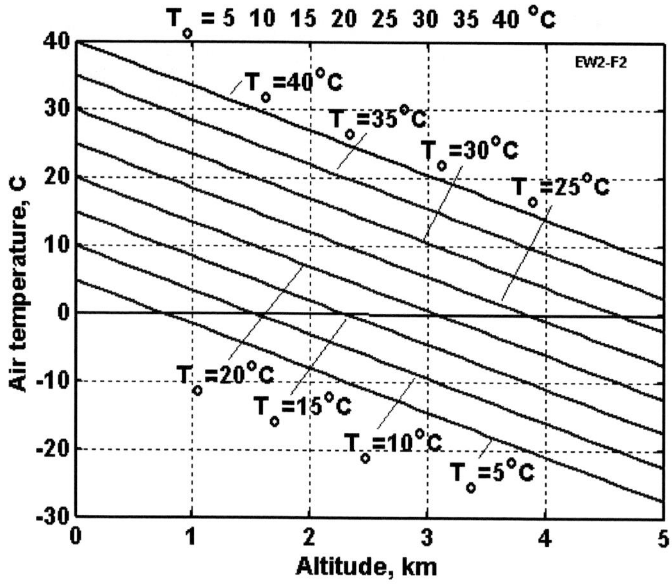

Figure 10. Air temperature versus the altitude for different temperatures.

After the encounter with the AB-Mountain range the atmospheric temperature is 5 – 7 C higher, because the air temperature increases when air pressure increases at low altitude. The analogy is to an air compressor; denser fractions are usually hotter. The water vapor is condensed and gives up its' heat to air. That additional air warming is important for countries which use the artificial AB-Mountains for protection from cold polar winds.

2. Altitude and wind speed. Wind speed, V, increases with altitude, H, as follows

$$\bar{V} = V / V_0 = (H / H_0)^{\alpha}, \tag{3}$$

where $\alpha = 0.1 - 0.25$ exponent coefficient depending upon surface roughness. When the surface is water, $\alpha = 0.1$; when surface is shrubs and woodlands $\alpha = 0.25$. The sub "0" means the data at Earth surface. The standard values for wind computation are $V_o = 6$ m/s, $H_o = 10$ m/s. The computation of this equation is succinctly presented in Figure 12. At high altitude the wind speed may be significantly more than equation (3) gives.

3. Water produced by AB-Mountains. Each linear meter of the mountain ridge (which may stretch for 5 km, 100 km, or even more) produces the water

$$W(H_m) = k_c V_0 \int_0^{H_m} m_w(T, H_m, h)\bar{\rho}(H)\bar{V}(H)dH$$

$$\text{where} \quad m_w = 0.00625\,[p(T_0)h - p(H_m)], \quad \text{for} \quad m_w > 0, \quad , \tag{4}$$

$$\text{if} \quad m_w < 0, \quad \text{than} \quad m_w = 0.$$

where W is water flow produced by 1 m of mountain ridge, kg/s/m; m_w is water in 1 m^3 [kg/m^3]; H_m is maximal height of artificial mountain, m; $k_c = 0.5 \div 1$ is collector (extraction) coefficient; $h \approx 0.5 \div 1$ is air humidity; V_0 is wind speed at H = 0, m/s.

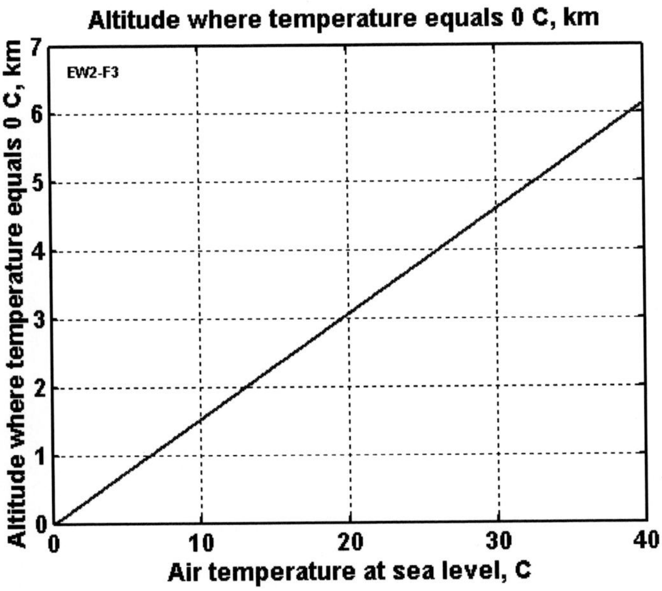

Figure 11. Altitude where temperature equals 0°C via air temperature at sea level.

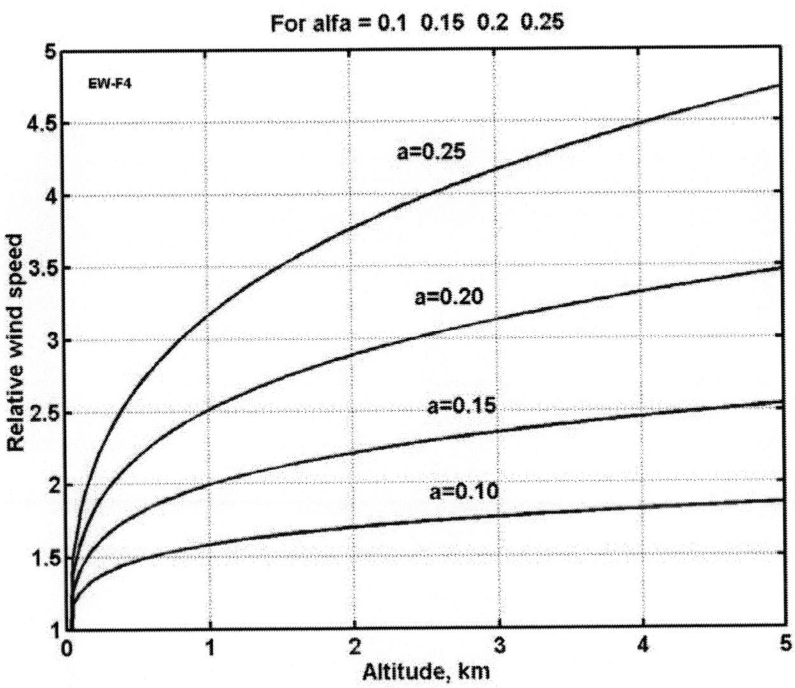

Figure 12. Relative wind speed versus altitude for $V_o = 6$ m/s, $H_o = 10$ m/s.

The computations of the equation (4) for $h = 0.3$, $k_c = 1$, $V_0 = 6$ m/s are presented in Figure 13.

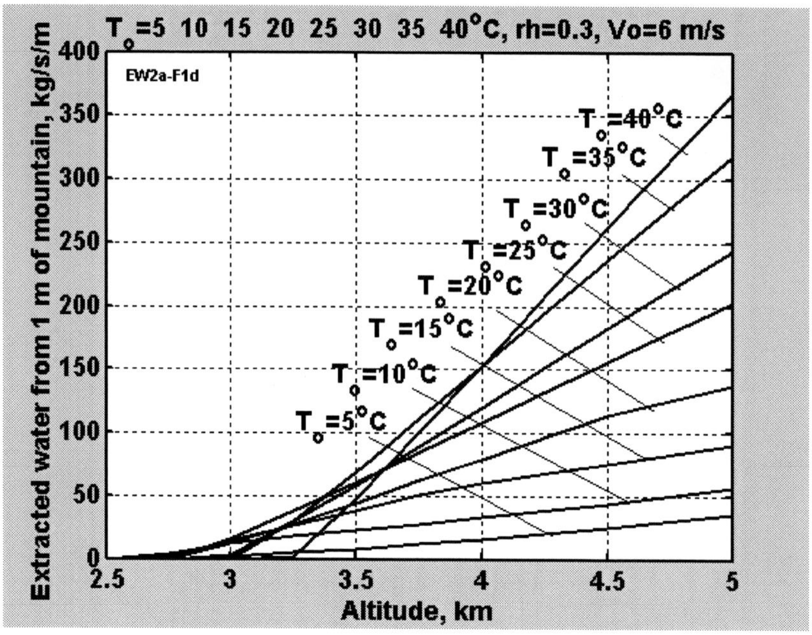

Figure 13. Amount of water flow may be extracted by 1 m of the AB-Dome artificial Mountain via mountain height for different air temperature T_o at H_o = 10 m, air speed V_o = 6 m/s, relative air humidity *rh* = *0.3* at sea level. Collector efficiency k_c = 1, coefficient increasing of air speed with altitude α = 0.15; H_m = 5 km.

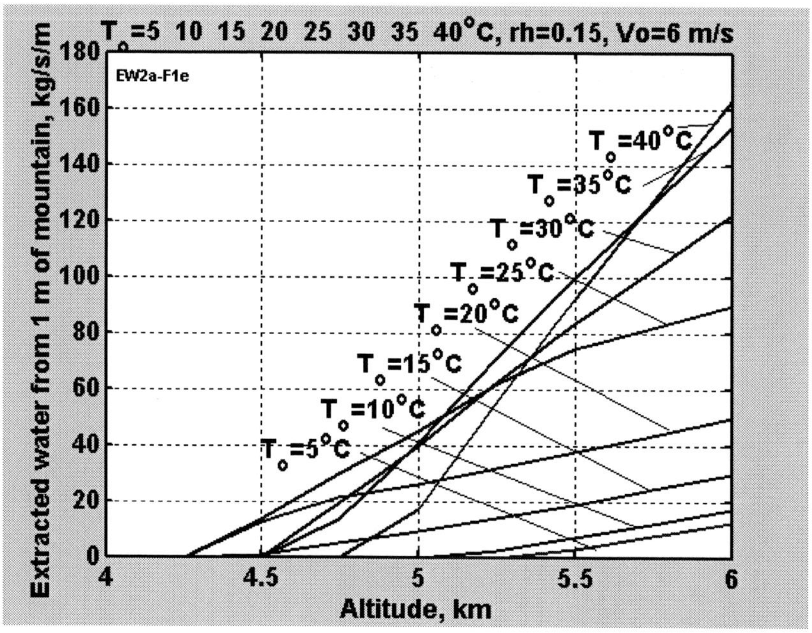

Figure 14. Amount of water flow may be extracted by 1 m of the AB-Dome artificial Mountain via mountain height for different air temperature T_o at H_o = 10 m, air speed V_o = 6 m/s, relative air humidity *rh* = 0.15 at sea level. Collector efficiency k_c = 1, coefficient increasing of air speed with altitude α = 0.15; H_m = 6 km.

Please notice: If relative humidity is $rh = 0,3$, the mountain begins to produce water over altitude 3 km, when the $rh = 0.3$. Every 100 - 300m ridge length of a 3-5-km altitude artificial mountain can produce a good sized river. However, in reality the water production after the mountain height over 3.5 – 5 km can not increase because all air water is condensed and precipitated. Many natural mountains over 3.5 – 5 km in elevation have dry landscapes.

The total water carried as vapor by atmospheric air is significantly more than we have here extracted. The perfect design of a water collector increases the extraction coefficient k_c. The computation of total water flow (for $k_c = 1$) is presented in Figure 14. The water non-extracted from atmosphere goes as clouds and rain after crossing the mountain ridge.

5. Energy produced by high altitude water. The water condensed at high altitude has huge energy because it has great mass and is located advantageously at very high altitude. For example, if artificial mountain has a height = 5 km, the water pressure at sea level is about 500 atm. We can easy convert this energy into electricity by conventional hydro-turbine and electric generators. And the higher the pressure, the smaller the installation needed.

Equation is below:

$$P = g\eta W H_m, \tag{5}$$

where P is water power, W/m; $g = 9.81$ m/s^2; η is efficiency coefficient; W is water flow, kg/s/m; (Eq. (4)); H_m is maximal altitude, m. The result equation is in Figure 15.

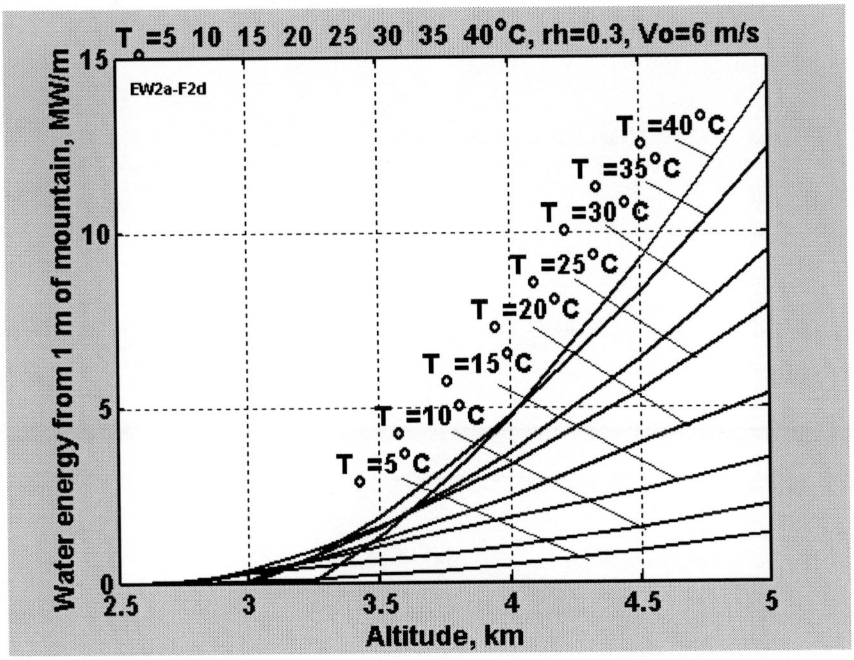

Figure 15. Water energy (MW) from 1 m of AB-mountain via dome height and air temperature T_o, air speed $V_o = 6$ m/s, relative air humidity $rh = 0.3$ at sea level $H = 10$ m. Collector efficiency $k = 1$, product of efficiency coefficients (tubes, hydro-turbine, electric generator) is $\eta = 0.8$, coefficient increasing of air speed with altitude $\alpha = 0.15$.

Again, please notice: That the artificial AB-Mountains produce gigantic levels of energy. A conventional large hydro-electric power station such as Hoover Dam may have about 2 GW of generating capacity. Only 4 km ridge length of 5 km tall artificial AB-Mountain can potentially produce energy comparable to the largest conventional hydroelectric stations in the world, in China and Brazil. (See for comparison http://en.wikipedia.org/wiki/List_of_the_largest_hydoelectric_power_stations) (And, most importantly, the AB-Mountain does not silt up its reservoirs like a conventional dam-made freshwater reservoir.) The huge amount of green water energy may be the most important profit potential obtainable from the offered AB-Mountain.

6. Hot mountain–derived wind. The water vapor condensing to water produces a lot of energy, about 2260 kJ/kg of water. The atmosphere absorbs this energy. The ground air temperature in the lee of the mountain is more than it is upwind of the mountain. The increase of air temperature may be estimated by equation received from equation of heat balance:

$$\Delta T = \frac{k_c \lambda m_{w0}(T_0)h}{c_p \rho_0}, \quad m_{w0} = 0.00625 p(T_0), \tag{6}$$

where ΔT is additional atmospheric temperature, C; $k_c \approx 0.7 - 1$ is extraction coefficient, m_{w0} is amount of water in 1 m^3 of air at temperature T_0, kg/m^3 (see Table 1); $h \approx 0.5 - 1$ is humidity; $c_p \approx 1$ kJ/kg/C is average air heat capability; $\rho_0 = 1.225$ kg/m^3 is standard air density. Note, the artificial mountain works better than a natural mountain because one has a smooth surface and good, indeed, selectable aerodynamic form (semicylinder).

The computations for $k_c = 1$ are presented in Figure 16.

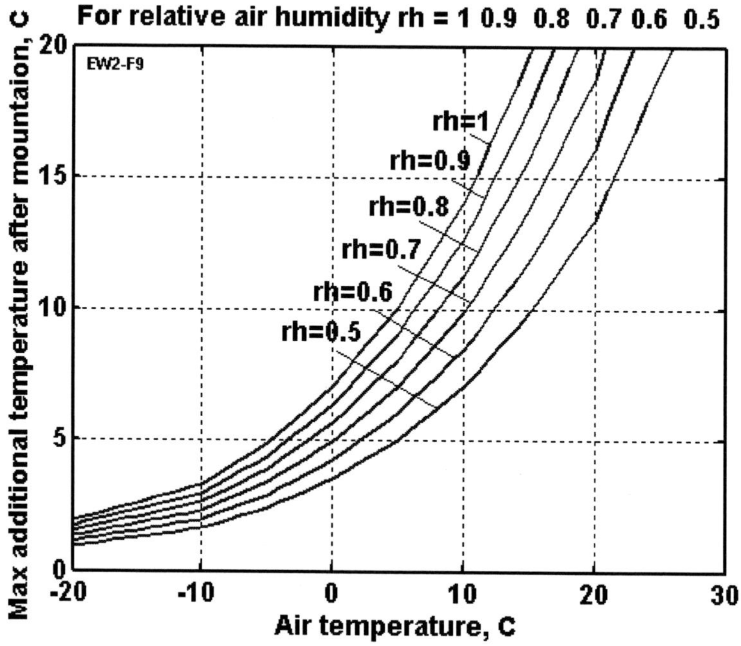

Figure 16. Maximal additional heating of atmosphere after mountain via temperature before mountain and relative humidity. Condenser coefficient equals 1.

As you see the artificial mountain can significantly increase temperatures of cold polar-derived winds. For example, if atmospheric temperature was 0 C, after an encounter with an artificial mountain it may be up to 7°C. If the initial air temperature was 10 C, after artificial mountain it may be up to 24°C. This effect is good relief for cold countries (Iceland-Scandinavia-Russia-Siberia, Canada, USA-Alaska), but it is not well for a hot area (Sahara). The cold water from AB-Mountain may be used for cooling the buildings of a city located behind the artificial Mountains. On the other hand, in the desert, we may place a new city before the mountains for wind cooling, and salt drying ponds behind the mountains, for free accelerated air-drying! Near the Red Sea, for example, 2.4 meters a year of salt water may evaporate from salt ponds. Such a rate could easily be doubled with hotter winds blowing for much of the time, doubling annual salt production.

Gigantic streams of energy flow through AB-Mountain ranges. If AB-Mountain length is 100 km only, water vapor power and air kinetic power is

$$P_v = c_{pm} SV\Delta T, \quad P_k = 0.5MV^2, \quad M = \rho VS, \tag{7}$$

where P_v is water vapor power, W; P_k is air kinetic power, W; c_{om} = 1.287 kJ/m^{3}K for air is the heat coefficient; V – average air speed, m/s; M – air flow mass, kg/s; S – cross-section of air flow, m^2; ρ – average air density, kg/m^3. If Mountain ridge is only 100 km, V = 5 m/s and ΔT = 10 C the water vapor power has gigantic value (1 - 3)×10^7 MW (10,000 one gigawatt power electric stations). The kinetic air power is only ≈ 2×10^4 MW. (2 x 10 gigawatts)—which is logical, because water is about a thousand times denser than air! But look at those figures—the whole energy use of mankind nowadays is only ~13,000 gigawatts! So, an AB-Mountain range can actually generate the energy that all mankind uses today, at an amazingly realistic cost (see below).

The control (of height, reflectivity, and position) of the AB-Mountain range manages this gigantic natural flow of energy and changes the weather and climate around the artificial mountain.

7. *Wind energy.* The AB-Dome has a high crest where there is a strong permanent wind. If we install the windmills at the top of AB-Dome, we get energy. This energy may be estimated by equations

$$N_1 = 0.5\eta\rho(H)D\cdot(\bar{V}V_0)^3, \tag{8}$$

where N_1 is windmill power from 1 m of the mountain ridge, W/m; η = 0.3 ÷ 0.6 is coefficient efficiency of wind rotor, ρ is air density at altitude H; D is rotor diameter, m; V is wind speed (see early), m/s.

Computations are presented in Figure 17.

As you see (compare with Figure 15) the wind energy in 500 times is less then the water energy and no reasons to install the complex and expensive windmills at top of AB-Mountain. But they may be useful on the low AB-Dome, which doesn't produce the water.

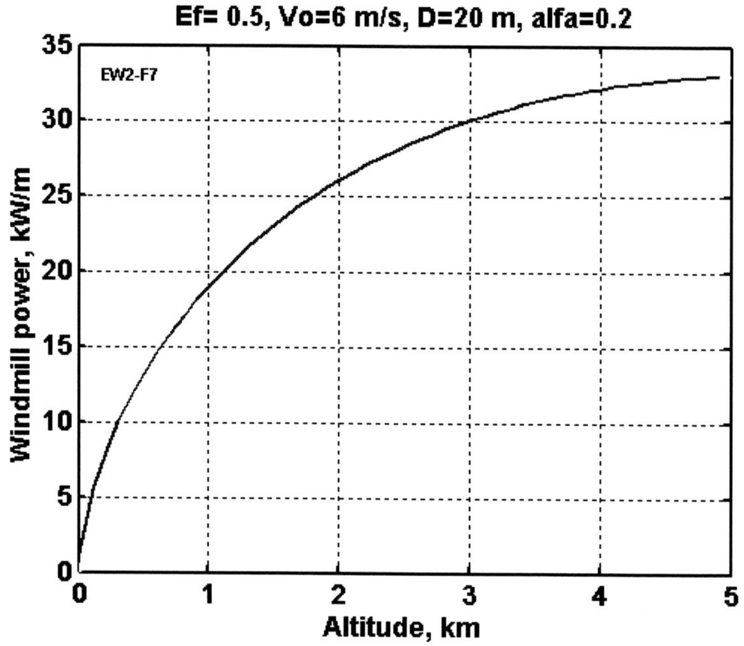

Figure 17. Windmill power [kW/m] via mountain height for air speed V_o = 6 m/s at sea level H_0 = 10 m, product of efficiency coefficients (windmill, electric generator) is η = 0.5, coefficient increasing of air speed with altitude α = 0.2, diameter of wind turbine D = 20 m.

8. Cooling of building in hot weather. The water from AB-Dome top has temperature about 0°C. That may be used for cooling of cities through cold fountains, cooling of buildings, dwelling, food storage in hot countries, even growing non-tropical crops in tropical countries! That can save much energy spent by conditions in summer time in hot weather. This energy may be estimated by expression:

$$O_1 = c_p W T_0 / \eta_c , \tag{9}$$

where Q_1 is possible energy from 1 m of AB-Mountain, W/m; c_p = 4.19 kJ/kg K is water heat capability; T_0 is air temperature at H = 0, C; W is water flow, kg/s/m; $\eta_c \approx 0.3$ is coefficient efficiency of condition.

Example. Let us to estimate the energy from cold water relative to the warm atmosphere. Assume the air temperature T_a = 35°C, the water temperature is 0°C, but while the water moves into delivery tubes, that warms approximately up T_w = 5°C. That means T_0 = 35 – 5 = 30°C. From Figure 13 we have the water flow W = 300 kg/m for T = 35 C. The total energy (for η_c = 1) is (Eq.(9)) Q_1 = 38 MW/m. That is *3* times more than the energy from high altitude water (P = 12 MW/m, Figure 15). The efficiency of condition is $\eta_c \approx 0.3$ and a comfortable room temperature T_r = 25°C. If we substitute these values (difference between air temperature and room temperature, $T_0 = T_r - T_w$ = 25 – 5 = 20 C) in Eq. (8), we may to save electric energy spent by air conditioners up to Q_1 = 84 MW/meter of mountain ridge. That is *7* times more then energy from high altitude water.

The cooling energy is sometimes more than P [Eq. (4)]. But that is difficult to realize in practice (except for cooling needs) because the difference of temperatures between water and air is small (≈ 20°C). The cold freshwater for cooling of building may be delivered by existing freshwater (heat transfer) systems. An on-land version of OTEC technology using solar hot water heaters and the cold water to maximize the temperature differences might tap these energy streams, but why? The hydroelectricity is far cheaper and less capital intensive. One use of such masses of cold water might be to lower the temperature of bodies of water to better hold oxygen for productive aquaculture.

9. The wind's dynamic pressure is computed by equation

$$p_d = \frac{\rho V^2}{2}, \qquad (10)$$

where p_d is wind dynamic pressure, N/m^2; ρ is air density, for altitude $H = 0$ the $\rho = 1.225$ kg/m^3; V is wind speed, m/s. The computation is presented in Figure 18.

The small overpressure of 0.01 atm forced into the AB-Dome or AB-Mountain to inflate it produces force p = 1000 N/m^2. That is greater than the dynamic pressure of very strong wind V = 35 m/s (126 km/hour). If it is necessary we can increase the internal pressure by some times if needed for very exceptional storms.

10. The thickness of the dome envelope, its sheltering shell of film and/or dense textile, is computed by formulas (from equation for tensile strength):

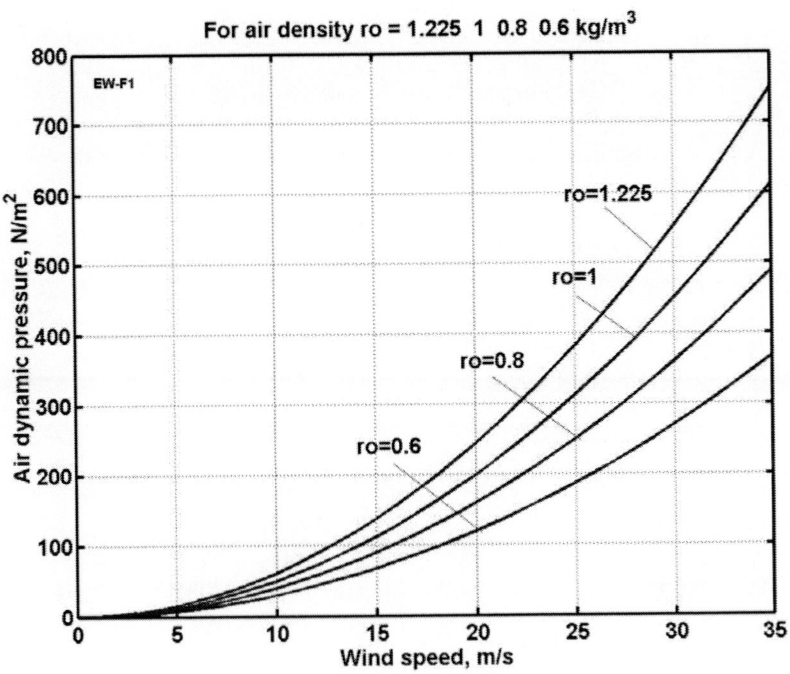

Figure 18. Wind dynamic pressure versus wind speed and air density ρ. The ro = 0.6 is for $H \approx 6$ km.

$$\delta_1 = \frac{Rp\bar{p}}{2\sigma}, \quad \delta_2 = \frac{Rp\bar{p}}{\sigma}, \tag{11}$$

where δ_1 is the film thickness for a spherical dome, m; δ_2 is the film thickness for a cylindrical dome, m; R is radius of dome (or AB-Dome cover between the support cable [1]), m; p is additional pressure into the dome (10÷1000), N/m^2; σ is safety tensile stress of film (up to 2×10^9), N/m^2.

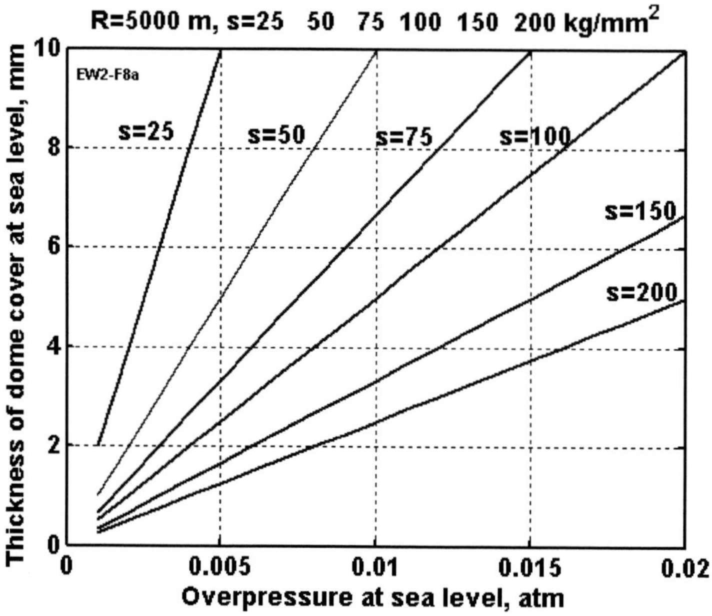

Figure 19. Thickness of AB-Mountain cover (without the support cables) via over pressure at $H = 0$ for different safety stress. Radius of AB-Dome is $R = 5$ km.

11. Cost of freshwater extractor. The cost, C [\$/kL], of produced freshwater may be estimated by the equation:

$$C = \frac{C_i / l + M_e}{M_{wy}}, \tag{12}$$

where C_i is cost of installation; l is live time of installation ($l \approx 10$), years; M_e is annual mountains; M_{wy} is annual amount of received freshwater, kL.

The retail cost of electricity for individual customers is \$0.18 per kWh at New York in 2007 (\$0.24 per kWh in 2008). Cost of other energy from other fuel is in [8] p.368. Average cost of water from river is \$0.49 - 1.09/kL in the USA, the water produced from sea costs about \$0.5 - 2/kL in Israel.

The estimations are in Macro-project section following.

12. Energy required by different methods of desalination. Below in Table 3 is some data about energy expense for different desalinizing means.

Table 3. Estimation of energy expenses for different methods of freshwater extraction

No	Method	Condition	Expense kJ/kL
1	Evaporation	Expense only for evaporation*	2.26×10^6
2	Freezing	Expense only for freezing, c.e. $\eta = 0.3$	1×10^6
3	Reverse osmosis	Expense only for pumping, 40 ÷ 70 atm	$(4\div7)\times10^3$

* This expense may be decreased by 2 -3 times when the water installation is connected with heat or nuclear electric station.

The AB-Mountain produces free freshwater and gives very cheap hydropower.

13. Using the AB-Dome for tourism, communication and wind energy. The relatively low AB-Dome can give good profit from tourism. As it shown in [9] p.93 for 4800 tourists in day and ticket cost $9 the profit would be about $15 million/year. (They would be attracted to the world-famous, towering facility for the talked-about sweeping horizon-to horizon view of a vast terrain exposed many kilometers below the viewpoint.)

The profit may be from tele-communication (TV, cell-telephone, military radars, etc). The best profit would be from high altitude wind electric station [14] (for wind speed $V = 13$ m/s at $H = 4$ km, wind rotor $R = 100$ m the power will be about 20 MW). In reality the wind speed at this altitude is more strong (up 35 m/s), stable, and rotor may to have radius up 150 m. That means the wind power may reach from 10 - 30 times more power. Since the generated power can go up as the cube of the wind speed, the installation would be well amortized.

14. Cost of material. The cost some materials are presented in Table 4 Attn. (2005-2007). Some of the differences in the tensile stress and density are the result of the different sources, models and trademarks.

MACRO-PROJECTS

AB-Mountain Ridge of Height 5 km and Length 100 km

Let us estimate the parameters, cost and profit of a system of AB-Mountains. Ridge of the height $H = 5$ km and length $L = 100$ km having simple primary cover (without control of transparency). Take the over pressure 0.01 atm (1000 N/m^2), safety tensile stress of cover σ =100 kg/mm^2 (10^9 N/m^2), specific weight of cover $\gamma = 1500$ kg/m^3.

Main cost of AB-Dome units. From Figure 18 we find the thickness of cover at ground is $\delta = 5$ mm, at 5 km is 3 mm. If we use the few support cable (decrease R in Eq. (9)[1]), the need cover thickness significantly decreases. We take the average $\delta = 3$ mm. The weight of S = 1 m^2 cover is $w = \sigma\gamma S = 4.5$ kg/m^2. The length of dome semi-cylinder $L_r = \pi R = 1.57\times10^4$ m, the weight of 1 m dome length is $W_1 = wL_r = 7\times10^4$ kg/m.

If cost of 1 kg cover is $1, then the cost of 1 m of the dome cover is $70,000/meter). We take the *total* cost of 1 m AB-Dome length (1 m of ready-made Dome Installation) c_d = $100,000/meter or $100 million/km. In this case the total cost of L = 100 km AB-Dome Installation is $C_d = c_dL = \$10^{10} = \$10\ B$ (where B is billion dollars). If maintenance is

\$1000/m/year, the total maintenance is \$100M/year (where M means million). The cost of cover is the main component in the AB-Mountain. If support cables are used the thickness (and cost) of cover may be decreased by some times (up to 10, [1]). The film may be reinforced by strong cheap artificial transparence fiber and this textile material will have very high tensile stress. The life time of the cover is about 20 years. (\$1 billion a year must be set aside for replacing it).

Profit from AB-Dome

a) *Profit from sale of electricity*. Let us take the average annual temperature outside of Dome $T_0 = 25$ C. From Figure 17 we find the average power of electricity is $P = 7$ MW/m. The Dome length is 100km = 10^5m. The total power is $7\times10^6\times10^5 = 7\times10^{11}$ W = 0.7 TW. The total (include oil, gas, coal, etc,) World Power usage was about 14 TW in 2005 (all World hydro-station power is equal to only 0.35 TW). The full year contains $t = 24\times 365 = 8{,}760$ hours. Assume the average wind blows 4 hours in day, and probability a needed direction is 0.25. Then useful year time is $t_u = 4\times0.25\times365 = 365$ hours. Annual energy of 1 m Dome is $E_1 = t_uP = 2.55\times10^6$ kWh/year/m. (2.55 billion kWh yearly for Mountain ridge 100 km).

Retail cost of electric energy in New York is \$0.18 1/kWh. We take price c_1 = \$0.1 1/kWh.

The profit from electricity of energy is $C_1 = c_1E_1$ = \$0.255 M/year/m. The length of Dome is $L = 10^5$ m. The total annual profit only from electricity is $C_e = C_1L$ = *\$25.5B/year. That is ~ 2.5 times more than the cost of the AB-Mountain installation.* (In reality, of course, increased supply usually decreases prices; but even at \$.01 a kilowatt hour, the equivalent of coal at \$82 a ton or gas at under \$3 per thousand cubic feet, the income of \$25 billion would pay back the costs of construction (of the dome, not the hydro works) in *only five months!* And at half even this price, or a half-cent a kilowatt-hour, the cleanness and controllability of hydroelectricity would seriously start to displace carbon fuels, in the sense that even coal and gas would have difficulty competing as simple thermal sources. So an interesting initiative for peace in the Mediterranean Basin region would be the EU's whole-hearted financing of such an AB-Mountain, the EU using the power and the desert regions to the south of the Mediterranean Sea taking the freshwater.)

The actual profile of the available water may be distributed around the year, depending on when prevailing winds cause a local 'rainy season'

b) *Profit from sale of water*. For temperature $T_0 = 25$°C the water flow from 1 m Dome is $q = 0.17$ m^3/s/m, 1m^3 = 1kL (Figure 15). Full year has $t = 24\times60\times60\times365 = 3.1536\times10^7$ s. Assume the average wind blows 4 hours in day, and probability a needed direction is 0.25. Then useful year time is $t_u = 4\times0.25\times365 = 365$ hours = 1.3×10^6 s. The annual water from 1 m Dome is $M_1 = qt_u = 2.2\times10^5$ kL/year/m. ($2.2\times10^5\times10^5 = 2.2\times10^{10}$ tonnes = 22 cubic kilometers of fresh water!) (For comparison, the Egypt is only allowed to remove from the Nile, about 55.5 cubic kilometers per year. (http://www.eoearth.org/article/Water_profile_of_Egypt). If the Egyptian government were to build the AB-Mountain rather than to buy from a foreign water vendor, for \$10 billion a year and \$1 billion a year maintenance it could produce and sell the water itself to farmers and use the revenue to pay for government employees in the cities. The annual water from 1 m Dome is $M_1 = qt =$

2.2×10^5 kL/year/m. If water cost is c = \$1/kL and length of Dome $L = 10^5$ m, the total water profit from Dome will be C_w = *\$22B/year. That is in 2.2 times more than the cost of the installation.*

The common annual income from electricity and water is \$47.5B/year. That is about 4.7 times more than the cost of the AB-Mountain installation.

Again, at one-twentieth or less these prices, (the likely price stagnation point) the AB-Mountain abundantly repays its construction, and in fact cultivates huge new demand for carbon replacement power and water cheap enough to irrigate with.

c) *The profit from cooling of buildings.* In point 8 (Eq,(8)) we computed the example which shows that cooling water has Q_1 = 38MW/m for air temperature T = 35°C. That may saves Q_2 = 84 MW/m electric energy spent by conditions. The year contains $t = 24\times365$ = 8,760 hours (working time t_u = 365 hours). Annual energy of 1 m Dome is $E_1 = t_uQ_1$ = 1.39×10^7 kWh/year/m. If we sale this energy by price c_1 = \$0.05/kWh the profit is $C_l = c_lE_l$ = \$0.7M/year/m. The length of Dome is $L = 10^5$ m. The total annual income is $C_e = C_lL$ *would,* theoretically, equal about \$70 B/year. *That is a benefit 7 times more than the installation cost.* That income is problematical; therefore we choose not to take it into account at this time. (The offered method of cooling is new and with great benefit but the poor tropical countries that need it most cannot yet pay for it. However, as a side benefit (i.e., cooling poor tropical countries such as India where extremes of 48 C (120°F) are not uncommon for weeks on end in summer, it could save many lives. It could also relieve inestimable human suffering and buy political peace with the poorest of the poor by giving them a real benefit from their country's decision to build an AB-Mountain range. There are no technical macro-problems in its application, the current water heating system may be used for cooling, and building cooling is a very important, even vital, macro-problem for a hot country or warm countries with normally hot summer weather regimes).

d) *Profit from harvest.* The surface into AB-Dome will have excellent control of the selected climate regime and may be used for harvest [1] three times in year (hydroponics). The area covered is $S = 10^9$ m^2. If profit is c = \$1/m^2, the common annual profit is $C_y = cS$ = *\$3B/year.*

e) *Profit from rent.* The area under Dome cover is a beautiful place for any city, because it has storm-free fine warm weather all of the time, may be protected from nuclear warhead assaults, chemical, and biological weapons in wartime [16]. If additional rent for ground into Dome is c = \$1/month/m^2, the annual profit will be $C_r = 12cS = 12\times1\times10^9$ = *\$12B/year.*

f) *There are a lot of other possibilities* to get the profit from offered AB-Mountain. For example, tourism at high altitude, communication, civil or/and military locators, fly gliding, cheap parachute jumping, non-gravity jumping, entertainments, etc. That can give additional profit.

As you see the AB-Mountain may just qualify as the macro-project in our troubled World that best gives incomparable rates of return, earning its cost back, and more, in one year (every year!). Whoever does not agree, can do the numbers with their own assumptions, to check our own.

Naturally, any national or private investor would have many questions: For example, how to build this gigantic AB-Dome. A.A. Bolonkin has many additonal inventions, which solve these macro-problems and is readily available for consultation with interested persons and groups.

We took the height 5 km. However, the artificial AB-Mountain having the height 3.5 – 4 km will have a closed efficiency, but lesser cost by 30 – 40%.

DISCUSSION

As with any innovative macro-project proposal, the reader will naturally have many questions. Author offer brief answers to the four most obvious questions our readers are likely to ponder.

(1) *How can snow and ice be removed from the dome?*

If water appears over film (rain), it flows down through a special tube into ground-based turbines. If snow (ice) appears atop the film, the control system passes the warm air between two cover layers. The warm air melts the snow (ice) and water flows down. The film cover is flexible and has a lift force of about 20 - 200 kg/m^2.

(2) *Storm wind.*

The AB-Dome has special semi-cylindrical form (Figure 8, 9). For internal pressure is 0.01 atm the Dome can resist a storm wind up to 40 m/s (144 km/hour). If wind is more powerful yet, we can increase the internal pressure up to the needed value.

(3) *Cover damage.*

The envelope contains a rip-stop cable mesh so that the film cannot be damaged greatly. Electronic signals alert supervising personnel of any rupture problems. The cover has internal and external cable (rope-ladders and rope cars, rope elevators) and workers can reach the any part of cover inside or out side for repair.

(4) *What is the design life of the film covering?*

Depending on the kind of materials used, it may be as much a decade (or up to 20 years and more). In all or in part, the cover can be replaced periodically.

(5) *How to build the AB-Dome?* The simplest method is to spread the section of cover on the ground's surface and to then turn on the air pump.

The lead author began this research as investigation of new method for receiving the cheap freshwater from atmospheric vapor. In processing research, he discovered that method allows producing huge amount energy, in particular, by transferring the atmospheric energy into electricity with high efficiency. The thin film (relative to volume contained, and absolutely) is very cheap. They are thrown out by the hundreds of tons every day and mess up the human environment, but properly employed they can conserve it as well. The theory of inflatable space towers [1]-[16] allows humanity to build very cheap high-height AB-Domes or towers, which can be used also for tourism, tele-communication, radio-location, producing wind electricity, outer space research [1-16].

5. CONCLUSION

One half of Earth's population is malnourished. *The majority of Earth is not suitable for unshielded human life*. The offered AB-Mountains can change the climate many regions, give them the water and energy. The increasing of agriculture area, crop capacity, carrying capacity by means of converting the deserts, desolate wilderness, taiga, permafrost into gardens are an important escape hatch from some of humanity's most pressing problems. The offered cheapest AB method may dramatically increase the potentially realizable sown area, crop capacity; indeed the range of territory suitable for human living. In theory, conversion of all Earth lands such as Alaska, North Canada, Siberia, or the Sahara or Gobi deserts into prosperous gardens would be the equivalent of colonizing another terraformed planet such as Mars. The suggested method is very inexpensive and may be utilized at the present time. We can start from small regions, such as small towns in bad climate regions and extended the practice over a large region—and what is as very important economically, earning financial profits for investors constantly!

Film domes can foster the fuller economic development of dry, hot, and cold regions such as the Earth's Arctic and even the ice-covered continent of Antarctica and, thus, increase the effective area of territory dominated by humans. The country can create the AB-Mountain barriers which will defend the country from cold North winds. Normal human health can be maintained by ingestion of locally grown fresh vegetables and healthful "outdoor" exercise. The domes can also be used in the Tropics and Temperate Zone. Eventually, they may find application on the Moon or Mars since a vertical variant, inflatable towers to nearby outer space, are soon to become available for launching spacecraft inexpensively into Earth-orbit or interplanetary flights [12].

The related macro-problems are researched in references [1]-[16].

Let us shortly summarize some advantages of this offered AB-Dome method of climate moderation:

1. The artificial AB-Mountains each give a lot of freshwater and hydropower and change a local climate (convert the dry climate to damp climate);
2. They can protect from cool or hot wind a very large region;
3. Covered region does not need large amounts of constant input water for irrigation;
4. Low cost of inflatable film AB-Dome per landscape reclaimed;
5. Control of inside ambient air temperature;
6. Usable in very hot and cold planetary regions;
7. Covered area is not at risk from weather;
8. Possibility of flourishing crops even with a sterile soil (hydroponics);
9. (9) 2 – 3 harvests in year; without farmers' extreme normal risks.
10. Rich harvests, at that.
11. Converting deserts, desolate wilderness, taiga, tundra, permafrost, and ocean surface into anthropogenic gardens;
12. Covering extant towns, cities with AB-domes;
13. Protection of city from external, tactical nuclear warhead assault, chemical and biological weapons [16];

14. Using the high artificial AB-Mountains for tourism, tele-communication, long-distance geographical location;
15. Using the AB-dome cover for illumination, pictures, movies and paid advertising.

We can make fine local weather and get new territory for living upon with an agreeable climate without daily rain, wind and low temperatures, and for agriculture. We can cover by thin film gigantic expanses of bad, unappealing dry and cold planetary regions. The countries having big territory (but bad poor-quality landscapes) may be able to use to increase their population and became powerful states in the centuries to come.

The offered method may be used to conserve a vanishing sea as the Aral Sea or Dead Sea. A closed loop water cycle saves this sea for a future generation, instead of bequeathing a salty dustbowl.

The author developed the same method for the ocean (sea). By controlling the dynamics and climate there, ocean colonies may increase the useful area another 3 times (after the doubling of useful land outlined above) All in all, this method would allow increasing the Earth's population by 5 – 10 times without the starvation.

The offered method can solve the supposed problem of global warming because AB-Dome units will be able to confine until use much carbonic acid (CO_2) gas, which appreciably increases a CO_2-stimulated crop harvest. This carbon dioxide gas will show up in yet more productive crops! The dome lift force reaches up 300 kg/m^2. The telephone, TV, electric, fresjwater and other tele-communications can be suspended to any dome's fabric and/or film cover.

The method can also help to defend valuable and important cities (or an entire strategic geographical region) from rockets, nuclear warheads, and military aviation. Details are briefly offered in a web-posted R&D paper [16].

References

(The reader may find some of these articles at the author's web page: *http://Bolonkin.narod.ru/p65.htm*, *http://arxiv.org*, search "Bolonkin", in the book "*Non-Rocket Space Launch and Flight*", Elsevier, London, 2006, 488 pgs. and in book "*New Concepts, Ideas, Innovations in Aerospace, Technology and Human Science*", NOVA, 2008, 430 pgs.)

[1] Bolonkin, A.A., *Irrigation without water (closed-loop water cycle). http://arxiv.org* search "Bolonkin" (2007).

[2] Bolonkin, A.A. and R.B. Cathcart, Inflatable 'Evergreen' Dome Settlements for Earth's Polar Regions. Clean. *Technologies and Environmental Policy*. DOI 10.1007/s10098.006-0073.4 (2006).

[3] Bolonkin A.A., C*ontrol of Regional and Global Weather*. 2006. *http://arxiv.org* search for "Bolonkin".

[4] Bolonkin A.A., *Cheap Textile Dam Protection of Seaport Cities against Hurricane Storm Surge Waves, Tsunamis, and Other Weather-Related Floods,* 2006. *http://arxiv.org*.

[5] Bolonkin, A.A. and R.B. Cathcart, Antarctica: *A Southern Hemisphere Windpower Station?* Arxiv, 2007.

[6] Cathcart R.B. and Bolonkin, A.A. *Ocean Terracing,* 2006. *http://arxiv.org.*

[7] Bolonkin, A.A. and R.B. Cathcart, *The Java-Sumatra Aerial Mega-Tramway,* 2006. *http://arxiv.org.*

[8] Bolonkin, A.A., "Optimal Inflatable Space Towers with 3-100 km Height", *Journal of the British Interplanetary Society* Vol. 56, pp. 87 - 97, 2003.

[9] Bolonkin A.A., *Non-Rocket Space Launch and Flight,* Elsevier, London, 2006, 488 pages.

[10] Macro-Engineering: *A Challenge for the Future*. Springer, The Netherlands, 2006, 318 pages. Collection of articles related to "Macro-engineering".

[11] Bolonkin A.A., Cathcart R.B., Inflatable 'Evergreen' Polar Zone Dome (EPZD) Settlements, 2006. *http://arxiv.org* search "Bolonkin".

[12] Bolonkin A.A., Inflatable Dome for Moon, Mars, Asteroids and Satellites, Presented as paper AIAA-2007-6262 by AIAA Conference "Space-2007", 18-20 September 2007, Long Beach. CA, USA

[13] Bolonkin, A.A. and R.B. Cathcart, "A Cable Space Transportation System at the Earth's Poles to Support Exploitation of the Moon", *Journal of the British Interplanetary Society* 59: 375-380, 2006.

[14] Bolonkin A.A., (2004b), Utilization of Wind Energy at High Altitude, AIAA-2004-5705, AIAA-2004-5756, International Energy Conversion Engineering Conference at Providence., RI, Aug.16-19. 2004. USA. *http://arxiv.org.*

[15] Bolonkin A.A., (2006a), Optimal Solid Space Tower, AIAA-2006-7717. ATIO Conference, 25-27 Sept. 2006, Wichita, Kansas, USA. http://arxiv.org, search "Bolonkin".

[16] Bolonkin A.A., Cheap Method of City Protection from Rockets and Nuclear Warheads. *http://arxiv.org* search "Bolonkin".

[17] Naschekin, V.V., *Technical thermodynamic and heat transmission.* Public House for High Universities, Moscow. 1969 (in Russian).

[18] Gleick, Peter; et al. (1996). in Stephen H. Schneider: *Encyclopedia of Climate and Weather*. Oxford University Press. *Encyclopedia of Desalination and water and Water Resources*.

[19] Encyclopedia of Earth - Surface water management.

[20] *http://Wikipedia.org. Some background material in this article is gathered from Wikipedia under the Creative Commons license.*

Chapter 2

THE GOLDEN GATE TEXTILE BARRIER: PRESERVING CALIFORNIA'S BAY OF SAN FRANCISCO FROM A RISING NORTH PACIFIC OCEAN*

ABSTRACT

Future climate regime change in the State of California may require construction of a seawater barrier separating the Pacific Ocean from San Francisco Bay and the Sacramento River-San Joaquin River Delta chiefly because Southern California is remarkably dependent on freshwater exported from Central California's Sacrament River-San Joaquin River Delta. We offer a new kind of seawater barrier, a macro-project built of impermeable tensioned textile materials stretched across the Golden Gate beneath or quite near to the famous bridge. The authors anticipate it might eventually substitute for a recently proposed "San Francisco In-Stream Tidal Power Plant" harnessing a 1.7 m tide at the Bay's entrance, if future climate conditions Statewide are at all conducive. First-glance physics underpin our proposed macro-project.

1. INTRODUCTION

While the public in Europe and the USA apparently have adopted a common outlook of technical resignation towards the world's evident post-Ice Age rising ocean, macro-engineers in Dubai assiduously sculpt the seashore and create new inhabitable islands. Indeed, by contemplating the future emplacement of a Strait of Hormuz Dam, Macro-engineering's foremost proponents have formulated a workable plan to isolate the Persian Gulf from the ocean and, thereby, making the Persian Gulf's water level controllable. It is true that macro-projects protecting major European cities from storm surges—such as London in the UK, St. Petersburg in the Russian Federation and Venice, Italy have been constructed or continue to be considered. The in-place infrastructures of The Netherlands are well known. In Asia, a movable textile seawater barrier facility is proposed to shield Palk Bay, located between India and Sri Lanka, from infrequent tsunami and yearly storm surges (Cathcart, 2006). To "correct oceanographic problems impairing the economic usefulness of coastal land", Richard B. Cathcart and Alexander A. Bolonkin, in Chapter 3 below, offer a deliberate "terracing" of the Mediterranean Sea within its vast Basin by its enclosure with a tensioned textile barrier in the Strait of Gibraltar that controllably excludes Mediterranean Sea evaporation compensation Atlantic Ocean seawater inflow. Such comprehensive and geographically large-scale corrective macroproject concepts stem, in part, from stimulation initiated by a summarizing technical book about Macro-engineering's burgeoning professionalism edited by Viorel Badescu, R.B. Cathcart and R.D. Schuiling, MACRO-ENGINEERING: A CHALLENGE FOR THE FUTURE (Springer, The Netherlands, 2007, 316 pages).

2. Good Cause for Concern by Coastal Americans

Before Macro-engineering and macro-management can be considered in the context of California's coast, a fundamental question must be asked: Is sea level rising along the USA's coast? After 1970, the world's popular news media discovered "global warming" and, since then, many prominent Californian legislators and politicians, such as Governor Arnold Schwarzenegger during 2007, as well as other citizens are nowadays convinced that anthropogenic carbon dioxide and other greenhouse gas global emissions are directly responsible for raising the Pacific Ocean's level of impingement on the State's coastline. Still, the tandem link between enhanced global greenhouse gas emissions and global sea level rise is still an unresolved matter of geoscientific dispute. As Earth's most recent Ice Age wanes, the world's atmosphere is likely to warm, some glaciers will become smaller and the ocean impacting the USA's coast will probably rise to an elevation noticeably above the present-day's level.

3. East Coast: Metropolitan New York City Sea Level Rise Impacts

Storm-induced seawater surges, caused by hurricanes and nor'easters, have in the past impacted the metropolitan New York City region. Future local sea level rise is expected to boost the destructive power of such storms to disrupt the function of vital urban infrastructure and to injure and kill people. Consequently, the possibility of protecting the Metropolitan Region with storm-surge barriers is being studied. Closable steel and concrete storm-surge barriers erected in three straits (between New York Bay and Lower New York Bay, Arthur Kill and near the Throngs Neck Bridge) could effectively, though only temporarily, seal off the Metropolitan Region from incoming Atlantic Ocean-generated storm-surges. See Chapter 5 in Part A of this book.

---

* Presented in http://arxiv.org in 2007.

4. Gulf Coast: Post-2005 New Orleans Region recovery

By the end of 2007, the US Army Corps of Engineerd presented to Congress a long-term, comprehensive plan for hydraulic macro-projects that will offer Gulf Coast residents and industry a predictably successful future defense against hurricane storm-surge damage. A complex integrated system of walls, barriers, offshore breakwaters, barrier island beach dune restoration, repaired or new levees, and massive mobile mechanical barrages that would close across inlets to exclude surging hurricane-stirred seawater and freshwater were considered by the US Army's Corps of Engineers. Americans have modified the Gulf Region's coast, either directly, by construction or dredging, or indirectly, as a consequence of profound inland landscape changes that then influence future sediment supply, freshwater runoff, or the Gulf Region's climate.

The commonest settlement/use practices that significantly alter the USA's coastline are the construction of coast-protection infrastructures such as jetties, groins, seawalls, bulkheads, revetments and the development of private and public property on and close to the shoreline. These oceanographic infrastructures constitute "Hard Structure Armoring". The alternative is "Beach Nourishment". Beach nourishment is the only coastal macro-management technique that adds sand to the littoral system nearest the seashore. Generally speaking, "Hard Structure Armoring" is monetarily the costliest—about 10% of initial infrastructure installation cost annually—to maintain in a robust state of physical readiness.

5. California, West Coast: San Francisco Bay/Delta Region

The U.S. Geological Survey published a useful geographical and historical survey of California's coastal land gains and losses during 2005. The present-day post-Ice Age sea level rise at the entrance to the 890 km^2 Bay of San Francisco is ~2.29 cm/decade. Future sea level rise, and the subsequent elevated storm-surges thereon, may severely affect San Francisco Bay sited shoreline cities, the levee-enclosed low-lying land above the intertidal flats and, very importantly, the adjacent Sacramento River-San Joaquin River Delta. Projected rising local sea level shore flooding can be viewed at Alex Tingle's animated website, GOTO: http://flood.firetree.net. On 16 April 2008, in Oakland, California, speakers examined various aspects of "Preparing for Rising Sea Levels in the Bay Area" based on maps released by the Bay Conservation and Development Commission (founded in 1965) during 2007. Those astounding maps illustrated the impact on landscapes around the San Francisco Bay if a 1 meter rise in local sea level ever happened during the 21st Century. All participants concluded exactly where land will recede because of seawater inundation without some aggressive, and quite expensive, near-term future macro-engineering interventions

Massive seawater intrusions into the largest bay on California's coast, its wetlands, associated surface freshwater storage systems and groundwater aquifers are undesirable prospective degradations of the San Francisco Bay/Delta. For example, it is imaginable that the Harvey O. Banks Pumping station feeding the California Aqueduct and the Tracy Pumping Plant feeding the Delta-Mendota Canal would have to cease operation, at least temporarily. Contamination of the Delta's freshwater with a permanent influx of saltwater would endanger a massive public freshwater supply that is pumped at considerable expense in dollars and energy to Southern California!

Projected sea level rises of 20-80 cm during the twenty-first century can only be expected to compound the vulnerability of subsided Delta Islands to levee failure and increase upstream backwater flooding. The Delta receives runoff from about 40% of California's land area (163,000 km^2) and about 50% of the State's total streamflow. A great earthquake at the infamous nearby San Andreas Fault—more powerful than the 7.1 Richter scale "Loma Prieta" earthquake of 17 October 1989—could cause many kilometers of Delta levees to collapse. Widespread Delta levee failure can be expected to divert river flow into many Delta polders—perhaps as much at 2.5 to 3.0 billion cubic meters—and, consequently, generate a strong flow of seawater from San Francisco Bay eastward into the unprotected Delta, suddenly contaminating the freshwater normally pumped to Southern California's burgeoning populace! The volume of San Francisco Bay is ~6.165 billion cubic meters. It has a mean depth of ~6 m.

In addition, a nearly inevitable increase of human population on San Francisco Bay's periphery is expected to require desalination plants providing a reliable future supplemental freshwater supply. Social fabrics and vital infrastructure will be stressed and strained negatively. Furthermore, future industrial and seaport explosive accidents may cause the generation of an infrastructure-devastating tsunami that might reach inland seaports such as Sacramento and Stockton, especially at high tide or during periods of extreme storm-surge. Though supercomputer models predict the 21st Century's climate for California will be dryer than today, it is worth noting that, during January 1862, the outflow of fresh water into the bays of San Francisco and into the Pacific Ocean through the Golden Gate was both large and persistent. Sea level at the Golden Gate was 17 cm above normal and for nearly two weeks fresh water flowed continually seaward through the Golden Gate as a strong plume of freshwater without tidal fluctuation. Fresh water covered the surface of the sub-embayments for two to three months (Engstrom, 1996). Peterson et al (1985) estimated that a freshwater inflow of ~120,000 m^3/second would have been necessary to overcome the tides [tidal range: maximum 2.65 m, ~2.0 billion cubic meters moving at ~2.5 m/second every 6 hours] at the Golden Gate during January 1862! In other words, an identical flood flow rate today could fill San Francisco Bay entirely with freshwater in only 14.27 hours!

What if California's climate becomes wetter than today? In that eventuality, more freshwater may be exported to Southern California. But if, as predicted, California becomes drier, then San Francisco Bay may need to be dammed at the Golden Gate to forestall saltwater intrusion into the Delta. The entrance to San Francisco Bay, the deepest in the bay (107 m depth), is an elongate seafloor depression with steep sidewalls of 10^0 to 17^0. The ocean shoals immediately west of the Golden Gate to a depth of ~38 m. Figure 1, below, is a computer simulation of the seafloor in the vicinity of the Golden Gate Bridge. Farther out to sea, a huge arcuate sandbar is present. It is a relict from the Ice Age. Damming of the Golden Gate may have a beneficial effect on the bar, perhaps stopping its elevational reduction and shoreward migration. Immediately after the 18 April 1906 earthquake, a tidal anomaly at the San Francisco Presidio's tide-gauge station was registered that indicated a 10 cm lowering of the sea level for a quarter of an hour followed by two to three tidal oscillations with maximum amplitude of only 5 cm.

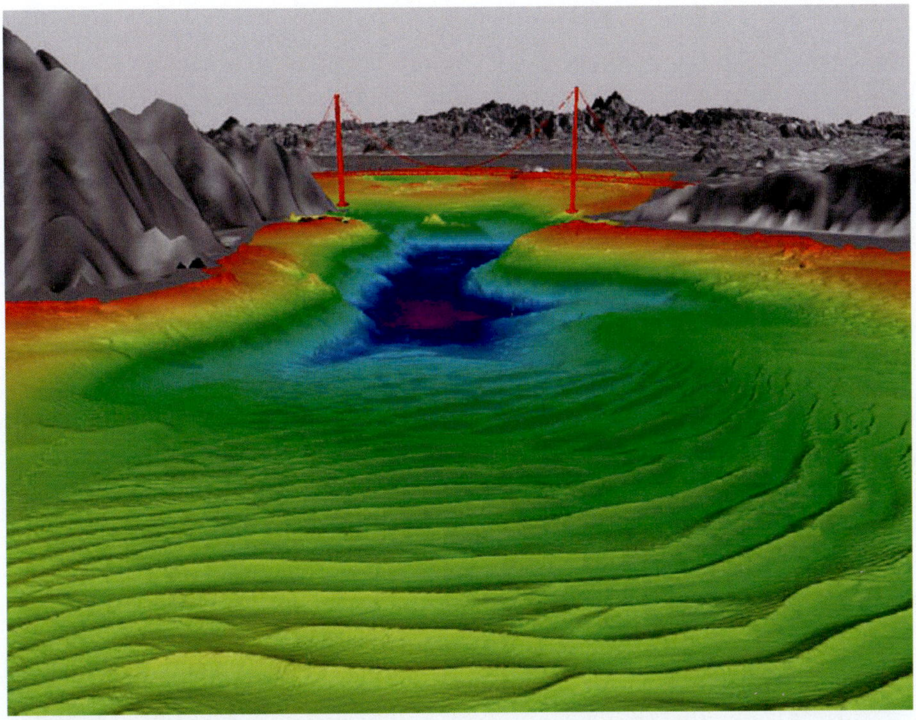

Figure 1. Seabed west of Golden Gate Bridge, viewed from Pacific Ocean

6. The Region's Planned, but Unbuilt, Hard Structure Armoring

Rivers first passed seaward through the Golden Gate approximately 500,000 years ago and post-Ice Age sea level first rose high enough to flow into the Bay of San Francisco about 15,000 years ago. The northern reach is a partially mixed sub-embayment dominated by seasonally varying freshwater inflows while the south reach is a tidally oscillating estuary. The Bay of San Francisco is an ocean-river mixing zone with a seaward flow equal to the sum of the river inflows less precipitation; at the present time only the freshwater inflow can be subjected to macro-management. During major droughts, freshwater inflows have been, effectively, nil; droughts have produced saltwater intrusions of the Sacramento River-San Joaquin River Delta. Anti-saltwater intrusion barriers were first proposed by nascent macro-engineers during the 1920s as a means of halting the contamination of Delta farmland irrigation freshwater as well as that used by nearby industries.

Since about 1950—the year when the term "coastal engineering" was first used in print in the Proceedings of the First Conference on Coastal Engineering—most of San Francisco Bay has shifted from being depositional to erosional as sediment supply diminished (when dams were built inland, especially in the Sierra Nevada mountain range) and existing currents and waves continued to remove sediment from the Region. Essentially all of San Francisco Bay has an anthropogenic character. Any future rise of the Pacific Ocean will, therefore, impact all existing and planned shoreline infrastructures. In 1990, landscape architect Steven Garey Abrahams proposed "Landscape Mobilis" which would consist of a fleet of waterborne landscapes—gardens, forests, meadows, crops, vineyards on barges towed around San Francisco Bay, visiting "receptacle parks" at strategic places along the shore.

When saltwater intrusions of the Delta first became a remarkable macro-problem for commercial Delta farming and industry during the early 1900s, macro-engineers of the time proposed physical barriers to block the inland migration of salty water. Many salt barrier macro-project plans were formulated, most of which involved the construction of low-level dams to separate freshwater upstream from the tidal seawater on the downstream side. Following the very low Delta outflow of 1923-1924, a 1929 macro-project plan for a barrier built at the Golden Gate was the greatest in geographical scope ever promoted by reclamation and hydraulic experts. Discussed by the respected macro-engineer Walker R. Young, had it been constructed, the dam would have become the longest "Hard Structure Armoring" of California's coast ever conceived and, until now, discussed! Figure 2, below, illustrates Young's concept.

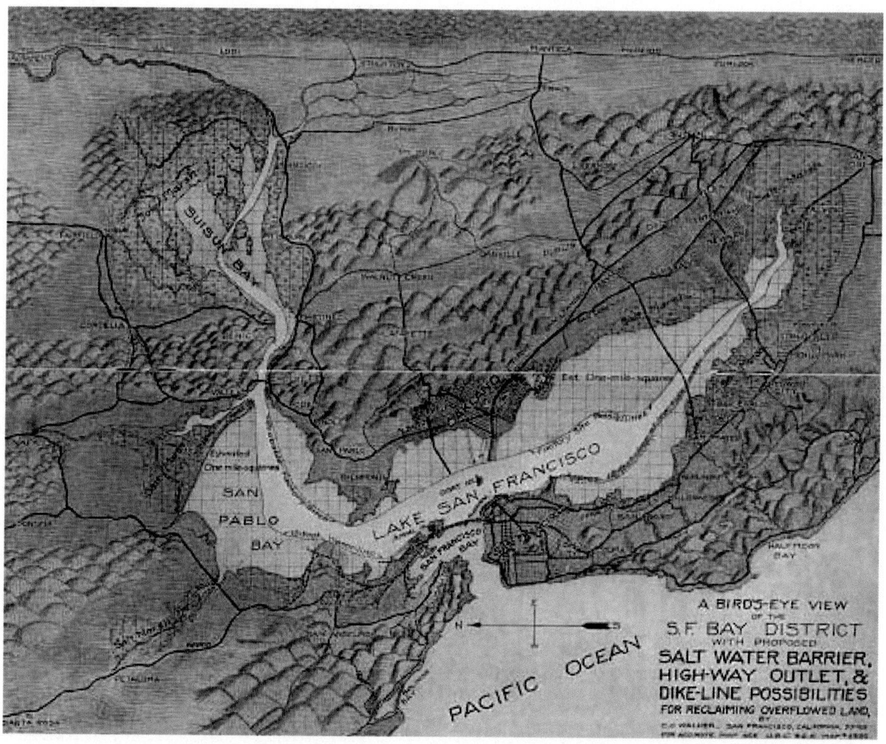

Figure 2. Subsequent to Young's imaginative speculated concept, from 1933 until his death, John Reber (1899-1960) campaigned for his plan to convert ~85% of the Bay of San Francisco into a freshwater lake by closing off the bay's northernmost and southernmost reaches and adding housing and industrial sites to the artificial shore. Similar macro-projects, with different geographic and economic goals of course, were offered by Norman Sper, who suggested that the Hudson River between New York Harbor and the Harlem River be filled with rock and soil, thereby making Manhattan Island no longer an island. Imagine "Tetrahedral City" invented by R. Buckminster Fuller (1895-1983) floating peacefully on Walker R. Young's 1929 vision of Lake San Francisco! Fuller's floating city, sometimes named "Triton City" was fully described by the Triton Fondation in a well-documented 1968 report, "TRITON CITY: A Prototype Floating Community", done for the US Department of Housing and Urban Development.

It would have been unnecessary to fabricate the Golden Gate Bridge, completed in 1938, if W.R. Young's 1927-29 speculative enclosure public work and super-highway had ever been constructed! [W.R. Young's engineering concept for a possible non-bridge S.F. Bay crossing was examined by State of California Govenor Clement Calhoun Young (1869-1947) during his 1927 to 1931 Adminstration.]

Walker Rollo Young (1885-1982), a 1908 University of Idaho graduate of engineering, became Chief Engineer at the USA's Bureau of Reclamation in 1945 after serving, from 1930 as Construction Engineer at Hoover Dam. Due to its site, W.R. Young's speculated "Golden Gate Dam and Highway" would have made the small embayment west of the discussed saltwater barrier a collection place for material being moved by wave and current action in the littoral cells to the north and south of the entrance to an areally smaller, enclosed Bay of San Francisco. Maintenance dredging would have become necessary. Basically, the so-called "Golden Gate Dam and Highway" would have shortened the fetch of winds driving waves into San Francisco Bay—such winds are strongest during Summer and during Winter storms—and since San Francisco Bay was to be a freshwater lake—one might say, a "real-estate lake"—the ecology of the Region would duplicate that endured by the same Region during January 1862!

7. Opportunity for Future Golden Gate Textile Barrier (GGTB)

The GGTB is worthy of investigation at this time because the City and County of San Francisco is considering the installation of a tidal in-stream hydroelectric permeable "curtain" spanning the Golden Gate near the present-day bridge connecting San Francisco and Marin Counties of California. Gulf Stream Energy, Inc. and Golden Gate Energy Company's Joint Application for Preliminary Permit for the San Francisco Bay Tidal Energy Project under P-12585-000 was filed on 26 April 2005 with the Federal Energy Regulatory Commission in Washington, D.C.

On 10 June 2006 a thorough technical examination of such a system for submarine power generation—that is, its system design, potential performance, construction and operational costs, as well as its economic and environmental benefits by the Electric Power Research Institute Inc. (EPRI) was publicly displayed. EPRI's experts concluded that ~35 MW could be extracted from the tidal stream "...without any negative impact on the environment" at a facility running North-South along the 122.47836^0 West meridian from 37.8111^0 North (on the San Francisco side of the seaway) to 37.8265^0 North (on the Marin County side of the strait), a distance of ~1,380 m where the mean depth is ~54 m. Since the City and County of San Francisco has an estimated average electrical power consumption of ~570 MW it is, therefore, possible that approximately 6% of the City and County's electrical supply could be developed by an underwater tidal energy "permeable curtain". Obviously, this nearly $100 million facility will not dam the Golden Gate nor will it impede the navigation of vessels entering or leaving the picturesque Bay of San Francisco, California via the strait. What it seems to indicate is (1) some influential people and organizations in San Francisco will tolerate an invisible hydropower plant at the Golden Gate and (2) future macro-engineering improvements there might also be tolerated if the Pacific Ocean should rise and/or the runoff passing through the Sacramento River-San Joaquin River Delta is greatly reduced by State-wide 21st Century climate change. The great seafloor pit below the Golden Gate Bridge may prove to be a boon since it can hold a great deal of sediment that may shift westwards from the central part of the Bay of San Francisco. And, if a Golden Gate Textile Barrier is emplaced, sand naturally migrating in the shallow Pacific Ocean's nearest littoral cell may improve the tsunami run-up on the facing cliffed-coastline that is present north and south of the entrance to San Francisco Bay. The GGTB would likely be equipped with penetrating ship-locks, perhaps emulating the design offered by Martin Cullen in his U.S. Patent Application 20050163570 issued 28 July 2005.

8. Seawater/Freshwater Textile Tension Structure

The typical textile barrier—more accurately, a membrane—is shown in FIGURE 3, below. Figure 3 includes the textile membrane seawater/freshwater barrier, floats, underwater pump/electricity generation, and the support cable. The hydropower dam can safely operate even when the water level on one side is as much at 10 m.

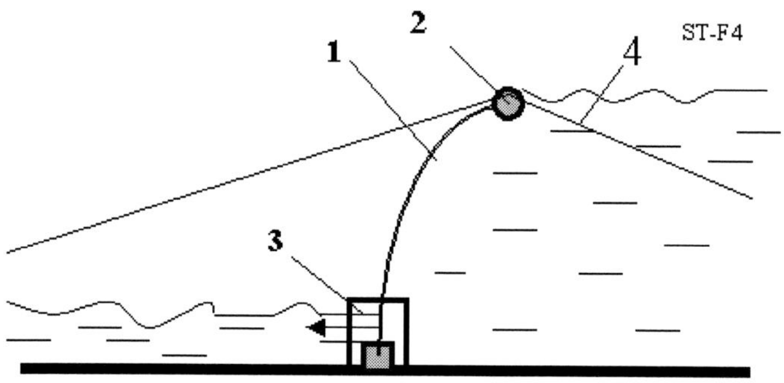

Figure 3. Textile Barrier. Notations: 1 - textile membrane, 2 - floats, 3 - pump/electric station, 4 - support cable.

8.1. Theory and Computation

1. Relative concentration of salt in water. When a river's freshwater inflows to a closed estuary that passes the freshwater only in one direction (namely, to the receiving ocean), the salt concentration of the seawater decreases. New relative concentration may be computed by equation

$$\bar{c} = e^{-Qt}, \tag{1}$$

where Q is a relative inflow of a river freshwater. $Q = q/W$ where q is inflow of freshwater, m^3; W is volume of like (gulf), m^3; t is time, sec. Computation for this equation is illustrated below

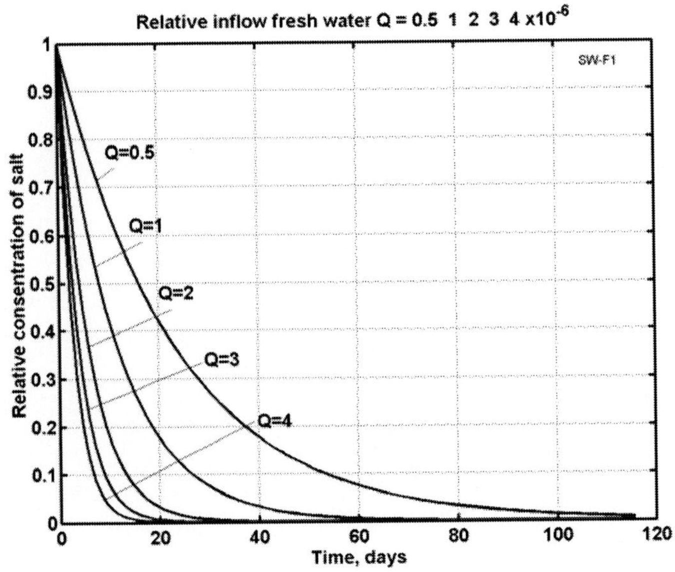

Figure 4. Relative concentration of salt via time (days) for different relative inflow of a river freshwater $Q = q/W$ where q is inflow of fresh water, m^3; W is volume of estuary, m^3.

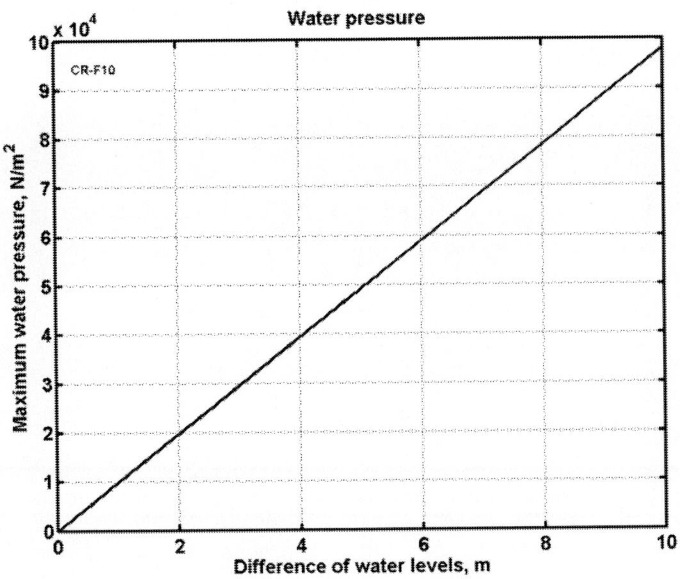

Figure 5. Water pressure via difference of water levels

2. Force P [N/m^2] for 1 m^2 of dam is

$$P = g\gamma h\,, \tag{2}$$

where g = 9.81 m/s2 is the Earth's gravity; γ is water density, γ=1000 kg/m3; h is difference between top and lower levels of water surfaces, m (see computation in Figure 5).

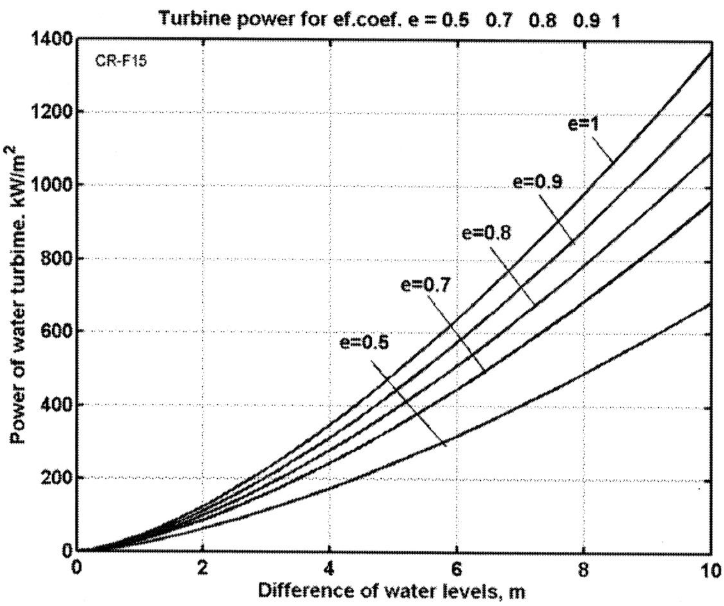

Figure 6. Specific power of a water turbine via difference of water levels and turbine efficiency coefficient

3. Water power N [W] is

$$N = \eta g m h, \quad m = \gamma v S, \quad v = \sqrt{2gh}, \quad N = \eta g \gamma h S \sqrt{2gh}, \quad N/S \approx 43.453 \eta h^{1.5}, \tag{3}$$

where m is mass flow across 1 m width kg/m; v is water speed, m/s; S is turbine area, m^2; η is coefficient efficiency of the water turbine, N/S is specific power of water turbine, kW/m^2.

Computation is presented in Figure 6.

4. Film thickness is

$$\delta = \frac{g\gamma h^2}{2\sigma}, \tag{4}$$

where σ is safety film tensile stress, N/m^2. Results of computation are in Figure 7. The fibrous material (Fiber B, PRD-49, see book Attachment) has maximum $\sigma = 312$ kg/mm^2 and specific gravity $\gamma = 1.5$ g/cm^3.

5. The film weight of 1m width is

$$W_f = 1.2\delta\gamma H, \tag{5}$$

Computation is illustrated by Figure 8. If our proposed textile barrier at the mouth of the Bay of San Francisco has a total length L m, we must multiple this result by L.

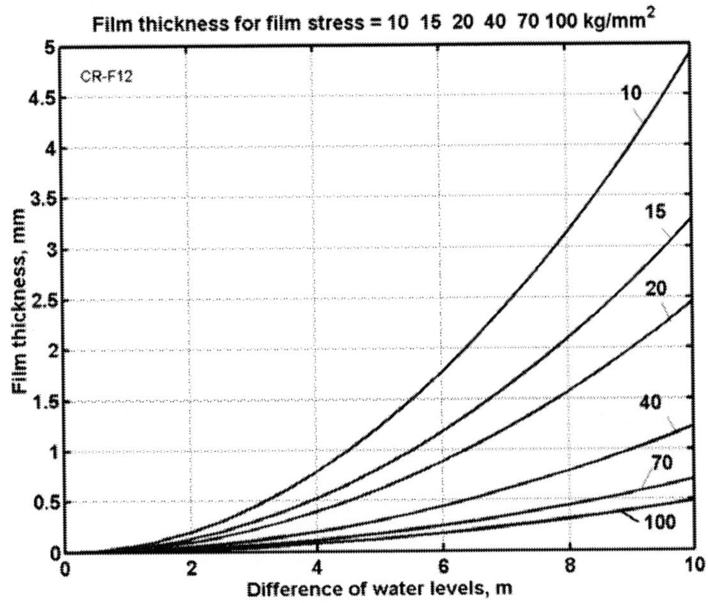

Figure 7. Film (textile) thickness via difference of water levels safety film tensile stress.

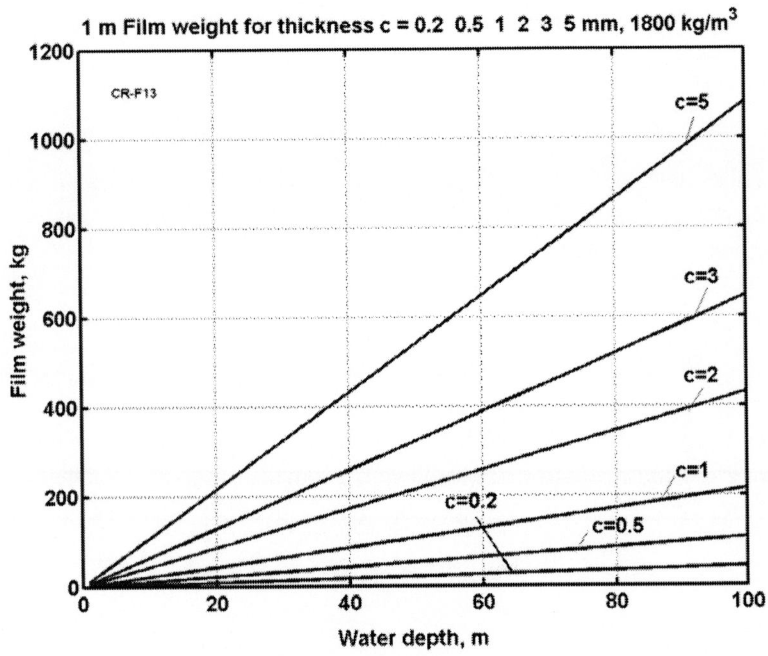

Figure 8. One meter of film weight via the dam depth and film thickness c, density 1800 kg/m^3.

6. *The diameter d* of the support cable is

$$T = \frac{Pl_2}{2}, \quad S = \frac{T}{\sigma}, \quad d = \sqrt{\frac{4S}{\pi}}, \tag{6}$$

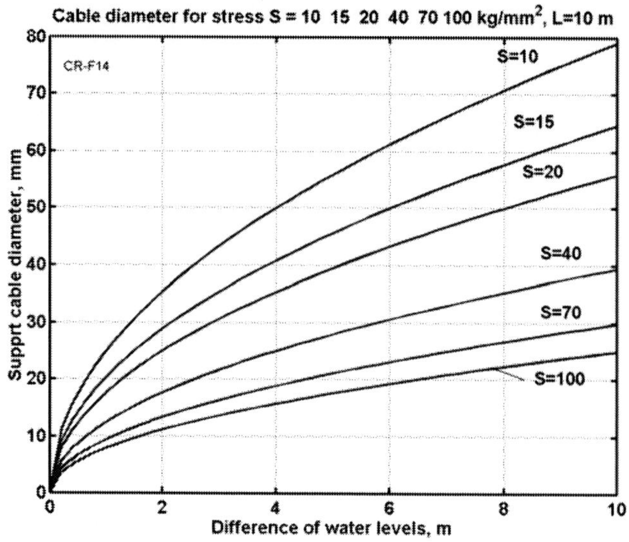

Figure 9. Diameter of the support cable via water level differences and the safety tensile stress for every 10 m textile dam.

where T is cable force, N; l_2 is distance between cable, m; S is cross-section area, m^2.

Computation is presented in Figure 8. The total weight of support cable is

$$W_c \approx 2\gamma_c HSL / l_2 , \quad W_a = \gamma_c SL, \tag{7}$$

where γ_c is cable density, kg/m^3; L is length of dam, m; W_a is additional (connection of seashore) cable, m. The cheapest currently marketed textile-suitable fiber has $\sigma = 620$ kg/mm^2 and specific gravity $\gamma = 1.8$ g/cm^3.

9. CONCLUSIONS

A drastic change in the State of California's climate—that is, a drying future period or a future wetter period compared to today's alleged "normal"—may necessitate the construction of a Golden Gate Textile Barrier. We have shown how this may be accomplished, but only in a preliminary Macro-engineering way.

Like San Francisco. 59 kb.

Golden Gate Bridge. 101kb.

REFERENCES

[1] Cathcart, R.B., "Sethusamudram ship channel macroproject: Anti-tsunami and storm surge textile arrestors protecting Palk Bay (India and Sri Lanka), *Current Science*, Vol. 91, No. 10, pages 1474-1476 (10 December 2006).

[2] Bolonkin A.A., *New Concepts, Ideas and Innovations*, Ch.5B, NOVA, 2008, 320p.

Chapter 3

LAKE TITICACA - PHYSICS OF AN INHERITED HYDROPOWER MACRO-PROJECT PROPOSAL*

ABSTRACT

Shared almost evenly by Peru and Bolivia, Lake Titicaca is situated on an Altiplano endorheic region of the northern Andes Mountains. Rio Desaguadero is the lake's only outlet. From 1908, several macro-engineers speculated on the creation of a second, completely artificial, outlet for Lake Titicaca's freshwater. Here we reconsider several 20th Century macro-project proposals, with the goal of examining and enhancing this technically interesting South American 21st Century Macro-engineering inheritance.

1. INTRODUCTION

Lake Titicaca is a lake located on the border separating the nations of Bolivia and Peru. It rests at ~3,810 m above mean global sea level, making it the highest commercially navigable lake in the world. By volume of water it is also South America's largest lake.

The lake is located at the northern end of the endorheic Altiplano basin high in the Andes Mountains on the border of Peru and Bolivia. The western part of the lake lies within the Puno Region of Peru, and the eastern side is located in Bolivia's La Paz Department.

The lake is composed of two nearly separate sub-basins that are connected by the Strait of Tiquina which is 800 m across at its narrowest point. The larger sub-basin, Lago Grande (also called Lago Chucuito) has a mean depth of 135 m and a maximum depth of 284 m. The smaller sub-basin, Lago Huiñaimarca (also called Lago Pequeño) has a mean depth of 9 m and a maximum depth of 40 m. The assumed average depth of the lake is 107 m.

Lake Titicaca is fed by rainfall and meltwater from nearby glaciers situated on the rugged mountains that directly abut the Altiplano basin. Five major river systems feed Lake Titicaca—in order of their relative flow volumes these are: Ramis, Coata, Ilave, Huancané, and Suchez. More than 20 other smaller rivers empty into Lake Titicaca, and the lake has 41 islands, some of which are densely populated by people.

---

* Presented in http://arxiv.org in 2007.

One of the islands seen from Lake Titicaca: in the distance, Amantani as seen from Taquile.

Having only a single season of free circulation, the lake is monomictic and water passes through Lago Huiñaimarca and flows out the single outlet at the Rio Desaguadero, which then flows south through Bolivia to isolated Lake Poopó. This only accounts for about 10% of the lake's water balance. Evapotranspiration, caused by strong winds and intense sunlight at altitude, balances the remaining 90% of the water input. It is very nearly a closed lake.

Lake Titicaca is, probably, a Quaternary Period remnant of the Mantaro mega-lake that filled the Basin to a maximum elevation of ~4,000 m. Nowadays, Lake Titicaca has a semi-regulated surface altitude of ~3810 m. Its freshwater volume is ~930 km^3 (UNESCO, 2003). During an extremely dry period between *circa* 4000 BC to *circa* 3000 BC, Lake Titicaca was ~75 m below its present-day altitude; by *circa* 2500 BC it began to fill again and rise to its current free water surface elevation. The presently existing higher elevation of the lake's free surface has submerged ancient artifacts and settlements. Indeed, the landscape surrounding the present-day Lake Titicaca indicates an ancient anthropogenic component—terraces, raised fields, sunken gardens and irrigated pastures.

Lake Titicaca constitutes a freshwater resource for approximately 2.5-3.0 million persons residing in the countries of Peru and Bolivia. However, about 95% of the freshwater removed from Lake Titicaca leaves by way of evaporation, while the remaining 5% is comprised of a discharge through the Rio Desaguadero amounting to ~20-35 m^3/second. During 1996, the two riparian states organized the Binational Autonomous Authority for Lake Titicaca (ALT). Since 2001, a dam near the International Bridge at the headwater of the Rio Desaguadero has been equipped with floodgates to provide a stable anthropogenic hydraulic capacity for Lake Titicaca. According to the ALT Master Plan, the water level of Lake Titicaca ideally ought to be maintained at range of between 3808 and 3811 m above mean global sea level. An increase in the human population around Lake Titicaca is forecast and freshwater usage will inevitably increase. A future resource for the needed freshwater may become possible with the reduction of natural evaporation from the lake's ~8,400 km^2 surface. In similar arid climates elsewhere as, for example, at Lake Eyre in Australia, a 50% reduction of free water surface evaporation could be achieved through deliberate spreading of an ultra-thin layer of organic molecules (in powder form) such as proposed by Robert Neville O'Brien in USA Patent 6303133, issued 16 October 2001. Under the sunny conditions prevailing at Lake Titicaca, after application biodegradation will occur in about 2.5 days; it will be technically necessary, then, for ~100 kg of O'Brien's patented powder to be broadcast three times per week, or 12 times per month—144 times per year.

Today, Peru's ~28,000,000 citizens obtain ~80% of their power from hydroelectric generating facilities. The nation's total operational base-load is nearly 6000 MW. (The country's largest hydropower installation is the Mantaro Complex in southern Peru. About 20% of Peru's hydroelectricity is generated there, nearly 1008 MW. The Mantaro Complex—built from 1973 to 1985—utilizes a ~1,000 m drop which occurs at the Mantaro River's first great bend as it flows generally eastward into the Basin of the famous Amazon River.) Peru has untapped hydropower resources on the eastern slopes of the Andes Mountains while thermal power stations are mostly concentrated along the densely populated coast affected directly by the Pacific Ocean.

2. Hydropower from Lake Titicaca for the Coast's Population

While the National Map of Peru was commenced on 10 May 1921 by plane-table and alidade methods, it remains incomplete at the start of the 21st Century (Mugnier, 2006). Accurate topographic mapping is, of course, essential to the planning of any macro-project, but especially a macro-project that relies on freshwater's long distance fall from the Andes Mountains to the Pacific Ocean!

Charles Reginald Enock (1868-1970), a British explorer of Peru, was first to explore and cursorily map the resources of Lake Titicaca and its surrounding watershed. He suggested a 120 km-long tunnel could convey some of the lake's liquid contents to the populated coastal zone adjacent to the Pacific Ocean and generate hydropower simultaneously. A vertical drop to sea level of ~20 m^3/second from a starting elevation of ~3810 m could theoretically produce 640 MW, close to 10% of Peru's existing installed electricity generation infrastructure! Although we cannot be certain, we suspect C.R. Enock may have been inspired by the North American engineer Alexis Von Schmidt (1821-1906) who, from 1865 onwards, proposed and promoted a freshwater aqueduct to the City and County of San Francisco drawing from California's Lake Tahoe. Whatever is the historical truth, the visionary macro-engineering idea was taken up again during the 1950s by the French engineer Marcel Mary. Translated into English and generalized, Mary offers the opinion that a diversion of Lake Titicaca to the Pacific Ocean by piercing at depth would provide a large head—perhaps as much as 3500 m—and would supply irrigation water to Peru's arid coast.

Trans-Andean railways, which depend on vast lengths of hard-to-excavate tunnel, are in a chaotic state at the present time. It is alleged by WIKIPEDIA that "In 2006, Ferrocarril Central Andino, work started to regauge the line from 914 mm to 1435 mm. There is also a proposal for a 21 km tunnel under the Andes" (WIKIPEDIA, accessed 29 January 2007).

Such a Macro-engineering proposal is made credible by the 53 km-long English Channel Tunnel and the planning work being done by Alp-Transit Gotthard for a 57 km-long tunnel between Erstfeld and Bodio, Switzerland that, at its deepest, will be 2 km underground! The key technology that makes it possible to dig deep and long tunnels efficiently is the hard-rock Tunnel-boring Machine (TBM), a machine that digs a tunnel by drilling out the heading to full size in one continuous industrial operation. During early 2007, a Robbins TBM began to bore Peru's 20 km-long Olmos Transandino Project. When finished, the Olmos Transandino Project will siphon freshwater from sources higher in the Andes Mountains to a reservoir created by Limon Dam. Approximately, 2.0 billion cubic meters of freshwater will be shifted annually from the Rio Huancabamba, a tributary of the Amazon River, to the Olmos Valley in the Pacific Ocean watershed. The water will be used to irrigate 150,000 ha of farmland and will also generate ~600 MW. The Robbins TBM will have to negotiate quartz porphyry and andesite bedrock geology. If Lake Titicaca were drained at depth, at its deepest point, freshwater could be made to fall ~3500 m to the Pacific Ocean.

3. Physics of a Base-of-Mountain Hydropower Station

Selection of the best course of the TBM-excavated tunnel will have to be done on the basis of on-site macro-engineering and geological studies. The studies will aim to predict, in the alternative suggested tunnel courses, the influence of the rock conditions on TBM operation, the amount of tunnel lining, the site and depth of the adits to the tunnel and the extent of the definitive field studies that must be done before commencement of tunnel construction. The economic benefits from the studies can be estimated as a considerable percentage of the total construction cost of the tunnel driving. As with other modern-day macro-projects, Macro-engineering has changed from its 20th Century incarnation; 21st Century Macro-engineering leaders of any macro-project must have a list of social groups to be met with, environmental impact statements to be filed, national and international laws to be complied with, and public concerns to be addressed. Yet the result—a legal and financial go-ahead—if properly done, is well worth the constraints of time and direct financial hardship: hydropower technology chosen openly, democratically, and consensually, rather than being dictated. Within a range of about 250-700 m, both Francis and impulse turbine units can be used.

Ordinarily, Lake Titicaca hydropower potential would remain worthless (on a significant geographical and economical scale) in the near-term future for a number of reasons: (1) the absence of Environmental Impact Statements; (2) formidable geological and geomorphic impediments such as infamously powerful earthquakes and rugged, even jagged, incidental terrain; (3) nearly non-existent traffic infrastructure such as highways, roads and railroads; (4) high to very high initial investment costs and long-period financial pay-backs; (5) the requirement for reliable long-distance aerial weather-resistant electric transmission lines over great distances; (6) the reluctance of international money lenders to consider low-interest loans to Peru and Bolivia; (7) volatile and inconsistent national political opinions regarding priorities of national, regional and centralized or decentralized energy system development and (8) the considerable on-going development of alternative energy resources such as natural gas fields.

Furthermore, we must assume that future global climate change during the 21st Century may instigate flexibility requirements for many existing and planned infrastructures! If, for example, the Altiplano climate becomes drier than today, then Lake Titicaca will be reduced in volume and the freshwater could be wasted through uncontrolled evaporation. On the other hand, if the Altiplano becomes wetter, Lake Titicaca will refill faster and pose a severe flood damage threat to all established infrastructures surrounding that body of water that is wedged and isolated in the Andes Mountains! Our offered technical suggestion creates a situation whereby—in either instance—Peru's coast-sited population will flourish and prosper! The map of Titicaca region is shown in Figure 1.

Most Relevant Data about Lake Titicaca

Data on Lake Titicaca is derived from different researchers from different time periods—in other words, these data must always be assumed to be somewhat "iffy" facts. Here, we show the most reliable factual generalities about Lake Titicaca.

1. Catchment total area 58,000 km^2.
2. Lake Titicaca's average surface area 8,372 km^2.
3. Water volume 893-930 km^3.
4. Average depth 107 m.
5. Maximum depth 283-351 m.
6. Modern Surface elevation 3810 m.
7. Primary source is 27 rivers. Main influx is Rio Ramis (flow 76 m^3/s).
8. Effluent Rio Desaguadero (flow 20-35 m^3/s).
9. Annual precipitation 703 mm/year.
10. Annual evapo-transpiration 652 mm/year (173 m^3/s).
11. Annual discharge 281 m^3/s.
12. Altitude of western ridge is 4000-4200 m.
13. Lake Titicaca's level oscillation is about 1 m/year up to ~6.47 m in 29 year period.
14. Used water by people settled in basin is 9.5 m^3/s (include irrigation 7.4 m^3/s).
15. Annual fish catch is 6,327 metric tons.
16. Economic losses of ~$890,000/decade are due to floods.
17. Usual cost of electricity in Peru is about $3/kilowatt-hour.

Lake Titicaca extends between 14^0 and 20^0 South Latitude and between 66^0 and 71^0 West Longitude. It is 176 km long and approximately 70 km wide.

Bolonkin-Cathcart Infrastructure Innovation

The conventional method to harness hydropower energy obtained from a very high-altitude lake is by drilling a difficult-to-complete tunnel, or several such tunnels, through the enveloping hard-rock mountains. But in our case, in particular, this common method is excessively expensive, dangerous and requires long excavation period. So far, the world practice of tunneling has no experience with very long tunnels, especially those incised in hard-bedrock mountain geological settings.

At this time, we offer an important technical innovation—to put hermetic steel or prefabricated reinforced concrete tube line, both emplaced by heavy-lift helicopters, over Andes Mountains (Figure 2). The water pumping station 2 lifts the freshwater from Lake Titicaca to the mountain ridge and, thence, into the conveying hermetic tube. Then, the water can safely flow to lower-elevation hydroelectric plant 4. The 90-95% energy spent for fluid pumping will be recuperated because the lower hydroelectric plant 4 will work off of a greater water level (and higher pressure). This technical innovation decreases our unconventional facility's monetary installation cost by hundreds of times!

Table. Annual flow in ten control stations of Lace Titicaca and Desaguadero River

River	Station	Average (m^3/s)	Maximum (m^3/s)	Minimum (m^3/s)
Ramis	Ramis	75.6	130.4	24.4
Huancané	Huancané	20.0	38.8	6.9
Suchez	Escoma	10.6	18.9	4.0
Coata	Maravilla	41.5	75.5	2.4
Ilave	Ilave	38.5	96.6	5.0
Desaguadero	International	35.5	186.5	–3.5
Desaguadero	Calacoto	51.9	231.6	6.2
Mauri	Abaroa	4.9	9.8	2.3
Caquena	Abaroa	2.8	5.6	0.9
Mauri	Calacoto	18.6	31.8	5.7
Desaguadero	Ulloma	77.1	282.7	19.7
Desaguadero	Chuquiña	89.0	319.3	20.0

The main tributary of the Lake Titicaca basin is the Desaguadero River, with an average annual discharge of 89m^3/s and a maximum of 319 m^3/S.

Computations and Estimations

1. Estimation of the producible power of proposed electricity station. The hydropower is computed by equation:

$$N = \eta g m H , \tag{1}$$

where N is power, W; η is coefficient efficiency of full system, $g = 9.81$ m/s^2 is Earth's gravity; m is water extension, m^3/s; H is difference of water levels, m.

Computation is presented in Figure 3.

As the reader can readily notice, the minimal water expense (20 m^3/s, without in any fashion changing Lake Titicaca's free surface level anthropogenically) produces about 640 MW while the maximal expense (35 m^3/s, without any change of the voluminous lake's free water surface level) produces 1070 MW. If we straight-forwardly simply agree to decrease Lake Titicaca's area by two times (100 m^3/s, with alteration of the lake's free surface water level, Figure 8), then the electricity generating station can permanently produce ~3060 MW.

That is a very powerful electric plant, just about equaling 50% Peru's current electricity base-load!

Map 21.2: Basin map

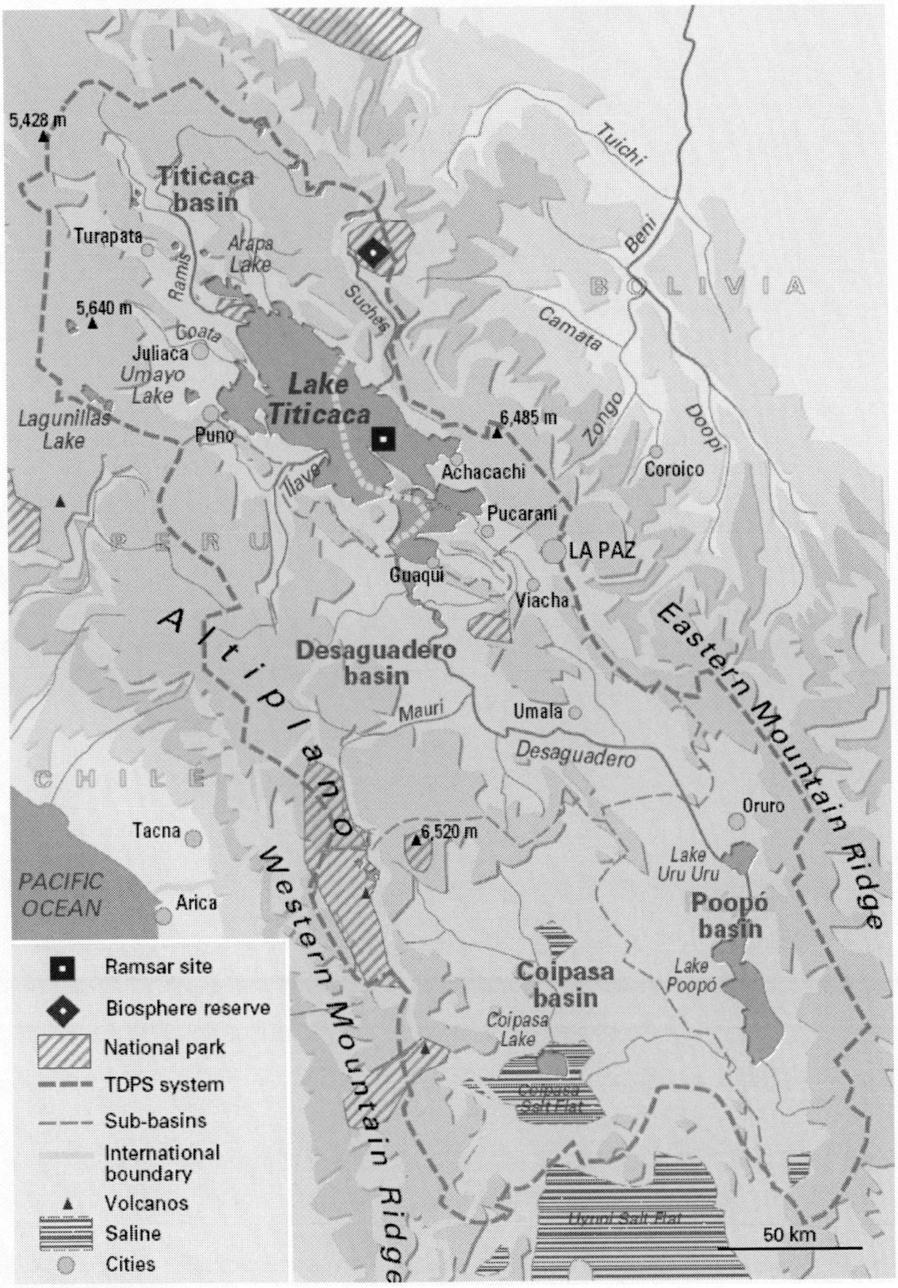

Source: Prepared for the World Water Assessment Programme (WWAP) by AFDEC, 2002.

Figure 1. Lake Titicaca.

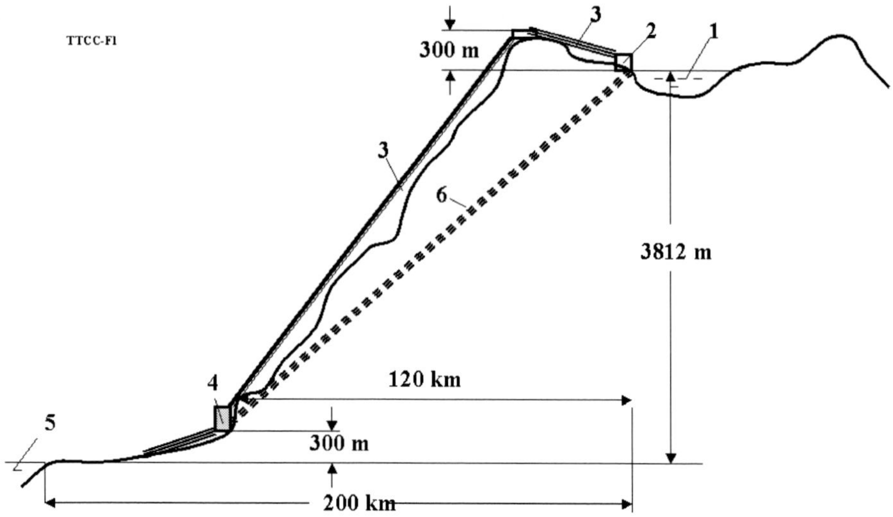

Figure 2. Sketch of proposed Lake Titicaca Electricity Generation Station. Notations: 1 – Lake Titicaca, 2 – freshwater pumping station, 3 - water tube, 4 - hydropower plant, 5 – Pacific Ocean, 6 – possible underground tunnel into a hard-bedrock mountain formation. In Peru, the shortest distance is between Puno to nearby maintain top is ~20 km, the linear distance to nearest mountain river (west of ridge) that drains to the Pacific Ocean is ~60 km and, thence, to the Pacific Ocean is ~200 km (see map in Figure 1). In Bolivia, the shortest distance from Pucarani to Rio Zongo (after passing over the mountain ridge) is ~50 km.

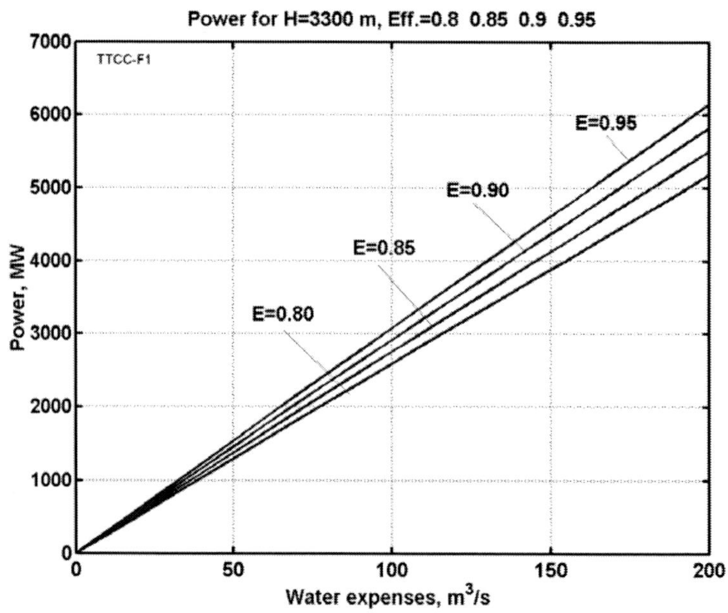

Figure 3. Full installation potential hydropower of proposed Andes-Pacific Ocean electricity plant via the water expense for different coefficients of efficiency.

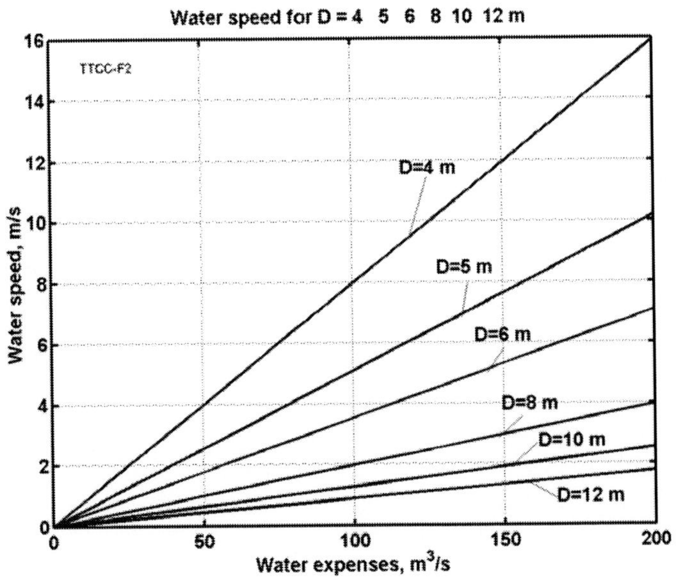

Figure 4. Freshwater speed in tube via water expenses for various tube diameters.

2. *The freshwater speed* into the hermetic tube can be estimated by equation

$$V = \frac{4m}{\pi D^2}, \tag{2}$$

where V is water speed, m/s; m is water extension, m^3/s; D is tube diameter, m.

Computation is presented in Figure 4. Greater tube diameter will promote less freshwater flow speed and, thus, reduced inside water losses.

3. *The loss of water pressure* (in m) is computed by equation

$$h = f\frac{L}{2g}\frac{V^2}{D}, \tag{3}$$

where h is loss of water pressure, m; f is friction coefficient; L is length of tube, m; V is water speed, m/s; D is tube diameter, m.

The water friction coefficient is

$$f = \frac{0.25}{\left[\log\left(\frac{k}{3.7D} + \frac{5.14}{R_e^{0.4}}\right)\right]^2}, \tag{4}$$

where R_e is Reynolds number, k is roughness.

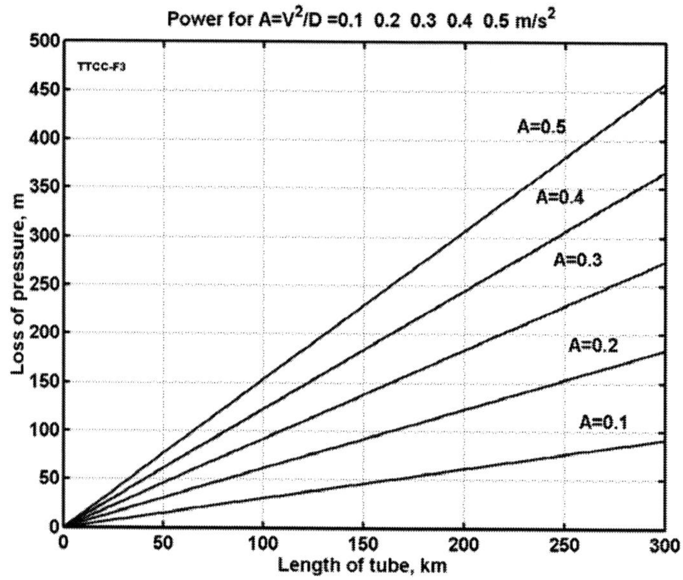

Figure 5. Loss of the water pressure into the closed tube system via the length of tube for the different ratios $A=V^2/D$, where V is water speed, D is hermetic tube diameter. Friction coefficient $f = 0.06$.

In our case we use the reinforced concrete or steel tubes. The friction coefficient for both is approximately 0.06.

The computation of equation (3) is presented in Figure 5. Loss amounts to about 200 m of water pressure over a flow distance of 200 km, or 130 m over a flow distance of 150 km.

4. Relative loss of water pressure. The relative loss of the tube's water pressure may be estimated by equation

$$\bar{h} = f\frac{V^2}{2gD}, \tag{5}$$

where $\bar{h} = h/L$ is relative loss of water pressure, m/km. The computation is presented in Figure 6.

The big hydraulic tube, with a large diameter, is more expensive to emplace but such a tube has the advantage of decreasing greatly the pressure loss and increases significantly the efficiency and the power of our proposed electric-generation station for Peru located west of the Andes Mountains.

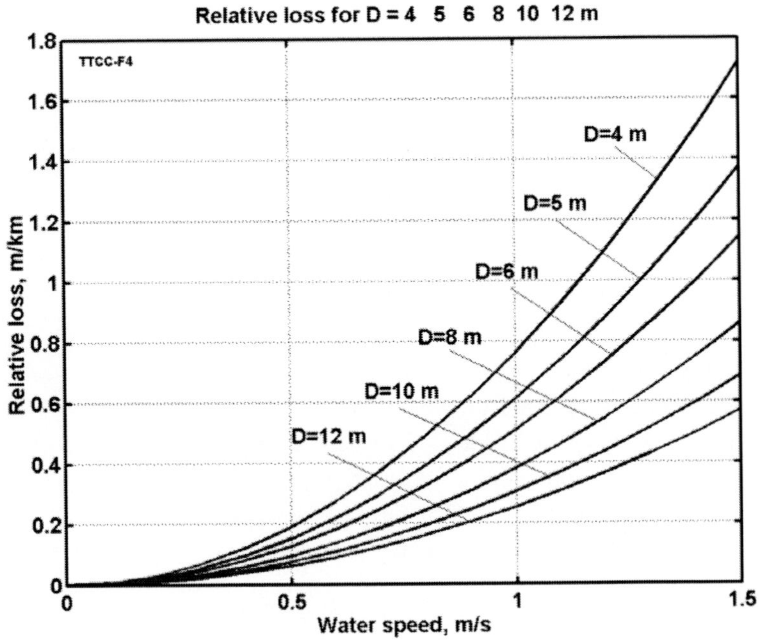

Figure 6. Relative loss (m/km) of the water pressure via the water speed for different tube diameters. Friction coefficient $f = 0.06$.

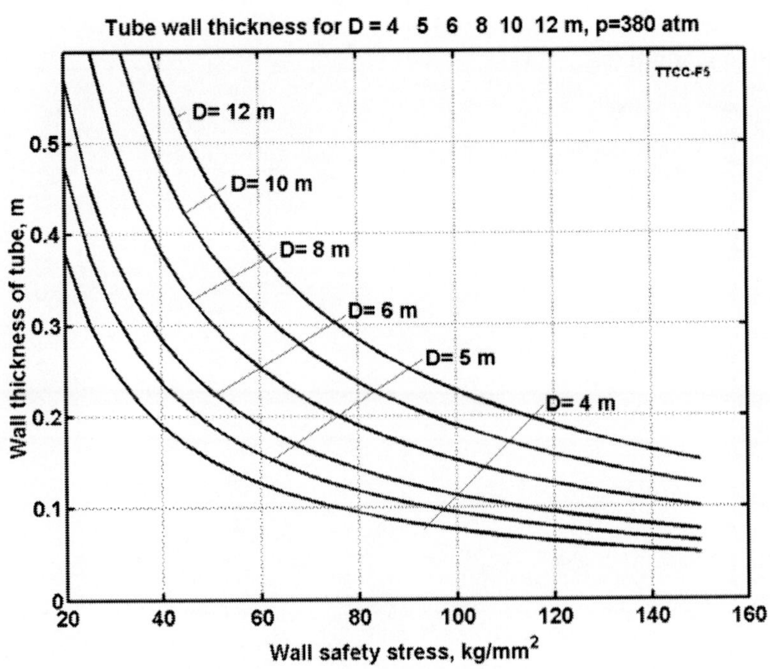

Figure 7. Tube-wall thickness via wall safety tensile stress for various tube diameters and water pressure 380 atmospheres.

5. Tubes. Tubes near the lower hydropower station have high pressure (up 380 atmospheres). They must be made from steel or something even stronger—perhaps with some composite fiber material. Such composed material has higher maximum stress (up 600 kg/mm^2, steel has only 120 kg/mm^2) and low specific-density (1800 kg/m^3, steel has 7900 kg/m^3). It may also be, or become, cheaper quite soon. Coefficient of safety is 3 to 5. Below, is the useful equation for computation of the tube wall's requisite thickness:

$$\delta = \frac{pD}{2\sigma}, \tag{6}$$

where δ is tube-wall thickness, m; p is water pressure, N/m^2; σ is safety tensile stress, N/m^2.

The result of computation is presented in Figure 7.

6. Over-expenditure of water. In case of permanent water over-expenditure, Lake Titicaca will be balanced at a new, human-selected lower mean free surface water level. The lake's area decreases and evaporation also decreases. The site-specific computations for Lake Titicaca are presented in Figure 8. The maximum of freshwater extended downward flow ought to be ~200 m^3/s and that is capable of producing a maximum of ~6100 MW. That is equivalent to all extant electric power plants of Peru! But, eventually, Lake Titicaca will vanish as a geographic feature. (It is possible to imagine that future global climate change might eventuate in a wetter regional climate allowing the constancy of this terminal lake.) The other way is to decrease evaporation by using some special anthropogenic floating coverings that create a thin layer on the freshwater's free surface and, thereby, retard or even prohibit evaporation or, ultimately, to simply cover the water surface with an extremely pliable and thin fabric film.

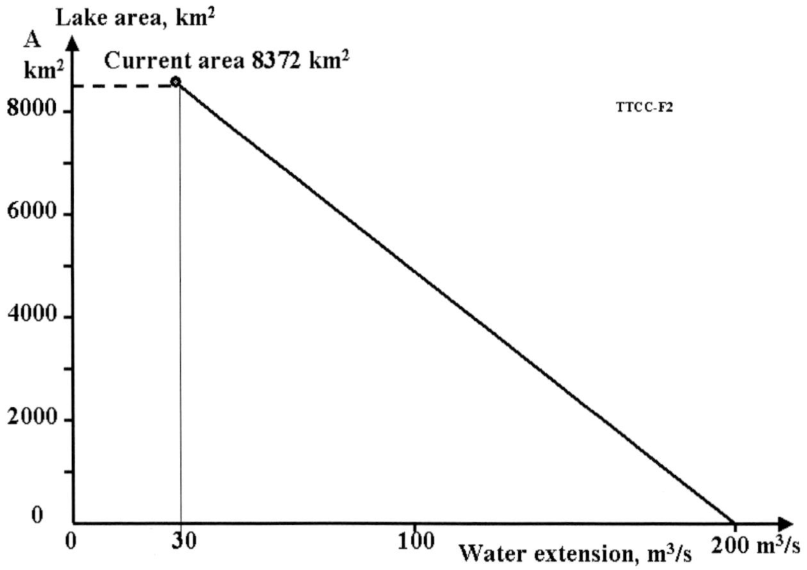

Figure 8. Decreasing of the Titicaca's lake surface area via average flow through the power plant. If the average flow is <30 m^3/s, we can decrease the loss (flow) in the Rio Desaguadero and retain Lake Titicaca's level. (Nowadays, there is a dam regulating the Rio Desaguadero.) Any permanent use of >30 m^3/s inevitably decreases the lake's mean free water surface area.

CONCLUSION

The inexpensive hydroelectric power plant, having generation capability of 640-1000 MW may be built near Lake Titicaca without any hazardous, ugly change to Lake Titicaca's area or volume (water expense up 20-35 m^3/s). For this to happen, the floodgates of the dam on the Rio Desaguadero near the International Bridge must be permanently closed. Average Rio Desaguadero flow is 70 m^3/second and Rio Desaguadero will become a river of smaller total flow emptying into Lake Uru Uru, which is north of isolated Lake Poopo.

If we choose to permanently utilize >30 m^3/s, then Lake Titicaca's free surface total area will decrease markedly. We can, however, receive, distribute and use more or less permanently a maximum up 6100 MW. Lake Titicaca will vanish, but our innovative hydroelectric station will still produce utilizable power. If, sometime, we find a truly economical method for halting or greatly reducing evaporation of the high-altitude lake, then it will certainly become possible to save Lake Titicaca in perpetuity and yet still harness a great quantity of electricity! The electricity will be fed into the dispersed main load centers by the national power grid while the freshwater can be used for irrigation and urban use in Peru's desertic coast region.

One more fact must be noted here: on 15 September 2007, close to the village named "Carancas" located near Lake Titicaca, a stony meteorite hit a dry riverbed in Peru. It dug a 14 m-wide crater in the landscape (16^0 39' 54" South by 69^0 02' 38" West). If that meteorite had by chance splashed into Lake Titicaca, it would very likely have caused a significant damaging tsunami!

Most probable geopolitical obstacle to macro-project's realization: Bolivia may not agree to such freshwater diversion because Lake Titicaca is a shared resource managed by ALT. However, Bolivia also has a known need for low-cost electricity and, therefore, we think/believe that an amicable United Nations Organization-sanctioned new, additional, international treaty agreement is contemplatable.

Typical Hydro-Station. 42 kb.

REFERENCES

[1] Baker, P.A. et al., "Holocene hydrologic variation at Lake Titicaca, Bolivia/Peru, and its relationship to North Atlantic climate variation", *Journal of Quarternary Science* 207: 655-662 (2005).

[2] Binational Autonomous Authority of Lake Titicaca, "Lake Titicaca Basin, Bolivia and Peru", Chapter 21, pages 466-480, IN THE 1ST *UN World Water Development Report: Water for People, Water For life* (UNESCO, 2003).

[3] Calaway, M.J., "Ice-cores, sediments and civilization collapse: a cautionary tale from Lake Titicaca", *Antiquity* 79: 778-790 (2005).

[4] Morris, Arthur Stephen, *Raised Field Technology: The Raised Fields Projects Around Lake Titicaca* (2004), page 1.

[5] Revollo, Mario M., "Management issues in the Lake Titicaca and Lake Poopo system: Importance of developing a water budget", *Lakes & Reservoirs: Research & Management* 6: 225-229 (September 2001).

[6] *WIKIPEDIA, "Trans-Andean Railways"*: *http://en/wikipedia.org/wiki/Trans-Andean_Railways* (2007).

[7] Chandler, David L., "Incoming!", *New Scientist,* 2 February 2008, page 40.

Chapter 4

PRESERVATION OF THE MEDITERRANEAN SEA DURING GLOBAL SEA LEVEL RISE WITH A GIBRALTAR STRAIT TEXTILE BARRAGE (GSTB)*

ABSTRACT

Artworks can improve humanity's ability to apply Macro-engineering's still-developing guiding principles which skirt or correct oceanographic problems impairing the economic usefulness of coastal land, the overhead airshed, and seawater's offshore temperature and salinity stability. A new form of Art, "Ocean Art", is here proposed which centers on deliberate terracing of appropriate regions of our world's ocean; a proposed example of macro-engineered useful "Ocean Art" is the technically possible 21st Century terracing of the Mediterranean Sea. "Ocean Art" is applicable worldwide to places that might be practically improved by its judicious employment. Such "Ocean Art" may constitute an entirely unique category of preventative planning solutions to anticipated future steric global sea level rise.

Keywords: Mediterranean Sea, sea level rise, tension textile barrage

1. BACKGROUND

Space Art proponents opt to construct various symbolic artifacts in Outer Space visible from Earth's surface. With plastic film and textile envelopes, Air Art advocates exploit the possibilities of compressed air or naturally generated atmospheric wind. Land Art results from different human interpretations of the significance of natural and anthropogenic sub-aerial landscapes.

Famed artist Christo installed 11 flamingo-pink floating plastic collar-mats covering approximately 600,000 m^2 of a saltwater lagoon's watery surface in Miami, Florida's Biscayne Bay for his temporary "Surrounded Islands Project" artwork of 1983. During 1969, Peter Hutchinson and Dennis Oppenheim installed artworks in the coastal waters off Tobago in the West Indies. "Wave Organ", constructed by Peter Richards and George Gonzales, and the seashore artworks of Andy Goldsworthy are also examples of what is, charitably, labeled Oceanographic Art since they are essentially decorative. However, Ocean Art is a form of seawater sculpting by aquatic terracing focused on the 70% of Earth's surface that constitutes our world's ocean; the modern-day originator of geographically large-scale intentional Ocean Art is the German architect Frei Otto, who seems to have first contemplated the concept during 1953. Ocean Art has a practical, commercially useful aspect that Oceanographic Art does not and, therefore, is of interest to 21st Century adherents of Macro-engineering.

Gibraltar Strait, background, as seen from the Rock of Gibraltar looking southward.

2. Inadvertent Terracing of the World Ocean

About 5,000 years ago, the Earth-atmosphere's methane concentration began to increase. Its generation source was human cultivation of rice in flatland paddies; about 2,000 years ago, humans commenced growing rice in watery paddies on terraced hillsides. Methane is a greenhouse gas and the anthropogenic contribution to the Earth-atmosphere causes global warming and a rise worldwide of sea level. Perhaps half of all living humans—i.e., during 2008, approximately 3.2 billion persons out of 6.4 billion persons—eat rice and the microorganisms living in anoxic rice field soils contribute between 10% and 25% of annual global methane emissions; by 2030, there might be five billion human consumers of rice. Artificial wetlands emplaced on shaped hillsides have in the past, and continue nowadays, to contribute to the world ocean's sea instability in terms of volume and area. Catastrophic

[*] Presented in http://arxiv.org in 2007.

methane releases, caused by submarine slumping during which submarine sliding of bottom material results in hydrate dissociation triggered by depressurization (sea level fall), have occurred in the Mediterranean Sea Basin. An anthropogenically-induced maintenance of the Mediterranean Sea's mean free surface level at one meter below the future prevailing future world ocean level is unlikely, even with a possible contributory permanent seawater-warming steric phenomenon, to trigger a large methane release from the seafloor during the 21st Century.

Especially subsequent to the Industrial Revolution, dams on rivers were constructed that retain large volumes of freshwater. There are probably 45,000 dams on land exceeding 15 m high that, in the aggregate, can store more than 6,500 km^3 of freshwater. (The Netherlands' famous Ijsselmeer did not directly terrace the ocean, but simply replaced a segment of seawater with an equal volume of freshwater; France's La Rance Tidal Barrage merely harnesses the daily tides.) Absent these reservoirs, the world's ocean would be higher than today. Anthropogenic global warming of the atmosphere caused, in part, by an uncontrolled but measured build-up of carbon dioxide gas during the past two centuries, is plausibly suggested to also cause a global sea level rise, which will initially be regionally differentiated.

3. Known World Ocean Terracing Factors

By 2100, the world's ocean could rise by 0.5-1 m relative to its present-day level thereby directly affecting the world's landmass coastlines. A 1.0 m rise represents the addition of approximately 4 x 10^{14} m^3 of freshwater to the world ocean—that amount was added to the world ocean in a few months about 8,200 years ago during the rapid final drainage of Lake Agassiz. Macro-engineers consider such a sporadic change as marginal in the mathematical sense of the word. All coastal nations should now be planning for a 0.5-1 m rise in world ocean level during the 21st Century. Wisely, the 21 signatories to the 1978 Barcelona Convention, an international agreement to protect the Mediterranean Sea, agreed to ban real estate development within 100 m of today's coastline during the group's 15th official meeting held during early-2008 in Almeria, Spain.

Humanity's activities (to make and earn a living) will "globalize" the Mediterranean Sea—its seawater, organic and inorganic contents, and coast. Mediterranean Sea Basin nations have several old and expensive ameliorative macroprojects options available (Cathcart, 2006). However, recent R&D and newer products derived from advanced material technologies—particularly, technical textiles exhibiting high-performance, purely functional, and precisely woven or non-woven fabrics—offers the prospect of cheap regional anti-sea level rise barrier macro-projects.

Commercial shipping interests and mass tourism organizers and related industries are the most obvious pro-amelioration seawater barrier macro-project constituencies examining the impacts of future sea level rise in the Basin in addition to the ecologically sustainable fisheries, recreational boating and yachting, harbor maintenance and the military communities.

The 21st Century seawater in the Mediterranean Sea has visibly and measurably different characteristics than during the 20th Century. Remarkable shifts in the thermohaline circulation and water mass characteristics have occurred because of mankind's activities on land, including large freshwater reservoir construction and the widespread regulation of river sediment deposition.

Peak evaporation from the Mediterranean Sea, driven by energy released from seawater, takes place during the wintertime. The Mediterranean Sea measurably warmed during the 20th Century; the causes and potential effects of the elevated seawater temperature on pelagic food webs and the Basin's present-day and future climate are not well known but could foster extinction of fish species. The regional atmosphere's response to an idealized 2 K cooling of the Mediterranean Sea has been estimated, but not the reverse, a warming. According to some authorities, no realistic climate change scenario for the twenty-first century with all forcings has been tested for the Mediterranean Sea. Unmonitored Basin seafloor volcanoes, which have in the past created new land for a short period and may create new land in the future—most probably at the Marsili seamount north of Sicily—are indicative of significant seafloor geothermal heating. Standard scientific explanations for imposed seawater warming and expansion within the Basin—the steric effect—rely mostly on the alleged enhanced atmospheric anthropogenic global warming theory. During the 21st Century, Mediterranean Sea Basin hurricanes possibly may become slightly more intense owing to seawater's steric warming, with higher boundary layer wind speeds, and increased vertical mixing of seawater masses. A Basin-wide sea level higher by 1 m may also lead to higher wintertime storm surges affecting the Basin's coast and its costly-to-repair/replace infrastructure.

4. Trans-Gibraltar Strait Bridge and Barrage Macroprojects

The Strait of Gibraltar connects the North Atlantic Ocean and the Mediterranean Sea. Long-span bridges are becoming longer because the use of ever-faster supercomputers permits useful calculations of the physical forces impinging these long-span structures with greater accuracy and because modern construction techniques can isolate suspension bridge tower bases from earthquakes. Macro-engineers mostly must meet managerial challenges such as financing and resource accumulation. Technical challenges remain, of course, but new materials—such as carbon fibers embedded in composite materials—are being developed as practical solutions to such impending macro-problems.

Three-dimensional, supercomputer-generated bathymetric maps and high-resolution geologic sections (based on sparker profiles and sea-bottom drill cores) are already available that generally illustrate the Strait of Gibraltar's geologic framework today. Anticipation of a Punta Malabata, Morocco to Punta Paloma, Spain Gibraltar Strait Tunnel, with construction tentatively slated to commence sometime during 2008, has provoked all macro-project site investigation efforts so far. Additional pre-planning site studies may be required if our proposed Gibraltar Strait Textile Barrage (GSTB) textile/plastic macroproject proposal is seriously considered. GSTB will be judiciously draped on a general alignment between Tarifa (36^0 01' North by 5^0 36' West) in Spain and Ksar e' Sghir (35^0 50' North by 5^0 32' West) in Morocco), creating an aerial and submarine artwork somewhat imitative of Christo's "Valley Curtain, Rifle, Colorado, 1970-1972", which was macro-engineered by Ernest C. Harris (1915-98). The GSTB will bow or "billow", like a ship's wind-filled sail, eastwards from the selected 20 km alignment because of marine (difference in sea elevations on a two-sided bottom-anchored membrane and natural currents such as tidal solitons coming from the North Atlantic Ocean) and aerial (seasonal winds) pressures acting directly upon the GSTB. Indeed, prevailing seasonal winds flowing along the Strait of Gibraltar will pile approximately 5-6 mm of seawater on the GSTB's westward face. GSTB macro-project planners will draw on the installation experience with heavy wire nets, floatation systems and their moorings derived from World War II anti-submarine net installations in strategic harbors and that documented experience offered by the 100 km-long World War I anti-submarine Otranto Strait Barrage (1915-19). One of the main factors influencing the GSTB's cost will be the sea-bottom cut-off wall to minimize seawater seepage—there is some slight possibility a submarine cut-off trench need not be dug on the seafloor, nor that an uninterrupted underwater grout curtain need ever be installed, to ensure proper functioning of the completed Gibraltar Strait Textile Barrage.

From its western approaches, the Gibraltar Strait Textile Barrage will have the characteristic of an architectural deception resembling an English Garden or zoo landscape architect's geotextile "ha-ha" (also known as a "sunken fence") in that—absent warning light-buoys and radar reflectors—ship navigators will visually misapprehend the true nature of the sea route ahead. Those mariners (such as private-sector fishermen and yachtsmen) piloting their boats and other types of watercraft without benefit of up-to-date navigational sea charts that indicate the GSTB's presence in the seascape will have no inkling via normal optical clues whatsoever that a 1 m drop in sea level obstructs the Strait of Gibraltar! Mariners without radar readouts using the eastern approaches will visually spy a 1 m-high tensioned fabric wall, which if made of clear or aquamarine-colored material might be almost invisible until closely sighted! Approximately 50,000 vessels of all types currently pass through the Strait of Gibraltar annually and the two ship traffic lanes are considered to be military chokepoints; at least one, and possibly two, Frei Otto-style tensioned fabric ship locks will be required to accommodate post-construction GSTB shipping traffic (Otto, 1967). Collapse of a blocking 1 m-high GSTB draped across the 20 km breadth of this oceanic gateway won't necessarily be a disaster for the Mediterranean Sea Basin nations; more or less, it will be similar to a strong storm surge event with a constant 106,000 m^3/s incoming "tide" rippling rapidly eastwards towards Greece and Turkey to eventually inundate (by 1.0 m) 2.5×10^6 km^2 of the Mediterranean Sea surface. A 1.0 m-high tsunami impacting the 1 m-high air-exposed eastern face of the GSTB will probably exert a momentary pressure of 10,000 kg/m^2, or about 200,000,000 kg overall; in other words, a new kind of flexible hydrostatic seawall installable at suitable locales will become effectively available to coastal macro-engineers.

Other than collision events caused by errant ships and the cycling pressure changes of small intra-Basin tides, the most significant prospective structural integrity maintenance threats facing a Gibraltar Strait Textile Barrage are tsunami generated within the Basin or in the Atlantic Ocean. For example, the tsunami caused by the 1 November 1755 Lisbon, Portugal seismic event caused a maximum 11 m wave run-up at Tarifa and 10 m at Tangier in Morocco; the Rock of Gibraltar's seaport was impacted with a maximum wave run-up of 2 m. The historically recorded past and the predicted future related to tsunami impacts on the GSTB-served region means that tsunami momentum effects must be very carefully considered in great detail. When a tsunami wave meets a semi-slack barrier the barrier takes some momentum from the wave and transfers it to the barrier if it can move, making it taut. A variable amount of momentum is bounced back to the ocean in various directions. If one GSTB is judged insufficient to withstand the forces applied (snap loading), then another paralleling GSTB can be installed since each is not as costly as a single Gibraltar Strait Bridge! Elongation of the GSTB's super-ropes under dynamic loading will dissipate some tsunami-deposited energy.

In the case of the GSTB, there is a quite interesting newly apprehended hydraulic effect only partially comprehended by GSTB proponents: any North Atlantic Ocean tsunami overtopping the Gibraltar Strait Barrage will encounter a sudden 1 m hydraulic descent. Could a sudden 1 m-high waterfall at the GSTB's eastern face effectively attenuate a potentially catastrophic tsunami run-up on 13,000 km of Mediterranean Sea Basin shore? The acute observation of the remarkable effect of a sudden topographical drop on tsunami propagation hydrodynamics may be as instigative subsequently as the contemplations of Benjamin Franklin (1706-90) on the spreading of oil on a UK freshwater pond as well as the ocean's seawater.

When a tsunami wave originating within the Mediterranean Sea Basin or in the Atlantic Ocean Basin impacts the eastern face of the GSTB, part of the wave's energy is transmitted through the GSTB, part is reflected from the GSTB and another part is absorbed in the various materials of which the GSTB is constructed. Tsunami overtopping of the impermeable tensioned fabric dam with zero freeboard will cause a North Atlantic Ocean seawater hydraulic flow (from supercritical to subcritical) that induces GSTB vibrations (near-critical flow induced vibrations, vortex shedding and suction on the air-exposed "downstream" face of the dam-artwork). In effect, the Mediterranean Sea close to the intact quivering/resonating GSTB will become a "stilling basin" that will dissipate the kinetic energy of the overtopping flow. (A 1 m overtopping results in a temporary 106,000 m^3/s flow whilest a 2 m overtopping results in a 300,000 m^3/s waterfall.) The optimum value of the drop height at the GSTB must be determined mathematically and by physical model testing since macro-engineers will wish to stabilize its geographical position and not cause any dangerous GSTB structural damage. Unfortunately, R&D reports on tensioned membranes vertically spanning water depths of limited extent are infrequent in the appropriate literature. The total area of the water-retaining fabric drape that is planned to comprise the GSTB is about 200 km^2 but only approximately 20,000 m^2 of it will actually be fully exposed to the air and material-degrading sunshine on its eastern face while under a continuous imposed 1 m seawater head (hydraulic pressure) on its submerged western face.

5. Seawater-impervious Membrane Architecture

Cables and membranes are the essential main components of Frei Otto's architecture proposals. Application of advanced technical textiles and super-ropes composed mainly of Kevlar—or, eventually, even carbon nanotubes—could permit safe emplacement and use of a pontoon bridge spanning the Strait of Gibraltar—in fact, a vehicle-carrying floating bridge imitating the span used by Xerxes in 480 BC to support his marching troops and their wartime baggage train as they crossed the Hellespont! Braided or stranded super-ropes could stabilize a pontoon bridge in a fixed geographical alignment for a long period, especially in a 1 m s^{-1} eastward flowing surface current refilling the evaporating Mediterranean Sea, which has a natural yearly seawater deficit of about 0.5 m.

The ultimate hydrostatic head supported by a textile (woven or a non-woven film) is the measure of the resistence to the passage of seawater through the material; the standard applicable for determining the resistence to seawater penetration is the hydrostatic pressure test. Several international organizations, as well as many national agencies generally accept the height of a seawater column given in metric units of distance as the applicable validation of a test method primarily intended for dense fabrics and films. Waterproof and watertight are synonyms in this report. In the past, the resistence to seawater penetration (in ship sails, ship hatch-covers etc.) has been technically achieved by coating woven textiles with various waterproofing materials; watertight textiles can now be achieved by dense weaving of strong fibers. Multi-axial multi-ply textiles are bonded by a loop system, consisting of one or more yarn layers stretched in parallel; yarn layers can have different spatial orientations and different yarn densities. The combination of multi-directional fiber layers has been found by scientific laboratory and commercial factory testing to be remarkably capable of distributing extraordinarily high strain forces; multi-axial multi-ply textile structures are dimensionally stable in any direction and exhibit isotropic distribution of stress forces with uniform strain behavior. Kevlar 29, 49, and 149 since 1971 is a Dupont, USA, trade name for aromatic polyamides with a Tensile Strength of >3 Gpa, a Failure Strain of 3% and a Material Density of ~1.4 g/cm^3 is a good example. All extremely strong materials able to perform as unitary form-active structures ought to be investigated for use in the proposed GSTB.

The characteristic strength of a structural textile or film material must have a low probability (~5%) of not being reached during the lifetime of the material's use in the GSTB and the characteristic load must not have more than a 5% probability of being exceeded during the design lifetime of the GSTB. Potentially, embedded fiber-optic electronics—detectors, reporters and automated alarm actuators—ought to be used to monitor in real-time the super-ropes as well as the draped barrier, which must fit tightly to Gibraltar Strait's sea bottom and sidewalls to successfully fulfill its macroproject functions, giving instant alerts to immediately responsible shore-based supervisors of all developing GSTB structural problems related to the GSTB's safe and efficient performance. Should the GSTB be separated from its two sidewalls and sea-bottom anchorage, swept away by an over-whelming tsunami after full loss of its structural integrity, then possibly it might be partly retrievable/salvageable and, if so, its quick post-failure replacement would "re-initiate" the artificial Mediterranean Sea reduction (by natural evaporation) GSTB macro-project in a manner timely. A clever and practical collapse design could even optimize recovery of a broken GSTB's components—in other words, the GSTB might be constructed with a design philosophy including the possibility of semi-controlled collapsibility, even pre-planned folding!

Let us now consider the very inexpensive-to-construct GSTB mathematically. The width of the Gibraltar Strait at the place designated previously (in Section 4) is 20,000 m. The maximum depth is 900 m, with an average depth of ~450 m. The GSTB will have a seawater surface difference of 1 m. If the top of the Gibraltar Strait Textile Barrage is partially supported by pontoons floating on the Atlantic Ocean, the installation may be utilized as a vehicular highway between Spain and Morocco. Sea-going ships arriving and departing the Mediterranean Sea Basin will bypass the GSTB by using sturdy fabric ship-locks built at each terminus of the bridge-dam. Our purpose in Section 5 is to estimate the needed materials (film, supporting super-ropes) as well as the hydropower output potential. A simple sketch of the GSTB is provided in Figure 1:

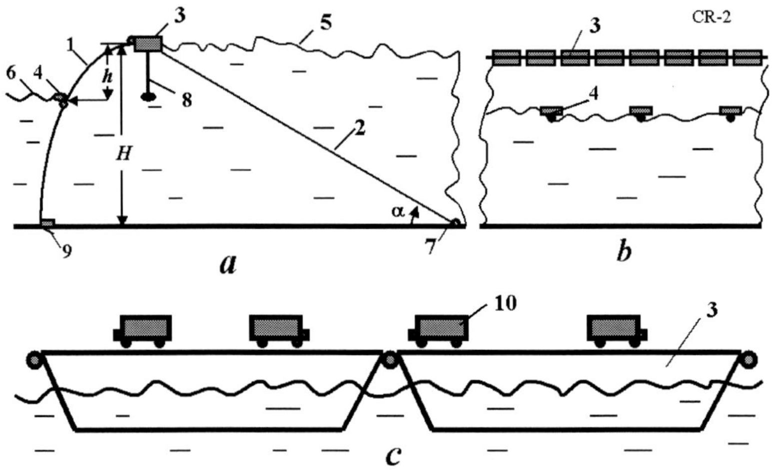

Figure 1. Kevlar or other suitable film-like flexible material forming the seawater barrier and the installation's associated hydropower station. (a) side-view, (b) front-view, (c) pontoon supported highway bridge joining Spain and Morocco. Notations: 1—flexible non-woven textile dam, 2—support cable, 3—pontoon, 4—hydroelectric turbine, 5—North Atlantic Ocean water level, 6—Mediterranean Sea water level, 7—anchor, super-rope spool, motor of support cable, 8—stabilizer, 9—stones, 10—tracks. Angle $\alpha = 30^0$.

Computation

1. Force P [N/m^2] for 1 m^2 of dam is

$$P = g\gamma h , \tag{1}$$

where g = 9.81 m/s2 is the Earth's gravity; γ is water density, γ=1000 kg/m3; h is difference between top and lower levels of water surfaces, m (see computation in Figure 2).

2. Water power N [W] is

$$N = \eta g m h, \quad m = \gamma v S, \quad v = \sqrt{2gh}, \quad N = \eta g \gamma h S \sqrt{2gh}, \quad N/S \approx 43.453 \eta h^{1.5}, \tag{2}$$

where m is mass flow across 1 m width kg/m; v is water speed, m/s; S is turbine area, m^2; η is coefficient efficiency of the water turbine, N/S is specific power of water turbine, kW/m^2.

Computation is presented in Figure 3.

3. Film thickness is

$$\delta = \frac{g\gamma h^2}{2\sigma} , \tag{3}$$

where σ is safety film tensile stress, N/m^2. Results of computation are in Figure 4.

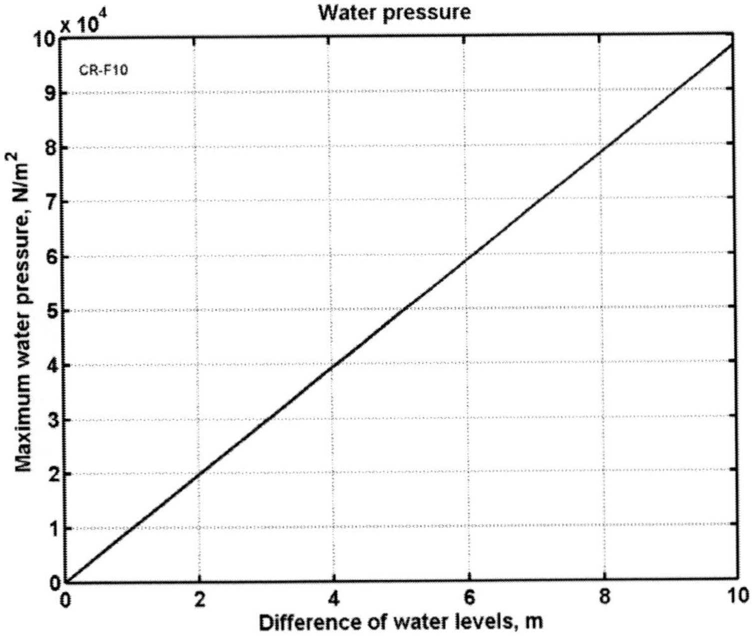

Figure 2. Water pressure via difference of water levels.

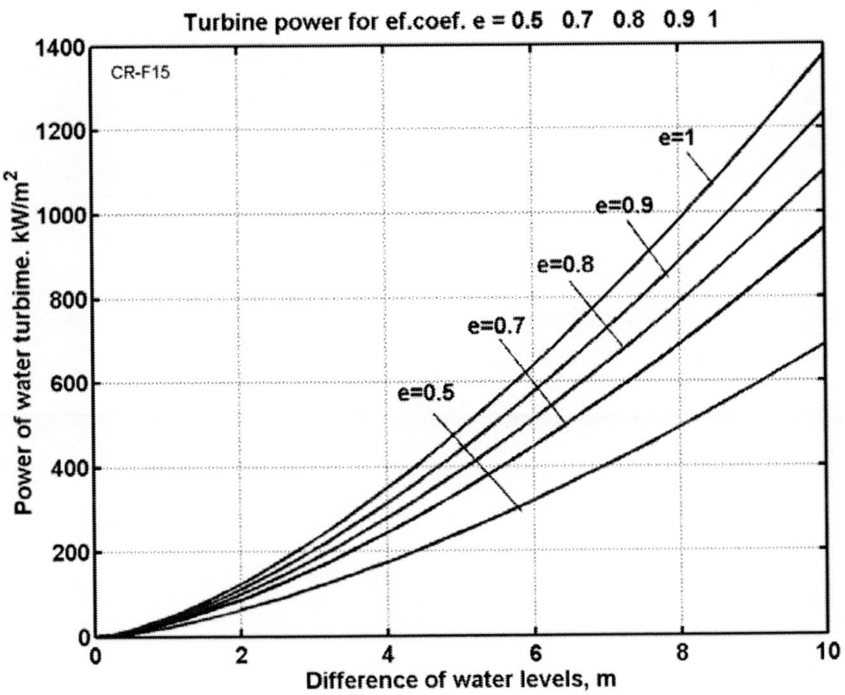

Figure 3. Specific power of a water turbine via difference of water levels and turbine efficiency coefficient.

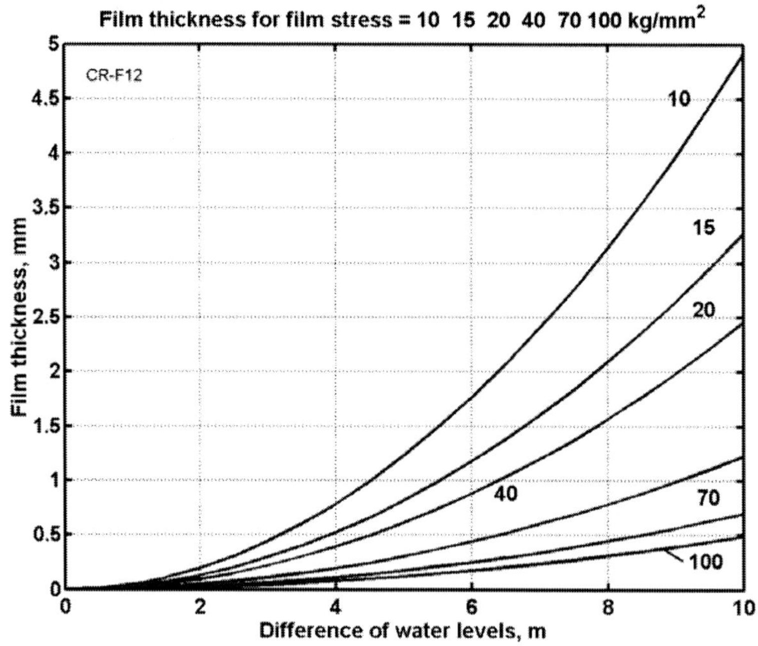

Figure 4. Film thickness via difference of water levels safety film tensile stress.

4. *The film weight* of 1 m width is

$$W_f = 1.2\delta\gamma H \quad , \tag{4}$$

Computation appear in Figure 5. If our dam has length L m, we must multiple the result by L.

5. *The diameter d* of the support cable is

$$T = \frac{Pl_2}{2}, \quad S = \frac{T}{\sigma}, \quad d = \sqrt{\frac{4S}{\pi}} \quad , \tag{5}$$

where T is cable force, N; l_2 is distance between cable, m; S is cross-section area, m^2.

Computation is presented in Figure 6.

The total weight of support cable is

$$W_c \approx 2\gamma_c HSL / l_2 , \quad W_a = \gamma_c SL, \quad , \tag{6}$$

where γ_c is cable density, kg/m^3; L is length of dam, m; W_a is additional (connection of banks) cable, m.

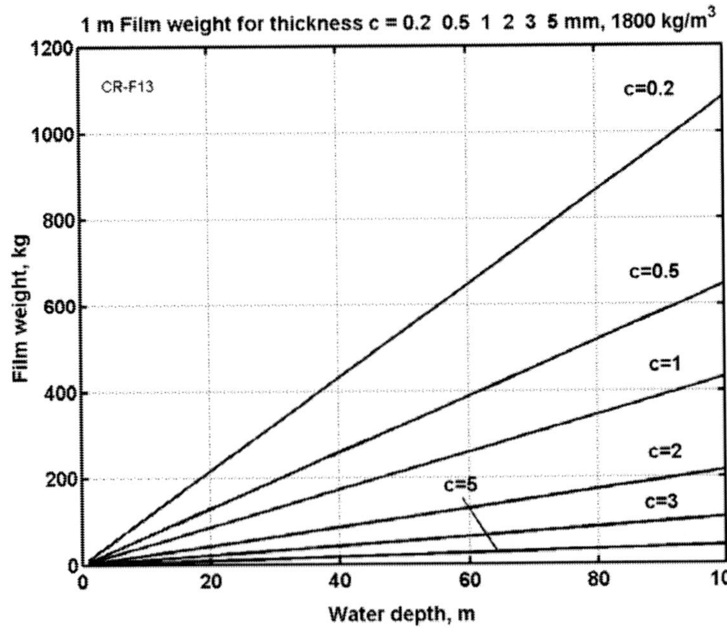

Figure 5. Film weight via the dam height and film thickness *c*, density 1800 kg/m^3.

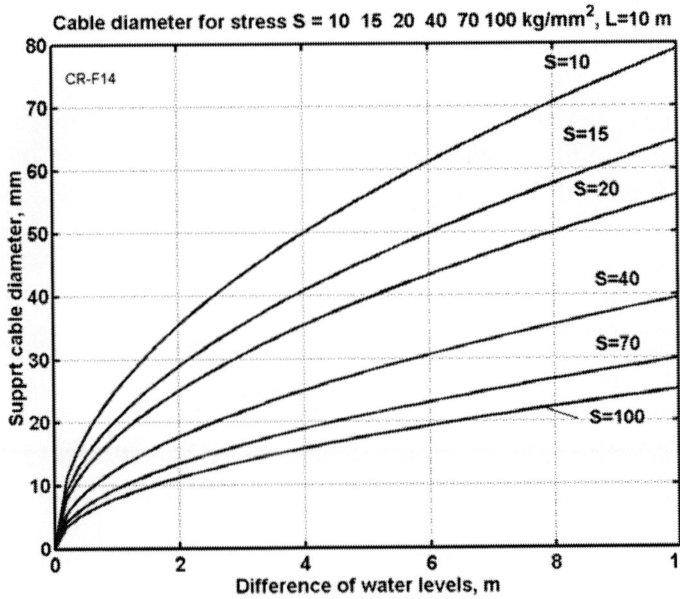

Figure 6. Diameter of the support cable via water level differences and the safety tensile stress.

Using the graphs above, we can estimate the relevant physical parameters of many interesting macroprojects.

Let us consider the GSTB emplaced between Spain and Morocco in the Strait of Gibraltar. The width of channel is 20000 m, the maximum depth is 900 m whilst the average depth is 450 m and the surface seawater level difference is 1m.

If the topmost part of the GSTB, on the Atlantic Ocean side, is partially supported by pontoons, the installation may be used as a fixed link super-highway, possibly even as a railroad bridge. Transiting ocean-going ships can use fabric sea-locks gates to pass through, across, the GSTB.

Our suggested Gibraltar Strait Textile Barrage very cheap to construct, especially when it is compared with a conventional concrete gravity dam installation as first proposed during 1928 by Herman Sorgel (1885-1952) in Germany.

Our purpose is to establish some kind of estimation of the needed film, supporting super-ropes, and energy development potential. A simple sketch of the GSTB is presented in Figure 1. The assumed difference of top and lower water levels (head) is 1 m. The estimation may be made using Figures 2 ÷ 6. However, we offer a clarifying computation that any reader can appreciate in this non-standard macro-engineering effort.

The force on 1 m^2 of film is $P = g\gamma h = 10\times1000\times1=10^4$ N/m^2, Eq.(1). The power 1 m of dam width for $\eta = 0.9$ is $N = \eta Pv \approx 4\times10^4$ W/m. For a dam with a length of 20,000 m the total power is $N_t = 800$ MW. That is power of middle-of-the-Gibraltar Strait electric station. The head is small (just 1 m) and the energy generation is not nearly as much.

Let us take the safety film and cable tensile stress $\sigma = 10$ kg/mm^2 $=10^8$ N/m^2. Using Eq. (3) we find the film thickness $\delta = 0.05$ mm, total film weight is $W_f = \delta\gamma HL = 0.05\times10^{-4}\times 1800\times450\times20000 = 810$ tonnes. Assume the support cables located $l_2 = 10$ m one from the other. The cross-section area of cable is $S = Pl_2/\sigma = 10^4\times10/10^8 = 10^{-3}$ m, diameter is $d = 36$ mm. The total cable weight is 3240 metric tons plus 32 tonnes comprised of the connecting 20,000 m super-rope. The total weight of our GSTB installation (without floating pontoons) is W = 810+3240+32 = 4182 tonnes. Floats can be also made from non-woven textile film and additional weight will be small ($\approx 5 \div 10\%$). We used conventional film, cables from cheapest available plastic. The artificial fiber has maximum tensile stress $\sigma = 500 \div 620$ kg/mm^2. Conventional strength factor has value 3 - 6. If we will use the artificial fiber with safety stress $\sigma = 100$ kg/mm^2, the total weight of installation decreases by 10 times (up to 418 tonnes). That is millions of times less (and much, much cheaper) then any conventional reinforced concrete gravity dam. The pontoons may be used to support the transfer of trains and/or motor vehicles. Conventional bypass channels may be used for surface shipping.

6. GSTB Hydropower Generation Opportunity

Adjacent to the super-rope sidewall seals and sea-bottom anchorages of our proposed Gibraltar Strait Barrage macroproject, the landscapes of Spain and Morocco offer nearly ideal conditions for the emplacement of Siphonic Hydropower facilities. Low-head hydroelectric power plants of a recently perfected type can manufacture electricity using gearless air turbines moved by seawater falling 1 m from the North Atlantic Ocean to the reduced/stabilized Mediterranean Sea. There can be an efficient recovery of economic energy in the form of electricity from low-head seawater. The monetary construction costs of these two electric power plants will be reasonable, even less than that of the GSTB; in total, this type of GSTB might be as little as 10% of the cost of a massive 1930s-style Herman Sorgel concrete gravity dam—that is, about USA \$10 billions instead of USA \$100 billions! All the machinery housed in these fixed power plants ought to be made secure from damaging environmental elements—possibly even small tsunami—and be readily examined and repaired by Spanish and Moroccan crews working on solid ground! Researched in the UK by Professor M.J. French since 1989, Siphonic Hydropower has evidently reached a near-perfection status and early marketing stage in its rigorous macro-engineering R&D. Our GSTB macroproject will present Siphonic hydropower developers with all the basic elements required for a potential hydroelectric development—a "limitless" world-ocean stream and an anthropogenic "drop" through which the North Atlantic Ocean's seawater can be used to convert the potential hydraulic energy into electrical energy. Where the available seawater head for the GSTB is only 1 m the losses in the Siphonic system (such as for aeration, due to pipe friction and due to upward drift of the air bubbles) would waste probably 0.4 m of the future available Atlantic Ocean head. Thus, the efficiency of the Siphonic system would be about 60%, without considering losses in the air turbines and electrical transmission losses. A Siphonic system of the size envisioned for the GSTB would justify the use of very efficient—that is, costly and precisely manufactured—air turbines, say at least 80%, so that overall efficiency could be approximately 50%. Thus, the electricity output could be ~500 MW. While not remarkably large, the power would be generated 24 hours/day, every day of the year regardless of weather conditions, forming a Mediterranean Sea Basin electrical generation and distribution system's reliable base-load. One megawatt is enough to power 1,000 European-style homes. Compressed air turbines energized by falling seawater will not affect the region's migratory birds, which is an advantage over wind farms where windmills are clustered at a site with persistent favorable winds.

7. Concluding Remark

We propose a unique Ocean Art macroproject—the combination of the Gibraltar Strait Textile Barrage and Siphonic Hydropower—that extends in scope a proposed protective macro-project to safeguard the Sethusamudram Ship Channel in Palk Bay being constructed by India—as an inexpensive total solution for the major known and forecast multiple environmental macro-problems of the Mediterranean Sea Basin. Proof and practical adaptation of this technology at the GSTB might its foster adoption elsewhere. We have also offered a new kind of textile-based hydrostatic seawall capable of successfully tolerating unpredictable incident 1 m-high tsunami. Joseph-Marie Jacquard (1752-1834) invented the automatic loom and his work with calculating machines has eventuated in today supercomputers. How fitting, then, that commercially available textiles/super-ropes and electronic computers are precisely the two industrial tools most needed to successfully resolve our *gedanken experiment*!

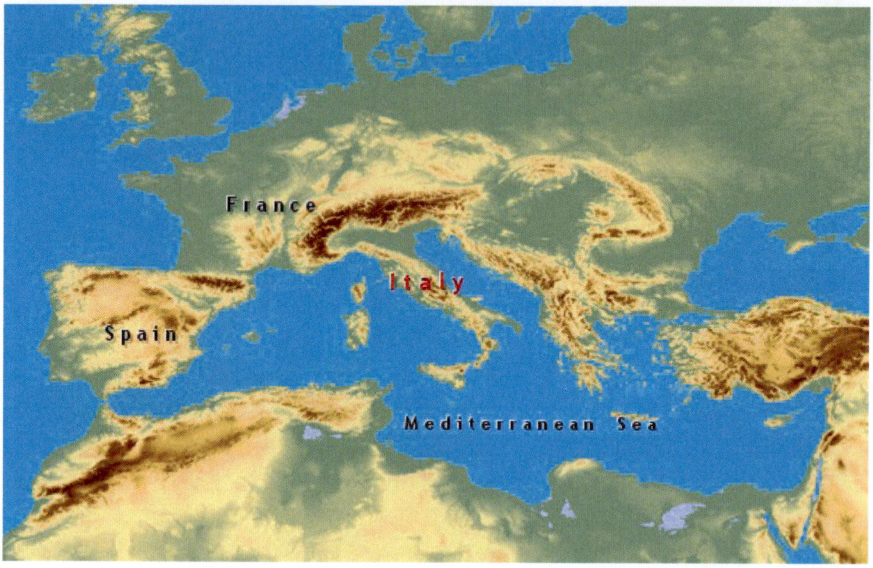

Mediterranean Sea

Gibraltar, 89 kb.

REFERENCE

[1] Cathcart, R.B. (March 2006) *"Gibraltar Strait Dam Macroprojects"* posted at a UK website: *www.daviddarling.info/Cathcart.html*.

[2] Bolonkin, A.A., "Air Cable Transport and Bridge", *Journal of Aircraft*, Vol. 40, No. 2, March–April 2003, pp.265-269.

Chapter 5

THE JAVA-SUMATRA AERIAL MEGA-TRAMWAY*

ABSTRACT

A mega-tramway based on the Indonesian islands of Sumatra and Java is proposed to span Sunda Strait. The Java-Sumatra Aerial Mega-Tramway (JSAMT) will be self-elevating and will regularly and cheaply launch passengers and cargoes via two conveyor belt-like facilities using standard winged shipping containers like those currently used by international trucking and ocean shipping industries that are volplaned across the Sunda Strait. The JSAMT will be a self-sustaining toll facility free of any requirement for international loans or funding guarantees for its construction. Its existence will remove any immediate need for an expensive to dig/maintain Nusantara Tunnel. We offer the formative basic technical specifications for the JSAMT and indicate some of the physical and cultural geographical facts underpinning our macro-engineering proposal; offshoots of a perfected and tested JSAMT may be installed at Palk Strait between India and Sri Lanka, the Gibraltar Strait and the Bering Strait by mid-21st Century.

1. INTRODUCTION

Sumatra is geographically defined by a 3 km-wide Malacca Strait and a 30 km-wide Sunda Strait; across the Sunda Strait is Java; further south Java is defined by the 18 km-wide Lomboc Strait. Approximately 245,000,000 people are citizens of the Republic of Indonesia; the nation's capital, Jakarta, situated on Java, may have ~21,000,000 residents by 2015. Jakarta Bay seaport facilities, which constitute a "mega-harbor" is being improved at considerable expense. Jakarta, Indonesia suffered a 7.5 magnitude earthquake, centered about 10-15 km east of the Capital city, at 5.968^0 South latitude by 107.655^0 East longitude on 8 August 2007. This "shaker" damaged some infrastructure locally but created no tsunami waves.

Among the natural hazards affecting those persons living/working on Sumatra and Java, tsunamis, earthquakes, volcanoes and annual forest fires are most remarkable[1]. The 27 August

---

* Presented in http://arxiv.org in 2007.

[1] Entries taken from figures are suffixed with "f" and tables with "t"

Add to list of index and check up the words:

AB Levitator
AB levitation
AB propulsion.
AB engine
AB solar sail
AB electric storage
AB reactor
AB Ramjet space propulsion
Primary ramjet
Cooling system
Electricity storage
Fly (flight) man
Flight car
Flight track
Levitation aircraft
Levitation space launch system

Artificial Magnetic field
Beam Space Propulsion
Interplanetary probe
Interstellar probe

Magnetic sail
Magsail
Electrostatic Magsail

High speed solar sail

Transfer of electricity into space
Wireless transferring of electricity

Plasma
Thermal velocity
Power density
Reaction rate
Spitzer plasma resistance
Deep of penetration
Bremsstrahlung loss energy
Wavelength
NOVA
OMERA
NIF

Atmospheric reentry
Parachute
Canopy

Optimal solid space tower

Electrostatic liner engine
Electrostatic space launcher
Electrostatic climber for space elevator
Suspended satellite system
Electrostatic levitation train
Electrostatic gun system

Micro (Mini)-Thermonuclear reactor
Radiation confinement
Self-magnetic confinement
Kinetic plasma pressure
Equilibrium pressure
Ion concentration
Electron concentration
Solar wind

Darreus Rotor
Inflatable dome settlement
Cathcart R.
Convection
Heat losses

Artificial Intelligence
Breakthrough in immortality
Brain information
Biological society
Electronic society
Computer
Immortality

1883 Krakatoa hydro-volcanic explosion and tsunami in the Sunda Strait are famous as is the haze caused by forest fires on Sumatra. During 2002, six of the ten members of the Association of Southeast Asian Nations vowed to fight fire pollution (smoke) in the region. Further development of comprehensive hazard mitigation is vitally necessary to reduce as much as possible the impact of natural and human-caused hazards. The region surrounding the Sunda Strait has a high potential to endure a possibly predictable volcanic eruption of Anak Krakatoa (volcanic ash cloud, air shockwaves, tsunami) and strong ground motion caused very strong earthquakes; in future, the Sunda Strait also may endure industrial hazards emanating from the Merak-Cilegon region and agricultural hazards in the Lampung and Ujung Kulon regions, where fallow land cleared for an anticipated season of crop planting, as well as natural forest felled and burned to create plantations, is subject to seasonal wildfires.

For these reasons, the authors concur that a Nusantara Tunnel bored beneath the Sunda Strait's seafloor is an inappropriate macroproject for the Republic of Indonesia to undertake at this time. As a viable alternative, we suggest the Java-Sumatra Aerial Mega-Tramway (JSAMT) based, in part, on the US Patent 6,494,143 awarded to Alexander Alexandrovich

---

Intelligent being
Electronic chip
Modeling of soul
Non-Biological Civilization
Entropy Second Law of Thermodynamics
Low of increasing Complexity
Rate of complexity
Fundamental Base law of Nature
Self-coping system
Natural selection
Electronic civilization
Electronic immortality
Robot
Super brain
Supreme mind
Human Soul
Mankind's existence

Johnson, A.
Organizing of research
Economical efficiency
Organizing scientific research,
planning of research,
funding research,
funding new ideas (concepts),
funding inventions and innovations,
estimating research cost,
assessment of research results,
research efficiency criteria,
innovation in organizing of scientific R&D
NIAC (NASA Institute for Advanced Concepts)
NASA (National Aeronautic and Space Administration)
DARPA (Defense Advanced Research Project Agency)
NSF (National Science Foundation) and Government Research Laboratories
SBIR - Small Business Innovation Research
Publications
Patenting
Recommendations
Final Recommendations

Bolonkin on 17 December 2002. If built, the JSAMT will be a truly remarkable 21st Century new kinetic aviation technology extension of the 260 m "Transporter Bridge" spanning the River Tyne since 1911 at Middlesbrough in England! Further, we conceive a cross-Palk Strait application of this particular technology will very likely be commenced due to the ongoing industrialization of the adjacent regions and construction of the Sethusamudram Ship Channel in a low-seismicity region.

2. Ground and Aerial Island Linkages

Indonesia's longest bridge, construction commenced during 2003 and is scheduled for completion by 2008, is the cable-stayed Suramadu Bridge; it will connect the island of Java and the island of Madura with a modern high-speed four-lane highway. Soon, Indonesians will experience the "zoomscape"—the localized human experience of architecture and landscape that has been fundamentally altered by the globalization of speedy transportation technologies — and come to appreciate their geographical surroundings in 21st Century synoptic apprehensions. The islands of Indonesia are connected by a network of hundreds of airports but the national highway and railway network remains fragmented; as a consequence, the logistics of intra-national commerce is complicated and, sometimes, uneconomical. Even common inter-modal standard shipping containers — like those stacked by the hundreds at Malaysia's seaports of Tanjung Pelepas and Port Klang — cannot yet be shifted throughout Indonesia with dispatch or economy. In other words, the Republic of Indonesia's future industrialization requires a national policy encouraging the timely initial organization of a semi-fixed aerial linkage capable of moving containerized people and cargoes over the Sunda Strait at low cost; the Java-Sumatra Aerial Mega-Tramway is a logical, and affordable, kinetic aviation technology to accomplish that task within a reasonable period of time.

The Berlin Airlift (27 June 1948 to 12 May 1949) was an extraordinary first use of a new tool of policy, a system of transporting supplies by air when ground routes (highways, railways and canals) were blockaded. The famous airlift's history is recounted excitingly in Jon Sutherland and Dian Canwell's THE BERLIN AIRLIFT: THE SALVATION OF A CITY (2007). Ultimately, about 2.1 million tonnes of supplies were ferried from western Germany to isolated Berlin in nearly 139,000 flights to Berlin through inbound routes ranging from 240 km to 450 km in length. (Weather forecasting for the two inbound routes and the single outbound air corridor was organized more thoroughly than any aviation operation previously devised.) What if an aerial mega-tramway connecting Berlin with a single site in western Germany had existed during 1948-49? Using calculations developed in SECTION 4, below, we have an answer for that alternate-history postulation: 2.1 millions metric ton of material could have been delivered in just 52500 standard containers of 40 tons each by two catapults (the first is located in Berlin and the second located in West Germany), 10 gliders (+2 in reserve, start weight of each glider is 240 metric tons) for $4 \times 40 = 160$ tonnes average capacity and flight frequency 0.1 hours (+20 minutes of loading, 16 minutes of flight, and +20 minutes of unloading), over a much shorter period of time 55 days instead of almost 320 days and 200 conventional transport aircraft! The Berlin Airlift advanced the logistics of airpower. We think the Java-Sumatra Aerial Mega-Tramway macroproject has the potential to truly revolutionize the Republic of Indonesia's industrial and commercial logistics for the suggested passengers and cargo transfer system after installation of universalized connection-disconnection devices atop the standard shipping containers; JSAMT will consist of a 40-200 meter-long cable path with ground-based engines that propel standardized winged shipping containers 35 km across the 30 km-wide Sunda Strait. The distance across Palk Strait is about the same. Foreseeably, such winged containers will carry passengers, with each container having seats, baggage stowage and a minimal cabin crew. While the new Airbus A380 will carry 555 persons, each winged container will likely safely carry about 100 persons so that loading/unloading logistics will be far, far simpler and efficient.

3. Java-Sumatra Aerial Mega-Tramway (JSAMT)

An aerial tramway is a type of aerial lift that uses comfortable cabins carrying passengers and pre-packaged cargoes; it constitutes the exact opposite of a water-traversing cable ferry that uses barges to haul passengers and cargoes across bodies of water even when water currents impose strong transverse flow. Naturally, suspended aerial tramway cabins can be violently jiggled by buffeting winds (head winds, trailing winds, tail-winds, cross-winds). JSAMT will utilize kinetic aircraft to accelerate the standard weather-tight winged shipping containers to subsonic speed of 250 m/s until its speed decays to a safe landing speed of 50-60 m/s on a paved airdrome on the other side of Sunda Strait. An acceleration of 3g will not discomfort passengers with normal health and the paved airfield runway length actually necessary to bring the winged container to a halt is only about 1.5 km. The flight path spanning the Sunda Strait will still be subject to the vagaries of the weather as well as other naturally hazardous flight conditions (Figure 1a and 1b). We anticipate a need for the internationally approved shipping containers—some of which will be loaded with passengers — to go no higher than 300 m - 500 m altitude above the Sunda Strait.

This system for Aerial Mega-Tramway—literally, a non-fixed aerial bridge—includes (Figure 1) a closed-loop cable and drive station located on the Earth's surface. The cable is supported in the air by columns equipped with rollers. Each drive station has engines located on the ground and works on any cheap energy. The system works in the following way. The subsonic load glider (winged container, aircraft) starts from a small conventional area (40 - 200 m, aerodrome, railway, see Figure 4), and is accelerated (with 0.7 - 3 g) in air up to a speed of 270–300 m/s (Mach number 0.9) by the drive station on the distance 1–6 km. The glider flies (for distance 30–70 km, Figures 2), gradually loses speed and increases attack angle of wing. When the speed drops is reduced to near landing speed, the glider then lands safely and normally.

For take-off, instead of an aerodrome, a short (40-200 m) railway may be employed (Figure 4). The glider can be started from special bogie (trolley) up to speed 50 m/s, take off and then it is accelerated in air up to a speed of 250-300 m/s (Mach number 0.9) by the cable. After acceleration the cable is disconnected and glider free flights.

The flight data are significantly improved if the kinetic aviation vehicle (glider) has variable wing area or variable swept wings. The flight altitude does not influence its range because the energy spent in climbing will be returned in gliding. The Java-Sumatra Aerial Mega-Tramway offers the following noteworthy advantages:

(a) load capability of kinetic aircraft—standard shipping containers—increases as a factor of two since the winged container has no fuel or engine;
(b) kinetic aircraft, the attachable/detachable wings, is significantly less expensive than convention aircraft with a pressurized cabin (no engines and expensive navigation devices);
(c) ground-based propulsion engine can operate on the cheapest available fuel;
(d) maximum flight-time is just a few minutes at most.

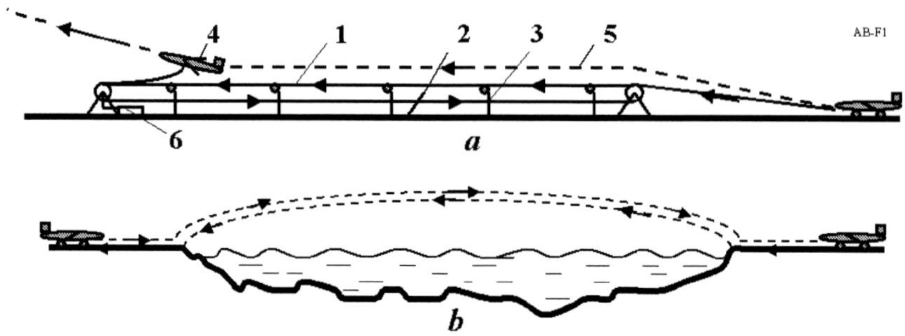

Figure 1. JSAMT. (*a*) Terminal. Drive system for acceleration of kinetic glider (acceleration station); (*b*) Aerial Mega-Tramway across water. Notation: 1 – closed-loop cable; 2 – Earth's ground; 3 – support column with roller; 4 – flight glider; 5 – trajectory of flight vehicle; 6 – drive station.

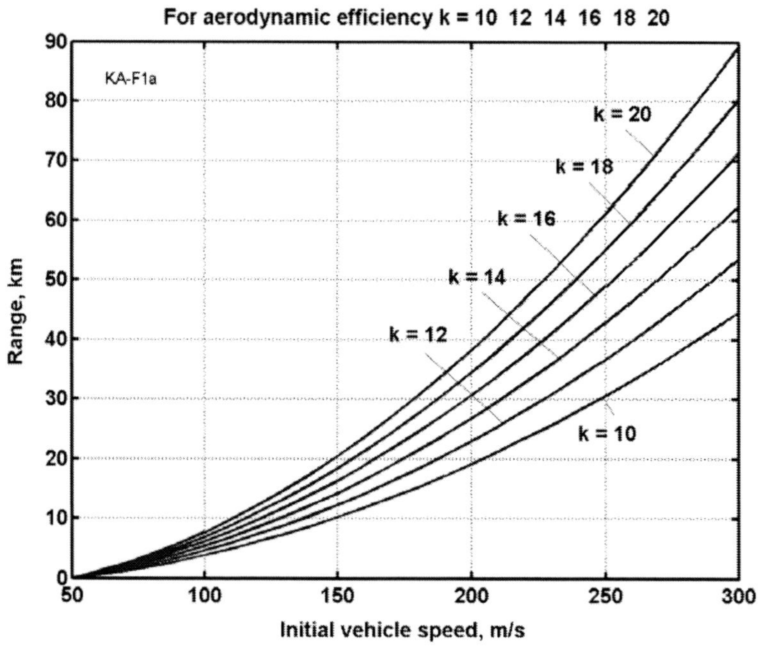

Figure 2. Range of the subsonic kinetic glider versus initial speed for different aerodynamicefficiency *K* = 10 12 14 16 18 20.

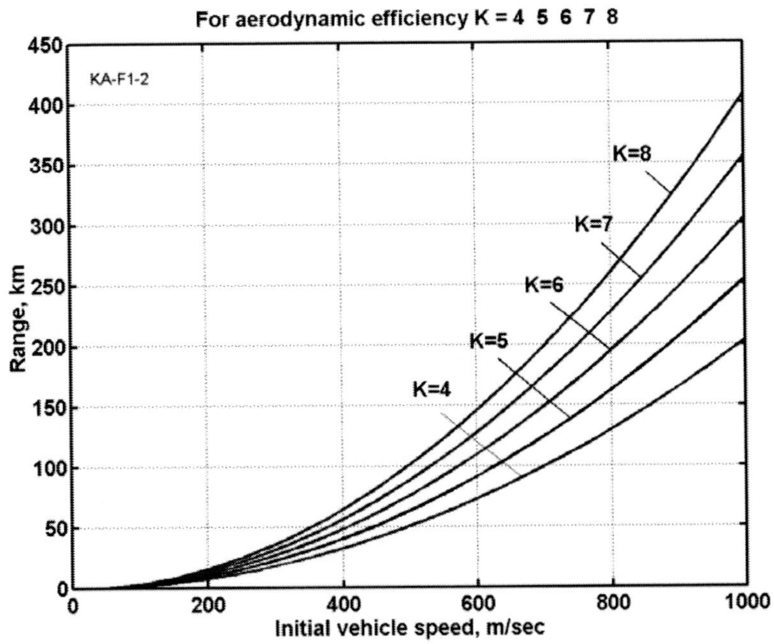

Figure 3. Range of the supersonic kinetic glider versus initial speed for different aerodynamic efficiency $K = 4\ 5\ 6\ 7\ 8$.

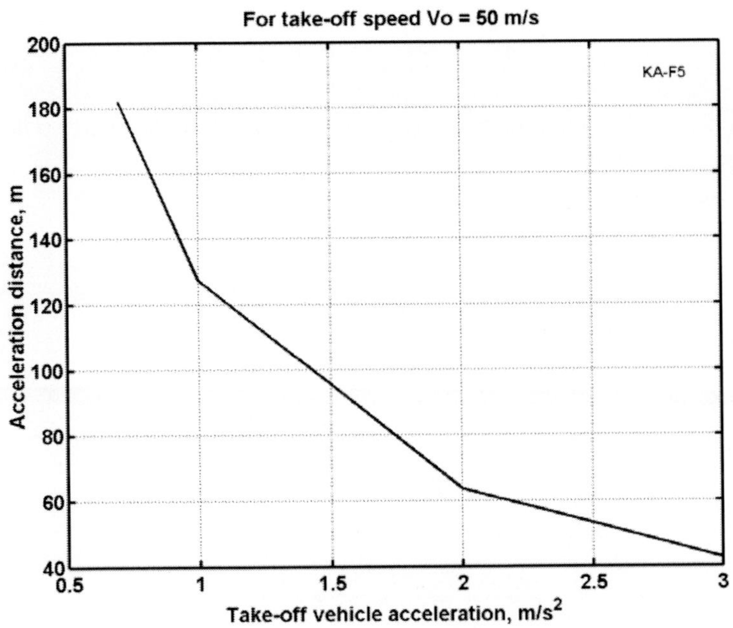

Figure 4. Aerodrome (railway) acceleration distance via a take-off acceleration of glider. Acceleration (3g) distance is 1500 m for a speed of 300 m/s for the subsonic vehicle and 17 km ($n = 3$) for a speed of 1 km/s for the supersonic vehicle (Figures 5, 6). That is not aerodrome length—that is acceleration in air.

Theory of Kinetic Vehicles and a General Estimation of Flight Data

1. The maximum range, *R*, of kinetic air vehicles is obtained from the kinetic energy of theoretical mechanics. It is equals

$$d\left(\frac{mV^2}{2}\right) = \frac{mg}{K}dR, \quad R \approx \frac{K}{2g}\left(V_1^2 - V_0^2\right), \tag{1}$$

where R is range [m]; K is the average aerodynamic efficiency (K = 10–20 for subsonic air vehicles and K = 5–8 for supersonic air vehicles. For example: the subsonic Boeing-747 has maximum K = 16, the recently retired European supersonic "Concorde" has maximum K = 7,5, supersonic aircraft XB-70 and YF-12 have $K = 7$, and Boeing 2707-300 has $K = 7.8$); g = 9.81 m/s^2 is gravity; V_1 is initial (after acceleration) speed [m/s]; $V_0 < V_1$ is final, near landing speed [m/s] (V_0 = 50–60 m/s); V is variable speed, $V_0 < V < V_1$ [m/s], $mg/K = D$ is air drag [N]; m is vehicle mass [kg]. Last equation in (1) is obtained from the first equation using integration. Results of our computations for subsonic (V < 300 m/s, M < 0.9, M is Mach number) and supersonic (M = 1 - 3) vehicles are presented in Figures 2 and 3. The range of a subsonic vehicle is 45–90 km for V_1 = 300 m/s; the range of a supersonic vehicle can reach up 400 km for V_1 = 1000 m/s.

2. Maximum acceleration distance can be calculated using the equation

$$S = \frac{V_1^2}{2gn}, \tag{2}$$

where n is overload, g. Computation results for both subsonic and supersonic modern aircraft are presented in Figure 4.

3. Average speed and flight-time are

$$V_a = \frac{V_1 + V_0}{2}, \quad T = \frac{R}{V_a}. \tag{3}$$

4. The trajectory of *horizontal turn* can be found from the following differential equation

$$\dot{V} = -\frac{gn}{K}, \quad \dot{\varphi} = \frac{L_1}{mV} = \frac{g\sqrt{n^2-1}}{V}, \quad \dot{x} = V\cos\varphi, \quad \dot{y} = V\sin\varphi, \quad or$$
$$V = V_1 - \frac{gn}{K}t > V_0, \quad \varphi = -\frac{K\sqrt{n^2-1}}{n}\ln\left(1 - \frac{gn}{K}t\right), \quad \dot{x} = V\cos\varphi, \quad \dot{y} = V\sin\varphi, \tag{4}$$

where L_1 is the projection of the vehicle lift force to a horizontal plane (vertical overload is 1); t is time [seconds]; φ is turn angle [rad].

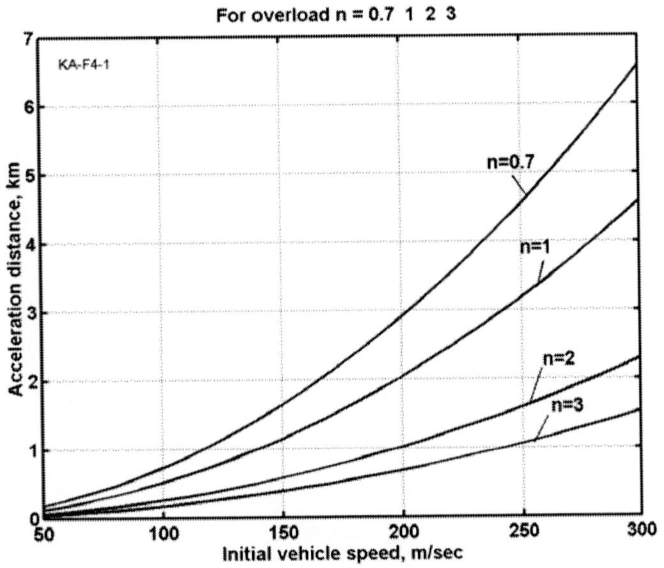

Figure 5. Full acceleration distance (air acceleration is included) of subsonic kinetic glider versus an initial speed and different horizontal overloads.

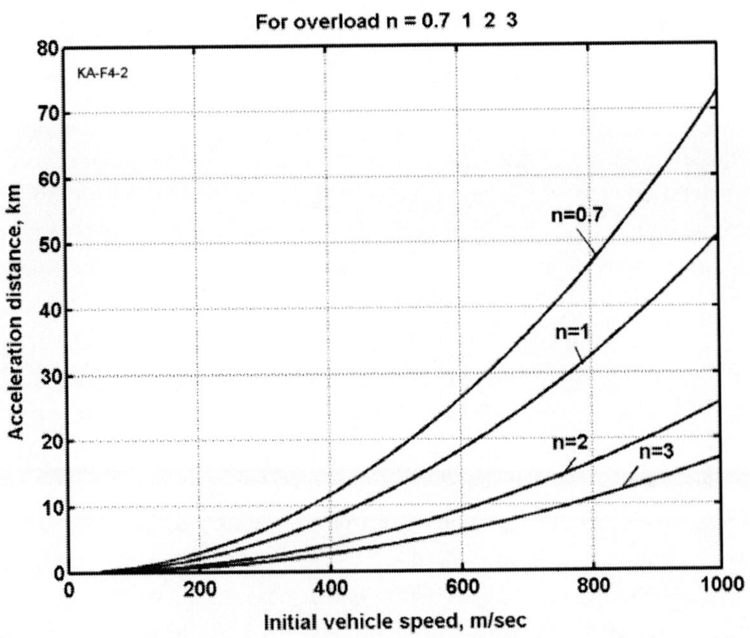

Figure 6. Full acceleration distance (air acceleration is included) of supersonic kinetic glider versus an initial speed and different horizontal overloads.

Results of computations for different overloads are presented in Figure 8. They show that the vehicle can turn back and return to its original aerodrome.

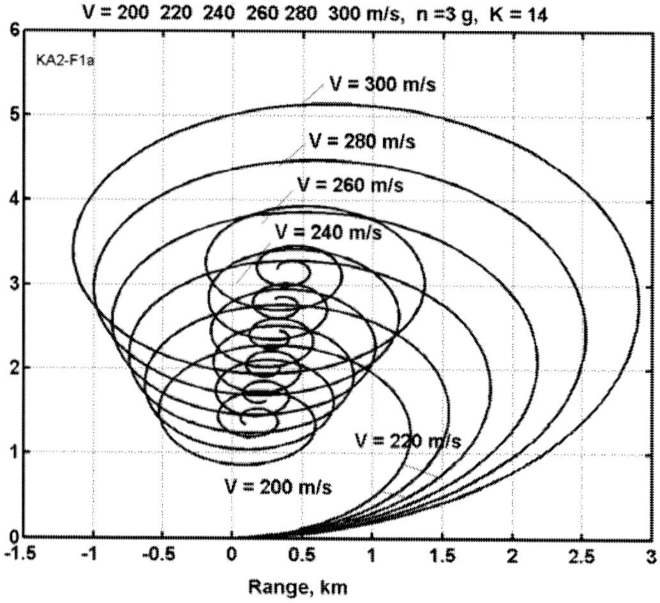

Figure 7. Horizontal deviation versus range of the subsonic kinetic vehicle for initial speed V = 200 220 240 260 280 300 m/s, horizontal overload n = 3g, aerodynamic efficiency K = 14.

5. JSAMT MACROPROJECT

Assume the mass of the flight vehicle is m = 15 tons (100 passengers and 4 members of crew); the acceleration is $a = 3g$ (this acceleration is acceptable for untrained people). The range is approximately L = 35 km (Sunda Strait, the seaway located between the Republic of Indonesia's major islands of Java and Sumatra as well as Palk Strait between India and Sri Lanka) (see Figure 3 or calculate using Eq. (1) for a final acceleration speed of 250 m/s and K = 14, range is 43 km). The needed acceleration distance is S = 1000 m (Figure 4). The starting area is only 40 m (Figure 3). The time of horizontal acceleration is $t = (2S/a)^{0.5}$ = 8.2 seconds. The flight time $t = 2\times L/(V_1+V_0)$= 467 s = 7,8 minutes. Assuming it uses a low-cost artificial fiber (σ = 600 kg/mm^2) widely produced by current industry, and a safety tensile strength of the drive vehicle cable is σ = 180 kg/mm^2 (the safety factor is 600/180 = 3.33), density γ = 1800 kg/m^3 (K_1 = 180/1800 = 0.1). Then the cross-section area of the vehicle cable around the vehicle will be $S_1 = 3m/\sigma$ = 250 mm^2, and the cable diameter is d = 18 mm. The mass of the cable is $M = S_1\gamma L_d$ = 450 kg. Here L_d = 1000 m is the maximum length of the drive cable.

The energy required for acceleration of the aircraft and the cable is $E = mV^2/2$. This is about E = 47 Mega Joules (1 Mega Joule = 10^6 J) if V = 250 m/s. The drag of the aircraft and cable is about D = 3 tons, which means $E = DL = 3\cdot10^4\cdot1000$ = 39 Mega Joules. If the launches are made launches are made every 0.1 hours and the ground-based engines must have a total power of about $P = E/t = 30\cdot10^6/6/60$ = 83 kW. If the engine efficiency is η = 0.3 the fuel consumption will be $F = E/\varepsilon/\eta = 39\cdot10^6/\varepsilon/0.3$ = 3.1 kg per flight. Here $\varepsilon = 42\cdot10^6$ [J/kg] is the energy capability of diesel fuel. This means that 0.031 kg of fuel is used for each passenger.

If tensile strength is $\sigma = 180$ kg/mm^2 = $1.8{\cdot}10^9$ N/m^2, $\gamma = 1800$ kg/m^3, then the total weight of the flywheels (as storage energy) will be about $M_w = 2E\gamma/\sigma = 2{\cdot}47{\cdot}10^{6}{\cdot}1800/1.8{\cdot}10^9$ = 94 kg.

Assume a cost of twenty million 2008 USA dollars for the installation, a lifetime of 20 years, and an annual maintenance cost of one million dollars. If 100 passengers are launched on every flight, there are 10 flights every hour for 350 days a year and the load coefficient is 0.75, then $N = 2\times100\times10\times24\times350\times0.75$ = 12,600,000 passengers will be launched per year. The launch cost per passenger is \$2,000,000/12,600,000 = \$0.16 plus fuel cost. If 0.031 kg of fuel is used for 1 passenger and the liquid fuel price is \$0.5 per kg, then the cost is \$0.016/person for liquid fuel. The total production cost will be about \$0.18/person. If each ticket costs but one USA dollar, then the profit could be 10.3 million USA dollars each year! The efficiency will be improved when the glider can take 200 or more passengers. For such a short distance, five—four working + one in reserve—gliders are a more-than-sufficient vehicle stock. Fuel prices change with time, but in any case the cost of delivery will be sometimes less than delivery by conventional aircraft. During January 2008, a barrel of oil sold on the world market for 100 USA dollars for the first time in the modern world's recorded history! The delivery time is less than any city transportation (bus, railway or subway)!

6. OTHER MACRO-PROJECT FACTORS

The public image of the JSAMT is at least as important as the macro-project's facility management during its physical creation. Considering that the JSAMT is a novel freight and passenger moving innovation, some serious research must be undertaken in order to artfully craft a favorable public image of JSAMT that will enlighten and attract the public in the Republic of Indonesia. For example, many Indonesians will find computer simulations of a trip across the Sunda Strait aboard the winged containers too much "zoomscape", perhaps even frightening. JSAMT will be a pioneering new aviation and cargo-handling technology that must be presented locally both as a patented invention familiar enough not to alarm people and innovative enough so as not to seem too unfamiliar! One of the world's tallest office buildings, the Petronas Towers in Kuala Lumpur, has excited the public of Malaysia since 1996. So, public relations image managers will need to tune their public regional and global presentations to investors, government authorities and future JSAMT users carefully, taking into account all likely negative reactions.

The development of spacious well-lit and secure container storage yards at both ends of the JSAMT will require cooperation at many levels of government. There will be no need for transit sheds on Java or Sumatra since the standard shipping container is weather-tight and functions as a mobile warehouse. Refrigerated containers will require a reliable source of electricity, of course. We view JSAMT as a unified/unitary transportation kinetic aviation system cooperating with railroad operators and truckers that has the potential to provide fixed linkages amongst, at least, the main islands of the Republic of Indonesia. JSAMT could reduce ship traffic in congested sea-lanes (Malacca Strait, Lomboc Strait and Sunda Strait) and help shippers avoid pirates plaguing slow-moving high-value cargo vessels. Once the JSAMT has been perfected, it may become adaptable to other Earth-surface sites such as the

Bering Strait or Gibraltar Strait. At Palk Strait, the Aerial Mega-Tramway could lessen the ship traffic environmental impacts on Palk Bay.

Even some suitable old aircraft could be cut-down and remodeled for use in the above proposed transport system for the Republic of Indonesia

Boeing 737.

Douglas C-133 Cargomaster.

REFERENCES

[1] Bolonkin, A.A., Non-Rocket Space Launch and Flight Elsevier, London, 2006.

[2] Bolonkin, A.A. and Cathcart, R.B., A Low-Cost Natural Gas/Freshwater Aerial Pipeline (2006). *http://arxiv.org*.

Chapter 6

ANTARCTICA: A SOUTHERN HEMISPHERE WINDPOWER STATION?*

ABSTRACT

The International Polar Year commenced in 2007. We offer a macro-project plan to generate a large amount of electricity on the mostly ice-encased continent of Antarctica by using sail-like wind dams incorporating air turbines. Electricity can be used to make exploration and exploitation efforts on Antarctica easier. We offer the technical specifications for the Fabric Aerial Dam and indicate some of the geographical facts underpinning our macro-engineering proposal.

INTRODUCTION

Antarctica is Earth's southernmost continent, overlying the South Pole. It is situated in the southern hemisphere, almost entirely south of the Antarctic Circle, and is surrounded by the Southern Ocean. At 14.4 million km^2, it is the fifth-largest continent in area after Asia, Africa, North America, and South America. Some 98% of Antarctica is covered by ice, which averages at least 1.6 kilometres in thickness.

On average, Antarctica is the coldest, driest and windiest continent, and has the highest average elevation of all the Earth's continents. Since there is little precipitation, except at the coasts, the interior of the continent is technically the largest cold desert in the world. There are no permanent human residents and there is no evidence of any existing or pre-historic indigenous population. Only cold-adapted plants and animals survive there, including penguins, fur seals, mosses, lichen, and many types of algae.

The Antarctic Treaty was signed in 1959 by twelve countries; to date, forty-five countries have signed the treaty. The treaty prohibits military activities and mineral mining, endorses scientific research, and protects the continent's pristinity. Ongoing experiments are conducted by more than 4,000 scientists of many nationalities and with different research interests. The International Polar Year, sponsored by the International Council of Science and the World

* Presented in http://arxiv.org in 2007.

Meteorological Organization, commenced during March 2007. The USA's Geological Survey collaborates on unique, comprehensive space imagery of Antarctica. GOTO: http://landsat.usgs.gov/.

Including all air flows within a 1 km layer above our Earth's solid surface, a technically possible wind energy resource base would likely be about 6 TW; probably less than 10% of that flux initially energized by the Sun could actually now be converted directly to power-line distributed electricity without causing major detrimental changes to the Earth-atmosphere. Scientific supercomputer modeling has shown that the extraction of kinetic energy from wind would somewhat alter Earth's turbulent transport in the atmospheric boundary layer (Archer and Jocobson, 2005).

Nevertheless, wind is a clean source of energy that has been utilized by humans for centuries to grind grains, pump water, propel sailing craft, and to perform other work, according to Wendy Williams and Robert Whitcomb's *Cape Wind: Money, Celebrity, Class, Politics, and the Battle for our Energy Future on Nantucket Sound* (Public Affairs: NY, 2007). Wind power, for many years deemed to be as fickle as the wind itself, can now be manipulated to become a steady, dependable electricity resource. Because wind farms can be interconnected with transmission cables, it has been found that large-scale installations of wind power generators could be utilized as a reliable baseload about 33-45% of the time (on average), according to Cristina L. Archer and Mark Z. Jocobson ["Supplying Baseload Power and Reducing Transmission Requirements by Interconnecting Wind Farms", *Journal of Applied Meteorology and Climatology*, 46: 1701-1717, November 2007].

Antarctica, during summer, of course!

Artists such as Tal Streeter, Howard Woody and Tsutomu Hiroi have flown kite-like sculptures and Jose M. Yturralde flew geometrical structures; Otto Piene's "Olympic Rainbow" consisted of five 600 m-long helium-filled polyethylene tubes displayed at the 20th Olympiad in Munich, Germany, during 1972. Piene coined the descriptive term "Sky Art" in 1969. "Wind farm" is the popular term used for an aggregation of wind turbines clustered at a site with persistent favorable air fluxes. Unfortunately, existing wind energy systems have deficiencies that limit their commercial applications: (1) wind energy is unevenly distributed geographically and has relatively low energy density. A single huge wind turbine cannot be placed on the ground. Instead, numerous small wind turbines must be used; (2) wind power is a function of the cube of wind velocity. At ground level, wind speed is low and rarely steady; (3) wind power system productivity is entirely dependent on prevailing weather, making it nearly impossible for productivity to be scheduled; (4) wind turbines of conventional design produce noise and are aesthetically unattractive.

Old-fashion ground-based wind energy installations.

Modern offshore wind turbines near Copenhagen, Denmark.

2. DESCRIPTION

It was not until *circa* 1840 that Antarctica was established to be an isolated continent. Its coastline is about 18,000 km in length. Dry katabatic (gravity-driven) winds blow coastward from the high interior icecap. The windiest place on Earth close to sea level is Cape Denison (67.02^0 South latitude by 142.58^0 East longitude) in Commonwealth Bay, Antarctica, where winds exceeding 50 m/second have been recorded regularly. Of all the continents, only on Antarctica does a single meteorological element (wind) have such an overwhelming influence on the climate. The dry katabatic winds blow with great constancy in direction, often moving at 20-40 m/sec over the smoothest icecap surface for hundreds of kilometers. As the katabatic winds leave the South Pole and approach to within about 100 km of the coastline, they tend to decrease in speed owing to drag over a rougher ice surface. Thereafter, Antarctica's winds generally blow off the continent's coastal escarpment toward the Antarctic Circle (Figure 1).

Worldwide, there are many macro-project R&D programs for the development of wind energy systems but most of them are ground or tower based; Australian macro-engineers have proposed Earth-stratosphere deployed kite-like electricity generators tethered to the ground by strongly anchored cables. We propose an innovative wind energy harnessing system, the Fabric Aerial Dam (FAD), which can operate successfully up to the lower boundary of the Earth's stratosphere; our defining challenge is to manage the transfer of the energy obtained to the consuming ground-based infrastructure. We propose that the first installation be made on the continent of Antarctica because electrical power is needed and because the wind energy harvesting system we propose can cause a "windbreak" or "shelterbelt" effect downwind that would be inappropriate for a densely populated region (Badescu and Cathcart, 2008).

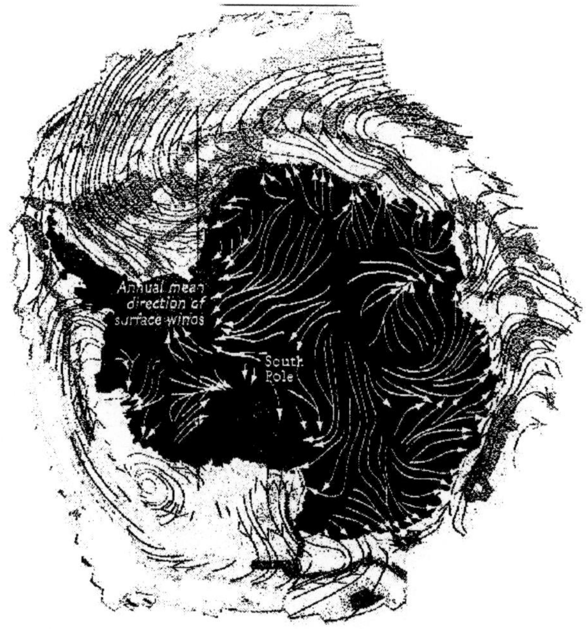

Figure 1. Map of annual mean direction of surface wind over ice.

In 1997, the first truly accurate map of Antarctica was produced by the Canadian Space Agency using its Radarsat-1 Earth-orbiting satellite. Antarctica's winter starts in April and ends during September; at the South Pole (elevation 2835 m), the sun rises on September 21 and sets on March 21. Antarctica has no indigenous human population; about 1,000 persons over-winter and 20,000 persons may work during summer on the icy continent. The lowest temperature ever measured was recorded in 1983 at Russia's Vostok Station (elevation of 3744 m) in East Antarctica—minus 89^0 C. Started in 2002, and virtually completed by 2006, Antarctic-1 is the continent's only "ice highway", connecting McMurdo on the coast to the Amundsen-Scott South Pole Base; Antarctic-1 was anticipated by an artist, the sculptor Rachel Weiss, more than twenty years ago! FAD, the Fabric Aerial Dam can be used as a beautifying decoration, as a projection screen for nighttime motion pictures (outdoor cinema), as a nighttime solar reflector and as a daytime partly focused heater for targeted small regions on the continent of Antarctica.

The Fabric Aerial Dam (FAD) is diagramed in Figure 2.

The FAD embodies a thin, possibly transparent, film 1, support cables 2, conic tunnels (3) funneling naturally flowing air to the turbines 4, electric generator 5, spool 6 for support cable, spool motor 8, film cable 9. The FAD will be installed perpendicular to the average main direction of the site's wind at an altitude where the wind is more than 1 m/s most days of the year. The FAD shields the downwind territory from extreme wind impacts and it can help to cause orographic precipitation. If storm a wind blows in opposed direction, the spools (6) can reel in the cable (2) and settle the flexible film aerial dam material safely on the ground for temporary storage. The same technique may be used when repairs to the FAD are found to be necessary.

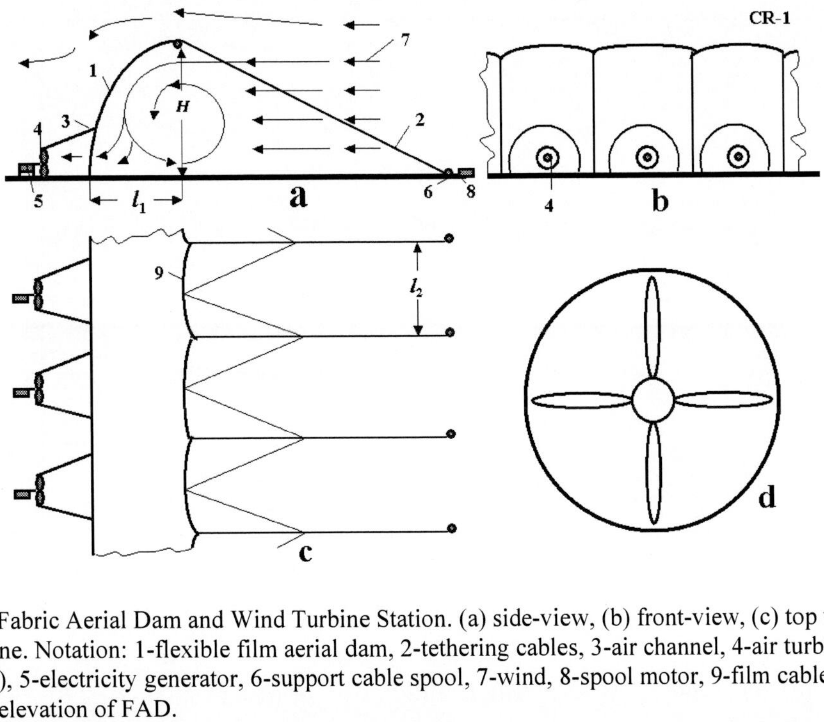

Figure 2. Fabric Aerial Dam and Wind Turbine Station. (a) side-view, (b) front-view, (c) top view, (d) wind engine. Notation: 1-flexible film aerial dam, 2-tethering cables, 3-air channel, 4-air turbine (propeller), 5-electricity generator, 6-support cable spool, 7-wind, 8-spool motor, 9-film cable. H - deployed elevation of FAD.

If the wind is less than 1 m/second, the FAD will tend to fall to ground but with a stronger wind the FAD will billow, taking off from the ground; if the wind is excessively strong, however, it may irreparably damage the FAD. We can calculate the minimum and maximum admissible to safely billow an installed FAD facility. Our purpose in doing so is to estimate the time (% or actual days in a year) when the FAD can operate properly. We assume the average wind speed at an altitude of H_0 = 10 m and 50 m is approximately V_0 = 8 and 12.7 m/s respectively.

The change of wind via altitude approximately is described by equation

$$\frac{V}{V_0} = \left(\frac{H}{H_0}\right)^{\alpha}, \qquad (1)$$

where V_0 is the wind speed at the original height, V the speed at the new height, H_0 the original height, H the new height, and α the surface roughness exponent.

We assume the surface roughness exponent, α, over Antarctica's ice is 0.10.

The result of our computations is shown in Figure 3 below.

Annual speed distributions vary widely from one site to another, reflecting climatic and geographic conditions. Meteorologists have found that the Weibull probability function best approximates the distribution of wind speeds over time at sites around the world where actual distributions of wind speeds are unavailable. The Rayleigh distribution is a special case of the Weibull function, requiring only the average speed to define the shape of the distribution.

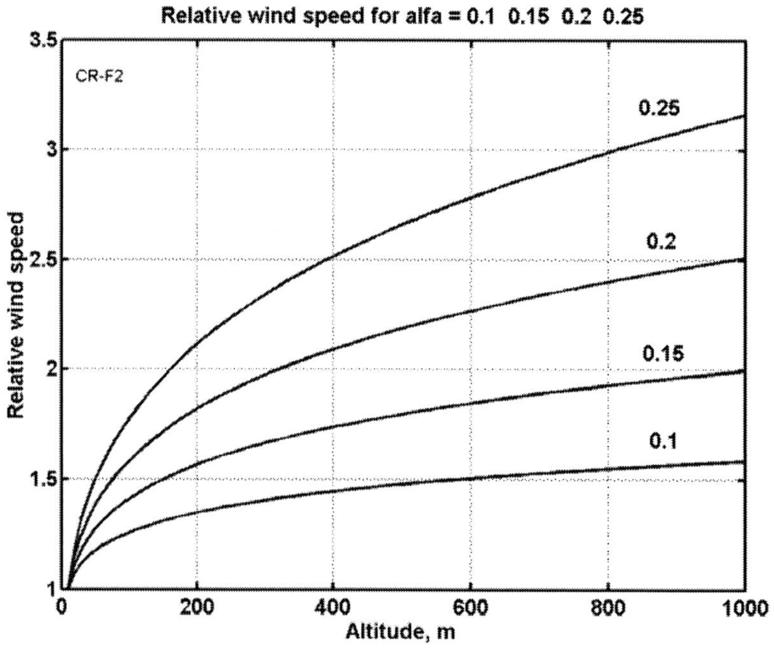

Figure 3. Relative wind speed via altitude and Earth Surface. For ice α = 0.1.

Equation of Rayleigh distribution is

$$f_x(x) = \frac{x}{\alpha^2}\exp\left[-\frac{1}{2}\left(\frac{x}{\alpha}\right)^2\right], \quad x \geq 0, \quad E(X) = \sqrt{\frac{\pi}{2}}\alpha, \quad Var(X) = \left(2 - \frac{\pi}{2}\right)\alpha^2, \tag{2}$$

where α is new parameter.

These data gives possibility to easy calculate the amount (percent) days (time) when air dam can operate in year (Figure 4). It is very important value for the estimation efficiency of offered devices.

Let us compute two examples:

1) Assume, the air dam has a minimum admissible wind speed of ~3 m/sec., the average annual speed in the given region is 6 m/s. From Figure 4, Eq. (2), we can then derive the ~15% probability of a wind of less than 3 m/sec. By the same means, we can compute the probability of a storm wind speed.
2) Assume, the average annual speed in given region is 6 m/s, the maximum admissible wind speed is 7 m/s. The probability that a wind speed will be less then 7 m/second is 55%, less then 8 m/second is 65% (Figure 4).

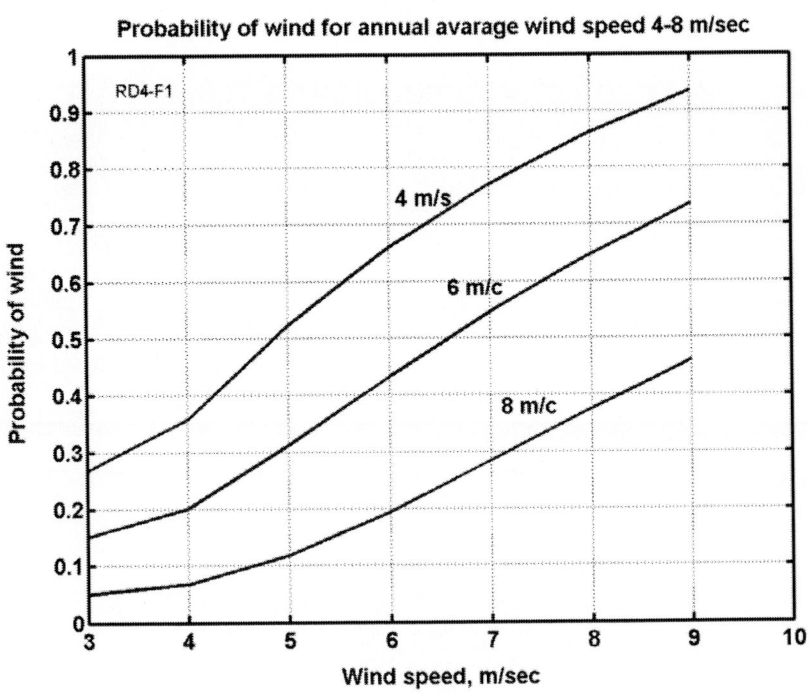

Figure 4. Probability of wind via wind speed and average wind speed in given place.

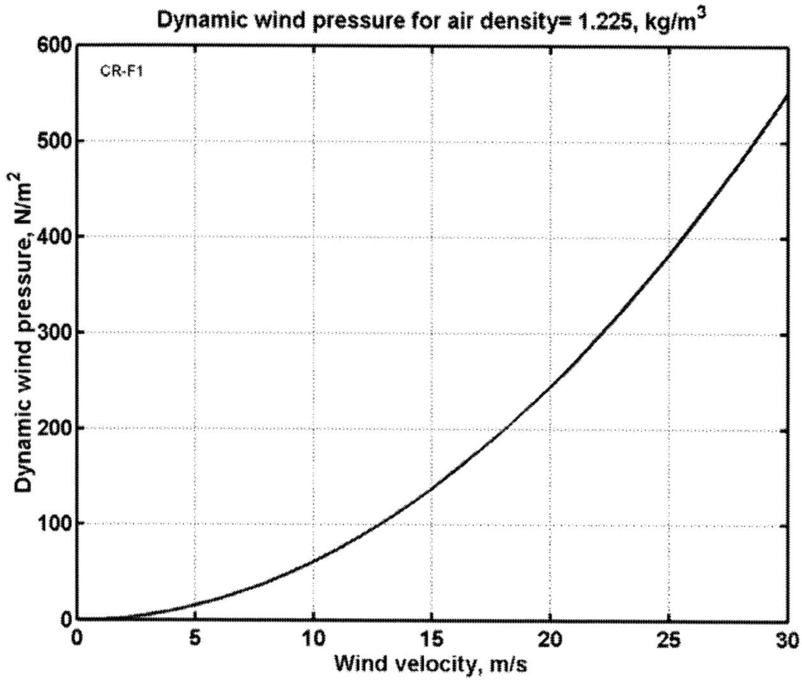

Figure 5. Wind dynamic pressure via wind speed.

3. Theory of Fabric Aerial Dams (FAD)

1. Dynamic pressure P of motion air (wind) can be computed by equation:

$$P = \frac{\rho V^2}{2}, \tag{3}$$

here P is wind dynamic pressure, N/m^2; ρ = 1.225 kg/m^3 (standard value) is air density; V is wind speed, m/s. Result of computation is presented in Figure 5.

2. The wind power can be computed by

$$N = \eta \frac{\rho A V^3}{2}, \tag{4}$$

where N is power, W; η is coefficient efficiency of air turbine, $\eta = 0.3 \div 0.5$; A is turbine area, m^2; a conical entry into turbine 4 is shown in Figure 2 can increase the effective area sometimes.

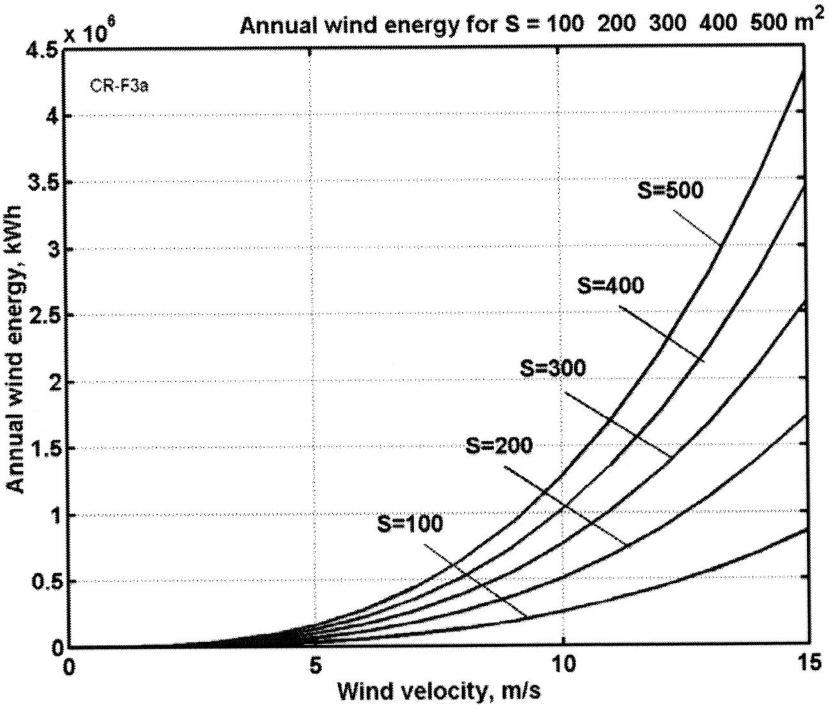

Figure 6. Annual wind energy via wind velocity and turbine area S.

The annual average wind speed near latitude 60^0 South is 12.7 m/s at height 50 m. If FAD has an off-the-ice cap base-top height of 50 m and is 1,000 km long, a coefficient efficiency of 0.5, the total power of FAD wind turbines may be more than 30 GW. In other words, a ring-shaped FAD 'enclosing' Antarctica theoretically might generate 450 GW of electricity." Something like our FAD has since been proposed by Laurie Chetwood in the UK, as reported in the March 2008 issue of the US periodical POPULAR SCIENCE (272: 28-29)

3. *Annual energy E* received from wind turbine is

$$E = 8.33N \text{ [kWh]} \tag{5}$$

Computation of this equation is presented in Figure 6.

4. *Requisite film thickness* δ is

$$\delta = \frac{T}{2\sigma}, \quad T = PH\ , \tag{6}$$

where T is wind force in 1 m width of the film, N/m; H is dam height, m (Figure 2); σ is safety tensile stress, N/m^2. Computational results for different values of σ are presented in Figures 7-8. Multiplier 1/2 accounts the force T is kept in two points (top and bottom).

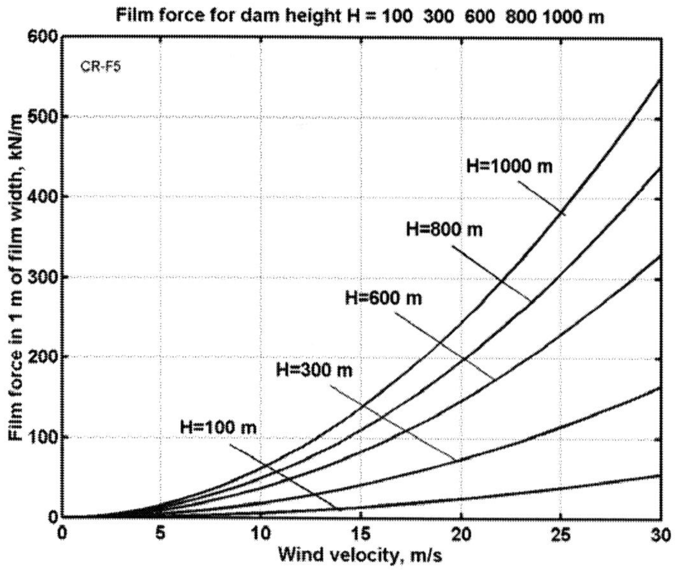

Figure 7. Film force active in one meter film width versus wind velocity and height of fabric air dam H.

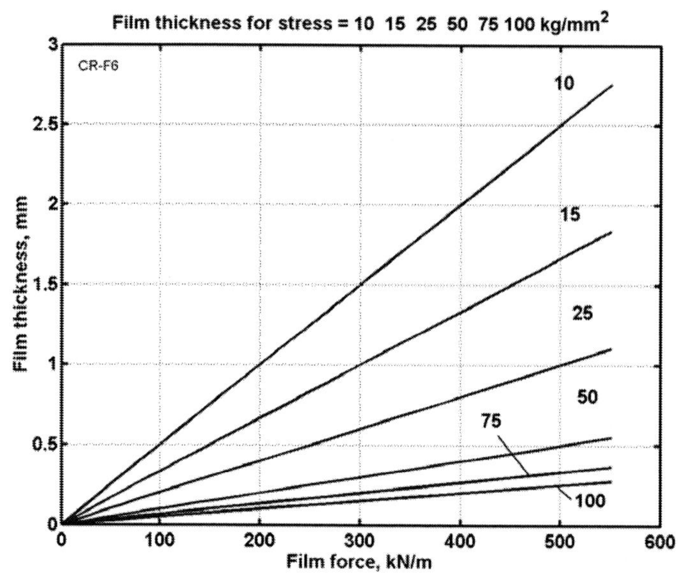

Figure 8. Film thickness via film force and safety film stress σ= 10, 15, 25, 50, 75, 100 kg/mm^2.

5. *The requisite diameter d of support cable* is

$$S = \frac{T l_2}{\sigma}, \quad d = \sqrt{\frac{4S}{\pi}}, \tag{7}$$

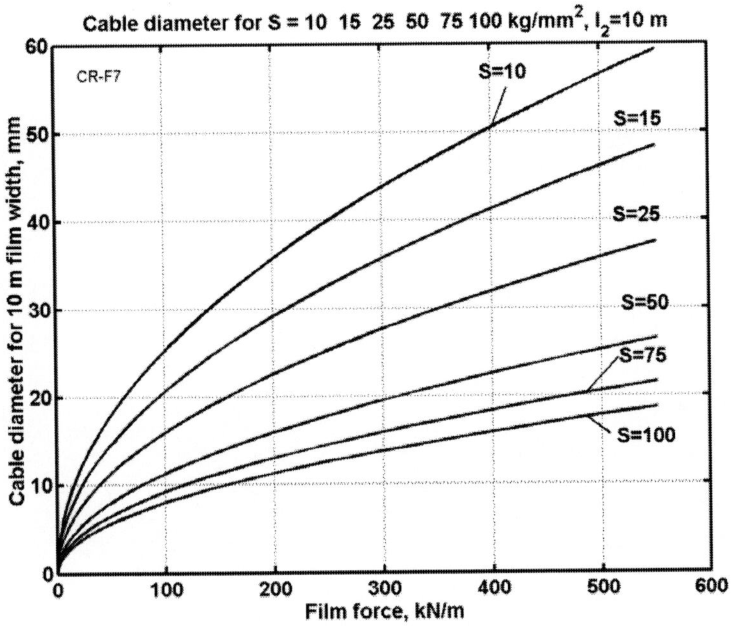

Figure 9. Cable diameter for 10 m film width via film force and safety cable tensile stress.

where S is cross-section area of cable [m^2]; l_2 is distance between cables, m. Results of computation for distance 10 m are presented in Figure 9. If distance between support cables is different from 10 m, the cross-section cable area must be changed in proportion.

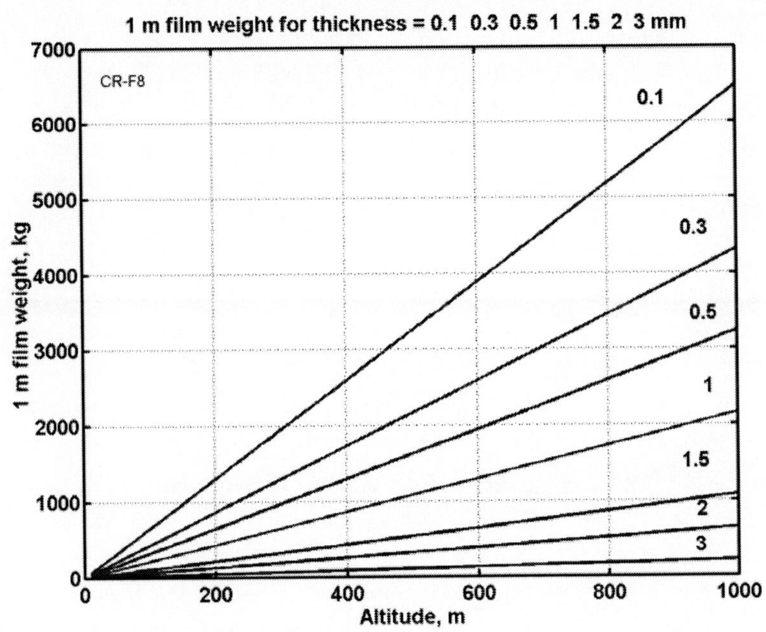

Figure 10. Film weight of width 1 m via dam height and film thickness.

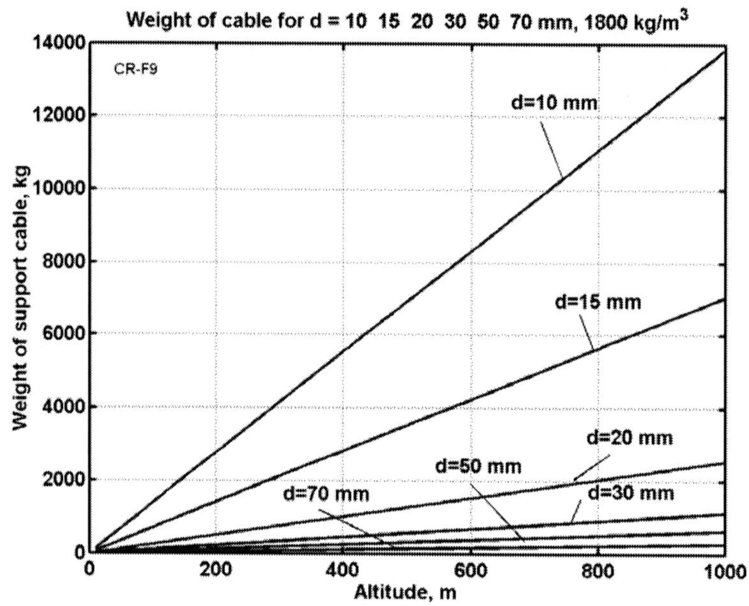

Figure 11. Weight of film support cable versus dam height and cable diameter for cable density of 1800 kg/m^3.

6. *Weight W_f of 1 m width film* is

$$W_f = 1.2H\delta\gamma\,, \tag{8}$$

here γ is specific density of the film (conventionally, for the most artificial fiber $\gamma = 1800$ kg/m^3). For Factor 1.2, take into curve form of air-blocking film. Result of computation is in Figure 10.

7. *The weight W_c of the single cable* for angle 30° to geographic horizon is

$$W_c = 2HS\gamma\,, \tag{9}$$

where γ is specific density of the cable (conventionally, for the most artificial fiber $\gamma = 1800$ kg/m^3).

Results of computation are shown in Figure 11.

4. Environmental Consequences

All of the anchor points for the FAD tie-down cables will be in thick ice. We propose the use of "ice-anchors" that are implanted using a simple heated water drill far smaller than the device currently being used to reach a deep lake beneath Antarctica. These anchor points can be removed and reset without environmental consequence.

Isolated Antarctica, which surrounds the South Pole, has no native land-based vertebrates save flightless penguins. Since 2003, penguins have lived in close proximity with big wind turbines at the Australian Mawson Research Station; neither the tall metal tower structure nor the whirring of the large-diameter propellers seems to disturb their normal activities. Located on the coast, the Mawson Research Station is already harnessing the katabatic winds we intend to harness further inland. Other seabirds, such as the Wandering Albatross, the Grey Headed albatross, Antarctic Fulmers, Cormorants, Antarctic and Giant Petrels will not be affected by FAD mainly because these flying birds do not range inland very far. FAD will carefully be sited beyond their normal ocean feeding movements and nesting activities.

5. Conclusion

We have shown conclusively that Fabric Aerial Dams that include air turbines generating electricity can be successfully deployed and operated almost continuously on the continent of Antarctica. Construction of our macroproject would open vast territories to exploration and exploitation. The FAD, resembling the sails of watercraft, can be laced with conductive wires to heat slightly, which would amplify the Teflon-coated textile's capacity to shed snow and ice. Electric land vehicles could travel on safe paths, such as the Antarctic-1 highway, while drawing their motive power from battery-recharge stations fed by FAD. It seems reasonable to anticipate the use of battery-powered vehicle

Imagine, if you will, cargo and passenger-carrying "Land Trains", such as a modern-day technical adaptation of R.G. LeTourneau, Inc.'s famous 1955 SNO-FREIGHTER (Model VC-22), perhaps traveling periodically on an Antarctic-2 highway encircling Antarctica approximately 100 km inland from the coastline to maintain the FAD. Various industrial activities, such as mining remote ore bodies or petroleum deposits, will also be enhanced made more economic and convenient by the ready availability of large amounts of clean energy (Green electricity) derived from wind-power.

At present, the Protocol on Environmental Protection to the Antarctic Treaty, which came into force on 14 January 1998, precludes all extraction of minerals except for scientific study. However, with the increase of the world's population, new geopolitical pressures may soon arise that will initiate a major change of the Antarctic Treaty to accommodate the globalized needs of humanity.

Therefore, we anticipate an early-21st Century reopening of the worldwide debate on the Convention on Regulation of Antarctic Mineral Resource Activities that terminated in 1988.

We anticipate that Antarctica will, eventually, produce an excess of electricity and we, therefore, expect that such excess will be exported via super-conductive undersea electric power cables to South America. Also, we forecast that FAD can be used to power the directed movement of Antarctica's gigantic tabular icebergs to freshwater-short arid lands such as Australia. In additional to ground wind the Antarctic has strong high atmospheric flows. They also can be utilized.

References

Archer, C.L. and Jacobson, M.Z., “Evaluation of global wind power”, *Journal of Geophysical Research,* Vol. 110, page D12110 (2005).

Bolonkin, A.A., and Cathcart, R.B., "Antarctica: a southern hemisphere wind power station?", *International Journal of Global Environmental Issues,* Vol. (2008).

Chapter 7

FLOATING CITIES, ISLANDS, AND STATES*

ABSTRACT

Many small countries are in real need of additional sovereign territory. Sometimes, they even build on rotting landfills and install various kinds of expensive artificial islands. The ocean covers 71% of our Earth's surface. Those countries (or persons of wealth) starting the early colonization of the ocean may obtain advantages through additional territory or creating their own independent state. An old idea is building a big ship. The best solution to this problem, however, is the provision of floating cities, islands, and states. The authors propose to employ floating cities, islands, and states as cheap floating platforms created from a natural ice field taken from the Arctic or Antarctic oceans. These platforms protected by air-film (bottom and sides) and a conventional insulating cover (top) and having a cooling system can exist for an almost unlimited period of time. They can be multiplied in number or increased size at any time, float even in warm-water subdivisions of the world-ocean, travel to different countries on continents and islands, serve as artificial airports, harbors and other marine improvements, as well as floating cities and industrial bases for virtually any use.

Authors have researched and computed the parmount parameters of these floating ice platforms, other methods of building such floating territory, compare them and show that the suggested method of realization is the least costly and most efficient means for sustainable ocean colonization by mobile humans.

Keywords: floating cities, ice floating platform, ocean colonization

INTRODUCTION

Short Information about Oceans, History Large Ship, and Ice Fields

An *ocean* is a major body of saline water, and a principal component of our planet's remarkable hydrosphere. Approximately 71% of the Earth's surface (an area of 361 million

---

* Presented in http://arxiv.org, April 2008.

square kilometers) is covered by ocean, a continuous body of seawater that is customarily divided into several principal named oceans and smaller named seas. More than half of this area is deeper than 3,000 meters. Average oceanic salinity is around 35 parts per thousand (ppt) (3.5%), and nearly all seawater has a salinity in the range of 31 to 38 ppt. Interestingly, the place furthest from the world-ocean—that is, the official "pole of inaccessibility" is in Asia (46^0 17' North latitude by 86^0 40' East longitude), according to Daniel Garcia-Castellanos and Umberto Lombardo's "Poles of Inaccessibility: A calculation algorithm for the remotest places on Earth", SCOTTISH GEOGRAPHICAL JOURNAL 123: 227-233 (September 2007).

The volume of Earth's ocean is approximately 1.3 billion cubic kilometers, and its average depth is 3,790 meters. The vast volume of the deep ocean (anything below 200 m) covers about 66% of our Earth's surface.

The total mass of the planetary hydrosphere is about 1.4×10^{21} kilograms, which is about 0.023% of the Earth's total mass. Less than 2% is freshwater, the rest is saltwater, mostly in the ocean.

Though generally recognized as several 'separate' oceans, these waters comprise one global, interconnected body of salt water often referred to as the "world-ocean" or "global ocean". That includes: Pacific Ocean, the Atlantic Ocean, the Indian Ocean, the Southern Ocean, and the Arctic Ocean.

Ocean colonization is the theory and practice of permanent human settlement of oceans. Such settlements may float on the surface of the water, or be secured to the ocean floor, or exist in an intermediate position. "Marine city" is defined at length at http://parole.aporee.org and the history of such facilities is briefly outlined in "Prototype cities in the sea", authored by Sandra Kaji-o'grady and Peter Raisbeck, for THE JOURNAL OF ARCHITECTURE 10: 443-461 (September 2005).

One primary advantage of ocean colonization is the expansion of livable area. Additionally, it might offer various other possible benefits such as expanded resource access, novel forms of governance (for instance mini-nations), and new recreational activities for athletic humans.

Many lessons learned from ocean colonization will likely prove applicable to near-term future outer space and other-planet colonization efforts. The ocean may prove simpler to colonize than interplanetary space and thus occur first, providing a proving ground for the latter. In particular, the issue of legal sovereignty may bear many similarities between ocean and outer space colonization with space station settlements; adjustments to social life under harsher extra-terrestrail circumstances would apply similarly to the world-ocean and to outer space; and many technologies may have uses in both environments.

Economy of world-ocean. Central to any practical attempt at ocean colonization will be the underlying global and local economic reality. To become self-sustaining, the colony will aim to produce output of a kind which holds a comparative advantage by occurring on or in the ocean. While it can save the cost of acquiring land, building a floating structure that survives in the turbulent open ocean has its own costs. Ocean-front land—say, land more than 100 meter from the coastline—can hold a very high value, especially in countries with no income taxes, so building space and selling it may prove popular. Tourists often visit warm locales during the winter; indeed, tourism drives the economies of many small island nations. The colony might also compete as an offshore center for financial transactions.

While importing food and fishing may compose the majority of ocean settlement food consumption, other possibilities include hydroponics and open-ocean aquaculture. Thus, an ocean settlement may be either a net importer or a net exporter of food products.

Such settlements or cities would probably import diesel and run conventional power plants as small islands everywhere do. However, other possibilities include solar power, nuclear plants, deep-sea oil deposits, and developing/farming a species of seaweed or algae for biofuel.

Ocean thermal energy conversion (OTEC) is another potential energy source. All that is required is tropical (warm) surface water and access to deep, very cold water. The difference in temperature is used to drive an electric generator via a turbine. (There is an added benefit in that the deep cold water usually is more fertile than surface water in the open ocean, and can support mariculture).

Similar communities already exist in the form of hotels, research stations, houseboats, houses on stilts, land below sea level behind dikes, vacation cruise ships, ocean oil rigs, etc.

Furthermore, humans migrating to small islands throughout the world has already occurred and is ongoing. Some wealthy persons have even bought private islands! Using current technology to create artificial islands is just an incremental step in continuing the spread of humanity.

Millennial projects. An artificial island is an island that has been constructed by humans rather than formed by natural means. They are created by expanding existing islets, construction on existing reefs, or amalgamating several natural islets into a bigger island.

An internet mailing list formed to attempt to organize it. The group incorporated as the "Living Universe Foundation." The list was still in existence as of 2007.

Some contemporary projects are much more ambitious. In 1994, Kansai International Airport was the first general aviation commercial airport to be built completely on an artificial island, followed by Chūbu Centrair International Airport in 2005 and the New Kitakyushu Airport and Kobe Airport in 2006.

Dubai is home to some of the largest artificial island complexes in the world, including the three Palm Islands macro-projects, The World and the Dubai Waterfront macro-projects, the last of which will be the largest in geographical extent.

The Israeli government is now planning for four artificial islands to be completed in 2013, off the coasts of Tel Aviv, Herzliya, Netanya and Haifa. Each island will house some 20,000 people and offer the employment opportunity of, probably, some 10,000 jobs at least. The islands should help with overcrowding in the coastal Israeli cities and might even be employed to do the same in Gaza.

A well thought out macro-engineering project proposal has also been presented in The Netherlands to create artificial islands, perhaps in the shape of a tulip, in the North Sea.

Under the United Nations Convention on the Law of the Sea treaty (UNCLOS), artificial islands have little legal recognition. Such islands are not considered harbor works (Article 11) and are under the jurisdiction of the nearest coastal state if within 200 nautical miles (370 km) (Article 56). Artificial islands are not considered islands for purposes of having their own territorial waters or exclusive economic zones, and only the coastal state may authorize their construction (Article 60). However, on the high seas beyond national jurisdiction, any "state" may construct artificial islands (Article 87).

Some attempts to create mini-nations have involved artificial islands such as Sealand and Republic of Rose Island. These were abject failures.

Big Ship Macro-Projects

America World City (originally named Phoenix World City) is a concept for a floating city proposed by John Rogers of World City Corporation. It is conceived as the first of three such behemoths serving United States of America ports and flying the U. S. flag. Rogers died in October, 2005.

Freedom Ship. Freedom Ship was a concept proposed by Norman Nixon. One has a design length of 1400 m, a width of 230 m, and a keel-to-mast height of 110 m, *Freedom Ship* would be more than four times longer than the retired *Queen Mary*. The design concepts include a mobile modern city featuring luxurious living, an extensive duty-free international shopping mall, and a full 160,000 m^2 floor set aside for various companies to showcase their products. *Freedom Ship* would not be a cruise ship, it is proposed to be a unique place to live, work, retire, vacation, or visit. The proposed voyage would continuously circle the globe, covering most of the world's coastal regions. Its large fleet of smallish commuter aircraft and hydrofoils would ferry its permanent residents and vetted visitors to and from shore.

The program statement has also announced that the propellers would be a series of 400 fully-rotational azipods; despite the high number of screws, the ship would still be the slowest in the world. Despite an initially stated in-service date of 2001, construction has not even begun as of 2008.

Net price estimates for the ship have risen from 6 billion US$ in 1999 to 11 billion US$ in 2002. Using the common formula for economic inflation will furnish the reader with an updated price estimate.

The Freedom Ship has little in common with any conventional ship; it is actually nothing more than a big barge. The bolt-up construction and the unusually large amount of steel to be incorporated into the ship meets the design engineer's requirements for stability and structural integrity and the cost engineer’s requirements of "economic feasibility" but the downside is a severe reduction in top speed, making the ship useless for any existing requirements. For example, it would be too slow to be a cruise ship or a cargo ship.

But what if this enormous barge was assigned a voyage that required slowly cruising around the world, closely following the shoreline, and completing one circum-navigation approximately every three years? If the designers then incorporated the following amenities into this big barge, what would result?

Figure 1. A side view of the proposed Freedom Ship. The largest existing ship in the world, the Knock Nevis, is approximately only one third of length of the Freedom Ship!

Figure 2. Freedom Ship (Front view).

- 18,000 living units, with prices ranging from $180,000 to $2.5 million, including a small number of premium suites currently priced at $44 million. 3,000 commercial units in a similar price range
- 2,400 time-share units
- 10,000 hotel units
- A World Class Casino
- A ferryboat transportation system that provides departures every 15 minutes, 24 hours a day, to 3 or more local cities giving ship residents access to the local neighborhood and up to 30,000 land-based residents a chance to spend a day on the ship.
- A World-Class Medical Facility practicing Western and Eastern medical doctoring as well as preventive and anti-aging medicine.
- A School System that gives the students a chance to take a field trip into a different country each week for academic purposes or to compete with local schools in numerous sporting events. For example; The Freedom Ship High School Soccer team plays a Paris High School team this week at home and an Italian team next week in Italy, while the Freedom Ship High School Band presents a New Orleans Jazz musical at a concert hall in London in the UK.
- An International Trade Center that gives on-board companies and shops the opportunity to show and sell their products in a different Country each week.
- More than 41 ha of outdoor Park, Recreation, Exercise and Community space for the enjoyment of residents and visitors.

Project Habakkuk or *Habbakuk* (spelling varies) was a World War II-hatched plan by the British to construct an aircraft carrier out of Pykrete (a 14% mixture of wood pulp and freshwater ice), for use against German U-boats in the mid-Atlantic, which was out of range of protective Allied land-based airplanes.

The *Habakkuk*, as proposed to Winston Churchill (1874-1965) by Lord Mountbatten (1900-1979) and Geoffrey Pyke (1893-1948) in December 1942, was to be approximately 610 m long and 91 m wide, with a deck-to-keel depth of 61 m, and bulkhead walls 12 m thick. It was intended to have a draft of 150 feet, and a displacement of 2,000,000 tons or more, to be constructed in timber and freshwater-rich wartime Canada from 280,000 blocks of ice.

The ice *Habakkuk* itself was never begun but experiments were conducted in the field.

Arctic and Antarctic (Southern) oceans. The ice fields of these oceans will be used for getting float platforms.

The amount of sea ice around the poles in winter varies from the Antarctic with 18,000,000 km² to the Arctic with 15,000,000 km². The amount melted each summer is affected by the different environments: the cold Antarctic pole is over land, which is bordered by sea ice in the freely-circulating Southern Ocean.

The Arctic Ocean occupies a roughly circular basin and covers an area of about 14,056,000 km². The situation in the Arctic is very different from Antactic sea (a polar sea surrounded by land, as opposed to a polar continent surrounded by sea) and the seasonal variation much less, consequently much Arctic sea ice is multi-year ice, and thicker: up to 3–4 meters thick over large areas, with ridges up to 35 meters thick. An *ice floe* is a floating chunk of sea ice that is less than 10 kilometers in its greatest dimension. Wider chunks of ice are called *ice fields.*

The North Pole is significantly warmer than the South Pole because it lies at sea level in the middle of an ocean (which acts as a reservoir of heat), rather than at altitude in a continental land mass. Winter (January) temperatures at the North Pole can range from about −43 °C to −26 °C, perhaps averaging around −34 °C. Summer temperatures (June, July and August) average around the freezing point (0 °C). In midsummer of South pole, as the sun reaches its maximum elevation of about 23.5 degrees, temperatures at the South Pole average around −25 °C. As the six-month 'day' wears on and the sun gets lower, temperatures drop as well, with temperatures around sunset (late March) and sunrise (late September) being about −45 °C. In winter, the temperature remains steady at around −65 °C. The highest temperature ever recorded at the Amundsen-Scott South Pole Station is −13.6 °C, and the lowest is −82.8 °C. However, this is by no means the absolute lowest recorded anywhere in the Earth, that being −89.6 °C at Antarctica's Vostok Station on July 21, 1983.

Descriptions and Innovations

The macro-engineering concept is to efficiently use a cheap floating platform taken from the ice fields in Arctic and Antarctica's Southern Ocean for the floating cities, island, and states. These cheap platforms protected by air-film (bottom and sides) and conventional insulating cover (top) and having cooling systems to deal with any leak-through heating can sustain the platform for an unlimited time. They can be increased in number or size at anytime, floated in warm oceans, travel to different continents and countries, serve as artificial airports, harbors and other marine improvements, as well as floating cities and industrial bases for virtually any use.

One possible means of construction is as follows: A scouting aircraft (helicopter) confirms a satellite-surveyed ice field as suitable and delives to it a small tractor with extensible wire–saw. The tractor saws up the ice platform to hew there from a platform of a specified size (including allowance for melting before insulation for example, 500×500×10 m)(Figure 4a) and an ice-braker ship tows this platform to open water. Here the platform is equipped with air-film covers, protected by from warm water on all sides. Platform is towed to a place where it will be provided for with final protection and other improvements; a suitable location for building the city or other floating improvement that it will come. One method of adding thermal protection of the ice is the following: The double film is submerged lower than the bottom of platform, moved under the platform (or the platform is moved over film) and filled with air. The air increases the lift force of platform and protects the bottom, sides and top of the platform from contact with warm water and air (Figure 4b, pointer 3). Simultaneously, the coolant fluid (it may be chilled air) flows throgh the cooling tubes 4 (Figure 4b) and keeps the ice at lower than melting, or indeed softening, point.

The top side of the platform may be covered with conventional heat protection and insulation means on top of which construction elements may be added. (film, ground, asphalt, concrete plates, houses, buildings, gardens, airdrome runways, and so on).

The other method allows us to custom-produce ice of any thickness and composition, including ices of low density (high lift force). Thin plastic tubes are located under the ice-bottom to be (which may be isolated from circulation by a film barrier) and and cold air (in the polar regions, or in winter, simple outdoor air) is blown throgh them. One freezes the water and produces an ice platform. The ice has a lower density and a high lift force (load capability) because the ice has internal channels (tubes) filled by air. We may evade spending energy for it in cold countries or in winter. The arctic (antarctic) winter air has temperature of -40 to about -50°C. In the Arctic Ocean, seawater is useful as a heat source (having 0°C) which can heat the outer air up to -3-5°C, turning an the air turbine, the turbine then turning the pump air ventilator. The corresponding estimation is in theoretical section. We can get the ice density of $\gamma = 500$ kg/m^3 having load capability of 500 kg/m^3 (the conventional ice has the lift force 80 kg/m^3). For decreasing the ice density, macro-enginers may use cork filler material or other such available low-density matrix fillers.

In second method, we can produce platform from *Pykrete* (also known as *picolite*). That is a composite material made of approximately 14% sawdust (or, less frequently, wood pulp) and 86% water by weight then frozen, invented by Max Perutz. Pykrete has some interesting properties, notably its relatively slow melting rate (due to low thermal conductivity), and its vastly improved strength and toughness over pure ice, actually closer to concrete, while still being able to float on water. Pykrete is slightly harder to shape and form than concrete, as it expands while freezing, but can be repaired and maintained from the sea's most abundant raw material.

The pykrete properties may be significantly improved by employing the cheapest available strong artificial fibers (such as basalt fibers, class or mineral wool, and others).

The composites made by mixing cork granules and cement have low thermal conductivity, low density and good energy absorption. Some of the property ranges of the composites are density (400–1500 kg/m^3), compressive strength (1–26 MPa) and flexural strength (0.5–4.0 MPa).

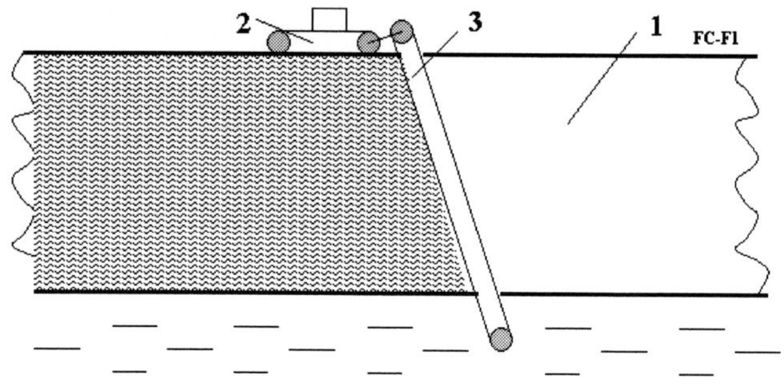

Figure 3. Cutting of floating platform from ice field. *Notations*: 1 – ice field in arctic (Antarctic) ocean; 2 – small tractor with band-saw or slicing wiresaw; 3 – mechanical band saw or slicing wiresaw.

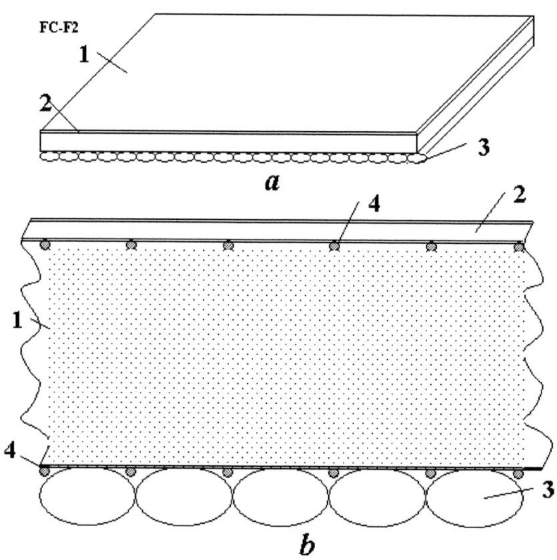

Figure 4. Ice platform prepared for floating city. (a) Common view, (b) Cross-section of platform. *Notations*: 1 – ice; 2 – top heat protection; 3 – low (bottom) heat protection and floating support (inflatable air balloon); 4 – cooling tubes.

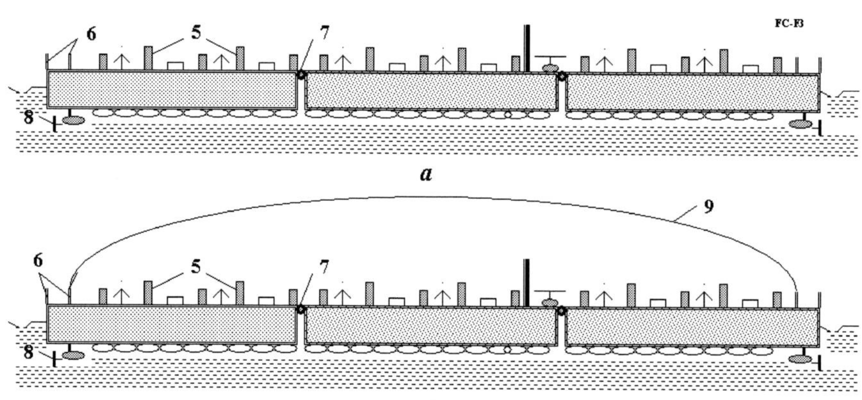

Figure 5. Floating city on ice platform: (a) Open floating city, (b) Floating city closed by film. *Notations*: 5 – city; 6 – protection from ocean waves in storm; 7 – turning connection (joint) of separated ice platform; 8 – fully-rotation azimuth thruster propellers; 9 – film dome.

The platform of floating city has protection (walls) 6 (Figure 5) against stormy ocean waves, joints 7 (Figure 5b) which decrease the platform stress in storms, propellers for maneuvering and moving. The platform may also have an over it a filmic AD-Dome (Figure 5b, pointer 9) such as is offered in [2. 11]. This dome creates a warm constant "deck" temperature, protects the floating city from strong winds and storms.

Summary of innovations:

1. Using a big natural ice platform for building large floating cities, islands and states.
2. Technology for getting these platforms from the natural ice fields (saw up).
3. Technology (artificial freezing without spending energy) for getting the artificial high lift force ice platform of any thickness (that means any load capability) from low density ice.
4. Composite material where ice is a matrix (base) and cork (or other material) as stuff.
5. Heat protection for natural ice fields by air film balloons.
6. Building the platforms from separated ice segments and connection them by joints.
7. It is offered the protection of the suggested platform by special double walls 6 (Figure 5) from ocean storm waves.
8. Protection of the suggested platform by the special transparent film 9 (dome), (Figure 5b) and creating a constant temperature in the floating city, plus protection from strong winds and storm.

Theory of Estimation and Computation

1. Material. The important values and characteristics of candidate materials for floating platforms are their price, lift force (in water), life time, strength, chemical stability in water, reliability, and so on. The water lift force of matter (L_f) is difference between density of water (d_w) and density of platform matter (d_m).

$$L_f = d_w - d_m , \tag{1}$$

where all value are in kg/m^3.

Air is the cheapest material, having the most lift force. However it needs a strong cover (in vessels, balloons) which can significantly increase the cost of the installation. The other lack (disadvantage) of using air is loss of lift force in case of damage to its container.

Ice is a cheap substance. It may be mined, rather than of necessity built, into a ready-made floating platform. But it has a small buoyancy force and low melting temperature, which are lower then the temperature of ocean seawater. We can decrease these disadvantages by using special air balloons under the platform, heat protection materials and barriers and a refreezing system. If we initially produce the platform by custom freezing, we can produce the custom-tailored, strong light ice having a high lift (buoyancy) force.

The other systems are metal or concrete constructions, filled by rock, foam plastic, aerocrete and so on. Their disadvantages are well known: A high cost and a huge procurement necessary for big installations (platforms).

Some materials and their properties are presented in Tables 1 and 2.

2. Computation of heat protection. We use in our project the cheap natural ice platform variant.

The heat loss flow per 1 m^2 by convection and heat conduction is (see [13]):

$$q = k(t_1 - t_2), \quad \text{where} \quad k = \frac{1}{1/\alpha_1 + \sum_i \delta_i / \lambda_i + 1/\alpha_2}, \tag{2}$$

where k is heat transfer coefficient, W/m^{2}K; $t_{1,2}$ are temperatures of the inter and outer multi-layers of the heat insulators, $^{\circ}$C; $\alpha_{1,2}$ are convention coefficients of the inter and outer multi-layers of heat insulators ($\alpha = 30 \div 100$), W/m^{2}K; δ_i are thickness of insulator layers; λ_i are coefficients of heat transfer of insulator layers (see Table 1 Attn.), m. The magnitudes of α are:

1) From water to metal wall $\alpha = 5000$ W/m^{2}K.
2) From gas to wall $\alpha = 100$ W/m^{2}K.
3) From heat isolator to air $\alpha = 10$ W/m^{2}K.

For example, let us estimate the heat flow from water to the bottom surface of ice platform protected by the small air balloons.

Assume the average thickness of air balloons is $\delta = 1$ m, the temperature of seawater at depth 10 m is 10°C. The heat flow from seawater to ice platform is

$$k \approx \frac{\lambda}{\delta} = \frac{0.0244}{1} = 0.0244 \ \frac{\text{W}}{\text{m}^2\text{K}};$$

$$q = k(t_2 - t_1) = 0.0244(10 - 0) = 0.244 \ \frac{\text{W}}{\text{m}^2\text{K}}.$$

That ($0.244 \ \frac{\text{W}}{\text{m}^2\text{K}}$) is a small value. This heat must be carried away (deleted) by cooling liquids or other fluids (i.e., cooled and force-driven gases or mixtures of gases such as Earth's air circulated especially for that purpose).

Estimate now the heat flow from outside air to the platform's top surface protected by 0.1 m wood gasket and 0.4 m humid soil. The air temperature is 25°C.

$$k \approx \frac{1}{\delta_1/\lambda_1 + \delta_2/\lambda_2} = \frac{1}{0.1/0.2 + 0.4/0.657} = 0.9,$$

$$q = k(t_2 - t_1) = 0.9 \cdot 25 = 22.5 \ \frac{\text{W}}{\text{m}^2\text{K}}.$$

If we change the wooden gasket for an asbestos plate of exactly the same thickness, then the heat flow decreases to q = 17 W/m^{2}K. Places where are situated houses, buildings and other structured constructions having concrete bases will have $q = 10 - 15$ W/m^{2}K. Using the wools or air protection significantly decreases the head loss through top platform surface. The average heat loss of top platform surface is about 15 W/m^{2}K. If we insert the air black gap 15 – 25 cm, this heat loss decreases to 1 – 2 W/m^{2}K. The side part of the floating platform may be protected the same as the bottom surface.

3. Freezing of platform. The freezing of 1 kg water requires energy

$$Q = c_p(t_2 - t_1) + \lambda_p \approx \lambda_p \,, \tag{3}$$

where c_p = 4.19 kJ/kg·C is energy needed for cooling water in 1°C; λ_p = 334 kJ/kg is energy needed for freezing 1 kg of water.

The energy needed for freezing may be received from the cold arctic air. That computed by equation

$$Q = c_{p,a}(t_2 - t_1) \,, \tag{4}$$

Here $c_{p,a}$ = 1 kJ/kg·K for air.

The computation shows for freezing 1 kg water we need about 22 kg air in temperature -20 C. Every 1 kg air heated from -20 C to -5 C in ocean water absorbs about 15 kJ/kg in heat energy.

4. Other heat flows. The radiation heat flow per 1 m^{2}s of the service area computed by equations (5):

$$q = C_r\left[\left(\frac{T_1}{100}\right)^4 - \left(\frac{T_2}{100}\right)^4\right], \quad \text{where} \quad C_r = \frac{c_s}{1/\varepsilon_1 + 1/\varepsilon_2 - 1}, \quad c_s = 5.67 \left[\frac{W}{m^2K^4}\right], \tag{5}$$

where C_r is general radiation coefficient, ε are black body rate (Emittance) of plates; T is temperatures of plates, °K.

The radiation flow across a set of the heat reflector plates is computed by equation

$$q = 0.5\frac{C'_r}{C_r}q_r \,, \tag{6}$$

where C'_r is computed by equation (5) between plate and reflector.

The data of some construction materials is found in Table 1, 5 Attn.

As the reader sees, the air layer is the best heat insulator. We do not limit its thickness δ.

The thickness of the dome envelope, its sheltering shell of film, is computed by formulas (from equation for tensile strength):

$$\delta_1 = \frac{Rp}{2\sigma}, \quad \delta_2 = \frac{Rp}{\sigma}, \tag{7}$$

where δ_1 is the film thickness for a spherical dome, m; δ_2 is the film thickness for a cylindrical dome, m; R is radius of dome, m; p is additional pressure into the dome, N/m^2; σ is safety tensile stress of film, N/m^2.

For example, compute the film thickness for dome having radius R = 100 m, additional air pressure p = 0.01 atm (p = 1000 N/m^2), safety tensile stress σ = 50 kg/mm^2 ($\sigma = 5\times10^8$ N/m^2), cylindrical dome.

$$\delta = \frac{100\times1000}{5\times10^8} = 0.0002\,m = 0.2\,mm \tag{8}$$

The dynamic pressure from wind is

$$p_w = \frac{\rho V^2}{2}, \tag{9}$$

Table 1. Estimation of different variants of floating platforms

#	Type of floating platform	Height, m	Cost,* $/m^2	Life time, Year	Load capasity, ton/m^2	Main-tains, $/m^{2}year	Draught m	Mass of platform, ton/m^2	Cooling Energy, W/m^2
1	Air-steel cylinder with steel walls	10 20	100 200	30-50	7 17	1 2	7.6 18.2	0.6 1.2	0
2	Steel cubs with net walls and air balloons	10 20	150 300	40-60	7 17	2 4	7.5 18	0.5 1	0
3	Steel cubs with net walls and foam plastic filler	10 20	150 130	40-60	7 17	1 2	7.6 18.2	0.6 1.2	0
4	Concrete empty cub with walls 0.1m, 100 $/ton	5 10	400 800	100-200	2.4 6	0.5 1	4 9.2	1.6 3.2	0
5	Aero Crete γ = 500 kg/m^3	10 20	220 440	100-200	4 7	1 2	9 17	6 17	0
6	Ice and 1 m air heat protection in bottom	5 10	2 3	∞	1 2	4 6	5 10	4 8	2 W/m^2 2 W/m^2
7	Air ice, γ = 500 kg/m^3 and 1 m air heat protect.	20 30	4 6	∞	9 15	5 10	19 30	10 15	2 W/m^2 2 W/m^2

* Only material.

where $\rho = 1.225$ kg/m^3 is air density; V is wind speed, m/s.

For example, a storm wind with speed $V = 20$ m/s (72 km/h), standard air density is $\rho = 1.225$ kg/m^3. Then dynamic pressure is $p_w = 245$ N/m^2. That is four time less than internal pressure $p = 1000$ N/m^2. When the need arises, sometimes the internal pressure can be voluntarily decreased, bled off.

MACRO-PROJECTS

The estimation of different variants of floating platforms is presented in Table 1.

The estimation cost of 1 m^2 of the platform in the contemplated "Freedom Ship" (the cost of cabins are included) is \$33,100/m^2 (2002). At the present time (2008) this cost has increased by a factor of two times more. Average cost of 1 m^2 of apartment in many cities is about \$1000/m^2 (USD).

DISCUSSION

Advantages and disadvantages of the speculated method.

Advantages

1. The offered method is cheapest by tens-to-hundreds of times relative to conventional shipbuilding operations, and beats nearly all but the remotest and most valueless land for cheapness as a construction substrate—yet the product may be relocated to within meters from some of the most valuable real estate on Earth—i.e., docked in Tokyo's Bay (Japan) or near the New York City island borough of Manhattan (USA) or close to China's economically booming Shanghai.
2. Unlimited areal enlargement of usable region is technically possible.
3. Easy increase of load capacity by additional freezing of new platform bottom area; and ease of restoring a damage sector.
4. High facility security attainable at a reasonable monetary cost.
5. Unlimited physical life-time. Well, at least not yet undetermined!

Disadvantages

1. Need for a permanent but small energy expenditure (in warm climates) for maintaining the ice at freezing temperatures.

RESULTS

It is a promising new method for obtaining a cheap ice platform suitable for many profitable engineering purposes, and for colonization the World Ocean.

REFERENCES

(The interested reader can find some of the lead author's works at *http://Bolonkin.narod.ru/p65.htm*, *http://Arxiv.org* Search: Bolonkin and books: Bolonkin A.A., *Non-Rocket Space Launch and flight:*, Elsevier, 2006, 488 pgs., Bolonkin A.A., "*New Concepts, ideas, and Innovations in Technology and Human life*", NOVA, 2008, 400 pg.)

[1] Bolonkin, A.A., *Irrigation without water* (closed-loop water cycle). *http://arxiv.org* search "Bolonkin"(2007).

[2] Bolonkin, A.A. and R.B. Cathcart, Inflatable 'Evergreen' Dome Settlements for Earth's Polar Regions. *Clean. Techn. Environ. Policy.* DOI 10.1007/s10098.006-0073.4.

[3] Bolonkin A.A., *Control of Regional and Global Weather.* 2006. *http://arxiv.org* search for "Bolonkin".

[4] Bolonkin A.A., *Cheap Textile Dam Protection of Seaport Cities against Hurricane Storm Surge Waves, Tsunamis, and Other Weather-Related Floods*, 2006. *http://arxiv.org.*

[5] Bolonkin, A.A. and R.B. Cathcart, *Antarctica: A Southern Hemisphere Windpower Station?* Arxiv, 2007.

[6] Cathcart R.B. and Bolonkin, A.A. *Ocean Terracing,* 2006. *http://arxiv.org.*

[7] Bolonkin, A.A. and R.B. Cathcart, *The Java-Sumatra Aerial Mega-Tramway,* 2006. *http://arxiv.org.*

[8] Bolonkin, A.A., "Optimal Inflatable Space Towers with 3-100 km Height", Journal of the British Interplanetary Society Vol. 56, pp. 87 - 97, 2003.

[9] Bolonkin A.A., *Non-Rocket Space Launch and Flight, Elsevier*, London, 2006, 488 ps.

[10] Macro-Engineering: *A Challenge for the Future.* Springer, 2006. 318 ps. Collection articles.

[11] Bolonkin A.A., Cathcart R.B., *Inflatable 'Evergreen' for Polar Zone Dome (EPZD) Settlements,* 2006. *http://arxiv.org* search "Bolonkin".

[12] Bolonkin A.A., Inflatable Dome for Moon, Mars, Asteroids and Satellites, Presented as *paper AIAA-2007-6262 by AIAA Conference "Space-2007"*, 18-20 September 2007, Long Beach. CA, USA.

[13] Naschekin, V.V., Technical thermodynamic and heat transmission, Public House of High University, Moscow, 1969 (in Russian).

[14] Koshkin H.I., Shirkevich M.G., Directory of elementary physics, Moscow, Nauka, 1982.

[15] *Wikipedia. S*ome background material in this article is gathered from Wikipedia under the Creative Commons license.

Chapter 8

SAVING OF DEAD SEA AND STABILIZATION THEREOF*

ABSTRACT

The author offers new, cheap methods for the extraction of freshwater and energy from the Earth's atmosphere, delivery of freshwater to the Dead Sea and stopping the evaporation of the Dead Sea. The suggested methods are fundamentally distinct from all existing methods that extract freshwater from the air and sea. All other industrial methods extract water from a saline water source (in most cases from seawater). These new methods extract freshwater from the atmosphere and may be used at any point in the Earth except for the Polar Zones. They do not require long-distance freshwater transportation. This inexpensive new method is very environment-friendly, definitely Green. The author's methods have three working versions: (1) In the first variant the warm (hot) atmospheric air is lifted by an inflatable tower to a high altitude and atmospheric water vapor is condensed into freshwater: (2) in the second version, the warm air is lifted by an artificial inflatable mountain; (3) In the third version, the Dead Sea is covered by thin film vapor barrier which stops evaporation.

In version (2)-(3) wind and solar heating of air are used to induce air flow.

The received freshwater may be used for irrigation, civil and industry needs and, in particular, for stabilization of the Dead Sea. The author offers another innovation - to connect the freshwater installation to the Dead Sea by cheap inflatable tubes.

The methods do not need energy, or need a small amount. Moreover, the freshwater in vertion (1)-(2) has a high pressure (300 or more atm.) owing to the advantage of the water column height and can be used for production of energy such as electricity and for transportation of water to a long distance. That way the freshwater cost is lower. In method (1) for increasing the productivity of the system, seawater is injected into the air and a solar air heater may be additionally used. The solar air heater produces a huge amount of electricity as a very powerful electricity generation plant. The offered electricity installation is 100 - 200 times cheaper than any common electric plant of equivalent output.

For stabilization of the Dead Sea level and converting Dead Sea region into a garden resort, almost Edenic in style, the author offers the cheap method – covering the Dead Sea and its shores by a thin film inflatable Dome (method (2)). That method creates a

---

* Presented to http://arxiv.org (May, 2008)

closed-loop cycle in Dead Sea region, stops the water loss of Dead Sea by vaporization, produces a fine climate within the Dome and increases the precipitation in the greater Dead Sea region.

The covering of the Dead Sea surface by a thin film (method) to reduce net evaporaton is cheaper by hundreds of times than the conventional method of desalination and delivering seawater by an undergraund pipe and cannel network some hundreds of kilometers.

Keywords: Extraction freshwater, method to obtain made freshwater, receiving energy from atmosphere, powerful renewal electric plant

INTRODUCTION

Dead Sea. The Dead Sea is a salt lake between Israel to the west, (including theWest Bank) and Jordan to the east. It is 420 metres below sea level, and its shores are the lowest point on the surface of the Earth on dry land. The Dead Sea is 330 m deep, the deepest hypersaline lake in the world. It is also the world's second saltiest body of water, after Lake Asal in Djibouti, with 30 percent salinity. It is 8.6 times saltier than the world-ocean's water. Israeli experts say that it is nine times saltier than the Mediterranean Sea (31.5% salt versus 3.5% for the Mediterranean). The Dead Sea is 67 kilometers long and 18 kilometers wide at its widest point. Surface area is 810 km^2 (North Basin). Average depth is 120 m. Water volume is 147 km^3. Shore length is 135 km. The present-day Dead Sea lies in the Jordan Rift Valley, and its main tributary is the Jordan River.The Jordan River is the only major water source flowing into the Dead Sea. There are no outlet streams. Some unknown quantity of groundwater—of various quality—does flow directly into the Dead Sea from the countryside encompassing the Dead Sea.

The northern part of the Dead Sea receives scarcely 100 mm of rain a year. The southern section barely 50 mm. The Dead Sea zone's aridity is due to the rain-shadow effect of the adjacent Judean Hills. The highlands east of the Dead Sea receive more rainfall than the Dead Sea itself.

The mountains of the western side, the Judean Hills, rise less steeply from the Dead Sea than do the mountains of the eastern side. The mountains of the eastern side are also reach higher elevations. Along the southwestern side of the lake is a 210 m (700 ft) tall halite formation called "Mount Sedom".

A rough Dead Sea, with salt deposits on cliffs. The mineral content of the Dead Sea is significantly different from that of ocean water. The exact composition of the Dead Sea water varies with season, depth, temperature and so on. The concentration of ionic species (in g/kg) of Dead Sea surface water in the early 1980s was found to be: Cl^- (181.4), Br^- (4.2), SO_4^{2-} (0.4), HCO_3^- (0.2), Ca^{2+} (14.1), Na^+ (32.5), K^+ (6.2) and Mg^{2+} (35.2). The total salinity was 276 g/kg.

These results show that w/w% composition of the salt, as anhydrous chlorides, was calcium chloride ($CaCl_2$) 14.4%, potassium chloride (KCl) 4.4%, magnesium chloride ($MgCl_2$) 50.8% and sodium chloride (common salt, NaCl) 30.4%. In comparison, the salt in the water of most oceans and seas is approximately 97% sodium chloride. The concentration of sulfate, SO_4^{2-}, ions is very low, and the bromide ion, Br^- concentration is the highest of all known waters on planet Earth.

Comparison between the chemical composition of the Dead Sea to other lakes and oceans show that the salt concentration in the Dead Sea is 31.5% (the salinity fluctuates somewhat). The unusually high concentration of salt results in a similarly high density of up to 1.24 kg/L, depending on temperature and salinity. Anyone can easily float in the Dead Sea because of natural buoyancy. In this aspect, the Dead Sea is similar to the Great Salt Lake in Utah, in the United States.

Left: A view across the Dead Sea from the Israeli side to the Jordanian side.
Right: Near Ein Gedi, salt build-up along the Dead Sea's tide-less shoreline.

The Dead Sea region has become a major center for human health research and special medical treatment. The mineral content of the waters, the very low content of pollens and other allergens in the air, the reduced ultraviolet component of solar radiation owing to the Dead Sea's extreme negative elevation, and the higher atmospheric pressure at this remarkable depth below worldwide sea level each have specific health effects. For example, persons suffering with reduced respiratory function from diseases such as cystic-fibrosis seem to benefit from the increased air pressure.

Sufferers of the skin disorder psoriasis also benefit from the ability to sunbathe for long periods due to its position below sea level and because the sun's harmful UV rays are markedly reduced. Furthermore, Dead Sea salt has been found to be beneficial to psoriasis sufferers.

In recent decades, the Dead Sea has been rapidly shrinking because of diversion of incoming water. From an elevation of 395 m below sea level in 1970 it fell 22 m to 418 m below sea level in 2006, reaching a drop rate of 1 m per year. Although the Dead Sea may never entirely disappear, because water vapor production slows down as surface area decreases and salinity increases, it is feared that the Dead Sea may substantially change all of its physical characteristics.

The Dead Sea Canal

The Dead Sea Canal is a proposed macro-project of building a canal from either the Mediterranean Sea (MDSC) or the Red Sea to the Dead Sea (RSDSC), taking advantage of the 400-meter difference in water levels between two segments of the world-ocean and the Dead Sea. The seawater flowing through the canal may help redress the drop in the level of the Dead Sea observed in recent years. The canal can also be used to generate hydroelectric power because of surface difference and maybe by salinity gradient power, and desalinate water by reverse osmosis.

History of canals. The idea was first proposed by the British Admiral William Allen in 1855 in an oeuvre called 'The Dead Sea - A new route to India'. At that time it was not yet known that the Dead Sea lies below sea level, and Allen proposed this canal as an alternative to the Suez Canal. Later, many civil and military engineers and politicians adopted the macro-project idea, also Theodor Herzl in his 1902 novel *Altneuland.* Most of the early proposals used the East bank of the Jordan River, but a modified form, using the West bank, was proposed after the separation of Transjordan from the Palestine Mandate. The idea was discussed at some length by Willy Ley in his 1954 book *Engineers' Dreams*, but it was then politically impractical; even the route west of Jordan would have to cross the 1949 armistice line twice.

The Dead Sea attracts over one million tourists a year. Notice the lady is not using any flotation device!

The idea was revived during the 1980s mainly for the purpose of electric power generation following the 1973 global oil crisis. The Mediterranean-Dead Sea Company studied various alternatives and recommended a route from the Mediterranean Sea through the Gaza Strip to a place near Masada. However, the macro-project did not commence due to unquenchable financial doubts by potential investors. The macro-engineering idea was re-examined again during the 1990s due to a freshwater crisis in the region. In addition to the Gaza Strip-Masada route, two other alternatives were considered, namely a Red Sea-Dead Sea canal and a northern route from the Mediterranean to the Bet She'an Valley, which was found to be the cheapest of the three proposals. At the present time, the Red Sea route is endorsed as a Jordanian project with Israeli and Palestinian support. Unfortunately, the Red/Dead route, in addition to being the least worthwhile in economic terms, may prove to be impractical due to chemical incompatibility of Red Sea and Dead Sea water.

The proposed *Two Seas Canal* would run from the Red Sea to the Dead Sea and provide electricity and potable water to Jordan, Israel and the Palestinian Authority.

Base 802834 (A00853) 9-01

Planimetric Map of Israel.

The water level in the Dead Sea is disappearing at a rate of almost one meter per year, and its surface area has reduced by about 30% in the last 20 years. This is largely due to the diversion of about 90% of the freshwater volume in the Jordan River. In the early 1960s, the river moved 1.3 billion cubic meters of freshwater every year from the Sea of Galilee to the Dead Sea. But dams, canals and pumping stations built by Israel, Jordan and Syria to divert water for crops and drinking have reduced the flow by more than 90% to only about 100 million cubic meters.

On May 9, 2005 Jordan, Israel and the Palestinian Authority signed an agreement to go ahead with a feasibility study for the Two Seas Canal. The agreement was signed on the Dead Sea's shore by Jordanian Water Minister Raed Abu Saud, Israeli Infrastructure Minister Binyamin Ben-Eliezer and Palestinian Planning Minister Ghassan al-Khatib.

The proposed canal would pump seawater 170 meters uphill from the Red Sea's Gulf of Aqaba in Jordan and then allow it to run downhill into the Dead Sea, which lies about 400 m below sea level. The macro-project will consist of about 175 km of canal, tunnel and piping along the Jordan-Israel border. It is expected to take about five years to build. Hydropower generated from the elevation difference is expected to be about 190 mega-watts. If it proves possible to generate power via salinity gradient power that too would be beneficial. Israeli experts say it is nine times saltier than the Mediterranean Sea (31.5% salt versus 3.5% for the Mediterranean).

The USD 5 billion macro-project includes a 550 MW power plant which will provide electricity for pumping the seawater in the initial stages and power a desalination plant which will provide around 850 million cubic meters of freshwater yearly to the three countries. Since hydroelectric generation will supply 190 megawatts, the remainder of the electric power needed annualy will be supplied using conventional energy sources.

A shorter and better route running from the Mediterranean Sea has been proposed by Israel in the past, but was discarded, as it would not suit Jordan's security and political needs.

The proposal has generated concern in Egypt which believes that the canal will increase seismic activity in the region; provide Israel with cooling water for its nuclear reactor near Dimona; turn the Negev Desert into a settlement region and increase the salinity of wells.

Unfortunately the Red/Dead route, in addition to being less worthwhile in economic terms than alternative canals to the Dead Sea, may prove to be impractical due to chemical incompatibility of Red sea and Dead sea water.

In the second version the canal would be about 180 kilometers long, extending from the Gulf of Aqaba down to the Dead Sea. It would carry up to 1,900 million cubic meters of water into the salt lake each year. Because the canal would cross a chain of hills, the water would first need to be pumped upwards 220 meters and then it would descend more than 600 meters and through the Wadi Arabah (or Arabah Valley) before finally reaching the Dead Sea.

Under the existing macro-project construction plan, the gradient would be used to generate electricity, much of which would be needed to power the pumping stations. The remainder could be used to run a large seawater desalination plant, which would produce up to 850 million cubic meters of drinking water a year for the Jordanians, Israelis and Palestinians dwelling in the West Bank.

The macro-project is being touted as a "peace canal" in a region shaken by political unrest. According to current estimates, construction would take about nine years and cost at least $5 billion.

The macro-project does in fact pose more than a few risks. The Wadi Arabah is an extension of the East African Rift and a tectonically active region. If there were a powerful temblor, the Canal could be badly damaged, realeasing saltwater and thus spoiling (by contamination) the groundwater.

In addition, experiments conducted by a team of scientists from the Geological Survey of Israel have shown that mixing the water from the two seas could lead to algae blooms, the precipitation of gypsum, and the water turning red.

Another fear is that the imported seawater from the Red Sea, which has a lower salinity and less density, might not mix completely with the salty water of the Dead Sea, causing a layering that would deprive the Dead Sea of its unique tourist-attracting physical characteristics.

Proponents of the Two Sea Canal point to the positive aspects related to the canal's construction, namely restoring the Dead Sea to its historic level, and making freshwater available for the benefit of neighboring countries. However, the transfer of massive volumes of unusual natural water from one sea to another, can bear drastic consequences on the unique natural characteristics of each of the two seas, as well as the desert valley which separates them, the Wadi Arabah. Some of these characteristics, especially in the Dead Sea region, are unique on a global perspective, and therefore crucially important for conservation. The environmental group Friends of the Earth Middle East has protested against the allegedly premature approval of the Two Seas Canal project by the Israeli government, without sufficient assessment of the project's impact on the natural environment of the area. The group lists several potential hazardous effects of the project on the unique natural systems of the Red Sea, the Dead Sea and the Wadi Arabah. These effects include:

1. *Damage to the unique natural system of the Dead Sea*, due to mixing its water with desalinated Red Sea water, which has a totally different chemical composition. This includes changes in water salinity, massive formation of gypsum, formation of volatile toxic compounds, change in water vapor rates, changes in the composition of bacteria and algae which inhabit the sea surface, chemical changes in the rocks which surround the water, and loss of unique health benefits that account for much of the tourist attraction to the Dead Sea area.
2. *Damage to the coral reefs of the Gulf of Aqaba*, due to water pumping.
3. *Damage to the natural landscape and ecosystem of the Arabah*, due to the construction, and the increase in humidity caused by the open canal segments.
4. *Damage to the aquifer of the Arabah*, due to contamination of groundwater with water from the Dead Sea. The alluvial deposits in Wadi Araba contain important supplies of fresh water. In the event that the pipeline ruptures (as might happen in the case of an earthquake), these aquifers will be irreparably damaged. This can bear fatal consequences to both the agriculture and ecosystem of the Arabah.
5. *Threats to archeological heritage*. The pipeline will cross areas of important cultural heritage, such as Wadi Finan, where the earliest copper mining and extraction in the world took place.

One of the plans which were suggested as a means to stop the recession of the Dead Sea is to channel water from the Mediterranean or the Red Sea, either through tunnels or canals (the Dead Sea Canal). Although a Mediterranean structure would be shorter, Israel is now committed to building a Red Sea canal in deference to Jordan's needs. The plan is to pump water 120 m up the Arava/Arabah from Aqaba or Eilat, tunnel under the highest point of the

Arava/Arabah valley, and then canalize the river of seawater as it falls 520 m to the Dead Sea. The downhill flow would be harnessed hydroelectrically, and the arriving seawater would run into a desalination plant to be constructed in Jordan. The salt water remaining after desalination would be deposited into the Dead Sea.

Jordan, Israel, and the Palestinian Authority signed an agreement On May 9, 2005. They want to begin feasibility studies on the macro-project, to be officially known as the "Two Seas Canal". The scheme calls for the production of 870 million cubic metres of fresh water per year and 550 megawatts of electricity. The World Bank is supportive of the project. However, several environmental groups have raised concerns about possible negative impacts of the project on the natural environment of the Dead Sea and Arava.

In 2007, the level of the Dead Sea fell by 1 metre.

Desalination

Desalination, desalinization, or desalinisation refers to any of several processes that remove excess salt and other minerals from water. Desalination may also refer to the removal of salts and minerals more generally, as in soil desalination, but the focus of this article is on water desalination.

Salty water is desalinated in order to be converted to freshwater suitable for consumption or irrigation, or, if almost all of the salt is removed, for human consumption. Sometimes the process produces table salt as a byproduct. It is used on many ships and patrolling nuclear submarines. Most of the modern interest in desalination is focused on developing cost-effective ways of providing fresh water for human use in regions where the availability of water is limited.

Large-scale desalination typically requires large amounts of energy as well as specialized, expensive infrastructure, making it very costly compared to the use of fresh water from rivers or groundwater. The large energy reserves of many Middle Eastern countries, along with their relative water scarcity, have led to extensive construction of desalination in this region. Saudi Arabia's desalination plants account for about 24% of our world's total capacity.

The world's largest desalination plant is the Jebel Ali Desalination Plant (Phase 2) in the United Arab Emirates. It is a dual-purpose facility that uses multi-stage flash distillation and is capable of producing 300 million cubic meters of water per year. The largest desalination plant in the United States is located in Tampa Bay, Florida, which began desalinating 95,000 m^3 of saltwater per day in December 2007. The Tampa Bay plant runs at around 12% the output of the Jebel Ali Desalination Plants. A January 17, 2008, article in the Wall Street Journal states, "World-wide, 13,080 desalination plants produce more than 12 billion U.S. gallons (45,000,000 m^3) of water a day, according to the International Desalination Association."

As of July 2004, the two leading methods were reverse osmosis (47.2% of installed capacity world-wide) and multi-stage flash (36.5%). The traditional process used in these operations is vacuum distillation—essentially the boiling of water at less than atmospheric pressure and thus a much lower temperature than normal. This is due to the fact that the boiling of a liquid occurs when the vapor pressure equals the ambient pressure and vapor pressure increases with temperature. Thus, because of the reduced temperature, energy is saved.

In the last decade, membrane processes have grown very fast, and most new facilities use reverse osmosis technology. Membrane processes use semi-permeable membranes and pressure to separate salts from water. Membrane systems typically use less energy than thermal distillation, which has led to a reduction in overall desalination costs over the past decade. Desalination remains energy intensive, however, and future costs will continue to depend on the price of both energy and desalination technology.

A number of factors determine the capital and operating costs for desalination: capacity and type of facility, location, feed water, labor, energy, financing and concentrate disposal. Desalination stills now control pressure, temperature and brine concentrations to optimize the water extraction efficiency. Nuclear-powered desalination might be economical on a large scale, and there is a pilot plant in the former USSR.

Critics point to the high costs of desalination technologies, especially for poor third world countries, the impracticability and cost of transporting or piping massive amounts of desalinated seawater throughout the interiors of large countries, and the byproduct of concentrated seawater, which some environmentalists have claimed "is a major cause of marine pollution when dumped back into the oceans at high temperatures".

It should be noted that typically the reverse osmosis technology that is used to desalinate water does not produce this "hot water" as a byproduct. Additionally, depending on the prevailing currents of receiving waters, the seawater concentrate byproduct can be diluted and dispersed to background levels within relatively short distances of the ocean outlet.

While noting that costs are falling, and generally positive about the technology for affluent areas that are proximate to oceans, one study argues that "Desalinated water may be a solution for some water-stress regions, but not for places that are poor, deep in the interior of a continent, or at high elevation. Unfortunately, that includes some of the places with biggest water problems" and "Indeed, one needs to lift the water by 2000 m, or transport it over more than 1600 km to get transport costs equal to the desalination costs. Thus, it may be more economical to transport fresh water from somewhere else than to desalinate it. In places far from the sea, like New Delhi (India) or in high places, like Mexico City (Mexico), high transport costs would add to the high desalination costs. Desalinated water is also expensive in places that are both somewhat far from the sea and somewhat high, such as Riyadh and Harare. In places close to the ocean, the dominant cost is desalination, not transport; the process would be therefore less expensive in places like Beijing (China), Bangkok (Thailand), Zaragoza, Phoenix, and, of course, coastal cities like Tripoli." For cities on the coast, desalination is being increasingly viewed as an untapped and unlimited water storage.

Israel is now desalinizing water at a cost of US$0.53 per cubic meter. Singapore is desalinizing water for US$0.49 per cubic meter. Many large coastal cities in developed countries are considering the feasibility of seawater desalination, due to its cost effectiveness compared with other water supply options, which can include mandatory installation of rainwater tanks or stormwater harvesting infrastructure. Studies have shown that desalination is among the most cost-effective options for boosting water supply in major Australian state capitals. The city of Perth (Australia) has been successfully operating a reverse osmosis seawater desalination plant since 2006, and the West Australian government has announced that a second plant will be built to service the city's needs. A desalination plant is to be built in Australia's largest city, Sydney, and Wonthaggi, Victoria in the near future.

The Perth desalination plant is powered partially by renewable energy from the Emu Downs Wind Farm. The Sydney plant will be powered entirely from renewable sources, thereby eliminating harmful greenhouse gas emissions to the environment, a common argument used against seawater desalination due to the energy requirements of the technology. The purchase or production of renewable energy to power desalination plants naturally adds to the capital and/or operating costs of desalination. However, recent experience in Perth and Sydney indicates that the additional cost is acceptable to communities, as a city may then augment its water supply without doing environmental harm to the atmosphere. The Gold Coast desalination plant will be powered entirely from fossil fuels and at a time when the coal fired power stations have significantly reduced capacity due to the drought. At a rate of over 4 kWh per cubic meter to produce this will be the most expensive source of water in Australia.

In December of 2007 the South Australian Government announced that it would build a seawater desalination plant for the city of Adelaide, Australia located at Port Stanvac. The desalination plant is to be funded by raising water rates to achieve full cost recovery. An online, unscientific poll showed that nearly 60% of votes cast were in favor of raising water rates to pay for desalination.

An informative January 17, 2008 article in the WALL STREET JOURNAL reports, "In November, Connecticut-based Poseidon Resources Corp. won a key regulatory approval to build a [US]$300 million water-desalination plant in Carlsbad, north of San Diego. The facility would be the largest in the Western Hemisphere, producing 50 million [U.S.] gallons [190,000 m^3] of drinking water a day, enough to supply about 100,000 homes... Improved technology has cut the cost of desalination in half in the past decade, making it more competitive... Poseidon plans to sell the water for about [US] $950 per acre-foot [1200 m^3]. That compares with an average [US]$700 an acre-foot [1200 m^3] that local agencies now pay for water." $1,000 per acre-foot works out to $3.06 for 1,000 gallons, which is the unit of water measurement that residential water users are accustomed to being billed in.

Reverse osmosis (RO) is a separation process that uses pressure to force a solution through a membrane that retains the solute on one side and allows the pure solvent to pass to the other side. More formally, it is the process of forcing a solvent from a region of high solute concentration through a membrane to a region of low solute concentration by applying a pressure in excess of the osmotic pressure. This is the reverse of the normal osmosis process, which is the natural movement of solvent from an area of low solute concentration, through a membrane, to an area of high solute concentration when no external pressure is applied. The membrane here is semipermeable, meaning it allows the passage of solvent but not of solute.

The membranes used for reverse osmosis have a dense barrier layer in the polymer matrix where most separation occurs. In most cases the membrane is designed to allow only water to pass through this dense layer while preventing the passage of solutes (such as salt ions). This process requires that a high pressure be exerted on the high concentration side of the membrane, usually 2–17 bar (30–250 psi) for fresh and brackish water, and 40–70 bar (600–1000 psi) for seawater, which has around 24 bar (350 psi) natural osmotic pressure which must be overcome.

The desalinated water is very corrosive and is "stabilized" to protect downstream pipelines and storages usually by adding lime and carbon dioxide to prevent corrosion of concrete or cement lined surfaces. Liming material is used in order to adjust pH at 6.8 to 8.1

to meet the potable water specifications, primarily for effective disinfection and for corrosion control.

Desalination processes are very effective barriers to pathogenic organisms, how ever disinfection is used to ensure a "safe" water supply.

Reverse osmosis units sold for residential purposes offer water filtration at the cost of large quantities of waste water. For every 5 gallons of output, a typical residential reverse osmosis filter will send around 10 - 20 gallons of water down the drain although it may be captured and used for watering plants and lawns.

DESCRIPTION AND INNOVATIONS

1. Delivery Tubes

Historically, for delivering water from the Mediterranean or Red Seas to the Dead Sea researchers have proposed underground tunnels and surface canals. Underground tunnels are very expensive, the surface canal cheaper, but that does not allow efficient use of the different sea levels and requires instead more energy for water pumping and maintenance cost. In effect that choice offloads cost from construction years to the operation years of the project. Underground tunnels require strong concrete linings and the surface canals need extensive sub-aerial site preparation and separation of water from ground and aquifers. Part of the incoming water will vaporize from the open canal surface, increasing the net cost of delivered water per unit of canal capacity. The historical project plan require also power and desalination plants costing about 1 billion USD and 550 MW capacity in power plants for desalination and delivery of water from the Mediterranean or Red Sea to Dead Sea.

The author published previously a series of methods [1]-[22] which are significantly cheaper and have significant advantages in comparison with currently proposed methods.

The author offers to deliver water to Dead Sea and for irrigation by *inflatable tubes made from artificial material or film.* That decreases the cost of building and delivering the system by tens of times (see in theoretical section and Table 3) and decreases the building period by a factor of some several times.

Material properties are in Table 5 Attn.The costs of some materials are presented in Table 6 Attn. (2005-2007). Some difference in the tensile stress and density are result the difference sources, models and trademarks.

2. High Height Tube Extractor and Electric Plant

The offered extractor is shown in Figure 1. The main part is a cheap inflatable high altitude tube 1 (up 3 - 5 km) supported by bracing wires 14. Tube is designed from inflatable toroids 13. They maintain the tube form. This tube has freshwater collector 2 and freshwater pipeline 3 inside. The tube can have a film solar heater 8 (optional) located on the ground. That consists of a transparent film cover over the ground (a black surface). The entering air flows between film and ground and is heated by solar radiation. That strongly increases the speed of incoming air flow (see computation) and therefore the total system productivity. If we additionally inject the saline (sea) water 7 into incoming air flow, we increase the water productivity yet more (see computation section). The installation can have a propeller 15 for increasing airflow in windless weather. The top end of the tube has an air turbine and electric generator 19 mounted there. The top end can have a wind turbine 20 (optional). The top end has also the observation desk 16 and elevator 18. The tube enter and exit 10 (Figure 1) have wind leafs (2, 4 Figure 2). The other parts are noted under Figures 1-2.

The installation works the following way. The wind leafs (3, 4 Figure 2) automatically are opened so as to use the wind dynamic pressure (and to draw off airflow for the top end of the tube). The airflow enters into the solar heater 8), is warmed and goes into the vertical intake tube (or directly to the entrance of the vertical intake tube when the solar heater is absent). At high altitude, the air expands, cools and water vapor condenses in the water collector. The pipeline going down from this (several kilometers high) collection region can deliver the freshwater (because of the altitude) under high water column pressure to a needed use. This high water pressure may be used for production of energy (electricity) or for delivery of water to far regions, by pipeline over great hills in between without need for pumps, because it is pre-pressurized. Moreover, we can install an air turbine and electric generator at the top of the tube and get electricity when we do not need a big amount of freshwater. In this case the solar energy of the solar heater is transferred to electric energy without a need for expensive solar cells (see computation).

3. The Inflatable Dead Sea AB-Dome

Our idea is a dome covering a large region (city, large wasteland, agricultural zone, or part or all of the Dead Sea) by a thin film with controlled clarity (reflectivity, and transmittal of solar spectrum)(optinal). The film is located at high altitude (up to 6 km) which include the Earth's troposphere where are the main climatic changes. The film dome is supported at altitude by a small additional air overpressure produced by ground ventilators and the film dome is connected to Earth's ground by cables. The closed area is also divided into sub-areas by film also having controlled clarity and albedo. That allows making different conditions (solar heating) in sub-areas and pumping hot, warm, cold, moist air from one sub-area to other sub-area, leading to many interesting engineering possibilities.

The film in question would be a cheap film having liquid crystal and conducting layers. The clarity of these is controllable by electric voltage assigned to various sectors by computer control. Such layers can pass or blockade the solar light (or parts of solar spectrum) and pass or blockade the Earth's re-radiation of heat. The outer and inside radiations have different wavelengths. That makes differential passage of desired and undesired radiations much easier to effect; the result is climate control within the dome. In conventional conditions about 50% of incoming solar energy reaches the Earth surface. The most part is reflected back to outer space by white clouds. In our closed system the clouds (and rain) will be formed at night when temperature is low, analogous to fog but in the higher colder regions of the dome.

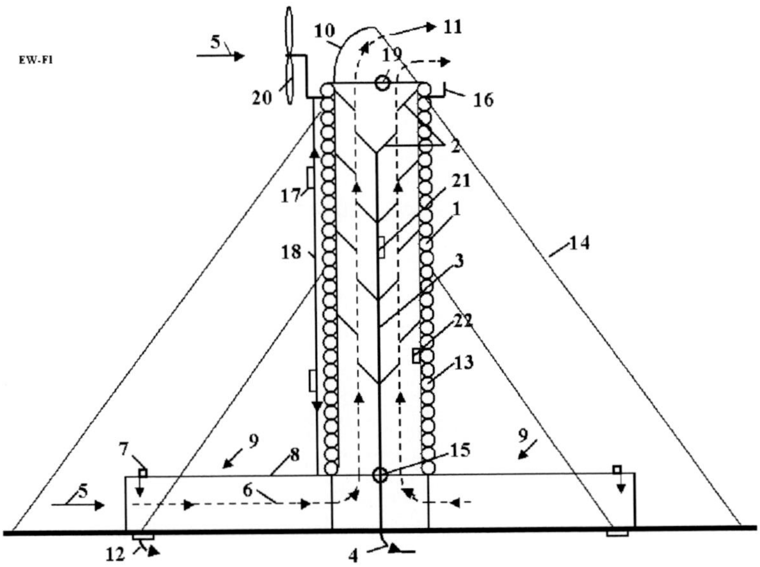

Figure 1. Inflatable freshwater-from-air extractor and electric power plant (side view). Notation: 1 - vertical tube, 2 - freshwater collector, 3 - freshwater pipeline, 4 - exit of freshwater, 5 - wind, 6 - air flow, 7 - injector of sea (saline) water (optional), 8 - transparent film and solar heater of air (optional), 9 - solar radiation (optional), 10 - air exit. 11 - air flow, 12 - collector of seawater (optional), 13 - inflatable toroid, 14 - support cable (bracing wire), 15 - ventilator (propeller) (optional), 16 - observation desk for tourists and communication installation (optional), 17 - passenger cabin, 18 - elevator, 19 - top propeller-electric generator, 20 - wind electric generator, 21, 22 - mobile cabins.

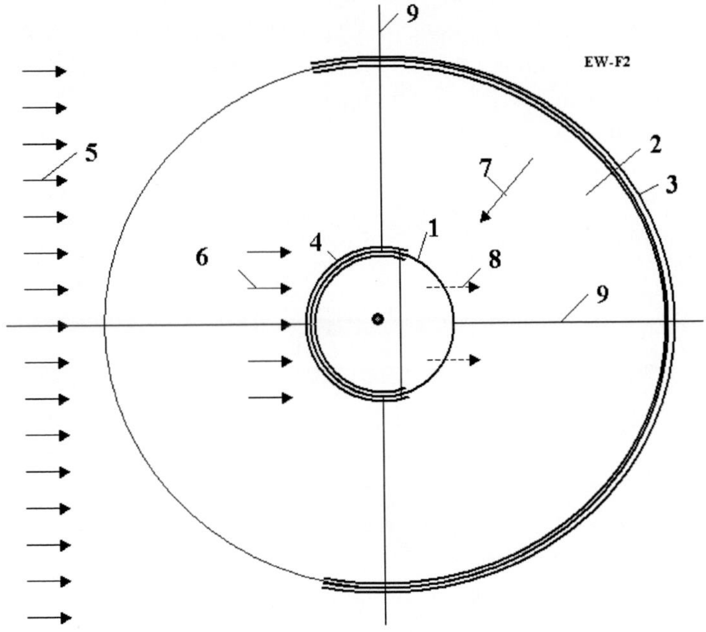

Figure 2. Inflatable extractor of freshwater from atmosphere (top view). Notation: 1 - vertical inflatable tube, 2 - solar heater (optional), 3 - wind leafs of solar heat, 4 - wind leafs of air exit, 5 - wind at ground, 6 - wind at altitude, 7 - solar radiation, 8 - exit tube air flow, 9 - tube support cable.

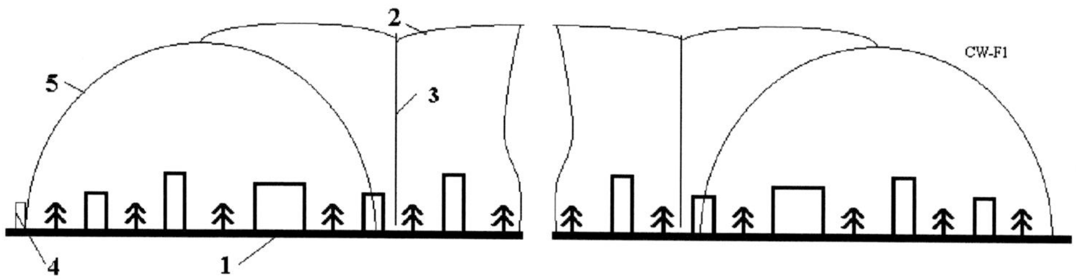

Figure 3. Film dome over city. Notations: 1 - city, 2 - thin film cover with controlled clarity, 3 - support cable, 4 - exits and ventilators, 5 - semi-cylindrical border section.

By this controllable means the many cold regions (Alaska, Siberia) may retain more solar energy and became a temperate climate or sub-tropic climate. That also means the Sahara desert can be reflected to coolness by day (and re-radiation controlled for warmth at night) and become a prosperous area with fine climate and with closed-loop water cycle.

It is worth noting that the water content in many crops is about a thousandth or so of the water required for the prolonged process of growing the crops. Usually that water is lost, but in a closed cycle dome all but a thousandth is recovered. *Therefore, only that tiniest fraction of the water needed would be replaced from outside the dome.*

The building of a film dome is very easy. We spread out the film over the ground, turn on the pumping propellers and the film rises by air overpressure to the needed altitude limited by the support cables. Local damage of the film is not a trouble because the additional air pressure is very small and the propeller pumps compensate for air leakage.

Included among the advantages of the suggested method are the possibility to paint pictures on the sky (dome), to show films on the sky by projector, to suspend illuminations, decorations, and air tramways from a dome ceiling that can serve as an engineering support. Long distance aircraft fly at altitude 8 - 11 km and our dome does not interfere with those flight levels. The support cable will have illumination for internal traffic safety and internal helicopters will thus avoid contact with them.

The offered dome can make our internal weather fine, and provide 'new' territory for living with a wonderful climate.

Our design for the dome is presented in Figure 3, which includes the thin inflated film dome. The innovations are listed here: (1) the construction is air-inflatable; (2) each dome is fabricated with very thin, transparent film (thickness is 0.1 to 0.3 mm) having controlled clarity without rigid supports; (3) the enclosing film has two conductivity layers plus a liquid crystal layer between them which changes its clarity, color and reflectivity under an electric voltage (Figure 4, Ch.1A); (4) the bound section of dome has a hemisphere form (#5, Figure 3). The air pressure is more in these sections and they protect the central sections from outer wind.

Figure 3 and Figure 4 Ch.1A illustrate the thin controlled transparency dome cover we envision. The inflated textile shell—technical "textiles" can be woven or non-woven (films)—embodies the innovations listed: (1) the film is very thin, approximately 0.1 to 0.3 mm. A film this thin has never before been used in a major building; (2) the film has two strong reinforcement nets, with a mesh of about 0.1×0.1 m and $a = 1 \times 1$ m, the threads are about 0.5 mm for a small mesh and about 1 mm for a big mesh. The net prevents the watertight and airtight film covering from being damaged by vibration; (3) the film incorporates a tiny electrically conductive wire net with a mesh about 0.1 x 0.1 m and a line width of about 100 μ (microns) and a thickness near 10 μ. The wire net is also an electric (voltage) control conductor. It can inform the dome supervisors concerning the place and size of film damage (tears, rips, etc.); (4) the film may be twin-layered with the gap — $c = 1$ m and $b = 2$ m—between covering's layers for heat saving. In polar regions this multi-layered covering is the main means for heat insulation and puncture of one of the layers won't cause a loss of shape because the film's second layer is unaffected by holing; (5) the airspace in the dome's covering can be partitioned, either hermetically or not; and (6) part of the covering can have a very thin shiny aluminum coating that is about 1μ for reflection of unnecessary solar radiation in equatorial or polar regions [15].

Our design for a Dead Sea or people-housing "Evergreen" dome is presented in Figure 5, which includes the thin inflated film dome. Air-supported construction derives from the balloon principle to shape a building; the air pressure inside the building exceeds the external air pressure to support the roof. Sunlight can penetrate special roofing materials, making the interiors brighter than others types of constructed buildings. The innovations are listed here: (1) the construction is air-inflatable; (2) each dome is fabricated with very thin, transparent film without rigid supports; (3) the enclosing film is a two-layered element with air between the layers to provide insulation; (4) the construction form is that of a hemisphere (semi-cylinder), or in the instance of a roadway/railway a half-tube, and part of it has a thin aluminum layer about 1 μ or less that functions as a gigantic collector of solar incident solar radiation (heating for Polar regions). Surplus heat collected may be used to generate electricity or furnish mechanical energy; and (5) the dome is equipped with sunlight controlling louvers [AKA, "jalousie", a blind or shutter having adjustable slats to regulate the passage of air and sunlight] with one side thinly coated with reflective polished aluminum of about 1μ. Real-time control of the sunlight's entrance into the dome and nighttime heat exit is governed by the shingle-like louvers (or controlled transparency of film).

4. Covering of Dead Sea by Thin Transparent Film

The cheapest method of stopping the evaporation of the Dead Sea is by covering the sea surface by a thin transparent film (Figure 6). The film has thickness 0.01 ÷ 0.02 mm. It is supported by floats and chains itself to the 'ground' by cables connected to anchors. The covering film does not reach the seacoast by about 100 m margin and tourists can swim and boat. Any ships may float by special dedicated routes, which are free from film. About 2% of the Dead Sea's surface is sufficient for these macro-engineering concerns.

It is worth mentioning, in passing, that the evaporation prevented by the surface film is worth a cubic kilometer per year over the highly insolated region of the Dead Sea—that is a cubic kilometer a year that need not be made up by elaborate macro-engineering schemes. Sometimes prevention is much cheaper than paying for a cure, particularly a chronic cure.

Even in the cooler regions of Lake Kinneret, (The Sea of Galilee) 170 square kilometers of exposed surface will probably evaporate over a hundred fifty to two hundred million cubic meters a year just from the surface of the reservoir, a loss, which avoided, again need not be made up.

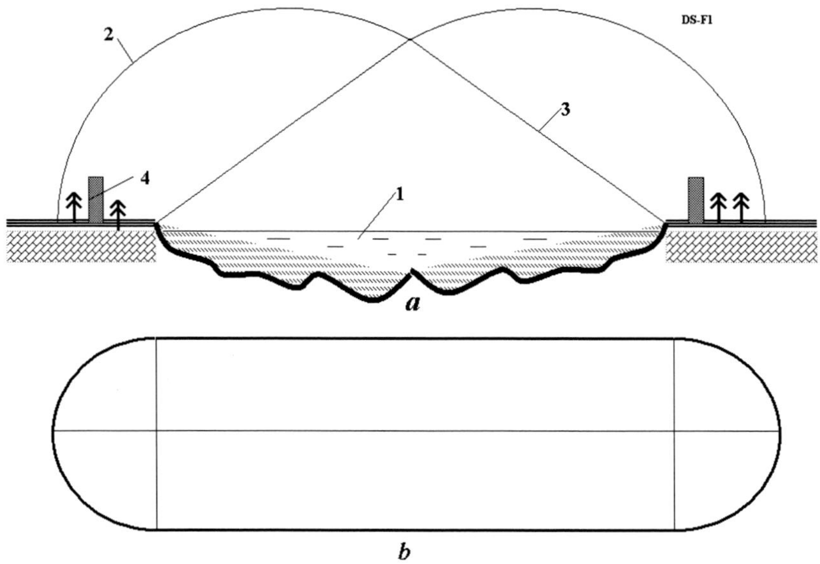

Figure 4-5. Artificial inflatable dome directly above the Dead Sea. Notations: (a) cross-section area of suggested bio-shell; (b) top view of cylindrical bio-shell. Notation: 1 – Dead Sea, 2 – version of transparent thin double film ("textiles") over sea; 3 – bracing; 4 – resort; 5 - enter; 6 - air pump (ventilator).

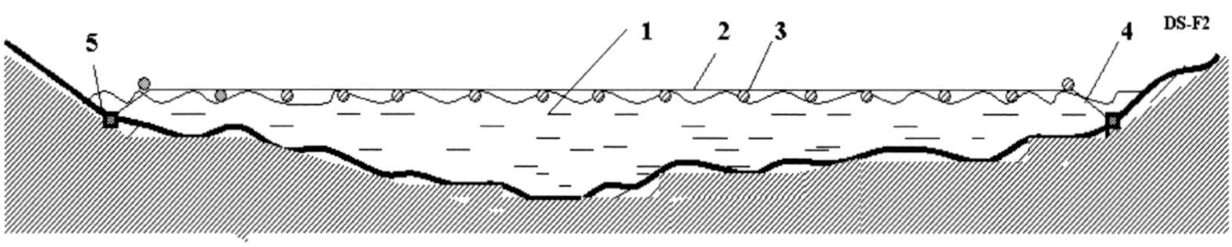

Figure 6. Covering the sea (lake) surface by thin film for stopping of vaporization. Notation: 1 – Dead Sea or other lake; 2 – thin transparent film; 3 – float; 4 – cable; 5 – anchor.

COMPUTATIONS AND ESTIMATIONS

Below reader can find the estimations and computations of textile tubes for delivery of water from Mediterranean and Red Seas to Dead Sea. The height of water tube track is about 200-800 m over sea level. The Dead Sea is located about 420m lower than Earth's global sea level, Lake Galilee about 200 meters below world sea level.

1. Estimation of the power for pumping water by textile tube. The needed power is computed by equation:

$$P = gmH/\eta\,, \tag{1}$$

where P is power, kW; $\eta \approx 0.8 \div 0.9$ is coefficient efficiency of full system; m is water extension, m^3/s; $g = 9.81$ m/s^2 is gravity; H is total difference of water levels (maximum ground height + friction loss of pressure in tube), m.

Computation is presented in Figure 7.

If a water pipeline has hydroelectric stations in a downhill part of water-line, then H is

$$H = h + \Delta H\ ,$$

where h is loss of pressure (in m) for water friction [m] of tube walls, ΔH is difference between the initial level and end level. Example, for Mediterranean Sea and Dead Sea $\Delta H = -420$ m. Minus of P means you are getting energy from the water flow. Coefficient efficiency of the line having the hydroelectric station is about $\eta = 0.8$.

2. *The water speed* in a hermetic tensioned textile tube can be estimated by equation

$$V = \frac{4m}{\pi D^2}\ , \tag{2}$$

where V is water speed, m/s; m is water extension, m^3/s; D is tube diameter, m.

Computation is presented in Figure 8. Greater tube diameter will promote less water flow speed and, thus, reduced inside water friction losses.

3. *The loss of water pressure* (in m) is computed by equation

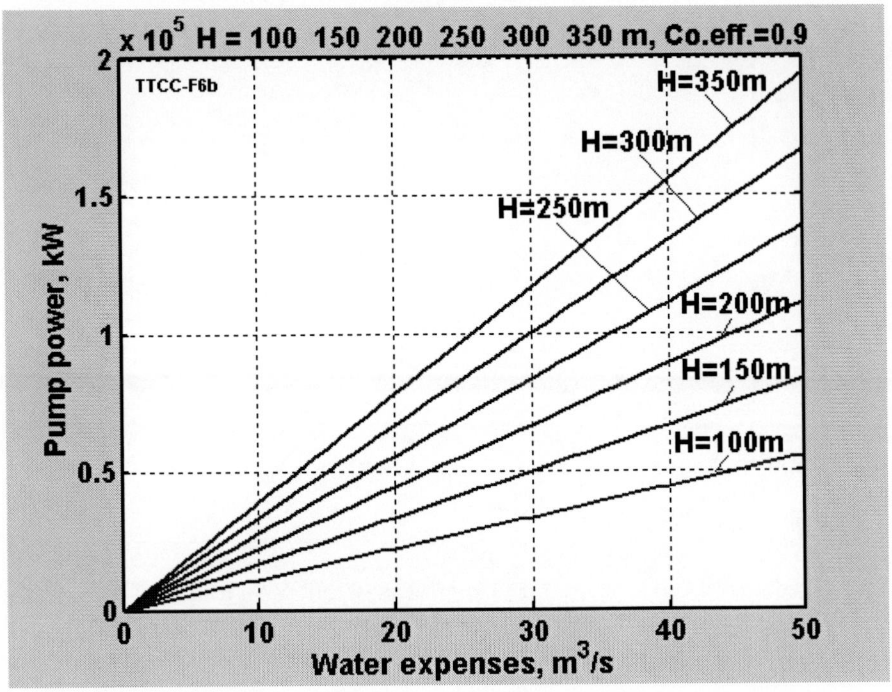

Figure 7. Required power of water pump station versus the water flow through (expense) for a different range of water levels in meters.

$$h = f\frac{L}{2g}\frac{V^2}{D}, \tag{3}$$

where h is loss of water pressure, m; f is friction coefficient; L is length of tube, m; V is water speed, m/s; D is tube diameter, m.

The water friction coefficient is

$$f = \frac{0.25}{\left[\log\left(\frac{k}{3.7D}+\frac{5.14}{R_e^{0.4}}\right)\right]^2}, \tag{4}$$

where R_e is Reynolds number, k is roughness.

In our case we use the textile tubes. The friction coefficient for both is approximately 0.06.

The computation of equation (3) is presented in Figure 8. Loss amounts to about 350 m of water pressure in distance 350 km, 200 m of water pressure in a distance of 200 km, and 150 m in a distance of 150 km.

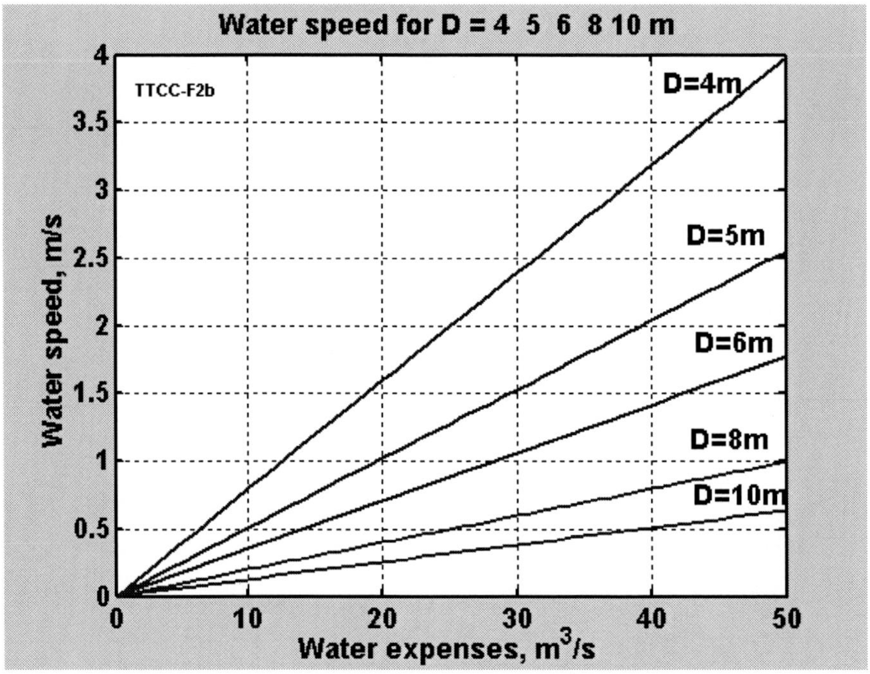

Figure 8. Water speed in textile tube via water flow through (expenses) for different tube diameters.

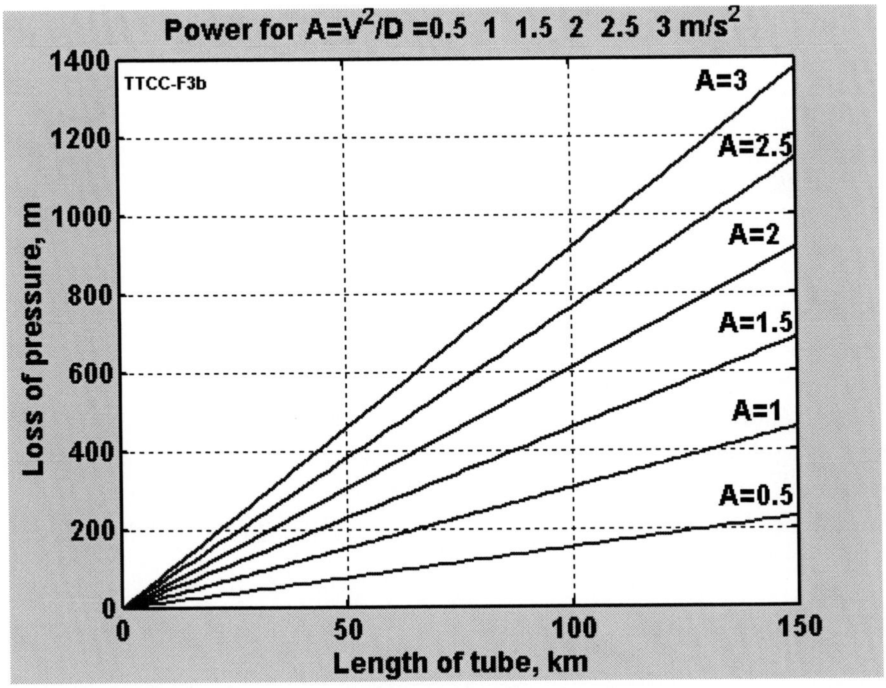

Figure 9. Loss of the water pressure into the closed tube system via the length of tube for the different ratios $A = V^2/D$, where V is water speed, D is hermetic tube diameter. Friction coefficient $f = 0.06$.

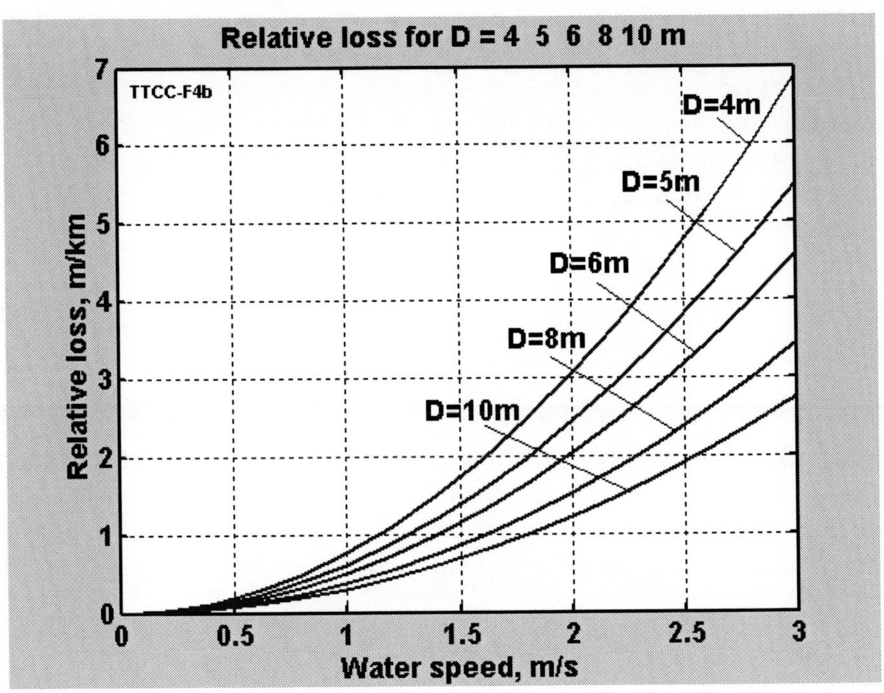

Figure 10. Relative loss (m/km) of the water pressure via the water speed for different tube diameters. Friction coefficient $f = 0.06$.

4. Relative loss of water pressure. The relative loss of the tube's water pressure may be estimated by equation

$$\bar{h} = f\frac{V^2}{2gD}, \qquad (5)$$

where $\bar{h} = h/L$ is relative loss of water pressure, m/km. The computation is presented in Figure 10.

The big hydraulic tube (i.e., with a large diameter) is expensive but such a tube has the advantage of decreasing greatly the pressure loss and increases significantly the efficiency and decreases the friction loss and pump power requirement.

The power needed for pumping seawater may be computed by equation

$$P = \frac{\pi\rho}{4\eta}D^2V(H+h), \qquad (7)$$

where P is power, W; ρ is seawater density, 1000 kg/m^3; D is tube diameter, m; V is water speed into pipe, m/s; H is altitude of final tube end, m; h is loss of seawater friction, m.

Example: For the pipe length L = 100 km and D = 4 m^2 and V =1.5 m/s, we have h = 160 m, H = 500 m; P = 12.4 MW.

5. Tubes. Tubes near the pump station have high pressure (up to 30 atmospheres.). They must be made from something even stronger—perhaps from some composite fiber material. Such composite material has higher maximum stress (up 600 kg/mm^2, steel has only 120 kg/mm^2) and low specific-density (1800 kg/m^3, steel has 7900 kg/m^3). It may be or become cheaper quite soon. Coefficient of safety is 3 to 5. Below, is the useful equation for computation of the tube wall thickness:

$$\delta = \frac{pD}{2\sigma}, \qquad (6)$$

where δ is tube-wall thickness, m; p is water pressure, N/m^2; σ is safety tensile stress, N/m^2.

The result of computation is presented in Figure 11. The textile tubes are supported by special chutes, troughs or inflatable pillows (To avoid chafing damage on the ground).

The installation can pump the amount of seawater given by:

$$M = \frac{\pi\rho}{4}D^2Vt, \qquad (9)$$

where M is mass (or volume, m^3/time) of seawater, kg/time; ρ is seawater density. kg/m^3 (or specific volume m^3/time), t is time (seconds, day, year).

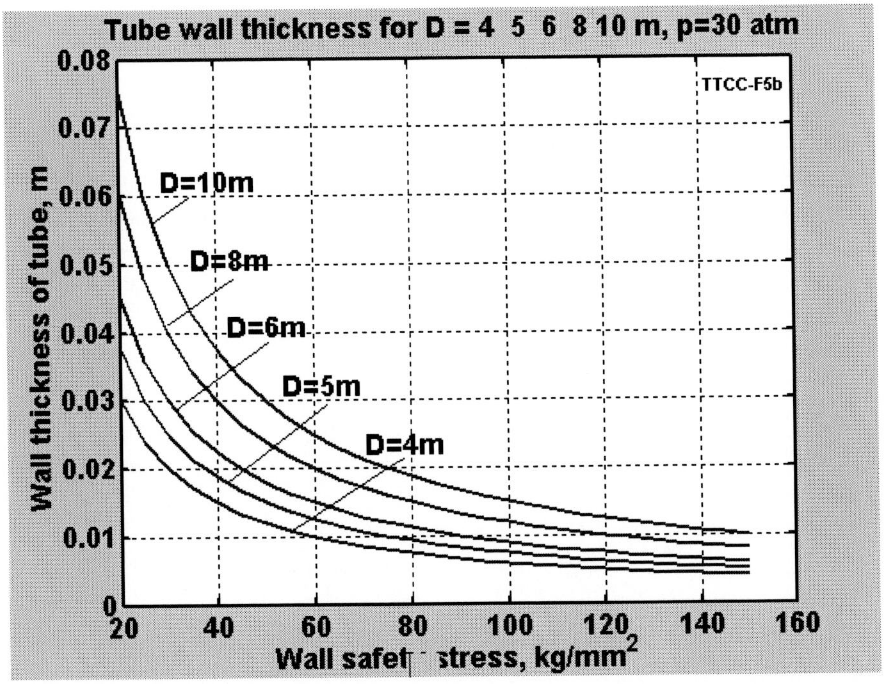

Figure 11. Tube-wall thickness via wall safety tensile stress for different tube diameters and water pressure 30 atm (300 m of altitude). If elevation is less the wall thickness will be proportionally less.

Example: For tube above (L = 100 km, D = 4 m) in daylight (t = 12 hours = 43200 sec) the deliverable seawater productivity will be 814,000 m^3/day.

6. Energy required by the different methods of desalination. Below in Table 1 is some data about expense of energy for different methods of desalination.

Table 1. Estimation of energy expenses for different methods of freshwater extraction

No	Method	Condition	Expense kJ/kL	Getting kJ/kL
1	Vapor	Expense only for vapor*	2.26×10^6	0
2	Freezing	Expense only for freezing, c.e. $\eta = 0.3$	1×10^6	0
3	Reverse osmosis	Expense only for pumping,	$(4\div7)\times10^3$	0
4	High Tower extr.	$t = 35$ C, $h = 0.7$, tube is black	0	30×10^3

* This expense may be decreased by 2 -3 times when the installation is connected with heat or nuclear electric station, utilizing otherwise wasted heat.

MACRO-PROJECTS

1. Current Conventional Macro-projects

There are some macro-projects looking toward Dead Sea stabilization at the present time, planned by conventional methods. Those are projected to build an electrically powered desalination plant (emitting 1.9 km^3 freshwater per year) at the Mediterranean Sea or Red Sea shoreline and connect it to the Dead Sea by underground tunnel and canal. They include the pump and hydroelectric stations. There are three versions of these plans:

1. *Red Sea (Arabah) – Dead Sea Canal (RSDSC).* Length 180 km. The water (1.9 km^3/year) is desalted, pumped to the height of 125 - 200m and through an underground tunnel and a canal delivered to Dead Sea. Cost is about 5 billions USD and building time 9 years. Needed power is 550 MW. Useful net power added to the grid after desalination is about 30 MW.
2. *Mediterranean Sea (Katif, Ashqelon)– Dead Sea Canal (MSDSC).* Length 100 km (20 km canal – 80 km tunnel). The water (1.9 km^3/year) is desalted, pumped to the height of about 200m and through an underground tunnel and canal, delivered to the Dead Sea. Cost about 5 billions USD and building time 5 years. Requested power is 550 MW. added to the grid after desalination is about 190 MW.
3. *(Northern) Mediterranean Sea – Jordan River Canal (MSJRC)* (or Haifa – Sea of Galilee). Length 80 km. The water (1.9 km^3/year) is desalted, pumped to the height of about 100m and through an underground tunnel and canal delivered to the Jordan River and from the Jordan River to the Dead Sea. Cost about 5 billions USD and building time 5 years. Required power is 450 MW. Useful net power added to the grid after desalination is about 0 MW.

Disadvantages:

1. Desalted water is expensive (0.6 – 1 $/kL USD). It is better to spend money for irrigation than as vapor from the Dead Sea. Israel could double its agricultural production.
2. In version (3) Syria and Jordan and a future Palestine can take this water without Israeli authorization for irrigation and the water will not reach the Dead Sea.

Offered Variants of Dead Sea Conservation Plans

Below we offer four different cheaper approaches toward saving the Dead Sea. For decreasing the building cost, of the freshwater cost and increasing net economic efficiency (including useful energy output), author offers the following main innovations:

1. Change the underground water tunnel and canal to inflatable tubes. That significantly decreases building cost, pumping energy and increases net useful energy.
2. Change the expensive desalination plant (cost about 1000M USD) by an AB-Tower (200M USD) that extracts high pressure freshwater and energy from the atmosphere. In contrast to a desalination plant the AB-Tower is cheaper, produces high pressure

(up to 270 atm net) freshwater output (water lines under such pressure do not need expensive pumping stations) and produces net energy (2000 – 6000 MW)—conventional desalination plants, by contrast, require abaout 300MW energy, ultimately derived from coal.

3. Cover with an AB-Dome (5 – 10 km height) the Dead Sea. Cost of an inflatable AB-Dome (Mountain) is 3.5 Billions of USD. Freshwater from atmosphere may be about 2 -10 km^3/year, energy about 6,000 MW. The climate will change (increasing precipitation) for a radius of about about 20 km around Dead Sea.
4. Cover the Dead Sea surface with a thin film (0.01÷ 0.02 mm), which stops the evaporation of the Dead Sea and allows a slow restoration effort.

2. Freshwater Pipelines

Three versions of the inflatable water lines are computed:

a) Red Sea (Arabah) – Dead Sea. Length is 180 km. Maximum hill transit height is 200m.
b) Mediterranean Sea (Ashqelon) – Dead Sea. Length is 110 km. Maximum hill transit height is 200m.
c) Mediterranean Sea (Haifa) – Galilee Like. Length is 80 km. Maximum hill transit height is 100m.

Data used for computation and results are:

- Flow is 60 m^3/s.
- Diameter of inflatable tube is 6 m.
- Water speed is 2.12 m/s.
- Pumping powers are 133 MW, 133 MW, 67 MW respectively.
- Losses of water pressure into tubes for coefficient of friction $C_f = 0.06$ are: 405m, 248m, 180m respectively.
- Useful difference of levels and getting energy are: 15m, 55m, 20m (11MW, 69MW, 16MW) respectively.
- Cost of inflatable pipelines only are $90, $60, $40M of USD, respectively.

3. AB-Tower–freshwater and Energy Extractor

Let us to make some estimations of the high altitude tube freshwater extractor. Our data is far from optimum. Our aim is to demonstrate the methods of estimation and some possibilities of the offered idea.

Take the radius of inflatable tube $R = 575$ m ($S = 10^6$ m^2), height of tube $H = 3$ km, air temperature on Earth's surface $t = 25$ °C, air relative humidity $h = 0.7$, wind speed $V = 6$ m/s. From equations and graphs of research [1] and the above we get:

- Amount of freshwater into 1 m^3 of air is m_w = 0.0052 kg/m^3 = 5.2 g/m^3 [Eq. (1)].
- The average wind speed at altitude H = 3 km for α = 0.15 is V = 14 m/s [1, Eq. (2)].
- The wind dynamic pressure at H = 0 is p_1 = 22 N/m^2 and the air wind rarefaction at H = 3 km is p_2 = 89 N/m^2 [1, Eq. (3)].
- For average coefficient of air friction C_f = 0.005 and average air density ρ = 1.1 kg/m^3 the air speed into tube is V_t = 12.6 m/s [1, Eq. (3)].
- The volume and mass of air flow are $v = SV = 125\times10^5$ m^3/s, $M_a = \rho v = 15.3\times10^3$ tons/s.
- The freshwater flow is $M_w = vm_w$ = 65 kL/s = 5.6×10^6 kL/day = 2.05 km^3/year (1 kL = 1 ton).

Energy estimation:

- Power from freshwater (H = 2500 m) is 1500 MW.
- Power from wind turbine on tube (one at top, R = 100 m, $A = \pi R^2$, η = 0.5, V = 12.6 m/s) is $N = 0.5\eta\rho AV^3$ =14.3 MW.
- Power from black-colored tube material (heating from Sun radiation, q = 500 W/m^2) is 1725 MW (only in solar daytime).
- Power from solar heater on Earth's surface (S_h = 1×1 km, q = 500 W/m^2) is 500 MW (only in solar day time).
- If the solar heater has area S_h = 5×5 km the power will be 12500 MW. That is equivalent to the power of some powerful arrays of electric stations (only in solar day time).
- If nighttime and no wind, we can turn on the lower ventilator. For V_a = 4 m/s the requested ventilator power is 40 MW [1, Eq.(8)]. But the energy obtained from high altitude freshwater is 475 MW.

We take the average useful energy without solar heater to equal 2000MW and with solar heater 6000 MW.

Cost of high tube extractor-generator. The cost of areas of thin film (the main construction material of inflatable tower) is about USD \$1/m^2. The full area of the tower (H = 3 km) is 68 km^2. The total cost of the tower's construction film is about \$70 million (M). The costs of ventilator-electric generators are \$15 M. Cost of hydroelectric generators are about \$20 M. Cost of R&D, design and building of tower is taken as \$20 M. Total cost of Installation (without Solar Heater) is USD\$125 M. The tower produces 5.6×10^6 kL/day of freshwater and the powerful electric station has an average power of more then 6,000 MW (with solar heater). Let us add a solar heater. The area of a solar heater (5×5 km) is 25 km^2 and the solar heater costs USD\$15 M (one increases the water by 30 - 50% and energy by 3 times). The total cost of the installation together with solar heater is USD\$140 M. Let us take the total cost as \$200 M of USD.

The 1 MW of a nuclear electric station costs about \$1 M. The offered installation (electric plant same power) is cheaper than the same nuclear electric station by 10 times. Our installation is safety, friendly to environment, and produces free energy and high pressure freshwater. The pressurized water may be delivereded long distances by tubes without pump stations and high pressure water may produce electricity far from the tower. The nuclear station requires a nuclear fuel complex and produces energy which costs as comparably to conventional heat cycle (fossil fueled) electric stations. Remainder, our estimation (data) is not optimal, and can be improved over that given above.

4. Transparent Inflatable AB-Dome over the Dead Sea

The Dead Sea has a size taken as 67×18 km. Let us take the size of our AB-Dome as70×20 km.

1. The average thickness of cover is

$$\delta = \frac{pR_c}{\sigma},$$

where p is average pressure, N/m^2; R_c is average radius of Dome between support cables, m; σ is safety stress of film, N/m^2. For p =100 N/m^2; R_c = 5000 m; σ = 10^9 N/m^2, we have $\delta = 10^{-3}$ m =1 mm.

2. Area of surface, S [m^2], volume, V [m^3], and mass, M [kg], of the AB-Dome are:

$$S = 0.5\pi DL,\ V = S\delta, \quad M = \gamma V,$$

where D is width of Dome, m; L is length of Dome, m; γ is cover density, kg/m^3. For γ = 1500 kg/m^3 we have $S = 2.2\times10^9$ m^2; $V = 2.2\times10^6$ m^3; $M = 3.3\times10^9$ kg.

3. Cost of cover for c = \$1/kg is \$3.3 billions (B) of USD. Building cost about 200 millions. Total cost of AB-Dome is about \$3.5 billions of USD.

5. Covering Dead Sea Water Surface by Thin Film

The author offers to stop evaporation of the Dead Sea by covering the 98% of sea surface with very thin (0.01 mm) transparent film. That is the cheapest method. We omit covering only 100 m of sea along the coast, where tourists swim and rowed or motorized small boats may move. In this case we save and gradually restore the Dead Sea because every year 75 mm of precipitation will be added to the Dead Sea level and 100 millions cubic meter of water is added from the Jordan River. This freshwater may be also used for irrigation although by the time it reaches the Dead Sea it often carries a less than favorable mixture of rural and urban effluents.

The computations show: Cover area is S = 800 km^2, film thickness is $\delta = 10^{-5}$ m, cover volume is $V = 8\times10^3$ m^3; cover mass is $M = 1.2\times10^7$ kg, for cost cover c = \$2/kg, the film cost is \$24 millions of USD. If covering work costs \$20 millions, the total cost is \$44 millions USD.

RESULTS

The results of our research and computation are presented in Table 4.

Table 4. Data of different methods of saving the Dead Sea

#	Project	Useful Electricity MW	Total cost US $M	Cost of Desalination Plant or Tower, $M	Cost of Canal (Tube)$M	Pump or Electric Station, MW	Time of Building, Years	To Maintain $M/year
1	Conventional RSDSC, d=6 m	30* (-300)**	5000	1000	3700	300	5 - 9	100
	Offered projects							
2	Inflatable tubes MSDST, d=6 m RSDST, d=6 m	 190 30	 1360 1390	 1000 1000	 60^{+} 90^{+}	 300 300	 1 1.5	 50 70
3	AB-Tower MSDST, d=6 m RSDST, d=6 m	 2000-6000 2000-6000	 260 290	 200^{***} 200^{***}	 60^{+} 90^{+}	 0 0	 1 1.5	 60 80
4	AB-Dome	30,000MW+ 20÷100 km^3 freshwater	3500	0	0	300 electric station	2	10
5	Thin film Sea Cover	0	44	0	0	0	0.5	1

* Useful energy; ** Energy requested by desalination plant; *** Cost of AB-Tower; $^{+}$ Cost of tube line.

CONCLUSION

Table 4 shows the offered innovations can significantly decrease the cost and increase efficiency of saving the Dead Sea per dollar spent.

Option 2 (inflatable tubes) decreases the installation cost by 3 times (no underground tunnel and canals), Building time decreases by 3 times, maintenance decreases by 30%.

Option 3 (Combine AB-Tower and inflatable tubes) decreases the Installation cost in 17 times, building time in 3 times, maintenance decreases by 20%.

Option 4 (Inflatable Dome) decreases the cost of installation only 30%, but produces a huge benefit in freshwater (up 20 km^3/year, profit 2 Billion (B)/year of USD for price 0.1\$/kL), energy (up 2000 ÷ 6000 MW, profit 1.75 ÷ 5.25B/year for price 0.1\$/kWh) and change climate (increasing precipitation) in distance 20 km.

Option 5 (thin film) produces a small amount of freshwater (up 6×10^7 m^3), no energy, but installation cost is less by 114 times than the conventional version 1.

Notes: Author began this research as an investigation of a possible new method for acquiring cheap freshwater from the atmosphere. In the process of research he discovered that this method allows producing huge amounts of energy, in particular, by transferring the solar energy into electricity with high efficiency (up to 80%). If direct to electricity solar cell panels are very expensive and have a typical efficiency about 15%, the thin film method of tapping solar power by turning it into fluid (air and rainwater) flows is very cheap. The theory of inflatable space towers [1]-[6] allows us to build very cheap high height towers, which can be used also for tourism, communication, radiolocation, producing wind electricity, space research [8].

The others important artifact and infrastructure technical innovations are the big AB-Dome [19] and the man-made inflated Mountains range composed of numerous aggregated AB-Dome units [20].

One half of Earth's population is malnourished. 2008 has been a year of worldwide food riots by the underfed poor. The majority of Earth is not suitable for unshielded human life. The offered AB-Domes and AB-Mountains can change the climate many regions, give them the water and energy. The increasing of agriculture area, crop capacity, carrying capacity by means of converting the deserts, desolate wilderness, taiga, permafrost into gardens are an important escape hatch from some of humanity's most pressing problems. The offered low-cost AB-Dome method of territorial improvement may dramatically increase the potentially realizable sowing area, crop capacity; indeed the entire range of territory suitable for human living. In theory, converting all nonproductive Earth land such as Alaska, North Canada, Siberia, or the Sahara or Gobi deserts into prosperous garden would be the equivalent of colonizing an entire new planet. The suggested method is very cheap and may be utilized at the present time, not in some distant future. We can start from small areas, such as stadium, small towns in bad regions and extended the practice over a large area—and what is as important, making money most of the way.

Film domes can foster the fuller economic development of dry, hot, and cold regions such as the Earth's Arctic and Antarctic and, thus, increase the effective area of territory dominated by humans. A country can create virtual mountain barriers, which will defend sub-Arctic or near Antarctica countries from cold polar winds. Normal human health can be maintained by ingestion of locally grown fresh vegetables and healthful "outdoor" exercise. The domes can also be used in the Tropics and Temperate Zone. Eventually, they may find application on the Moon or Mars since a vertical variant, inflatable towers to outer space, are soon to become available for launching spacecraft inexpensively into Earth-orbit or interplanetary flights [21].

The related problems are researched in references [1]-[21].

Let us shortly summarize some advantages of this offered AB Dome method of climate moderation:

(1) The artificial Mountains give a lot of freshwater and energy and change a local climate (convert the dry climate to more pleasant moist climate);
(2) They protect from cool or hot wind over a large region;
(3) Covered area does not need large amounts of constant input water for irrigation;
(4) Low cost of inflatable film Dome per area reclaimed;
(5) Control of inside temperature;
(6) Usable in very hot and cool regions;
(7) Covered area is not at risk from weather;
(8) Possibility of flourishing crops even with a sterile soil (hydroponics);
(9) 2 – 3 harvests in year; without farmers' extreme normal weather risks. (Ice storms, wind storms, drought, airborne blights, etc)
(10) Rich harvests, at that.
(11) Converting deserts, desolate wilderness, taiga, tundra, permafrost, and ocean into gardens;

(12)Covering towns and, cities by offered domes;
(13)Protection of city from external attack by tactical nuclear warheads, chemical and biological weapons [16];
(14)Using the high artificial mountains for tourism, communication, triangulation and navigation services, and so on;
(15)Using the dome cover for illumination, pictures, films and advertising.
(16)These forms of hydropower do not silt up their reservoirs, or erode their water ways.

We can almost manufacture perpetually fine local weather, get new territory for living with an agreeable climate without daily rain, wind and low temperatures, and for agriculture. We can cover by thin film gigantic expanses of near useless, dry and cold regions. The countries having big territory (but bad land) such as Canada and the Sahara states, Mongolia and Australia, may be able to use to increase their population and became powerful states in the centuries to come.

The offered method may be used to conserve a vanishing sea as the Aral or Dead Sea. A closed loop water cycle saves this sea for a future generation, instead of bequeathing a salty dustbowl.

The author developed the same method for the ocean (sea). By controlling the dynamics and climate there, ocean colonies may increase the habitable area of the planet another 3 times (after the doubling of useful land outlined above) All in all, this method would allow increasing the Earth's population by 5 – 10 times without starvation.

The offered method can solve the problem of global warming because AB domes will be able to confine until use much carbonic acid (CO_2) gas, which appreciably increases a harvest. This carbon will show up in yet more productive crops! The dome lift force reaches up 300 kg/m^2. The telephone, TV, electric, water and other communications can be suspended on the dome cover.

The offered method can also help to defend cities (or an entire given region) from rockets, nuclear warheads, and military aviation. Details are offered in paper [19].

REFERENCES

(The reader may find some of these articles at the author's web page: *http://Bolonkin.narod.ru/p65.htm*, *http://arxiv.org*, search "Bolonkin", in the book "*Non-Rocket Space Launch and Flight*", Elsevier, London, 2006, 488 pgs. and in book "*New Concepts, Ideas, Innovations in Aerospace, Technology and Human Science*", NOVA, 2008, 430 pgs.)

[1] Bolonkin A.A., (2006), *"Extraction of Freshwater and Energy from Atmosphere"*, *http://arxiv.org* search "Bolonkin".

[2] Bolonkin, A.A., (2002), "Optimal Inflatable Space Towers of High Height". COSPAR-02 C1.1-0035-02, *34th Scientific Assembly of the Committee on Space Research* (COSPAR), The World Space Congress – 2002, 10–19 Oct 2002, Houston, Texas, USA.

[3] Bolonkin, A.A., (2003), "Optimal Inflatable Space Towers with 3-100 km Height", *JBIS*, Vol. 56, No 3/4, pp. 87–97, 2003.

[4] Bolonkin A.A.,(2004a), "Kinetic Space Towers and Launchers", *JBIS*, Vol. 57, No 1/2, pp. 33–39, 2004.

[5] Bolonkin A.A., (2004b), Utilization of Wind Energy at High Altitude, AIAA-2004-5705, AIAA-2004-5756, *International Energy Conversion Engineering Conference at Providence.*, RI, Aug.16-19. 2004. USA. *http://arxiv.org*.

[6] Bolonkin A.A., (2006a), Optimal Solid Space Tower, AIAA-2006-7717. *ATIO Conference,* 25-27 Sept. 2006, Wichita, Kansas, USA. http://arxiv.org, search "Bolonkin".

[7] Bolonkin A., Cathcart R., (2006b). *A Low-Cost Natural Gas/Freshwater Aerial Pipeline. http://arxiv.org*, search "Bolonkin".

[8] Bolonkin A.A., (2006c) *Cheap Textile Dam Protection of Seaport Cities against Hurricane Storm Surge Waves, Tsunamis, and Other Weather-Related Floods. http://arxiv.org*, search "Bolonkin".

[9] Bolonkin A.A., (2006d) *Non-Rocket Space Launch and Flight*, Elsevier, 2006, 488 ps.

[10] Bolonkin A., Cathcart R., (2006e). *The Java-Sumatra Aerial Mega-Tramway, http://arxiv.org*.

[11] Bolonkin A., Cathcart R., (2006f). *Inflatable Evergreen Polar Zone Dome (EPZD) Settlements. http://arxiv.org*, search "Bolonkin".

[12] Bolonkin, A.A. and R.B. Cathcart, (2006g) "A Cable Space Transportation System at the Earth's Poles to Support Exploitation of the Moon", *Journal of the British Interplanetary Society* 59: 375-380.

[13] Bolonkin A., (2006h). *Control of Regional and Global Weather. http://arxiv.org*, search "Bolonkin".

[14] Bolonkin A., Cathcart R., (2006i). *Antarctica: A Southern Hemisphere Windpower Station? http://arxiv.org*, search "Bolonkin".

[15] Bolonkin, A.A. and R.B. Cathcart (2006j), "Inflatable 'Evergreen' dome settlements for Earth's Polar Regions", *Clean Technologies and Environmental Policy* DOI 10.1007/s10098-006-0073-4.

[16] Cathcart R., Bolonkin A., (2006k),*The Golden Gate Textile Barrier: Preserving California Bay of San Francisco from a Rising North Pacific Ocean. http://arxiv.org*. Search: "Bolonkin".

[17] Cathcart R., Bolonkin A., (2006*l*), Ocean Terracing, *http://arxiv.org. Search: "Bolonkin"*.

[18] Book (2006m),: *Macro-Engineering - A challenge for the future*. Collection of articles. Eds. V. Badescu, R. Cathcart and R. Schuiling, Springer, (2006). (Collection contains Bolonkin's articles: Space Towers; Cable Anti-Gravitator, Electrostatic Levitation and Artificial Gravity).

[19] Bolonkin A.A., (2006n). *AB Method of Irrigation without Water* (Closed-loop water cycle) *http://arxiv.org*, search "Bolonkin".

[20] 20. Bolonkin A.A., (2007a). *Cheap Artificial AB-Mountains, Extraction of Water and Energy from Atmosphere and Change of Regional Climate. http://arxiv.org*, search "Bolonkin".

[21] Bolonkin A.A., (2007b). *Inflatable Dome for Moon, Mars, Asteroids and Satellites. Presented as paper AIAA-2007-6262 by AIAA Conference "Space-2007",* 18-20 [September 2007, Long Beach. CA, USA. *http://arxiv.org*, search "Bolonkin".

[22] Bolonkin A.A., (2007c). *Cheap AB-Protection of a City from Rockets and Nuclear Warheads. http://arxiv.org*, search "Bolonkin".

[23] Gleick, Peter; et al. (1996). in Stephen H. Schneider: Encyclopedia of Climate and Weather. Oxford University Press.

[24] *Encyclopedia of Desalination and Water and Water Resources.*

[25] *Encyclopedia of Earth - Surface water management.*

[26] *The MEH-Method (in German): Solar Desalination using the MEH method,* Diss. Technical University of Munich.

[27] *Wikipedia.* Some background material in this article is gathered from Wikipedia under the Creative Commons license.

PART III. SCIENCE RESEARCH AND TECHNICAL PROGRESS

ABSTRACT

At the present time the USA's Federal Government spends big tax monies for Scientific Research and Development (R&D). How to best organize this vast governmental activity, how to best estimate its ultimate utility and profitability (real and potential), how to best increase efficiency of innovation and production, how to best estimate the worth of new discoveries and innovations, how to properly fund R&D of new concepts and innovations, and how to correctly estimate their results - all these macro-problems are important for successful planning of scientific research, new macro-systems—these are all complex and pressing questions that require answers if further industrial progress and scientific improvements are ever to be! The authors consider these major-system problems and offer many remarkable innovations in organization, estimation, suggestions for entirely new research efficiency criteria, development, new methods for assessments of new ideas, innovations in science and industry, and new methods in patenting technology.

The author A.A. Bolonkin worked for many years within the USA's Federal Government entities (scientific laboratories of NASA, Air Force), and USSR and USA universities and industry.

Keywords: Organizing scientific research, planning of research, funding research, funding new ideas (concepts), funding inventions and innovations, estimating research cost, assessment of research results, research efficiency criteria, innovation in organizing of scientific R&D

1. INTRODUCTION

Since the beginning of the 20th Century, science discoveries and improving technology have held the main role in human progress. Humanity has amassed more knowledge than during all previous centuries. People researched aerodynamics, flight dynamics and the design of aircraft. Trained people developed rocket theory and traveled into outer space and landed living people on the Moon by 20 July 1969. Organized research focused on nuclear physics began the exploration of nuclear energy and the creation of powerful computers, which help with the further fast-paced study of Nature. Astronomy's devices allow humans to see and study extra-Solar System planets, possibly even worlds inhabited with forms of life, located millions of light-years beyond our homeland, the Earth.

The power and influence any modern ecosystem-State in our world has is defined by its science, technology, and industry capabilities. The USA is a world leader because, for many years the USA's industry and national government spent more money than any other country on R&D, science-based technical innovations. For example, the USA funds space research more than all other countries combined. Until recently, the all the main scientific advances in space, aviation, and computers originated in the USA.

If the USA's citizens still desire to continue to be science and technology's world leader, they must continue this practice and further refine this public and private policy. However, it is possible when the country will get moving is when it has competitors and takes part in a peaceful competitive struggle. Men on our nearby Moon became possible because the former USSR launched the first satellite (4 October 1957), commencing humankind's Space Age, and the USA's leaders at that critical time clearly understood the USA had temporarily lost global leadership in the important field of science and technology. Only in 1969, after the first manned flight to the Moon, did the USA return to undoubted leadership in space exploration and exploitation development. That program effectively ended in 1972. However, before collapse (1991) the USSR launched more satellites than all the rest of the World's space-faring ecosystem-States together, including the USA! The USA decided to restore this program only when China, that is the PRC, publicly announced its 21st Century program for a manned Moon exploration.

The second very important side of scientific R&D is the efficient use of available funding. The financing of any project is limited everywhere, every time. Unlimited funding is inconceivable. The right organization of scientific funding and research is a very important element of scientific progress. That includes: Organization and careful selection of the most feasible prospective ideas and innovations for research, selection of a "can do" principal investigator - scientists who are the authors or enthusiasts of this idea, its champion, a real hard-headed estimation of the macro-project cost, potentially reachable results, and practical application perspectives.

All these problems are very complex investigations. However, there are common criteria that help to solve these problems of selection and comprehensive organization, and which can save a lot of taxpayer money and achieve practical success in short period of time.

The investigation of these macro-problems is impossible without consideration of current systems of research and frank criticism of its disadvantages. The authors suggest new criteria and new forms of organizing science funding that were tested and/or applied in limited particular cases herein, and which show a high specific efficiency. They also offer new criteria for evaluation of science results which allows more evenly for an observer to estimate the honesty of finished scientific work reports by specialists and to separate pseudo-scientific or non-honest works from real ones.

For customers, leadership and management is also very important for correct estimation of the cost of an offered research, a capability of principal investigator, group, or organization to do this research. Unfortunately, the practice shows mistakes occur very often and they can easily cost millions of USA Dollars (and the EU's Euros)! Herein, A.A. Bolonkin suggests a straight-forward set of simple rules that will permit avoidance of the big strategic mistakes and big awkward and embarrassing tactical slips in the planning of future research efforts.

The human presence is very important in the selection and distribution of limited monetary funding. In many organizations we observe and comment on the situation when large government money distribution—money shifted from all national taxpayers—is channeled by the directives of just one man. As a result, he or she begins to give money to his/her friends, to his/her colleagues or worse - to take bribes. Such a person keeps elementary information about the activities of his/her organization secret. A.A. Bolonkin, herein, offers a method for the best selection to foil this insidious practice, making it exceedingly difficult to initiate or, if revealed, to continue.

2. Support of New Concepts

The monetary support of new aviation and extra-terrestrial space concepts is the basic element of humankind's ongoing scientific and technical progress. All useful things, which we see around us everyday, were developed from new concepts, ideas researched in the rather recent past. This fact is gracefully, eloquently, and comprehensively outlined in Robert Friedel's A CULTURE OF IMPROVEMENT: TECHNOLOGY AND THE WESTERN MILLENNIUM (MIT Press, 2007). But, let us consider the state of affairs now existing.

Science and technology are very complex and have a very high level presence globally nowadays. The production of new valid concepts and ideas, and the effort to fully substantiate them, can ONLY be done nowadays by highly educated people, not by tinkerers and private-sector putterers. The USA has hundreds of thousands of conventionally-trained scientists of every stripe possible. New concepts and ideas are generated only by a very few talented (genius-level) people supported by skilled workers. They are but a small percentage of every thousand scientists. That requires (from them) very much time and hard work. That is not going to be fully paid time in government or company laboratories. The Government and private laboratories develop ONLY known concepts and ideas because their purpose is to get the maximum profit in the shortest time; that means to produce and substantiate new ideas a scientist can only use his own private time. There are a lot of scientists, but most of them do conventional perfectional researches of well-known ideas and small improvements of them, to ensure a good career path. All countries are funding science and research, *but they do not usually fund new ideas or concepts*. Rather, they assimilate known new technology, often developed in other ecosystem-countries. The net funding for radically new concepts and ideas are close to zero in the world as a percentage of gross funding. Break-through funding, practically-speaking, almost does not exist!

In all countries the composers, writers, artists receive a royalty for performance of their musical compositions, books and artworks. Why must scientists gift their hard work to the world, as they labor on new concepts, ideas, theories, and equations for computations? It would be just as if companies making millions from a newly invented method of computation would pay a small ($1000) royalty for author, without the bizarre legal structure where the only people with assured income from innovations are the readily-despised lawyers. Oddly, in the USA at every known level of governmental over-sight, administration and law-formulations, most of the professionalized politicians are derived from the class of persons known as “attorneys”—them and realtors!

3. Studies of Innovation

The development of any new concept and idea can be presented in four essential stages (Figure 1). Efficiency, E, is possible profit, P, divided by cost, C, of realization.

$$E = P/C. \tag{1}$$

The innovation development has four stages:

1. The first stage is discovery of new concepts or idea. That stage includes an appearance of new idea and INITIAL RESEARCH of its possibilities and main conditions that are requisite for its practicability, initial proof of reality. A person can be only author of a new concept or idea if he/she made initial research and showed that this idea may become a future technical reality. A person who ONLY gave the idea (point 0 in Figure 1) is NOT its author because it is easy to produce a lot of ideas that are beneath or beyond realization. For example, the fantast Jules Verne (1828-1905) penned his famous book about the first manned flight to the Moon using a truly huge metal cannon cast *in situ* in the ground of Florida, USA. Is he the author of the idea for manned flight to Moon employing a big gun? No. Even primitive research shows that a human cannot tolerate the acceleration that is caused by this method, where the vehicle is a cannonball.

    The first stage is ONLY theoretical; strong individual and talented enthusiast in own time without any support because unknown concept or idea cannot be in government or company plan.
2. The second stage started after publication or public announcement of the primary idea during a scientific conference. Other researchers join the investigation of the new idea and make more detailed researches. Most of this new idea research is theoretical, and only a small part may be experimental.
3. The third stage includes the production of appropriate experimental examples, an early form of materialization.
4. The fourth stage is actual production of marketable versions of the idea.

We show the development of one innovation (curve 1 in Figure 1). However, any concept exhausts itself and its inherent efficiency possibilities over time. The new concept (idea) appears which promises even more efficiency (curve 2 in Figure 1). Conventionally, in initial time a new concept has less efficiency than a highly over-developed old idea, but as refinements occur in the future, the innovation efficiency becomes significantly more than the old idea.

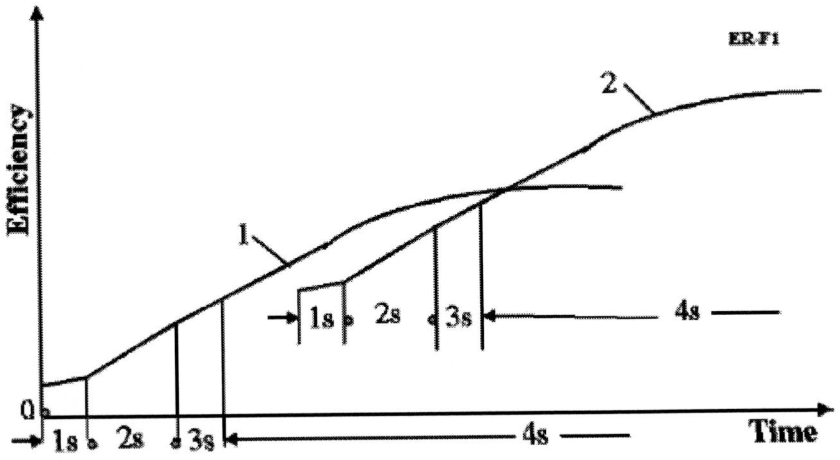

Figure 1. Four-stage innovation development.

For example, the original idea of a vehicle was startlingly original: People had the idea to connect a vehicle to a horse. Later they invited a motorized vehicle. Then they developed aircraft. At present, humanity is developing space transiting vehicles. People laughed at the first automobiles; the first airplanes were captured collapsing in amusing old movies; the first rockets tended to explode. What American can ever forget the USA's "Flopnik"? But as they matured, they opened—literally—new worlds of possibility.

4. Criteria of Scientific Works

There are two main simple criteria which allow recognition of the difference between true scientific research and that of some pseudo-scientific works by educated or merely clever persons:

1. Author in special paragraph or article conclusion must enumerate: *What is new (unknown before!) he/she offered and/or made in work offered?* That may be a new concept, an idea, multiple innovations, new mathematical models (equations), new non-conventional result of computation, new design of old or well-known macro and micro object (show its advantages), et cetera.
2. The author must DETAIL all his computation (equations and their receiving!) and initial data, which ALLOW to repeat (check up) his new equation and computation. If he offered new project, he must estimate the its cost.

   If offered idea, research and innovation are close to old or known idea or research, the author must enumerate all difference of his idea, innovations and results from earlier works (What NEW he/she offers/made in his/her work). If nothing on the list is actually new, that means the presented work is just idle talk.

   If author does not give the proof of the new equations, full data for computation, he deprives other scientists the means to check his equations and computations and the value of these equations and computations are virtually zero. That means the author(s) is afraid to let his/her work to endure a thorough examination.

The work offered the new macro-projects must contain the estimation of their cost. Without this estimation, the value of scientific work is very low.

In last time it appears very much "scientific" works which are presented as results of funded scientific research for government organizations. That means the burdened taxpayers pay for these works. The Scientific Committee of Auditing "Science", a member of the organization "Citizens Against Government Waste" (CAGW)*[1] applied these simple criteria which show: that is scientific or pseudo-scientific work? Those criteria also allow conventional or especially well educated people to recognize pseudo-scientific works (see details in http://auditing-science.narod.ru and http://www.geocities.com/auditing.science/, http://NASA-NIAC.narod.ru).

There is the third criterion which is applied ONLY to works funded by Government:

3. If this work is funded by Government (taxpayers), the sum of money received by any author (or a research organization) must be made public!

Note, sometimes the author(s) announce: this work was supported by (Government or funded by Government) organization. But if they did not show the exact monetary sum of "support", that means the reader can understand this work was done without spending any taxpayer money.

The sum allows other scientist (and interested people) to estimate the difference between the real cost and payment for the presented work.

Most taxpayer-funded works run by Government departments and agencies do not satisfy this simple criterion. Why? Most likely, because these so-called "researches" are really worthless pseudo-scientific products! The grant is received on the quiet (by backstairs influence). Especially, in this instance, the readers can see it in NIAC (see below) where former director Mr. Robert Cassanova and this defunct group, probably, stole more than 50 millions of USA taxpayer money (see details in http://auditing-science.narod.ru and http://www.geocities.com/auditing.science/, http://NASA-NIAC.narod.ru).

4. Organization of Scientific Works

Government Relations

Currently, the most important First Stage is the most difficult situation. No Federal or reliable private-sector funding, no extraneous technical support of any kind. This work can be done ONLY by individual enthusiasts and at one's own expense in time and money. Funding of the new perspective concept or idea is needed AFTER its initial theoretical research by an

---

[1] Citizens Against Government Waste is the nation's largest taxpayer watchdog group with over one million members and supporters nationwide. It is a nonpartisan, nonprofit organization dedicated to eliminating waste, fraud, abuse, and mismanagement in government. CAGW has helped save taxpayers $825 billion!

encompassing system of awards and prizes. For example, what NIAC should have been was an agency funding this difficult stage where seed-money is hard to come by (see below in section on NIAC).

Recommendations:

There is only one solution of this macro-problem – the United States of America's Federal Government must install a series (3 - 5) of special national Government prizes (awards of about $100K should be sufficient) in every important scientific field (space, energy, computing, biology, physics, et cetera.) for new-concept scientific researches that are:

(1) Given ONLY for new concepts and ideas developed by author and published or presented in sufficient qualifying detail at a scientific conference or on the Internet (stage 1 in Figure 1).
(2) The awards must be given ONLY to qualified individuals.
(3) The competition must be OPEN, advertised widely in public notices. ALL contenders and their work and proposals announced BEFORE any awards.
(4) The awarding Committee must be from *independent* well-known scientists in given field.

The same awards may be also in stage two (developing new concept or idea by non-author of this idea if the author of idea is awarded; or non-author make significant innovations which develop or solve problems important for progress this idea). In stage three the grants can be given ONLY for experiment or model.

5. NIAC (THE "NASA INSTITUTE FOR ADVANCED CONCEPTS")

The non-experienced reader objects - there exists NIAC (NASA Institute for Advanced Concepts) that must support new concepts and ideas in aerospace.

But millions of American tax dollars were awarded by NIAC Director Mr. *Robert Cassanova* for theoretical works before they were ever presented to an established scientific society! As a result, the applicant received money before researching and presented "research" that was more an exploration of an idea with potential for revolutionary discovery than an actual development of the idea itself.

In other places awards are given for well-known published scientific works in OPEN competition. It is impossible, for example, that the Nobel Prize for Physics would be given for merely *promising* to create a epoch-making discovery.

Mr. Cassanova (NIAC) announced that every proposal is reviewed by 6 reviewers (3 internal + 3 external reviewers), but he refuses to identify or present these reviews. Why?

He did not send the most obvious and really revolutionary proposals to any reviewers. He was afraid, apparently, to show them even to his marionette NIAC Research Council (Chairman Mrs. *Robert Whaterhead*, *Dava Newman* (MIT), *T. Wang, C. Bowden*, *L. Goff*, et al.).

What kinds of proposals are awarded money supports by Mr. Cassanova?

An important part of the answer to this question can be easily found by the reader at a website: *http://NASA-NIAC.narod.ru* and others.

Overview: The NIAC spent more 50 millions USA dollars during eight years of its existence, but they did not really put forth any really new concepts or ideas! Most NIAC final "research" reports are idle talk (no scientific results, no pre-production models, no correct scientific report, the final reports content a lot of scientific mistakes, and so on). For example, the final reports don't have any scientific results: Space Elevator (award about one million dollars), Bio Suit (awards about one million dollars), Chameleon Suit (award about 1 million dollars), Weather Control (awards about one million dollars), Winglee M2P2 MagSail (award about two million dollars), Cocoon vehicle (work contains only scientific mistakes), anti-matter sail (empty useless non-scientific seven page work), and so on ad nausea (see Final Reports in http://NASA-NIAC.narod.ru).

For example, Mr. Robert Cassanova awarded four million of dollars to the following persons: *Howe S., Colozza A., Nock K., Cash W., Dubowsky S.* He also awarded three or four times millions of taxpayer contributions to these persons: *Hoffman R. Maise G., McCarmack E., Rice E., Slough J. Kammash N., Winglee R., Newman D.*

The Science Committee of the organization *"Citizens Against Government Waste" (CAGW)* awarded NIAC and Mr. Cassanova the "Pseudo-Nobel Prize-2005" (and "Pseudo-Nobel Prize-2006" for wasting millions of taxpayer dollars by pseudo-scientific works (GOTO: http://www.geocities.com/auditing.science or http://auditing-science.narod.ru).

Recommendations:

The President and Congress of the United States of America, needs to, and must, thoroughly investigate the NIAC situation and remove, NASA and USRA leaders who allow any abuse and corruption on their watch. The Science Committee of CAGW stands ready to present to a Special Investigation Commission the documents that confirm the statements presented and outlined in this article.

In this saddening and costly national situation, it is the best decision, to stop the wasteful and ineffective financing of NIAC and pass their functions to Independent Committee created from well-known scientists, or NASA can create its own Committee from eminent volunteer scientists or to pass selected managerial functions to the National Science Academy, or National Science Foundation and to send awards only to finished scientific works in OPEN competition, or pass these vital functions to the growing and historically relevant and important International Space Agency Organization (http://www.international-space-agency.org or http://www.isa-hq.net) which would be better suited, and able, to stimulate, enable, and promote advanced space launch, propulsion, power, orbital, and planetary grant disbursements, R and D and implementation. This is based on an ever-increasing need for global cooperation, collaboration, common effort, and universal viewpoint. The International Space Agency's Directives, Charter, Purpose, Goals, and Certificate of Incorporation reflects this reality far better than the USRA or NIAC directives or charters. The many millions in Government-dispensed tax monies and private sector money and other relevant resources would be better used under the management and oversight of the International Space Agency Organization.

The CAGW Science Committee has available already an offer to NASA for a detailed plan on how to improve the work of NIAC, making it more open and its product more useful.

This plan includes three conventional conditions:

(1) Independent selection Committee having widely-known E-mail address.

(2) Open competition with publication of all nominated scientific works on Internet, including assessments made by scientists before any funding awards.
(3) Awarding ONLY actually achieved, not speculation about, scientific works not supported from other sources.

Discussion

The CAGW Science Committee considered, in detail, seven of about two hundred awards made by Mr. Robert Cassanova (GOTO: http://www.geocities.com/auditing.science or http://auditing-science.narod.ru). Amazingly, 90% of the "final reports" are just idle talk giving the impression to readers that there are NO talented scientists in the USA! That means, obviously, that the system of funding and awarding of scientific works is wrong. Mr. Cassanova is a university system employee and he evidently tries strenuously to fund his friends and protégés within his system of work. However, universities take the funded money and do not pay them over to professors who receive their fixed salary. Often, a professor is overloaded by lectures, direct work with talented students and ordinary classroom examinations. Such a person does not have time or the possibility to make serious research that requires huge efforts and much time. That's why he/she wrote the idle talk report, pseudo-scientific work!

The USA found the best solution of this problem – one sends scientists to government research centers or laboratories and they work full time 1-2 years on a problem there, shielded from busywork. Government centers and laboratories must directly invite the needed scientists without going through favored groups such as National Research Council (NRC) and ORAU (Oak Ridge Associated Universities). That would save much money and stop favoritism toward friends and weak scientists-- often non professionals in a given field of study. The Laboratory scientists know well the talented scientists in his field and they must solve what scientists must be invited.

Conclusion

The best way is to withdraw this function and this money from NASA-NIAC-USRA, pass them to a special government committee (or the National Academies, or ISA) including famous and reputable scientists and to award the published works (researches) containing new concepts, ideas, inventions, and innovations. Make it an open competition!

In 2007, after critics in international press spoke, after many letters from scientists in Government, Congress, the NASA stopped funding the NIAC and discharged Mr. Robert Cassanova. Now, the Federal Government must request the organizations presented the pseudo-scientific works (and Mr. Cassanova) to return the taxpayer's wasted money.

6. Fellowship and NRC (National Research Council)

Government created the good initiative – temporary attract the talented scientist for solving the difficult scientific problems (Research Association-ships). The NRC decided to use it for self profit. That received the right to select of candidates (main aim – to be the moderator at sinecure). That allows to take many money to self (NRC employees and NAS), promotion friends (Fellowships), create from NRC the charitable organization for untalented scientists but useful people.

Example, A.A. Bolonkin knows a well-known scientist – he has, so far, had a 30 year-long experience with the person, authored more than 170 scientific articles and books and tens inventions in given field. He developed the new method, contacted with Government laboratory. The laboratory gave an excellent review in his proposal. He sent the application to NRC. NRC program administrator Mr. *E. Basques* informed applicant: *NRC did not present his proposal to the (2007) NRC Pier Review, as applicant has a low a scientific score (7.4). The NRC deprived him to apply his proposals in during one year??!!* He asked Mr. Basques: send to him detail computation of his score; explain - why his score is so very low; how much years of experience, scientific works and inventions he must has for admission to any NRC review; how much years of experience, scientific works and inventions the selected candidate have; who is chair of NRC and NRC Advisory Committee. Mr. Basques answered, that such information are secret!

Very early in the game, NRC was accepting three different proposals from just one applicant in one Panel review and had four Reviews in every year. That was true competition which allows the talented active scientists to promote new ideas and develop America's technology. But now Mr. E. *Basques* accepts ONLY one application per year from one applicant including the candidates who he did *not admit* to review! He converts the NRC, scientific COMPETITION to *charitable organization* for untalented, dull scientists, his friends and other such useful persons.

We call your attention to the following abnormal economic situation. The Air force, Army, Navy, NASA and other well-known USA government scientific laboratories are staffed with leading scientists in their various fields. Laboratories can estimate and select the new ideas, concepts and innovations. They do not need a skewing mediator (NRC) for selection of proper, potentially very productive research candidates. The NRC mediator produces ONLY additional expenses (up to 50%) and imposes on such laboratories and facilities the good friends of a NRC moderator, but the bad scientists contracted make few useful or worthwhile discoveries.

We have same situation, when the mediator (NRC) stands between seller (scientist) and customer (Government Research Laboratory). When the laboratory wants to hire the scientists, the moderator stops buying, request the big moderation payment and sale (imposed) customer the other, own, bad goods.

Conclusion: The Associate-ship Government Program is a truly excellent and economically useful idea, but Government Research Laboratories do not need a NRC moderator to function successfully. They know best the specialists in their active investigational fields than any over-paid, biases current NRC bureaucrat and they can select the best scientists without NRC moderating, thus saving millions of taxpayer dollars and, at the same time, greatly accelerate America's further technical progress.

7. PUBLICATIONS

There are well-known organizations such as the American Institute of Aeronautics and Astronautics. One performs great works, organizes aerospace conferences and publishes a series of aerospace journals. But it doesn't have support from government and NASA and it became a strictly commercial organization. For example, the cost of participation in AIAA conferences is very high. That means only employees of government and big organizations can take part in scientific forums. But, almost by definition, they will display only conventional R&D plans of the type the system currently favors. The new revolutionary ideas and researches are made by talented individuals, enthusiasts in their free time. They can make a revolutionary research, but they do not have much money (some thousands of dollars) for payment of trip, hotel and conference fee. *Literally, the USA loses these revolutionary researches.*

Editors of AIAA journals do not get a salary for their arduous efforts. That means they want to see their name in every copy of journal, but they do not want to work as editor. They pass an article to a reviewer, and pass the review to author. That function can be done via computer. Some of them have allegedly converted their journal to essentially a private edition for their friends and protégés. For example, all 20 revolutionary researches which were published in the recent comprehensive book *"Non-Rocket Space Launch and Flight"*, Elsevier, London, 2006, offered for publication in AIAA *"Journal of Power and Propulsion" (JPP)*, but all were rejected by editor-in-chief *Vigor Yang* as researches were written in a non-American style and having poor English diction. What is "American style" he cannot explain, poor English-- the readers can see the book and decide: Is it a sufficiently important reason in refuse revolutionary innovations? From notes of Vigor Yang, it is seen he has poor knowledge of extant aerospace and vehicle propulsion systems. For some last years the "JPP" have not published any revolutionary ideas, but published many articles having serious scientific mistakes. The same situation with AIAA *"Journal of Spacecraft and Rockets"* (Editor-in-Chief *Vincent Zoby*).

It is a bad situation, that the USA has only a single journal about power and propulsion system or spacecraft and American authors must publish new ideas and researches in journals abroad.

It is bad that commercial publishing houses do not want to publish scientific literature, because it is not profitable. As a result, the scientific literature (and text-books) are very expensive and prohibitive not only for students, but for scientists.

It is bad that there is no free scientific Internet library (which would pay the government back by factors of 1000 in terms of net scientific development generated) to enable individuals of talent and enthusiasts to pursue their researches by using open sources of data and other information.

It is bad that the AIAA requests about $1000 for every publication in its journal and sells every scientific article for $10.

Recommendations:

(1) The USA must have minimum two rival journals in every scientific field. (These may be Internet journals). Every journal must have an Appeal Commission where author

can complain if he/she does not agree with editor's *clearly stated reasons* for article rejection.

(2) Every National Conference must have a small fund for supporting the individuals presenting revolutionary research and give them possibility to address a meeting.
(3) The US Government and the NASA must support with appropriate funding the points 1-2 above (scientific journal and scientific conferences), the AIAA (and all big old Scientific Societies), the scientific publishing houses, the free scientific Internet library.
(4) The AIAA (and all big old Scientific Societies) must freely publish on the Internet all manuscripts presented in AIAA Scientific Conferences. (Paper copies, of course, are its business and may be charged for as the publishers pleases).
(5) The government must create the free Internet Library of the technical, matematic, physic textbooks.

The Government and country lose more on the obstacles outlined above, which stop the generation and filtering and developing new ideas, than the output of the most talented individual researchers of this generation. The loss is incalculable and should stop immediately.

8. Patenting

The USA's magnificient Constitution proclaims a support of science and, as well, time-constrained protective patenting. Unfortunately, the USA's PTO (Patent and Trademark Office) had become a powerful means to extract money from inventive people. The Payment for PTO equals some thousands dollars and is prohibitive for individuals. The patenting approval process continues for at least 1-2 years. If the inventor complains, the PTO can sabotage submitted for review inventions. A.A. Bolonkin personally knows of an instance when an inventor paid for invention but the PTO did not award a patent! The PTO creates a lot of rules that permit the pumping of money from people and that allows the sabotaging of the patenting process.

Recommendations:

It must be category "Announced Invention". These are inventions written in PTO style and presented without PTO examination in special Internet websites. No, or an extremely small, fee (less \$2) may be for publication this invention on Internet. If author think the company used his invention without his permission, he applies to a Patent Special Court and requests compensation. That compensation cannot be more 5% of users income from this single invention.

(1) The Government must give permits to 3-5 competitive companies for giving patents. These companies compete with the cheapest patenting (who wishes to receive patent).
(2) Now, the PTO has rates tailored to big Companies and to small Business. It must be a special rate for individuals and the FULL payment (application, patenting, and maintenance) must be not more \$100 for all these stages. The maintenance fee is

usually what kills the ability of an individual to finance his own patents; sometimes corporations count on that and wait him out.

(3) There must be a new category of "important patents for Defense of the USA". If a Special Committee recognized a patent application as necessary (important) for Department of Defense or the US national security, the applicant should have a right to a free patent (he receives only a author certificate, the Government gets all patent rights), all American organizations or companies can use this patent but they must pay its author 1% and the PTO 1% from the gross-cost of the products incorporating this patent.

(4) All income received by PTO must be used for support of individual inventors programs, *not as an income center for the bureaucracy itself.*

9. FINAL RECOMMENDATIONS

Current system organization and funding of science researches is not efficient, especially for NRC, PTO, NIAC, NASA, DARPA, DOD, AF, SBIR and the NSF. They need re-organization. Main components of this reformation must be the following:

(1) The unwise and wasteful practice of advance funding of primary theoretical researches must be stopped and changed to OPEN competitions in any given field and in given topics.

(2) Government must install 3-5 annual Government Prizes (about $100K) in every important field of science (space, aviation, computer, physics, biology, energy, etc.) for important THEORETICAL achievements made by individuals. Practical results will flow from these if such are forthcoming from enthusiasts; but the way forward must be pointed out. It takes genius to do it, and genius needs its physical as well as spiritual rewards!

(3) The company using new methods of computation must pay small ($1000 or less) royalties to the authors for every licensing use.

(4) Must be also the additional form free registration inventions without PTO (PTO examination and PTO fee).

(5) NASA must be divided into at least two independent rival organizations.

(6) The main method funding of research must not be funding through Universities but it must be the work of University scientists done during 1-3 years 'sabbatical' as Fellow researchers in big Government laboratories. The NRC must be closed and Government laboratory straight invite the needed scientists.

(7) NASA, DARPA, Government laboratories must engage a head and main specialists of every project in OPEN competitions, preferably the authors of project (proposal) and scientists who made the main contributions in the project idea or concepts.

(8) The Government must support adequate scientific journals, publishing houses, free Internet scientific libraries; individual scientists should be aided to presented important researches to scientific national conferences.

(9) Government must make special small rates apply (<$100) to individual inventors, free patenting of important for DOD and National defense inventions and to use all

PTO profit for support of individual inventor programs important for DOD and the USA.

References

GOTO: *http://NASA-NIAC.narod.ru.*
GOTO: *http://auditing-science.narod.ru or http://www.geocities.com/auditing.science.*

APPENDIX

SUMMARY

Here there are values useful for calculations and estimations of aerospace projects.

1. System of Mechanical and Electrical Units

The following table contains the delivered metric mechanical and the electromagnetic SI units that have been introduced in this text, expressed in terms of the fundamental units *meter*, *kilogram*, *second*, and *ampere*. From these expressions the *dimensions* of the physical quantities involved can be readily determined.

Length............. 1 meter = 1 m
Mass................1 kilogram = 1 kg
Time................1 second = 1 s
Electric current.... 1 ampere = 1 A

Force.........1 newton = 1 N = 1 kg·m/s^2
Pressure.....1 N/m2 = 1 kg/m·s^2
Energy.......1 joule = 1 J =1 N/m = 1 kg·m^2/s^2
Power........1 watt = 1 W =1 J/s =1 kg·m^2/s^3

Rotational inertia................1 kilogram·meter^2 = 1 kg·m^2
Torque.............................1 meter·newton = 1 kg·m^2/s^2
Electric charge....................1 coulomb = 1 C = 1 A·s
Electric intensity..................1 N/C = 1 V/m = 1 kg·m/s^3·A

Electric potential..................1 volt = 1 V = 1 J/C = 1 kg·m^2/s^3·A
Electric resistance.................1 ohm =1 Ω =1 V/A = 1 kg·m^2/s^3·A^2
Capacitance........................1 farad = 1 F =1 C/V = 1 C2/J = 1 s^4·A^2/kg·m^2
Inductance..........................1 henry = 1 H =1 J/A2 = 1 Ω·s = 1 kg·m^2/s^2·A^2

Magnetic flux......................1 webwer = 1 Wb = 1 J/A = 1 V.s = 1 kg·m^2/s^2·A

Magnetic intensity................1 tesla = 1 Wb/m2 =1 V.s/m2 =1 kg/s^{2}·A=N/mA
Reluctance.........................1 ampera-turn/weber = 1 A/Wb =1 s^{2}·A^{2}/kg·m^{2}
Magnetizing force.................1 ampere-turn/meter = 1 A/m
Kelvin is fundamental unit of temperature
Candela is fundamental power-like unit of photometry

Fundamental Physical Constants

Speed of light in vacuum	$c = 299\,792\,459 \sim 3\times10^{8}$ m/s
Magnetic constant (permeability)	$\mu_o = 4\pi\times10^{-7}$ N/A^2
Electric constant $1/\mu_o c^2$	$\varepsilon_o = 8.854\,187\,817...\times10^{-12}$ F/m
Plank constant	$h = 6.626\,068\,76...\times10^{-34}$ J s
	$h/2\pi = 1.054\,571\,596...\times10^{-34}$ J s
Standard gravitational acceleration	9.806 65 m/s^2
Standard atmosphere (atm)	101 325 N/m^2
Thermochemical kilocalorie	4184 J
Speed of light in vacuum (c)	$2.997\,935\times10^{8}$ m/s
Electronic charge (e)	1.60210×10^{-19} C
Avogadro constant (N_A)	6.0225×10^{26}/kmol
Faraday constant (F)	9.6487×10^{7} C/kmol
Universal gas constant (R)	8314 J/kmol
Gravitational constant (G)	6.67×10^{-11} N·m^2/kg^2
Boltzmann constant (k)	1.3806×10^{-23} J/K
Stefan-Boltzmann Constant (σ)	5.670×10^{-8} W/K^4·m^2
Rest energy of one atomic mass unit	931.48 MeV
Electron-volt (eV)	1.60218×10^{-19} J

Rest masses of particles

	(u)	(kg)	(MeV)
Electron	$5.485\,97\times10^{-4}$	9.1091×10^{-31}	0.511 006
Proton	1.002 2766	$1.672\,52\times10^{-27}$	938.26
α-particles	4.001 553	6.6441×10^{-27}	3727.3

Detonation energy of 1 kiloton of high explosive is 10^{12} cal. 1 cal = 4.19 J.

Standard periodic table

Group → / ↓ Period	1	2	3	4	5	6	7	8	9	10	11	12	13	14	15	16	17	18
1	1 H																	2 He
2	3 Li	4 Be											5 B	6 C	7 N	8 O	9 F	10 Ne
3	11 Na	12 Mg											13 Al	14 Si	15 P	16 S	17 Cl	18 Ar
4	19 K	20 Ca	21 Sc	22 Ti	23 V	24 Cr	25 Mn	26 Fe	27 Co	28 Ni	29 Cu	30 Zn	31 Ga	32 Ge	33 As	34 Se	35 Br	36 Kr
5	37 Rb	38 Sr	39 Y	40 Zr	41 Nb	42 Mo	43 Tc	44 Ru	45 Rh	46 Pd	47 Ag	48 Cd	49 In	50 Sn	51 Sb	52 Te	53 I	54 Xe
6	55 Cs	56 Ba	*	72 Hf	73 Ta	74 W	75 Re	76 Os	77 Ir	78 Pt	79 Au	80 Hg	81 Tl	82 Pb	83 Bi			
7																		

* Lanthanides	57 La	58 Ce	59 Pr	60 Nd	61 Pm	62 Sm	63 Eu	64 Gd	65 Tb	66 Dy	67 Ho	68 Er	69 Tm	70 Yb	71 Lu
** Actinides	89 Ac	90 Th	91 Pa	92 U											

This common arrangement of the periodic table separates the lanthanides and actinides from other elements. The Wide Periodic Table incorporates the f-block; the Extended Periodic Table incorporates the f-block and adds the theoretical g-block.

Element categories in the periodic table

Metals						Metalloids	Nonmetals			Unknown
Alkali metals	Alkaline earth metals	Inner transition elements		Transition elements	Other metals		Other nonmetals	Halogens	Noble gases	
		Lanthanides	Actinides							

Atomic number colors show state at standard temperature and pressure (0 °C and 1 atm)

Solids **Liquids** **Gases** **Unknown**

Borders show natural occurrence

Primordial | Undiscovered

Astronomical Data of Solar system

Space body	Distance from sun (10^6 km) Mean Aphelilion Perithelion	Period of Revolution (day)	MeanMassMeanGravity radiusOrbitaldensityon surf.Speed (km)10^{24} kgMg/m^{3}m/s^{2}km/s
Sun	-- -- --	--	696 000 -- 1.41274--
Mercury	57.9 69.8 46.0	88.0	2 4203.1675.463.7248.8
Venus	108.1 109.0 107.5	224.7	6 2614.8704.968.6935.0
Earth	149.5 152.1 147.1	365.2	6 3715.9755.529.7829.8
Mars	227.8 249.2 206.6	687.0	3 3890.6394.123.7224.2
Jupiter	777.8 815.9 740.7	4 333	69 90019001.3323.0113.0
Saturn	142615081348	10 760	57 500 568.80.719.149.65
Uranus	286830072737	30 690	23 700 86.91.569.676.78
Neptune	449445374459	60 100	21 500 102.92.4715.05.42
Pluto	590873704450	90 740	2 900 5.375.50 8.04.75
Moon	0.384 from Earth	27.322	1 7370.07353.34 1.621.02

Astronomy. Short Information

A typical galaxy contains hundreds of billions of stars, and there are more than 100 billion (10^{11}) galaxies in the observable universe. Astronomers estimate that there are at least 70 sextillion (7×10^{22}) stars in the observable universe. That is 230 billion times as many as the 300 billion in the Milky Way.

In astronomy the distance is measured by light year (ly) or parcek (pc).

1 light year = 9.46×10^{12} km = 0.307 pc.

The nearest stars:	Distance (pc)
1. Proxime Centauri	1.29 pc
2. Bernard's star	1.82
3. Wolf 359(CN Leo)	2.39
4. Lalande 21185	2.56
5. Sirius A	2.74

The nearest star to the Earth, apart from the Sun, is Proxima Centauri, which is 39.9 trillion (10^{12}) kilometres, or 4.2 light-years away. Light from Proxima Centauri takes 4.2 years to reach Earth. Travelling at the orbital speed of the Space Shuttle (5 miles per second—almost 30,000 kilometres per hour), it would take about 150,000 years to get there.

Density of gases at normal pressure and temperature 0 ^{o}C in kg/m^3

Air	1.293
Hydrogen	0.08988
Helium	0.1785

Densities of various materials

Material	ρ in kg/m³	Notes
Interstellar medium	$10^{-25} - 10^{-15}$	Assuming 90% H, 10% He; variable T
Earth's atmosphere	1.2	At sea level
Aerogel	1 – 2	
Styrofoam	30 – 120	
Cork	220 – 260	
Water	1000	At STP
Plastics	850 – 1400	For polypropylene and PETE/PVC
The Earth	5515.3	Mean density
Copper	8960	Near room temperature
Lead	11340	Near room temperature
The Inner Core	~13000	As listed in Earth
Uranium	19100	Near room temperature
Iridium	22500	Near room temperature
The core of the Sun	~150000	
Atomic nuclei	$\sim 3 \times 10^{17}$	As listed in neutron star
Neutron star	$8.4 \times 10^{16} - 1 \times 10^{18}$	
Black hole	2×10^{30}	Mean density inside the Schwarzschild radius of an earth-mass black hole (theoretical)

Parameters of Earth atmosphere (relative density and temperature)

H km	/ o	T °K	H km	/ o	T °K	H km	/ o	T °K	H km	/ o	T °K
0	0	288.2	5	0.601	255.6	20	0.0725	216.7	50	0.000375	274
1	0.908	281.6	7	0.482	242,6	25	0.0332	216.7	60	0.000271	253.4
2	0.822	275.1	10	0,338	223.1	30	0.0146	230.4	100	0.32×10^{-6}	208 .2
3	0.742	268.6	12	0.255	216.7	35	0.00676	244.0	200	0.295×10^{-9}	122 7
4	0.669	262.1	15	0.159	216.7	40	0.00327	257.7	300	0.273×10^{-10}	1358

Specific impulse of liquid fuel (nozzle 100:0.1, seconds):
Oxygen – kerosene 372
Oxygen – hydrogen 463
AT-NDMG 350
Specific impulse of solid fuel (nozzle 40:0.1, seconds): 228–341.

Maximum energy and the specific impulse of the particle reactions.
$V = (2E/m)^{0.5} = 1.384\times10^4 (E/N)^{0.5}$ [m/s].
N is number of nucleons.

Reaction	Energy (*E*), eV	Particle speed, m/s
Burning		
Carbon + O_2 = CO_2	0.093	4224
Hydrogen + O_2 = HO_2	0.14	5178
Dissociation of gases		
H_2	4.48	20,714
O_2	5.1	5,209
N_2	9.76	8,171
Ionization		
H_2^+	2.65	15,931
O_2^+	6.7	5,625
H	13.6	51,039
Nuclear reaction		
Uranium	~200 MeV	$12,750\times10^3$
$^3H + {}^2H = {}^4He + n$	17.5 MeV	$25,892\times10^3$
Annihilation	938 MeV	-

Heat of combustion (MJ/kg):

Benzene 44	Mazut 30–41	Natural gases 42–47	Wood 15-30
Diesel fuel 43	Spirit 27.2	Hydrogen 120	Peat 6 –11
Kerosene 43	Coal 15–27	Acetylene 48	gunpowder 3

Energy Density

Storage type	Energy density by mass (MJ/kg)	Energy density by volume (MJ/L)	Peak recovery efficiency (%)	Practical recovery efficiency (%)
Mass-energy equivalence	89,876,000,000			
Binding energy of Helium-4 nucleus	683,000,000	$8.57\text{x}10^{24}$		
Nuclear fusion of hydrogen (energy from the sun)	645,000,000			
Nuclear fission (of U-235) (Used in nuclear power plants)	88,250,000	1,500,000,000		
Natural uranium (99.3% U238, 0.7% U235) in fast breeder reactor	24,000,000			50%
Enriched uranium (3.5% U235) in light water reactor	3,456,000			30%
Hf-178m2 isomer	1,326,000	17,649,060		
Natural uranium (0.7% U235) in light water reactor	443,000			30%
Ta-180m isomer	41,340	689,964		
Liquid hydrogen	143	10.1		
Compressed gaseous hydrogen at 700 bar	143	5.6		

Energy Density (Continued)

Storage type	Energy density by mass (MJ/kg)	Energy density by volume (MJ/L)	Peak recovery efficiency (%)	Practical recovery efficiency (%)
Gaseous hydrogen at room temperature]	143	0.01079		
Diesel fuel/residential heating oil	45.8	38.7		
Jet A aviation fuel / kerosene	42.8	33		
Biodiesel oil (vegetable oil)	42.20	33		
Specific orbital energy of Low Earth orbit	33 (approx.)			
Anthracite coal	32.5	72.4		36%
Wood	6–17	1.8–3.2		
Liquid hydrogen + oxygen (as oxidizer) (1:8 (w/w), 14.1:7.0 (v/v))	13.333	5.7		
TNT	4.184	6.92		
compressed air at 300 bar (at 12°C), without container	0.512	0.16		
Lithium ion battery with nanowires	2.54-2.72?	29		95%
Lithium thionyl chloride battery	2.5			
Fluoride ion battery	1.71.7-4.2	2.8-5.8		
Regenerative Fuel Cell (fuel cell with internal Hydrogen reservoir used much as a battery)	1.62			
Capacitor by EEStor (claimed capacity)	1.0			
Sodium-sulfur battery		1.23		85%
Liquid nitrogen	0.77	0.62		
Lithium ion battery-predicted future capability	0.54–0.9	0.9–1.9		95%
Lithium ion battery-present capability	0.23–0.28			
Lithium sulphur battery	0.54-1.44			
Kinetic energy penetrator	1.91.9-3.4	30-54		
5.56 × 45 mm NATO bullet	0.4-0.8	3.2-6.4		
Zn-air batteries	0.40 to 0.72			
Flywheel	0.5			81-94%
Ice	0.335	0.335		
Zinc-bromine flow battery	0.27–0.306			
Compressed air at 20 bar (at 12°C), without container	0.27	0.01		64%
NiMH Battery	0.22	0.36		60%
NiCd Battery	0.14-0.22			80%
Lead acid battery	0.09–0.11	0.14–0.17		75-85%
Compressed air in fiber-wound bottle at 200 bar (at 24°C)	0.1	0.1		
Commercial lead acid battery pack	0.072-0.079			
Vanadium redox battery	0.09	0.1188		70-75%
Vanadium bromide redox battery	0.18	0.252		81%

Energy Density (Continued)

Storage type	Energy density by mass (MJ/kg)	Energy density by volume (MJ/L)	Peak recovery efficiency (%)	Practical recovery efficiency (%)
compressed air in steel bottle at 200 bar (at 24°C)	0.04	0.1		
Ultracapacitor	0.0206	0.050		
Supercapacitor	0.01		98.5%	90%
Capacitor	0.002			
Water at 100 m dam height	0.001	0.001		85-90%
Spring power (clock spring), torsion spring	0.0003	0.0006		

Data for Estimation and Computation

Table 1. [1], p.351.[2], p.73. Heat Transfer

Material	Density kg/m^3	Heat transfer $\lambda = W/m^{o}C$	Heat capacity $kJ/kg^{o}C$
Concrete	2300	1.279	1.13
Baked brick	1800	0.758	0.879
Ice	920	2.25	2.26
Show	560	0.465	2.09
Glass	2500	0.744	0.67
Steel	7900	45	0.461
Air	1.225	0.0244	1
Asphalt	2110	0.6978	2.09
Asbestos plate	770	0.1162	0.810
Oak	800	0.207	1.758
Humid soil	1700	0.657	2.01
Mineral wool	200	0.0465	0.921
Dry sand	1500	0.326	0.795
Glass wool	200	0.037	0.67
Slag wool	250	0.0698	-
Aluminum	2670	204	0.921
Water	1000	0.5513	4.212
Sold rubber	1200	0.169	1.382
Aerocrete	-	0.07-0.32	-
Foam plastic	-	0.043-0.058	-
Reinforced concrete	-	1.55	-

Table 2. [13], p. 465. Emittance, ε (Emissivity)

Material	Temperature, T °C	Emittance, ε
Bright Aluminum	50 ÷ 500 °C	0.04 - 0.06
Bright copper	20 ÷ 350 °C	0.02
Steel	50 °C	0.56
Asbestos board	20 °C	0.96
Glass	20 ÷ 100 °C	0.91 - 0.94
Baked brick	20 °C	0.88 - 0.93
Tree	20 °C	0.8 - 0.9
Black vanish	40 ÷ 100 °C	0.96 – 0.98
Tin	20 °C	0.28

Sources:

1. Naschekin, V.V., Technical thermodynamic and heat transmission. Public House High University, Moscow, USSR. 1969 [in Russian].
2. Koshkin H.I., Shirkevich M.G., Directory of elementary physics, Moscow, Nauka, 1982.

Table 3. Maximum partial pressure of water vapor in atmosphere for given air temperature

t, C	-10	0	10	20	30	40	50	60	70	80	90	100
p,kPa	0.287	0.611	1.22	2.33	4.27	7.33	12.3	19.9	30.9	49.7	70.1	101

Table 4. Properties of various good insulators (recalculated in metric system)

Insulator	Resistivity	Dielectric strength	Dielectric Tensile strength	
	Ohm-m.	MV/m. E_i	constant, ε	kg/mm^2
Lexan	10^{17}–10^{19}	320–640	3	5.5
Kapton H	10^{19}–10^{20}	120–320	3	15.2
Kel-F	10^{17}–10^{19}	80–240	2–3	3.45
Mylar	10^{15}–10^{16}	160–640	3	13.8
Parylene	10^{17}–10^{20}	240–400	2–3	6.9
Polyethylene	10^{18}–5×10^{18}	40–680*	2	2.8–4.1
Poly (tetra- fluoraethylene)	10^{15}–5×10^{19} 40–280**	2	2.8–3.5	
Air (1 atm, 1 mm gap)	-	4	1	0
Vacuum (1.3×10^{-3} Pa, 1 mm gap)	-	80–120	1	0

*For room temperature 500 – 700 MV/m. E = 700 MV/m for t < 15 C.

** 400 – 500 MV/m.

Sources: Encyclopedia of Science & Technology (NY, 2002, Vol. 6, p. 104, p. 229, p. 231) and Kikoin, I.K., (Ed.), Tables of Physical Values. Atomuzdat, Moscow, 1976 (in Russian)., p. 321.

Note: Dielectric constant ε can reach 4.5 - 7.5 for mica (E is up 200 MV/m), 6 -10 for glasses (E = 40 MV/m), and 900 - 3000 for special ceramics (marks are CM-1, T-900) [17], p. 321, (E =13 -28 MV/m). Ferroelectrics have ε up to 10^4 - 10^5. Dielectric strength appreciably depends from surface roughness, thickness, purity, temperature and other conditions of materials. Very clean material without admixture (for example, quartz) can have electric strength up 1000 MV/m. As you see, we have the needed dielectric material, but it is necessary to find good (and strong) isolative materials and to research conditions which increase the dielectric strength.

Table 5. Material properties

Material	Tensile strength	Density g/cm3	Fibers	Tensile strength	Density g/cm3
Whiskers	kg/mm2			kg/mm2	
AlB_{12}	2650	2.6	QC-8805	620	1.95
B	2500	2.3	TM9	600	1.79
B_4C	2800	2.5	Allien 1	580	1.56
TiB_2	3370	4.5	Allien 2	300	0.97
SiC	1380-4140	3.22	Kevlar or Twaron	362	1.44
Material			Dynecta or Spectra	230-350	0.97
Steel prestressing strands	186	7.8	Vectran	283-334	0.97
Steel Piano wire	220-248		E-Glass	347	2.57
Steel A514	76	7.8	S-Glass	471	2.48
Aluminum alloy	45.5	2.7	Basalt fiber	484	2.7
Titanium alloy	90	4.51	Carbon fiber	565	1.75
Polypropylene	2-8	0.91	Carbon nanotubes	6200	1.34

Source: [22]-[27], Howatsom A.N., Engineering Tables and Data, p.41.

The cost some material is presented in Table 2 (2005-2007). Some difference in the tensile stress and density are result the difference sources, models and trademarks.

Table 6. Average cost of material (2005-2007)

Material	Tensile stress, MPa	Density, g/cm3	Cost USD $/kg
Fibers:			
Glass	3500	2.45	0.7
Kevlar 49, 29	2800	1.47	4.5
PBO Zylon AS	5800	1.54	15
PBO Zylon HM	5800	1.56	15
Boron	3500	2.45	54
SIC	3395	3.2	75
Saffil (5% SiO_2+Al_2O_3)	1500	3.3	2.5
Matrices:			
Polyester	35	1,38	2
Polyvinyl	65	1.5	3
Aluminum	74-550	2.71	2
Titanum	238-1500	4.51	18
Borosilicate glass	90	2.23	0.5
Plastic	40-200	1.5-3	2 - 6
Materials:			
Steel	500 - 2500	7.9	0.7 - 1
Concrete	-	2.5	0.05
Cement (2000)	-	2.5	0.06-0.07
Melted Basalt	35	2.93	0.005

Table 7. Estimation of energy expenses for different methods of freshwater extraction

No	Method	Condition	Expense kJ/kL	Getting kJ/kL
1	Vapor	Expense only for vapor*	2.26×10^6	0
2	Freezing	Expense only for freezing, c.e. $\eta = 0.3$	1×10^6	0
3	Reverse osmosis	Expense only for pumping,	$(4\div7)\times10^3$	0
4	High Tower extr.	$t = 35$ C, $h = 0.7$, tube is black	0	30×10^3

* This expense may be decreased in 2 -3 times when the installation is connected with heat or nuclear electric station.

Approximately cost of some material in USD (2008)

Cost of coal is 82 – 94$/ton

Oil per barrel $120 world market-price (May, 2008). 1 barrel = 138.97 liters.

Car gas (benzene) $3.2/gallon, (May, 2008, USA retail). 1 gallon = 3.785 liters (US).

Electricity $0.25/kWh (retail, USA)

Freshwater $0.6 – 1/kL (prime cost)

Approximately cost of some big macro-projects and main material components (2006-2008, USD, some are designed cost)

Macro-projects:

1. Oil line 467 km - $2.2B. (4.7M/km).
2. Oil line 120 km $0.7B. (5.8M/km).
3. Oil line (Byrgas-Aleksandropolus) (Black-Sea) 280 km - $1.2B. (4.3M/km).
4. East oil line 4188 km - $11.2B (2.67M/km),(2008).
5. Oil line Azerbaijan – Turkey 1767 km - $4B. (2.26M/km).
6. Gas line “Blue Stream” (Russia – Turkey) (under water of Black Sea, deep 2150m) - $3.2B.
7. Gas line 530 km (gas capacity 30B cub, m gas per year) more $1B.
8. MagLev (Magnetic Levitation Highway) in Shanghai, China, 30 km – 1.2B.
9. MagLeb 1 km – $24.6M/km.
10. Highway system 8 lines - $50M/mile (USA).
11. Airport Hong Kong - $20B (1998).
12. Sea bridge 25 km (Jersey (England)-France) - $2B (project).
13. Bridge (China) 36 km - $1.55B.
14. Canal Caspian Sea -Black Sea 750 km, caring capacity 32M tons, - $6.5B.
15. Tunnel English-France 50 km, - $12B.
16. Tunnel Rasia-USA 100 km $10-12B (Bering Str., project).
17. 1 km railway in Siberia (permafrost) $11M (1 km conventional railway $0.8-1.3M/km).
18. Wind electric plant 50MW cost $80-100M.
19. Solar electric plant 250MW cost $300M.
20. Nuclear electric plant (“Belene”, Bulgaria) 2000MW cost $6B, building 7 years.
21. Floating nuclear electric plant 70MW cost $200M, building 4 years.
22. Nuclear reactor RBEP-1000 cost $1B.

Materials:

23. Steel tube for gas (diameter 1420mm, wall thickness 19 mm, mark 17G1C) $712/ton, $470/m (2006).
24. Steel tube for gas (diameter 750mm, thickness 7-9 mm, mark 17G1C) $440/ton, (2006).
25. Steel tube for gas (diameter 273mm, thickness 5-6 mm, mark 17G1C) $1000/ton, (2006).
26. Plastic tube for cold water (diameter 125 mm, thickness 11.4 mm, mark PN1) $8/m.

Aircraft (2007):

27. C-17 Clobemaster (military-transport) $250M.
28. A-380 big passenger aircraft $320M.
29. Mig-29K (military fighter) $35M.
30. Su-34 (attack plane) $40M.
31. Tu-204 (cargo) $40M.
32. Typhoon (fighter, English) $120M.

NON-CONVENTIONAL MATERIALS

Artificial Fiber and Nanotubes

Artificial fiber and cable (film) properties. Cheap artificial fibers are currently being manufactured, which have tensile strengths of 3-5 times more than steel and densities 4-5 times less than steel. There are also experimental fibers (whiskers) that have tensile strengths 30-100 times more than steel and densities 2 to 5 times less than steel. For example, in the book [172] Ch.12A, p.158 (1989), there is a fiber (whisker) C_D, which has a tensile strength of $\sigma = 8000$ kg/mm^2 and density (specific gravity) of $\gamma = 3.5$ g/cm^3. If we use an estimated strength of 3500 kg/mm^2 ($\sigma = 7\cdot 10^{10}$ N/m^2, $\gamma = 3500$ kg/m^3), than the ratio is $\gamma/\sigma = 0.1\times10^{-6}$ or $\sigma/\gamma = 10\times10^{6}$.

Nanotubes come close to being the best fiber that can be made from graphite (see section "Nanotubes" in book Attachment).

For example, whiskers of Carbon nanotube (CNT) material have a tensile strength of 200 Giga-Pascals and a Young's modulus over 1 Tera Pascals (1999). The theory predicts 1 Tera Pascals and a Young's modules of 1-5 Tera Pascals. The hollow structure of nanotubes makes them very light (the specific density varies from 0.8 g/cc for SWNT's (Single Wall Nano Tubes) up to 1.8 g/cc for MWNT's, compared to 2.26 g/cc for graphite or 7.8 g/cc for steel). Tensile strength of MWNT's nanotubes reaches 150 GPa.

In 2000, a multi-walled carbon nanotube was tested to have a tensile strength of 63 GPa. Since carbon nanotubes have a low density for a solid of 1.3-1.4 g/cm^3, its specific strength of up to 48,000 kN·m/kg is the best of known materials, compared to high-carbon steel's 154 kN·m/kg.

The theory predicts the tensile stress of different types of nanotubes as: Armchair SWNT - 120 GPa, Zigzag SWNT – 94 GPa.

About 60 tons/year of nanotubes are produced now (2007). Price is about \$100 - 50,000/kg. Experts predict production of nanotubes on the order of 6000 tons/year and with a price of \$1 – 100/kg to 2012.

Commercial artificial fibers are cheap and widely used in tires and countless other applications. The authors have found only older information about textile fiber for inflatable structures (Harris J.T., Advanced Material and Assembly Methods for Inflatable Structures, AIAA, Paper No. 73-448, 1973). This refers to DuPont textile Fiber B and Fiber PRD-49 for tire cord. They are 6 times strong as steel (psi is 400,000 or 312 kg/mm^2) with a specific gravity of only 1.5. Minimum available yarn size (denier) is 200, tensile module is 8.8×10^6 (B) and 20×10^6 (PRD-49), and ultimate elongation (percent) is 4 (B) and 1.9 (PRD-49). Some data are in Table 5 Attn.

Industrial fibers have up to σ = 500 - 600 kg/mm^2, γ = 1500 -1800 kg/m^3, and σ/γ = 2,78x10^6. But we are projecting use in the most present projects the cheapest films and cables applicable (safety σ= 100 - 200 kg/mm^2).

Aerogel

Aerogel is a low-density solid-state material derived from gel in which the liquid component of the gel has been replaced with gas. The result is an extremely low density solid with several remarkable properties, most notably its effectiveness as a thermal insulator.

Silica aerogel is an especially good conductive insulator because silica is a poor conductor of heat—a metallic aerogel, on the other hand, would be a less effective insulator. Carbon aerogel is a good radiative insulator because carbon absorbs the infrared radiation that transfers heat. The most insulative aerogel is silica aerogel with carbon added to it.

Since it is 99.8% air, it appears semi-transparent. The world's lowest-density solid is a silica nanofoam at 1 mg/cm^3, which is the evacuated version of the record-aerogel of 1.9 mg/cm^3. The density of air is 1.2 mg/cm^3. It has remarkable thermal insulative properties, having an extremely low thermal conductivity: from 0.03 W/mK down to 0.004 W/mK, Its melting point is 1,473 K (1,200 °C or 2,192 °F).

Carbon aerogels. Due to their extremely high surface area, carbon aerogels are used to create supercapacitors, with values ranging up to thousands of farads based on a capacitance of 104 F/g and 77 F/cm^3. Carbon aerogels are also extremely "black" in the infrared spectrum, reflecting only 0.3% of radiation between 250 nm and 14.3 μm, making them efficient for solar energy collectors.

Super-alloys

A *superalloy*, or *high-performance alloy*, is an alloy that exhibits excellent mechanical strength and creep resistance at high temperatures, good surface stability, and corrosion and oxidation resistance. They can operate up to 1100 °C (1373 K).

Superalloys are metallic materials for service at high temperatures, particularly in the hot zones of gas turbines. Such materials allow the turbine to operate more efficiently by withstanding higher temperatures. Turbine Inlet Temperature (TIT), which is a direct indicator of the efficiency of a gas turbine engine, depends on the temperature capability of 1st stage high pressure turbine blade made of Ni base superalloys exclusively.

One of the most important superalloy properties is high temperature creep resistance. Other crucial material properties are fatigue life, phase stability, as well as oxidation and corrosion resistance.

Metal Foam

A metal foam is a cellular structure consisting of a solid metal - frequently aluminum - containing a large volume fraction of gas-filled pores. The pores can be sealed (closed-cell foam), or they can form an interconnected network (open-cell foam). The defining characteristic of metal foams is a very high porosity: typically 75-95% of the volume consists of void spaces. The strength of foamed metal possesses a power law relationship to its density; i.e., 20% dense material is more than twice as strong as 10% dense material (Ceteris paribus).

Closed-cell foams retain the fire resistant and recycling capability of other metallic foams but add an ability to float in water (for densities less than 1g/cc).

Meta-material

A metamaterial (or meta-material) is a material which gains its properties from its structure rather than directly from its composition. To distinguish metamaterials from other composite materials, the *metamaterial* label is usually used for a material which has unusual properties.

Mylar (boPET)

Biaxially-oriented polyethylene terephthalate (boPET) polyester film is used for its high tensile strength, chemical and dimensional stability, transparency, reflective, gas and aroma barrier properties and electrical insulation.

A variety of companies manufacture boPET and other polyester films under different trade names. In the USA and the UK, the most well-known trade names are *Mylar* and *Melinex.*

The orientation of the polymer chains is responsible for the high strength and stiffness of biaxially oriented PET film, which has a typical Young's modulus of about 4 GPa. biaxially oriented PET film has excellent clarity, despite its semicrystalline structure.

Biaxially oriented PET film can be aluminized by evaporating a thin film of metal onto it. The result is much less permeable to gasses (important in food packaging) and reflects up to 99% of light, including much of the infrared spectrum.

Mylar is base material for audio or video magnetic recording tapes.

Five layers of metallized boPET film in the NASA's spacesuits make them radiation resistant and help to keep astronauts warm.

Kapton

Kapton is a polyimide film developed by DuPont which can remain stable in a wide range of temperatures, from -269 °C to +400 °C (4 K - 673 K). Kapton is used in, among other things, flexible printed circuits (flexible electronics) and Thermal Micrometeoroid Garments, the outside layer of spacesuits.

According to a NASA internal report, Space Shuttle "wires were coated with an insulator known as Kapton that tended to break-down over time, causing electrical short-circuits and, potentially, fires. "The NASA Jet Propulsion Laboratory has considered Kapton as a good plastic support for solar sails because of its long duration in the space environment" (J. L. Wright, *Space Sailing*, Gordon and Breach, 1992).

Nylon

Nylon is one of the most common polymers and is technically a synthetic linen.

Property are: density 1.15 g/cm3; electric conductivity 10^{-12} S/m; melting point 190-350°C; tensale stress up 50 kgf/mm2.

Characteristics:

- Variation of luster: nylon has the ability to be very lustrous, semilustrous or dull.
- Durability: its high tenacity fibers are used for seatbelts, tire cords, ballistic
- cloth and other uses.
- High elongation
- Excellent abrasion resistance
- Highly resilient (nylon fabrics are heat-set)
- Paved the way for easy-care garments
- High resistance to:
  - insects, fungi and animals
  - molds, mildew, rot
  - many chemicals
- Melts instead of burning
- Used in many military applications

Kevlar

Kevlar is the registered trademark for a light, strong para-aramid synthetic fiber, related to other aramids such as Nomex and Technora. Typically it is spun into ropes or fabric sheets that can be used as such or as an ingredient in composite material components.

Currently, Kevlar has many applications, ranging from bicycle tires and racing sails to body armor because of its high strength-to-weight ratio—famously: *"...5 times stronger than steel on an equal weight basis..."* A similar fibre called Twaron with roughly the same chemical structure was introduced by Akzo in 1978, and now manufactured by Teijin.

When Kevlar is spun, the resulting fibre has great tensile strength (ca. 3 000 MPa), and a relative density of 1.44. When used as a woven material, it is suitable for mooring lines and other underwater application objects.

There are three grades of Kevlar: (i) Kevlar, (ii) Kevlar 29, and (iii) Kevlar 49. Typically, Kevlar is used as reinforcement in tires and rubber mechanical goods. Kevlar 29's industrial applications are as cables, in asbestos replacement, brake linings, and body armour. Kevlar 49 has the greatest tensile strength of all the aramids, and is used in plastic reinforcement for boat hulls, airplanes, and bicycles. The ultraviolet light component of sunlight degrades and decomposes Kevlar, a problem known as UV degradation, and so it is rarely used outdoors without protection against sunlight.

For a polymer Kevlar has very good resistance to high temperatures, and maintains its strength and resilience down to cryogenic temperatures (-196°C); indeed, it is slightly stronger at low temperatures. At higher temperatures the tensile strength is immediately reduced by about 10-20%, and after some hours the strength progressively reduces further. For example at 160°C about 10% reduction in strength occurs after 500 hours. At 260°C 50% reduction occurs after 70 hours. At 450°C Kevlar sublimates.

Applications

Armor. Kevlar is well-known as a component of some bulletproof vests and bulletproof facemasks. The PASGT helmet and vest used by US military forces since the early 1980s both have Kevlar as a key component, as do their replacements. Other military uses include bulletproof facemasks used by sentries. Civilian applications include Kevlar reinforced clothing for motorcycle riders to protect against abrasion injuries and also Emergency Service's protection gear if it involves high heat (e.g., tackling a fire), and Kevlar body armor such as vests for Police officers, security, and S.W.A.T.

Rope and cable. The fibre is used in woven rope and in cable, where the fibres are kept parallel within a polyethylene sleeve. Known as "Parafil", the cables have been used in small suspension bridges such as the bridge at Aberfeldy in Scotland. They have also been used to stabilise cracking concrete cooling towers by circumferential application followed by tensioning to close the cracks.

Building construction. A retractable roof of over 5,575 square meter of are, made of *Kevlar,* was a key part of the design of Montreal's Olympic stadium for the 1976 Summer Olympics held in Canada. It was spectacularly unsuccessful: completed ten years late and replaced ten years later in May 1998 after a series of problems.

Composite materials. Aramid fibres are widely used for reinforcing composite materials, often in combination with carbon fibre and glass fibre. The matrix for high performance composites is usually epoxy resin. Typical applications include monocoque bodies for F1 racing cars, and helicopter rotor blades.

GENERAL REFERENCES

(Parts of these works are in available from the Internet: http://Bolonkin.narod.ru/p65.htm, http://arxiv.org, search term: "Bolonkin", in book "*Non-Rocket Space Launch and Flight*", Elsevier, 2006, 488 pgs, and in book "*New Concepts, Idias, and Innovations in Aerospece and Technology*", NOVA, 2008, 420 pgs)

[1] Bolonkin, A.A., (1958a). Design of Hydro-Aircraft with Underwater Wing. *Report of Aircraft State Construction Bureau named Antonov*, Kiev, Ukraine, 1959 (in Russian), 120 p.

[2] Bolonkin, A.A., (1959a). Investigation of Aircraft AN-12 in Take-off. *Report of Aircraft State Construction Bureau named Antonov*, Kiev, Ukraine, 1959 (in Russian), 60 p.

[3] Bolonkin, A.A., (1959b). Research of Optimal Parameters of VTOL Aircraft, *Report of Aircraft State Construction Bureau named Antonov*, Kiev, 1959 (in Russian), Part 1 40 p., part 2 35 p.

[4] Bolonkin, A.A., (1959c). Research of Optimal Parameters of High Altitude Aircraft. *Report of Aircraft State Construction Bureau named Antonov*, Kiev, Ukraine, 1959 (in Russian), 60 p.

[5] Bolonkin, A.A., (1959d). Computation of High Thrust Aircraft with unsteady polar. *Report of Aircraft State Construction Bureau named Antonov*, Kiev, Ukraine, 1959 (in Russian), 50 p.

[6] Bolonkin, A.A., (1960a). Method of Estimation the Control of Aircraft Interface. *Presiding of Kiev High Engineering Aviation Military School* (KVIAVU), Kiev, Ukraine, Issue #72, 1960.

[7] Bolonkin, A.A., (1962a), *Theory of Flight Models*, Moscow, Association of Army, Air Force, and NAVY, 328p. 1962, (in Russian).

[8] Bolonkin, A.A., (1964a). Optimization of parameters of variation problems (Ukrainian. Russian and English summaries). *Dopoviti Akad. Nauk Ukrain. RSR,* 1964, #5, p. 580-582. Math.Rev. #6352.

[9] Bolonkin, A.A., (1964b) The calculus of variations and a functional equation of Bellman, and an interpretation of Langrange's undetermined multipliers. (Ukrainian. Russian and English summaries). *Dopovidi Akad. Nauk Ukrain., RSR* 1964, #10, p. 1290-1293. M.R. #5136.

[10] Bolonkin, A.A., (1964c). The extension principle and the Jacobi condition of the variation calculus. (Ukrainian. Russian and English summaries). *Dopovidi Akad. Nauk Ukrain. RSR* 1964. #7. p. 849-853. M.R. #5117.

[11] Bolonkin, A.A., (1965a), "Theory of Flight Vehicles with Control Radial Force". Collection Researches of Flight Dynamics, Mashinostroenie Publisher, Moscow, pp. 79–118, 1965, (in Russian). Intern.Aerospace Abstract A66-23338# (Eng).

[12] Bolonkin, A.A., (1965b), Investigation of the Take off Dynamics of a VTOL Aircraft. Collection Researches of Flight Dynamics. Moscow, Mashinostroenie Publisher, 1965, pp. 119-147 (in Russian). International Aerospace Abstract A66-23339# (English).

[13] Bolonkin, A.A., (1965c), Optimization of Trajectories of Multistage Rockets. Collection Researches of Flight Dynamics. Moscow, 1965, p. 20-78 (in Russian). International Aerospace Abstract A66-23337# (English).

[14] Bolonkin, A.A., (1965d). A method for the solution of optimal problems (Russian). Collection Complex Systems Control, pp.34-67. Naukova Dumka, Kiev, 1965. M.R. #5535.

[15] Bolonkin, A.A., (1965e). Special, Sliding, and Impulse Regimes in Problems of Flight Dynamics (Russian). Collection Complex Systems Control, pp.68-90. Naukova Dumka, Kiev, 1965. M.R. #5535.

[16] Bolonkin, A.A., (1966a). Boundary-value problems of Optimal Control. Military *Aviation Engeenering Academy (VVIA) named Zhukovskii*, Issue #1131, 1966, p. 103-128. (Russian).

[17] Bolonkin, A.A., (1967a). Special extreme, *Report presented to Symposium of Applied Mathematics*, Gorkii, USSR, 1967. (Russian).

[18] Bolonkin, A.A., (1968a). Impulse solution in control problems. (Russian). *Izv. Sibirsk. Otdel. Akad. Nayk USSR*, 1968, No. 13, p. 63-68. M.R. 7568. (Russian).

[19] Bolonkin, A.A., (1968b). Solution of Problem the Linear Optimal Control with one Control, *Journal "Prikladnaya Mechanica"*, Vol. 4, #4, 1968, p.111-121. (Russian).

[20] Bolonkin, A.A., (1969a). Solution of discrete problems of optimal control on the basis of a general minimum principle (Russian. English summary). *Vycisl. Prikl. Mat. (Kiev) Vyp. 7 (1969)*, 121-132. Mathematical Review 771.

[21] Bolonkin, A.A., (1969b), Special extreme in optimal control. Akademia Nauk USSR, Izvestiya. Tekhnicheskaya Kibernetika, No 2, Mar-Apr.,1969, p.187-198. See also *English translation in* Engineering Cybernetics, # 2, Mar- Apr.1969, p.170-183, (English).

[22] Bolonkin, A.A., (1970a). A certain method of solving optimal problems. *Izv. Sibirsk. Otdel. Akad. Nauk SSSR*. 1970, no.8, p. 86-92. M.R. #6163.

[23] Bolonkin, A.A., (1970b). A certain approach to the solution of optimal problems. (Russian. English summary). *Vycisl. Prikl. Mat.* (Kiev). Vyp. 12 (1970), p. 123-133. M.R. #7940.

[24] Bolonkin, A.A., (1971a), *Solution Methods for boundary-value problems of Optimal Control Theory.* Translated from Prikladnaya Mekhanika, Vol. 7, No 6, 1971, p. 639-650, (in English).

[25] Bolonkin, A.A., (1971b). Solution of Optimal Problems. *Collection "Mathematical Problems of Production Control"*, Moscow State University (MGU), Issue #3, 1971, p. 55-67.

[26] Bolonkin, A.A., (1972a), *New Methods of Optimization and their Applications*, Moscow Highest Technology University named Bauman, 1972, p.220 (in Russian).

[27] Bolonkin, A.A., (1982a), *Installation for Open Electrostatic Field*, Russian patent application #3467270/21 116676, 9 July, 1982 (in Russian), Russian PTO.

[28] Bolonkin, A.A., (1982b), *Radioisotope Propulsion*. Russian patent application #3467762/25 116952, 9 July 1982 (in Russian), Russian PTO.

[29] Bolonkin, A.A., (1982c), *Radioisotope Electric Generator*. Russian patent application #3469511/25 116927. 9 July 1982 (in Russian), Russian PTO.

[30] Bolonkin, A.A., (1983a), *Space Propulsion Using Solar Wing and Installation for It*, Russian patent application #3635955/23 126453, 19 August, 1983 (in Russian), Russian PTO.

[31] Bolonkin, A.A., (1983b), Getting of Electric Energy from Space and Installation for It, Russian patent application #3638699/25 126303, 19 August, 1983 (in Russian), Russian PTO.

[32] Bolonkin, A.A., (1983c), *Protection from Charged Particles in Space and Installation for It*, Russian patent application #3644168 136270, 23 September 1983, (in Russian), Russian PTO.

[33] Bolonkin, A. A., (1983d), *Method of Transformation of Plasma Energy in Electric Current and Installation for It*. Russian patent application #3647344 136681 of 27 July 1983 (in Russian), Russian PTO.

[34] Bolonkin, A. A., (1983e), *Method of Propulsion using Radioisotope Energy and Installation for It*. of Plasma Energy in Electric Current and Installation for it. Russian patent application #3601164/25 086973 of 6 June, 1983 (in Russian), Russian PTO.

[35] Bolonkin, A. A.,(1983f), *Transformation of Energy of Rarefaction Plasma in Electric Current and Installation for it*. Russian patent application #3663911/25 159775, 23 November 1983 (in Russian), Russian PTO.

[36] Bolonkin, A. A., (1983g), *Method of a Keeping of a Neutral Plasma and Installation for it*. Russian patent application #3600272/25 086993, 6 June 1983 (in Russian), Russian PTO.

[37] Bolonkin, A.A., (1983h), *Radioisotope Electric Generator*. Russian patent application #3620051/25 108943, 13 July 1983 (in Russian), Russian PTO.

[38] Bolonkin, A.A., (1983i), *Method of Energy Transformation of Radioisotope Matter in Electricity and Installation for it*. Russian patent application #3647343/25 136692, 27 July 1983 (in Russian), Russian PTO.

[39] Bolonkin, A.A., (1983j). *Method of stretching of thin film. Russian patent application* #3646689/10 138085, 28 September 1983 (in Russian), Russian PTO.

[40] Bolonkin, A.A., (1984a). Method Solution of Optimal Problem, Collection "*Research of Computer and Control System*", Irkutsk, USSR, 1984, pp.94-98. (in Russian).

[41] Bolonkin, A.A., (1987a). Method of Deformation of Extreme. Collection "*Methods and Programs of Optimal Problems in Networks*", Irkutsk State University and West Siberian Technological Institute, 1987, (in Russian).

[42] Bolonkin, A.A., (1987b). Method for Solution of Optimal Problems having equalities and non-equalities limitations. Collection "*Methods and Programs of Optimal Problems in Networks*", Irkutsk State University and West Siberian Technological Institute, 1987, (in Russian).

[43] Bolonkin, A.A., on any Set" (in Russian), Collection "*Design of Micro-processors*", Irkutsk (USSR), 1988 9 p.

[44] Bolonkin, A.A., (1988b). Deformation Method in Control Problems" (in Russian), Collection "*Design of Micro-processors*", Irkutsk (USSR), 1988 10 p.

[45] Bolonkin, A.A., (1990a). "Aviation, Motor and Space Designs", Collection *Emerging* Technology in the Soviet Union, 1990, Delphic Ass., Inc., pp.32–80 (English).

[46] Bolonkin, A.A., (1991a), *The Development of Soviet Rocket Engines*, 1991, Delphic Assoc. Inc., 122 p. Washington, (in English).

[47] Bolonkin, A.A., (1991b), New Approach to Finding a Global Optimum. *New American's Collected Scientific Reports.* Vol. 1, 1991, The Bnai Zion Scientists Division., New York.

[48] Bolonkin, A.A., (1992a), "*A Space Motor Using Solar Wind Energy (Magnetic Particle Sail)*". The World Space Congress, Washington, DC, USA, 28 Aug. – 5 Sept., 1992, IAF-0615.

[49] Bolonkin, A.A., (1992b), "*Space Electric Generator, run by Solar Wing*". The World Space Congress, Washington, DC, USA, 28 Aug. –5 Sept. 1992, IAF-92-0604.

[50] Bolonkin, A.A., (1992c), "*Simple Space Nuclear Reactor Motors and Electric Generators Running on Radioactive Substances*", The World Space Congress, Washington, DC, USA, 28 Aug. – 5 Sept., 1992, IAF-92-0573.

[51] Bolonkin, A.A., *The Twenty - First Century: The advent of the non-biological civilization.* GOTO: *http://Bolonkin.narod.ru.*

[52] Bolonkin, A.A. (1994a), "*The Simplest Space Electric Generator and Motor with Control Energy and Thrust*", 45th International Astronautical Congress, Jerusalem, Israel, 9–14 Oct., 1994, IAF-94-R.1.368.

[53] Bolonkin, A.A., Khot N., (1994b), *Optimal Structural and Control Design. 45th International Actronautical Congress.* Jerusalem, Israel. October 9-14, 1994, IAF-94-I.4.206.

[54] Bolonkin, A.A., (1994c), *Method for Finding a Global Min*imum, AIAA/ NASA/USAF/SSMO Symposium on Multi-disciplinary Analysis and Optimization, Panama City, Florida, USA, Sept. 7-9, 1994.

[55] Bolonkin, A.A. and Khot, N., (1995a), Design and Optimal Control in Smart Structures. Conference "*Mathematics and Control in Smart Structures*", 26 Feb.-3 March 1995, San Diego, CA, USA.

[56] Bolonkin, A.A., Khot N., (1995b), *Optimum Structural Vibration Control with Bounds on Control Forces*, 1995 ASME Design Technical Conference, 15th Biennial Conference on Vibration and Noise, September 17-21,1995, Boston, MA, USA.

[57] Bolonkin, A.A., (1995c), *The Twenty - First Century - the beginning of human immortality*. GOTO: *http://Bolonkin.narod.ru.*

[58] Bolonkin, A.A., Khot N., (1996a), *Optimal Bounded Control Design for Vibration Suppression. Acta Astronautics,* Vol.38, No. 10, pp803-813, 1996.

[59] Bolonkin, A.A. and Khot N., (1996b), *Minimum Weight of Control Devices with Bounded LQG Control.* The World Space Congress -96, June 1-6, 1996, Albuquerque, MN, USA.

[60] Bolonkin, A.A. and Khot N., (1996c), *Design of Smart Structures with Bounded Controls, Smart Structures and Materials,* Feb. 25-29, 1996, San-Diego, CA.

[61] Bolonkin, A.A. and Khot N., (1997a), Design of Structure control System using Bounded LQG, *Eng. Opt.*, 1997, Vol. 29, pp. 347-358.

[62] Bolonkin, A.A., (1997b). *Eccentric internal combustion engine*. Patent application US PTO 08/892,665 of 07/14/97.

[63] Bolonkin, A.A., (1998a). *Inflatable flight vehicles*. Patent application US PTO 09/271,700 of 1/26/98.

[64] Bolonkin, A.A., (1998b).*Protection from Tornado*. Patent application US PTO 09/236,749 of 1/25/98.

[65] Bolonkin, A.A., (1998c). *Uniblate Rotor*, Patent application US PTO 09/203,130 of 2/01/98. Patent US 6,234,422 of 5/22/01.

[66] Bolonkin, A.A., (1999a). *Method of space launch and hypersonic launch system*. Patent application US PTO 09/344,235 of 6/25/99.

[67] Bolonkin, A.A. (1999b), A High Efficiency Fuselage propeller ("Fusefan") for Subsonic Aircraft, 1999 World Aviation Congress, AIAA, #1999-01-5569.

[68] Bolonkin, A.A. and Gilyard, G.B., (1999c), Estimated Benefits of Variable-Geometry Wing Camber Control for Transport Aircraft. *NASA Center for AeroSpace Information (CASI), NASA/TM-1999-206586; H-2368; NAS 1.15:206586, 19991001; October 1999*

[69] Bolonkin, A.A., (1999d). The twenty-first century: the advent of the non-biological civilization and the future of the human race, *Journal "Kybernetes"*, Vol. 28, No.3, 1999, pp. 325-334, MCB University Press, 0368-492 (English).

[70] Bolonkin A.A., (1999e). *Science, Soul, Heaven and Supreme Mind, http:// Bolonkin.narod.ru* (Russian, English).

[71] Bolonkin, A.A., (1999f). *Breakthrough in Immortality. http://Bolonkin.narod.ru* (Russian, English).

[72] Bolonkin, A.A., (1999g). *Human Immortality and Electronic Civilization*. Electronic book, 1999. WEB: *http://Bolonkin.narod.ru*, *http://Bolonkin.narod.ru/p101.htm* (English), *http://Bolonkin.narod.ru/p100.htm* (Russian).

[73] Bolonkin, A.A., (1999h). *Bolonkin Rotor Engine*. Patent application US PTO 09/290,465 of 04/12/99. Patent US 6,248,921 of 10/09/01

[74] Bolonkin, A.A., (1999i). *Hypersonic Space Launcher (Method and Instellation)* Patent application US PTO 09/13,08 of 1/26/98; 09/344,235 of 06/25/99; 1/057,819 of 01/18/02.

[75] Bolonkin, A.A., Gilyard G.B., (2000a), *Optimal Pitch Thrust-Vector Angle and Benefits for all Flight Regimes, NASA Center for AeroSpace Information (CASI), NASA/TM-2000-209021; NAS 1.15:209021; H-2402, 20000301; March 2000.*

[76] Bolonkin, A.A., (2000b). *Eccentric Engine*. Patent application US PTO 09/575,256 of 05/22/2000. Patent US 6,526,937 B1 of 3/4/03.

[77] Bolonkin, A.A., (2001a). *Method and Installation for Space Trip*. Patent application US PTO 09/789,959 of 2/23/01.

[78] Bolonkin, A.A., (2001b). *Method and Installation for Space Launch*. Patent application US PTO 09/873,985 of 6/04/01.

[79] Bolonkin, A.A., (2001c). "*Method Transportation of Vehicles and Installations for It*". Patent application US PTO 09/893,060 of 6/28/01. Patent US 6,434,143 B1 of 12/17/02.

[80] Bolonkin, A.A., (2001d). *Method and Installation for getting of Energy*. Patent application US PTO 09/945,497 of 9/06/01.

[81] Bolonkin, A.A., (2001e). "*Method for Gas and Payload Transportation at Long Distance and Installations for It*", Patent Application USPTO # 09/978,507 of 10/18/01.

[82] Bolonkin, A.A., (2001f). Cable Launcher. Patent application US PTO 09/974,670 of 10/11/01.

[83] Bolonkin, A.A., (2002a), "*Non-Rocket Space Rope Launcher for People*", IAC-02-V.P.06, 53rd International Astronautical Congress, The World Space Congress – 2002, 10–19 Oct 2002, Houston, Texas, USA.

[84] Bolonkin, A.A, (2002b), "*Non-Rocket Missile Rope Launcher*", IAC-02-IAA.S.P.14, 53rd International Astronautical Congress, The World Space Congress – 2002, 10–19 Oct 2002, Houston, Texas, USA.

[85] Bolonkin, A.A., (2002c), "*Inexpensive Cable Space Launcher of High Capability*", IAC-02-V.P.07, 53rd International Astronautical Congress, The World Space Congress – 2002, 10–19 Oct 2002, Houston, Texas, USA.

[86] Bolonkin, A.A., (2002d), "*Hypersonic Launch System of Capability up 500 tons per day and Delivery Cost $1 per Lb*". IAC-02-S.P.15, 53rd International Astronautical Congress, The World Space Congress – 2002, 10–19 Oct 2002, Houston, Texas, USA.

[87] Bolonkin, A.A., (2002e), "*Employment Asteroids for Movement of Space Ship and Probes*". IAC-02-S.6.04, 53rd International Astronautical Congress, The World Space Congress – 2002, 10–19 Oct 2002, Houston, Texas, USA.

[88] Bolonkin, A.A., (2002f), "*Optimal Inflatable Space Towers of High Height*". COSPAR-02 C1.1-0035-02, 34th Scientific Assembly of the Committee on Space Research (COSPAR), The World Space Congress – 2002, 10–19 Oct 2002, Houston, Texas, USA.

[89] Bolonkin, A.A., (2002g), "*Non-Rocket Earth-Moon Transport System*", COSPAR-02 B0.3-F3.3-0032-02, 02-A-02226, 34th Scientific Assembly of the Committee on Space Research (COSPAR), The World Space Congress – 2002, 10–19 Oct 2002, Houston, Texas, USA.

[90] Bolonkin, A.A., (2002h) "*Non-Rocket Earth-Mars Transport System*", COSPAR-02 B0.4-C3.4-0036-02, 34th Scientific Assembly of the Committee on Space Research (COSPAR), The World Space Congress – 2002, 10–19 Oct 2002, Houston, Texas, USA.

[91] Bolonkin, A.A., (2002i). "*Transport System for Delivery Tourists at Altitude 140* km". IAC-02-IAA.1.3.03, 53rd International Astronautical Congress, The World Space Congress – 2002, 10-19 Oct. 2002, Houston, Texas, USA.

[92] Bolonkin, A.A., (2002j), "*Hypersonic Gas-Rocket Launch System.*" AIAA-2002-3927, 38th AIAA/ASME/SAE/ASEE Joint Propulsion Conference and Exhibit, 7–10 July 2002. Indianapolis, IN, USA.

[93] Bolonkin, A.A., Cloutier D., (2002k). Search, Observation, and Attack Problems, *Technical Report* AFRL-MN-EG-TR-2003-1717, 21p. 2002.

[94] Bolonkin, A.A., (2003a), "Air Cable Transport", *Journal of Aircraft*, Vol. 40, No. 2, March–April 2003, pp.265-269.

[95] Bolonkin, A.A., (2003b), "Optimal Inflatable Space Towers with 3-100 km Height", *Journal of the British Interplanetray Society*, Vol. 56, No 3/4, pp. 87–97, 2003.

[96] Bolonkin, A.A., (2003c), "Asteroids as Propulsion Systems of Space Ships", *Journal of the British Interplanetary Society*, Vol. 56, No 3/4, pp. 97–107, 2003.

[97] Bolonkin, A.A., (2003d), "Non-Rocket Transportation System for Space Travel", *JBIS*, Vol. 56, No 7/8, pp. 231–249, 2003.

[98] Bolonkin, A.A., (2003e), "Hypersonic Space Launcher of High Capability", *Actual problems of aviation and aerospace systems*, Kazan, No. 1(15), Vol. 8, 2003, pp. 45–58. Kazan Aviation Institute.

[99] Bolonkin, A.A., (2003f), "Centrifugal Keeper for Space Stations and Satellites", *JBIS*, Vol. 56, No 9/10, pp. 314–327, 2003.

[100] Bolonkin, A.A., (2003g), "Non-Rocket Earth-Moon Transport System", Advances in Space Research, Vol. 31/11, pp. 2485–2490, 2003, Elsevier, London.

[101] Bolonkin, A.A., (2003h), "Earth Accelerator for Space Ships and Missiles". *Journal of the British Interplanetary Society*, Vol. 56, No. 11/12, 2003, pp. 394–404.

[102] Bolonkin, A.A., (2003i), "Air Cable Transport and Bridges", TN 7567, International Air and Space Symposium – The Next 100 Years, 14–17 July 2003, Dayton, Ohio, USA.

[103] Bolonkin, A.A., (2003j), "Air Cable Transport System", *Journal of Aircraft*, Vol. 40, No. 2, March-April 2003, pp. 265-269.

[104] Bolonkin, A.A., Sierakowski R., (2003k). *Design of Optimal Regulators*. The manuscript is accepted as paper AIAA-2003-6638 by 2nd AIAA "Unmanned Unlimited" Systems, Technologies, and Operations- Aerospace, Land, and See Conference and Workshop and Exhibit, San Diego, California, USA, 15-18 Sep 2003. GOTO: *http://arxiv.org* and search "Bolonkin".

[105] Bolonkin, A.A., Murphy R., (2003-l). Geometry-Based Feasibility Constraints for Single Pursuer Multiple Evader Problems, 2nd AIAA "Unmanned Unlimited" Systems, Technologies, and Operations – Aerospace, Land, and sea Conference and Workshop and Exhibit, San Diego, California, 15-18 Sep 2003, AIAA-2003-6638.

[106] Bolonkin, A.A., (2004a), "Kinetic Space Towers and Launchers ', *JBIS*, Vol. 57, No 1/2, pp. 33–39, 2004.

[107] Bolonkin, A.A., (2004b), "Optimal trajectory of air vehicles", *Aircraft Engineering and Space Technology*, Vol. 76, No. 2, 2004, pp. 193–214.

[108] Bolonkin, A.A., (2004c), "Long Distance Transfer of Mechanical Energy", International Energy Conversion Engineering Conference at Providence RI, Aug. 16–19, 2004, AIAA-2004-5660.

[109] Bolonkin, A.A., (2004d), "Light Multi-Reflex Engine", *Journal JBIS*, Vol. 57, No 9/10, pp. 353–359, 2004.

[110] Bolonkin, A.A., (2004e), "Optimal trajectory of air and space vehicles", *AEAT*, No 2, pp. 193–214, 2004.

[111] Bolonkin, A.A., (2004f), "Hypersonic Gas-Rocket Launcher of High Capacity", *Journal of the British Interplanetary Society*, Vol. 57, No 5/6, pp. 167–172, 2004.

[112] Bolonkin, A.A., (2004g), "High Efficiency Transfer of Mechanical Energy". International Energy Conversion Engineering Conference at Providence RI, USA. 16–19 August, 2004, AIAA-2004-5660.

[113] Bolonkin, A.A., (2004h), "Multi-Reflex Propulsion System for Space and Air Vehicles", *JBIS*, Vol. 57, No 11/12, 2004, pp. 379–390.

[114] Bolonkin, A.A., (2004i), Utilization of Wind Energy at High Altitude, AIAA-2004-5705, AIAA-2004-5756, International Energy Conversion Engineering Conference at

Providence., RI, Aug.16-19. 2004. USA. GOTO: http://arxiv.org and search "Bolonkin".

[115] Bolonkin, A.A., Cloutier D., (2004j). Search and Attack Strategies, AIAA Conference Guidance, Navigation, and Control, Rhide Island, 16-19 August, 2004, Tr. #20201.

[116] Bolonkin, A., Cloutier D., (2004k). Search for Enemy Targets, *Technica. Report* AFRL-MN-EG-TR-2003-1716, June 2002. 49 p.

[117] Bolonkin, A.A., (2004k), Twenty-first century – the beginning of human immortality, *Journal "Kybernetes"*, Vol. 33, No.9/10, 2004, pp. 1535-1542, Emerald Press, *www.emeraldinsight.com/ISSN 0368-492X.htm* (English).

[118] Bolonkin, A.A., (2004L), Method of Recording and Saving of Human Soul for Human Immortality and Installation for It. US PTO, disclosure document #567484 of December 24, 2004.

[119] Bolonkin, A.A., (2005a) "*High Speed Catapult Aviation*", AIAA-2005-6221, Atmospheric Flight Mechanic Conference – 2005, 15–18 August, 2005, USA.

[120] Bolonkin, A.A., (2005b), "*Kinetic Anti–Gravitator*", AIAA-2005-4504, 41 Propulsion Conference, 10–12 July 2005, Tucson, Arizona, USA.

[121] Bolonkin, A.A., (2005c), "*Electrostatic Solar Wind Propulsion System*", AIAA-2005-3857, 41 Propulsion Conference, 10–13 July 2005, Tucson, Arizona, USA.

[122] Bolonkin, A.A., (2005d), "*Sling Rotary Space Launcher*", AIAA-2005-4035, 41 Propulsion Conference, 10–13 July 2005, Tucson, Arizona, USA.

[123] Bolonkin, A.A., (2005e), "*Electrostatic Utilization Asteroids for Space Flight*", 41 Propulsion conference, 10–12 July 2005, Tucson, Arizona, USA, AIAA-2005-3857.

[124] Bolonkin, A.A., (2005f), "*Guided Solar Sail and Electric Generator*", 41 Propulsion conference, 10–12 July, 2005, Tucson, Arizona, USA, AIAA-2005-3857.

[125] Bolonkin, A.A., (2005g), "*Problems of Levitation and Artificial Gravity*", 41 Propulsion conference, 10–12 July 2005, Tucson, Arizona, USA, AIAA-2005-3365.

[126] Bolonkin, A.A., (2005h), "*Radioisotope Sail and Electric Generator*", 41 Propulsion conference, 10–12 July, 2005, Tucson, Arizona, USA, AIAA-2005-3653.

[127] Bolonkin, A.A., Murphy R., (2005i). Geometry-Based Parametric Modeling for Single Pursuer Multiple Evader Problems. Journal JGCD, v.28, #1, 2005

[128] Bolonkin, A.A. (2006a), Book "*Non-Rocket Space Launch and Flight*", Elsevier, London, 2006, 488 ps. Contents can be accessed: GOTO: *http://Bolonkin. narod.ru/p65.htm*.

[129] Book contains more than 20 of the author's revolutionary concepts and ideas.

[130] Bolonkin, A.A., (2006b), *Beam Space Propulsion*, AIAA-2006-7492, Conference Space-2006, 18-21 Sept;, 2006, San Jose, CA, USA. GOTO: *http://arxic.org* and search "Bolonkin".

[131] Bolonkin, A.A., (2006c), *Electrostatic AB-Ramjet Space Propulsion*, AIAA/AAS Astrodynamics Specialist Conference, 21-24 August 2006, USA. AIAA-2006-6173. Journal "Aircraft Engineering and Aerospace Technology", Vol.79, #1, 2007. GOTO: *http://arxiv.org* and search "Bolonkin".

[132] Bolonkin, A.A., (2006d), Electrostatic Linear Engine, AIAA-2006-5229, 42nd Joint Propulsion Conference, 9-12 June 2006, Sacramento, USA. *Journal "Aircraft Engineering and Aerospace Technology*", Vol.78, #6, 2006, pp.502-508.

[133] Bolonkin, A.A., (2006e), *High-Speed Solar Sail*, AIAA-2006-4806, 42nd Joint Propulsion Conference, 9-12 June 2006, Sacramento, USA. GOTO: *http://arxiv.org* and search "Bolonkin".

[134] Bolonkin, A.A., (2006f), *A New Method of Atmospheric Reentry for Space Shuttle*, AIAA-2006-6985, MAO Conference, 6-9 Sept. 2006, USA. GOTO: *http://arxiv.org* and search "Bolonkin".

[135] Bolonkin, A.A., (2006g), *Suspended Air Surveillance System*, AIAA-2006-6511, AFM Conference, 21-29 Aug. 2006, Keystone, USA. *http://arxiv.org*, search "Bolonkin".

[136] Bolonkin, A.A., (2006h), Optimal Solid Space Tower, AIAA-2006-7717. ATIO Conference, 25-27 Sept. 2006, Wichita, Kansas, USA. *http://arxiv.org*, search "Bolonkin".

[137] Bolonkin, A.A., (2006i), *Theory of Space Magnetic Sail Some Common Mistakes and Electrostatic MagSail.* Presented as paper AIAA-2006-8148 to 14-th Space Planes and Hypersonic System Conference, 6-9 November 2006, Australia. *http://arxiv.org*, search "Bolonkin".

[138] Bolonkin, A.A., (2006j) *Micro -Thermonuclear AB-Reactors for Aerospace*. Presented as paper AIAA-2006-8104 in 14th Space Plane and Hypersonic Systems Conference, 6-8 November, 2006, USA. *http://arxiv.org*, search "Bolonkin".

[139] Bolonkin, A.A., Cathcart, R.B., (2006k). *A Low-Cost Natural Gas/Freshwater Aerial Pipeline.* GOTO: *http://arxiv.org* and search "physics/0701061".

[140] Bolonkin, A.A., (2006l) *Cheap Textile Dam Protection of Seaport Cities against Hurricane Storm Surge Waves, Tsunamis, and Other Weather-Related Floods. http://arxiv.org*, search "Bolonkin".

[141] Bolonkin, A., Cathcart R.B., (2006m). *The Java-Sumatra Aerial Mega-Tramway.* GOTO: *http://arxiv.org* and search "physics/0701099".

[142] Bolonkin, A., Cathcart R.B., (2006n). *Inflatable Evergreen Polar Zone Dome* (EPZD) Settlements. GOTO: *http://arxiv.org* and search "physics/0701098".

[143] Bolonkin, A.A. and R.B. Cathcart, (2006) "A Cable Space Transportation System at the Earth's Poles to Support Exploitation of the Moon", *Journal of the British Interplanetary Society* 59: 375-380.

[144] Bolonkin, A.A., (2006o). *Control of Regional and Global Weather. http://arxiv.org*, search "Bolonkin".

[145] Bolonkin, A.A., Cathcart R.B., (2006p). *Antarctica: A Southern Hemisphere Windpower Station?* GOTO: *http://arxiv.org* and search "physics/0701055".

[146] Bolonkin, A.A., (2006q). *AB Levitator and Electricity Storage*. http://arxiv.org, search "Bolonkin", AIAA-2007-4612.

[147] Bolonkin, A.A. (2006r). *Simplest AB-Thermonuclear Space Propulsion and Electric Generator. http://arxiv.org*, search "Bolonkin".

[148] Bolonkin, A.A., (2006s). *Wireless Transfer of Electricity in Outer Space. http://arxiv.org*, search "Bolonkin".

[149] Bolonkin, A.A., (2006t). *Electrostatic AB-Ramjet space propulsion for interplanetary flight, AEAT*, vol. 79, No. 1, 2007, pp. 3 - 16.

[150] Bolonkin A.A., (2006u). *Method of Recording and Saving of Human Soul for Human Immortality and Installation for it*. Patent application US PTO 11613380 of 12/20/06.

[151] Bolonkin, A.A. and R.B. Cathcart (2006v), "Inflatable 'Evergreen' dome settlements for Earth's Polar Regions", *Clean Technologies and Environmental Policy* DOI 10.1007/s10098-006-0073-4.

[152] Bolonkin, A.A., (2006w). New AB-Thermonuclear Reactor for Aerospace, AIAA-2006-7225, *http://arxiv.org*. Search: "Bolonkin".

[153] Cathcart, R.B. and Bolonkin, A.A., (2006x), The Golden Gate Textile Barrier: Preserving California Bay of San Francisco from a Rising North Pacific Ocean. GOTO: *http://arxiv.org* and search "physics/0702030".

[154] Cathcart, R.B and Bolonkin A.A., (2006y), Ocean Terracing. GOTO: *http://arxiv.org* and search "physics/0701100".

[155] Bolonkin, A.A., (2006z). Method of recording and saving of human soul for human Immortality and Installation for it. Patent application #11/613,380 of 12/20/06. USPTO.

[156] Book (2006z2), *Macro-Engineering - A challenge for the future*. Collection of articles. Eds. V. Badescu, R.B. Cathcart and R.D. Schuiling, Springer, 2006. (Collection contains two of A.A. Bolonkin's articles: Space Towers; Cable Anti-Gravitator, Electrostatic Levitation and Artificial Gravity).

[157] Bolonkin, A.A., (2007a). AB Levitrons and their Applications to Earth's Motionless Satellites, *http://arxiv.ru*, search: Bolonkin. AIAA Joint Propulsion Conference, Paper AIAA-2008-8150, 21-23 July, Hartford, Connecticut, USA.

[158] Bolonkin A.A., (2007b). Passenger life-saving in a badly damaged aircraft scenario, AIAA-2007-5844, *http://arxiv.ru*, search: Bolonkin.

[159] Bolonkin, A.A., (2007c). Optimal Electrostatic Space Tower (Mast, New Space Elevator), AIAA-2007-6201, *http://arxiv.ru*, search: Bolonkin.

[160] . Bolonkin, A.A., (2007d). Electrostatic Space Climber, AIAA-2007-5838, 43-rd Joint Propulsion Conference, Cincinati, Ohio, USA, 9-11 July, 2007. *http://arxiv.ru*, search: Bolonkin.

[161] Bolonkin, A.A., (2007e). Inflatable Dome for Moon, Mars, Asteroids and Satellites, AIAA-2007-6262, *http://arxiv.ru*, search: Bolonkin.

[162] Bolonkin, A.A., (2007f). Human Immortality and Electronic Civilization, 3-rd Edition, Lulu, 2007, 66 pgs., (Enflish and Russian), *http://www.lulu.com* search "Bolonkin".

[163] Bolonkin, A.A., (2007g). Memories of Soviet Political Prisoner, Lulu, 2007, 66 pgs., (Enflish and Russian), *http://www.lulu.com* search "Bolonkin". English is translated from same Russian book, New Your, 1991.

[164] Bolonkin, A.A., (2007h). Extraction Water and Energy from Atmosphere. *http://arxiv.org*. Search: "Bolonkin".

[165] Bolonkin, A.A., (2007i). AB Irrigation without Water (Closed-loop water cycle). *http://arxiv.ru*, search: Bolonkin.

[166] Bolonkin A.A., (2007j). AB Method of Irrigation without Water (Closed-loop water cycle). *http://arxiv.org*, search term: "Bolonkin".

[167] Bolonkin A.A., (2007k). Thermonuclear Reflect AB-Reactor, *http://arxiv.org*, search term: "Bolonkin".

[168] Bolonkin A.A., (2007-l). AB ferromagnetic Space Propulsion. *http://arxiv.org*, search term: "Bolonkin". AIAA Joint Propulsion Conference, Paper AIAA-2008-5151, 21-23 July, Hartford, Connecticut, USA.

[169] Bolonkin, A.A., (2008a). Floating Cities, Islands, States. *http://arxiv.org*, search term: "Bolonkin".

[170] Bolonkin, A.A., (2008b). AB-Tubes and Quasi-Superconductivity at Room Temperature, *http://arxiv.org*, search term: "Bolonkin".

[171] Bolonkin, A.A., Krinker M., Magnetic Aircraft, *http://arxiv.org*, search term: "Bolonkin". AIAA Joint Propulsion Conference, Paper AIAA-2008-4610, 21-23 July, Hartford, Connecticut, USA.

[172] Calasso, F.E., (1989), Advanced Fibers and Composite, Gordon and Branch Scientific Publisher, New York, 1989.

[173] Carbon and High Performance Fibers, (1995), Directory, NY 1995, Chapman and Hall, 6th ed., New York.

[174] Concise Encyclopedia of Polymer Science and Engineering, (1990), Ed. J. I. Kroschwitz, 1990. New York.

[175] Clarke, A.C.: Fountains of Paradise, Harcourt Brace Jovanovich, New York, 1978.

[176] Dresselhous, M.S., (2000), Carbon Nanotubes, Springer, New York, 2001.

[177] Fedorov, V.D., (1981), Basis of Rocket Flight, Moscow, Nauka.(Russian).

[178] Handbook of Physical Quantities, Ed. Igor S. Grigoriev, 1997, CRC Press, USA.

[179] Harris, J.T. (1973), *Advanced Material and Assembly Methods for Inflatable Structures*. AIAA, Paper No. 73-448.

[180] Johnson, A., *Space Research: Organizing for Economical Efficiency*. Presented as paper AIAA-2006-7224 in Conference "Space-2006", 19-21 September 2006, San Diego, California, USA.

[181] Johnson, A., Space research: problems of efficiency. *Journal "Actual Problems of Aviation and Aerospace System"*, No.1, 2007. *http://www.kcn.ru/tat_en/science/ans/journals/rasj_cnt/07_1_10.html*

[182] Kikoin, I.K., (1976), Table of Physical Magnitudes, Moscow, Atomic Publish House. (Russian).

[183] Landis, G.A., (2004), Interstellar Flight by Particle Beam, *Acta Astronautica*, Vol. 55, pp.931 - 934.

[184] Landis, G., (2000). "Dielectric Films for Solar- and Laser-pushed Lightsails," *AIP Conference Proceedings Volume 504*, pp. 989-992; Space Technology and Applications International Forum (STAIF-2000), Jan. 30 - Feb. 3, Albuquerque NM.

[185] Landis, Geoffrey A., Cafarelli Craig, (1999). The Tsiolkovski Tower Re-Examined, *JBIS*, Vol. 32, pp. 176 -180, 1999.

[186] Landis, Geoffrey A. and Cafarelli, Craig, (1995), "The Tsiolkovski Tower," paper IAF-95-V.4.07, 46th International Astronautics Federation Congress, Oslo Norway, 2-6 Oct. 1995. There is also an errata to this paper, which was published in JBIS: Landis, Geoffrey A., (2005),"Correction," *Journal of the British Interplanetary Society*, Vol. 58, p. 58.

[187] Landis, Geoffrey A., (1998), "*Compression Structures for Earth Launch*," paper AIAA-98-3737, 24th AIAA/ASME/SAE/ASEE Joint Propulsion Conf., July 13-15, Cleveland OH.

[188] Nishikawa, K., Wakatani, M., *Plasma Physics*, Spring, 2000.

[189] Omidi, N. and Karimabadi, H., (2003), "*Electrostatic Plasma Sail*", AIAA 2003-5227, 2003. 39th AIAA/ASME/SAE/ASEE Joint Propulsion Conference and Exhibit 20-23 July 2003, Hurtisville, Alabama. Contains a principal scientific mistakes.

[190] Regan, F.J., Anandakrishnan S.M., *Dynamics of Atmospheric Re-Entry*, AIAA, 1993

[191] Shortley, G., Williams D., Elements of Physics, Prentice-Hall, Inc., Englewood Cliffs, New Jersey, 1971, USA.

[192] Smitherman, D.V., Jr., Space Elevators, NASA/CP-2000-210429, 2000.

[193] Tsiolkovski, K.E.: "Speculations about Earth and Sky on Vesta," Moscow, Izd-vo AN SSSR, 1959; Grezi o zemle I nebe (in Russian), Academy of Sciences, U.S.S.R., Moscow, p.35, 1999.

Some Print Publications by Richard B. Cathcart

"Asteroid impact in the Black Sea: A Black Scenario", pp. 1-8 *in* S.J. Rzoska and V.A. Mazur (Eds), Soft Matter under Exogenic Impacts (Springer, 2007) [with R.D. Schuiling and V. Badescu].

"Tapping earth's upper-mantle methane gas resource at a nuclear drilling initiative area, Palk Bay, India/Sri Lanka", *CURRENTS SCIENCE* 92: 729-732 (25 March 2007).

"Sethusamudram ship channel macroproject: Anti-tsunami and storm surge textile arrestors protecting Palk Bay (India and Sri Lanka", *CURRENT SCIENCE* 91: 1474-1476 (10 December 2006).

"Asteroid Impact in the Black Sea: Death by Drowning or Asphyxiation?" *NATURAL HAZARDS* 40: 327-338 (2007) [with Dr. R.D. Schuiling, Dr. Viorel Badescu et al.].

"A Cable Space Transportation System at the Earth's Poles to Support Exploitation of the Moon", *JOURNAL OF THE BRITISH INTERPLANETARY SOCIETY* 59: 375-380 (2006) [with A.A. Bolonkin].

"Use of class A and class C stellar engines to control sun movement in the galaxy", *ACTA ASTRONAUTICA* 58: 119-129 (2006) [with Dr. Viorel Badescu].

"Environmental thermodynamic limitations on global human population", *International Journal of Global Energy* 25: 129-140 (2006) [with Dr. Viorel Badescu].

"The Hormuz Strait Dam Macroproject—21st Century Electricity Development Infrastructure Node (EDIN)?", *Marine Georesources and Geotechnology* 23:25-37 (January-June 2005).

"Nautical Jugs or Not?", *Current Science* 88:1211-1212 (25 April 2005).

"Palk Bay: A future industrial complex site?", *Current Science* 87: 849-850 (10 October 2004).

"Architectural Ecology: A Tentative Sahara Restoration", *International Journal of Environmental Studies* 61:145-160 (April 2004) [with Dr. Viorel Badescu].

"Geo-engineering Gone Awry: A New Partial Solution of Fermi's Paradox", *Journal of the British Interplanetary Society* 57: 209-215 (2004) [with Dr. Milan Cirkovic].

"Palk Strait power station: A future fixed link on Adam's Bridge?" *Current Science* 85: 430-431 (25 August 2003). [India]

"Unnatural Envelopment: Fieldwork on the Active Tectonics of J.G. Ballard's 'Build-Up' (1957)" *Journal of Geoscience Education* 50: 176-181 (March 2002).

"Stellar Engines for Kardashev's Type II Civilization" *Journal of the British Interplanetary Society* 53: 297-306 (September-October 2000) [with Dr. Viorel Badescu].

"Tibetan power: A unique hydro-electric Macroproject servicing India and China" *Current Science* 77: 854-855 (10 October 1999). [India]

"Land Art as global warming or cooling antidote" *Speculations in Science and Technology* 21: 65-72 (1998).
"Taming Mars with a Tent and Tunnel: Creation of a Biosphere-City" *SST* 21: 117-131 (1998).
"Greenhouse atmospherics: Mega-deaths or Macro-engineering?" *SST* 20: 17-20 (1997).
"Seeing is Believing: Planetographic Data Display on a Spherical TV" *Journal of the British Interplanetary Society* 50: 103-104 (1997).
"Gauthier's 'Linear City'" *Environmental Conservation* 23: 286 (1996).
"Mitigative Anthropogeomorphology: a revived 'plan' for the Mediterranean Sea Basin and the Sahara" *Terra Nova: The European Journal of Geosciences* 7: 636-640 (1995).
"Macro-engineering and terraforming: building modernized and additional functional regions" *Speculations in Science and Technology* 14: 34-40 (1991). [Abbreviated as SST.]
"Is the Trans-Channel Icebridge a viable alternative to the Channel Tunnel?" *SST* 10: 63-65 (1987).
"Improving the status of Rodoman's electronic geography proposal" *SST* 9: 37-39 (1986).
"What if We Lowered the Mediterranean Sea?" *SST* 8: 7-15 (1985).
"Macro-engineering Transformation of the Mediterranean Sea and Africa" *World Futures* 19: 111-121 (1983).
"Mediterranean Basin-Sahara Reclamation" *SST* 6: 150-152 (1983).
"A Megastructural End to Geologic Time" *Journal of the British Interplanetary Society* 36: 291-297 (1983).
"Radioactive Waste Element Liquefying Device for Geologic Fault Fusion" *SST* 4: 103-104 (1981).
"How a Global Warming Could Change the Geographical Future" *The Futurist* 14: 28 (October 1980).
"Meteorite Mining" *Future Life* #27 (June 1981), page 13.
"On the road to Antarctica" *Future Life* #23 (December 1980), page 13.

Works in Russian Language

[194] Tsiolkovski, K.E. (1959),"*Speculations about Earth and Sky on Vesta*", Moscow, Izd-vo AN SSSR, 1959; Grezi o zemle i nebe, Academy of Sciences, USSR, Moscow, p. 35, 1999.
[195] Mayboroda, A., Zero Gravity on Earth, "*Yuniy Tehnik*" ("Young Technician"), No. 10, October 1988, Moscow.
[196] Pokrovskii, G.I., (1964), *Space Tower*, TM, ("Technology for Youth"), No. 10.
[197] Poliakov, G., (1977), *Space Necklace of Earth*, TM, ("Technology for Youth"), No. 4.
[198] Yunitskii A., (1982), "*General Planetary Transport System*", "TM" ("Technology for Youth"), No. 6,) (in Russian). (see last four Russian works in: *http://www.ipu.ru/stran/bod/ing/sovet2.htm*, Pictures: *http://www.ipu.ru/stran/bod/ing/sovet_ris.htm*)

NIAC Reports in USA

[199] *http://auditing-science.narod.ru* or http:*//www.geocities.com/auditing.science/*

[200] Christensen C., "Ultralight Solar Sail for Interstellar Travel", *http://NASA-NIAC.narod.ru*

[201] Hose, S. D., "Antimatter Drive Sail for Deep Space Missions", *http://NASA-NIAC.narod.ru*

[202] Landis, G. A., "Advanced Solar and Laser Pushed Lightsail Concepts", *http://NASA-NIAC.narod.ru*

[203] Miller, D. W., "Electromagnetic Formation Flight", *http://NASA-NIAC.narod.ru*

[204] Zubrin, R., "The Magnetic Sail". *http://NASA-NIAC.narod.ru.*

[205] About NIAC: GO TO: *http://auditing-science.narod.ru* or *http://www.geocities.com/auditing.science/*, *http://NASA-NIAC.narod.ru*

Some of A.A. Bolonkin's Popular Publications

[206] Personal site: GOTO: *http://Bolonkin.narod.ru.*

[207] Bibliography (about the author discussing his ideas) publication in Russian press and Internet in 1994 - 2004 (*http://www.km.ru*, *http://pravda.ru*, *http://n-t.ru*, ets. Search: Bolonkin)

[208] Bolonkin, A.A., Our children may be a last people generation, *Literary newspaper*, 10/11/95, #41 (5572), Moscow, Russia (Russian).

[209] Bolonkin, A.A., Stop the Earth. I step off. *People Newspaper*, Sept., 1995. Minsk, Belorussia (Russian).

[210] Bolonkin, A.A., End of Humanity, but not End of World. *New Russian Word*, 3/6/96, p.14, New York, USA (Russian).

[211] Bolonkin, A.A., *Natural Human Purpose is to be God. http://Bolonkin.narod.ru* (Russian).

[212] Bolonkin, A.A., *American and Russian Science*.(English). *http://Bolonkin.narod.ru.*

[213] Bolonkin, A.A., Locate the God into Computer-Internet Network (in Russian). *http://Bolonkin.narod.ru.*

[214] Blekherman, A., *Short biography of Dr. A. Bolonkin*. (English, Russian) *http:// Bolonkin.narod.ru*

[215] Dr. Sci. M. Krinker, World Space Congress-2002 (about 9 Bolonkin's scientific works presented to Congress). *http://Bolonkin.narod.ru* (Russian).

[216] Ruduyk, B., New ideas of Dr. Bolonkin, *Newspaper "Fact",* Ukraine (in Ukraine). *http://Bolonkin.narod.ru.*

[217] Kurolenko, N., Electronic Society, *News paper "Kievskie Vedomosti*", C. Kiev, Ukraine, 27 May, 2002. *http://Bolonkin.narod.ru.*

[218] Bay, E., Our soul is only set our knowledge (about ideas of A. Bolonkin). *Newspaper "Literary Newspaper".* 10 July, 2002. Moscow (in Russian).

[219] Golovkov, A., Time for thinking, Magazime *"Ogonek",* 1988, Moscow, Russia (in Russian). *http://Bolonkin.narod.ru.*

[220] Levin, V., "*Take-off and Landing*" (article from magazine "*Vestnik*" about Bolonkin. 1992)(Russian).

[221] Bolonkin, A.A., Peculiarities of Soviet and American Sciences. (Russian). *http://Bolonkin.narod.ru*

[222] Bolonkin, A.A., Memoirs of Soviet Political Prisoner (English). *http://www.lulu.*com search "Bolonkin", *http://Bolonkin.narod.ru.*

[223] Getmanskii, Igor, Price of Immortality, Moscow, EKCMO, 2003, 480 ps., (Fantasy, in Russian).

INDEX

A

B

C

E

F

G

H

I

J

N

O

P

S

T

U

V

W

X

Y

Z